# SAP PRESS e-books

Print or e-book, Kindle or iPad, workplace or airplane: Choose where and how to read your SAP PRESS books! You can now get all our titles as e-books, too:

- By download and online access
- For all popular devices
- And, of course, DRM-free

Convinced? Then go to www.sap-press.com and get your e-book today.

# SAP® Cloud Integration Cookbook

**SAP PRESS**

SAP PRESS is a joint initiative of SAP and Rheinwerk Publishing. The know-how offered by SAP specialists combined with the expertise of Rheinwerk Publishing offers the reader expert books in the field. SAP PRESS features first-hand information and expert advice, and provides useful skills for professional decision-making.

SAP PRESS offers a variety of books on technical and business-related topics for the SAP user. For further information, please visit our website: *www.sap-press.com*.

Bilay, Singh, Singh, Gutsche, Krimmel
Cloud Integration with SAP Integration Suite: The Comprehensive Guide (4th Edition)
2024, 900 pages, hardcover and e-book
*www.sap-press.com/5760*

Martin Koch, Siegfried Zeilinger
Configuring SAP Business Technology Platform: The Practical Guide for Administrators
2025, 450 pages, hardcover and e-book
*www.sap-press.com/6080*

Aron, Gakhar, Vij
SAP Integration Suite
2021, 343 pages, hardcover and e-book
*www.sap-press.com/5326*

Martin Koch, Siegfried Zeilinger
Security and Authorizations for SAP Business Technology Platform
2023, 355 pages, hardcover and e-book
*www.sap-press.com/5627*

Martin Koch, Siegfried Zeilinger
Cloud Connector for SAP
2023, 352 pages, hardcover and e-book
*www.sap-press.com/5683*

Martin Koch, Thorsten Reisinger, Marc Urschick

# SAP® Cloud Integration Cookbook

Advanced Cloud Integration with SAP Integration Suite

Rheinwerk

Publishing

**Editor** Meagan White
**Copyeditor** Doug McNair
**Cover Design** Graham Geary
**Photo Credit** iStockphoto: 471406880/© Bill Oxford
**Layout Design** Vera Brauner
**Production** Kelly O'Callaghan
**Typesetting** III-satz, Germany
**Printed and bound in** Canada, on paper from sustainable sources

**ISBN 978-1-4932-2765-5**
1st edition 2026

**© 2026 by:**
Rheinwerk Publishing, Inc.
2 Heritage Drive, Suite 305
Quincy, MA 02171
USA
info@rheinwerk-publishing.com
+1.781.228.5070

**Represented in the E.U. by:**
Rheinwerk Verlag GmbH
Rheinwerkallee 4
53227 Bonn
Germany
service@rheinwerk-verlag.de
+49 (0) 228 42150-0

**Library of Congress Cataloging-in-Publication Control Number:** 2025041038

# Contents at a Glance

# Contents

# 3   Mappings: Groovy, XSLT, and Message Mappings   101

# 4    Adapters

# 5    Loosely Coupled Interfaces

# 6    Artifact Reusability

# PART I
# Cloud Integration Recipes

# Chapter 1
# Introduction

*SAP Integration Suite is the central platform for seamlessly connecting systems, applications, and processes—both in the cloud and on-premises. It offers companies a powerful set of tools and services for implementing integrations quickly, securely, and with future-proof capabilities. It thus forms the backbone of a modern, networked IT landscape in line with the clean core approach.*

*Cloud Integration* is a central component of SAP Integration Suite, and this chapter will teach you the basics, structure, and benefits of this technology. It'll help you learn from the best and gain practical knowledge from real-world projects. That's what we're aiming for in this book.

Cloud Integration is a powerful but complex tool within SAP Integration Suite. The platform is constantly evolving and offers a huge range of possibilities—from simple Representational State Transfer (REST) integration to hybrid architectures with Event Mesh, cloud connector, and asynchronous message processing via Java Message Service (JMS).

There are already several books that explain the basics of SAP Integration Suite, and they provide a solid understanding of the system, introduce features, and are heavily based on official SAP documentation. However, anyone who works on projects knows that theory is good but practice beats everything. So we wrote this book because we worked on customer projects with Cloud Integration for many years and realized that a crucial piece of the puzzle was missing: a resource that not only shows how to implement something technically but also shows why certain decisions should be made, what pitfalls to avoid, and which strategies will prove themselves in the long term. We wanted to create a book that reflects the everyday life of SAP developers, architects, and consultants, not dry feature lists. This book tackles real challenges, provides solutions, and distills the knowledge we've gained from countless projects into a concise, accessible format. So, forget about excerpts from the product documentation—this is a practical guide for anyone who deals with complex integration scenarios on a daily basis.

We've deliberately focused on depth—this book isn't an introduction; it's an in-depth exploration. It's designed for those who've already gained some experience with SAP Integration Suite and now want to learn how to solve real problems like the following:

- How to structure modular integration flows (*iFlows*)
- How to implement Groovy scripting sensibly and securely

- What to consider when using JMS queues or Event Mesh
- Which strategies have proven effective in the areas of security, monitoring, and deployment

We've prepared real-life scenarios for you, not artificial examples. These scenarios include the integration of *SAP SuccessFactors* with third-party systems, hybrid interfaces via the cloud connector, and the use of *event-based architectures* for *decoupling* systems. We don't just show the technical structure; we also explain the background, decision-making processes, and alternatives.

This publication is distinct from others because it directly addresses crucial yet often overlooked aspects. It covers setting up an end-to-end continuous integration/continuous delivery (CI/CD) process for SAP integrations, reusability and maintenance of integration artifacts, and secure handling of sensitive data in productive operations. It also provides detailed coverage of infrastructure topics such as authentication via OAuth 2.0, the use of SAP Cloud Identity Services, and the use of SAP Cloud ALM for monitoring and error analysis.

Our goal was to write a book that's on par with the professionals. This isn't some glossy demo; it's a real workbook written by practitioners for practitioners. You'll learn how to use Cloud Integration well, and you'll understand how to use it in a maintainable, expandable, secure, and future-proof way. This book is for anyone who wants to shape integration at the highest level, not just be part of it.

## 1.1   Objectives and Target Group

This book is for pros who work with Cloud Integration and want to go beyond the usual tutorials and documentation. It's got the technical depth, architectural know-how, and practical insights you need to design, build, and operate strong and future-proof integration scenarios in the SAP ecosystem.

The book's goals are to do the following:

- Teach you how to use Cloud Integration in depth, including how to design it into modules, manage its interfaces, and make it work better.
- Provide real-world examples and practical integration patterns that go beyond theoretical knowledge.
- Make it so that projects that use different systems can be used again, be made larger, and have a simple design.
- Prepare developers and architects for more advanced topics. These include event-driven architectures, secure message handling via JMS queues, and end-to-end monitoring.

This book also addresses several key roles in the SAP integration landscape, each of which has specific needs and challenges. Here's a list of roles that specifies why this book is particularly valuable for each one:

- **SAP developers and integration architects**
  They are the hands-on builders of iFlows and system landscapes. They need to create robust, maintainable solutions that can scale with business needs, not just understand basic iFlows. This book provides concrete implementation techniques, modularization strategies, and architectural patterns to help reduce complexity and technical debt.

- **Project teams that implement complex integration projects**
  Large projects inevitably entail a variety of systems, demanding deadlines, and stringent performance and reliability standards. Teams must deal with versioning, transport, testing, and operation of interfaces under real-world constraints. This book will support them with structured approaches, reusable templates, and project-proven patterns that accelerate development while ensuring quality and consistency.

- **SAP consultants and solution architects**
  Consultants design blueprints, enforce best practices, and advise clients on integration strategy. They need a solid foundation of what works in practice—not just what's theoretically possible. This book helps them define reusable integration concepts, establish governance models, and make architectural decisions based on real-world experience.

- **Experienced developers seeking advanced topics**
  Cloud Integration developers who know the basics often reach their limits when working on asynchronous communication, error handling, or event-based patterns. This book will take them to the next level by diving into advanced topics such as the following:

  - Designing loosely coupled architectures with Event Mesh
  - Reliably processing messages with JMS queues
  - Using APIs and scripting (Groovy and JavaScript) for powerful integration logic
  - Troubleshooting and tuning performance in productive systems

  Gaining a working knowledge of these topics is essential for building robust and future-proof integrations, and they are rarely covered in such depth elsewhere.

## 1.2   Overview of Cloud Integration

To put it succinctly, *SAP Integration Suite* is the digital backbone of modern SAP landscapes. It plays a pivotal role in hybrid system landscapes by orchestrating all integration scenarios across on-premises and cloud-based systems. In a business environment where companies are increasingly relying on hybrid models that combine traditional

SAP systems, cloud solutions, and third-party software, having a robust, scalable, and strategically managed integration platform is crucial. Consequently, SAP Integration Suite is a top priority for further development at SAP itself.

In Section 1.2.1, we'll start with a look at the product roadmap. In Section 1.2.2, we'll dive into the different communication types, and in Section 1.2.3, we'll give you an overview of the different SAP integration suite capabilities. Finally, in Section 1.2.4, we'll present SAP Cloud Transport Management.

### 1.2.1 Product Roadmap

The *product roadmap* is comprehensive, dynamic, and closely integrated with the innovation areas surrounding SAP Business Technology Platform (SAP BTP), event-driven architectures, and AI-supported integration. Those who are interested can follow the current and planned further development of the suite at any time in *SAP Discovery Center* (*https://discovery-center.cloud.sap/*), which provides roadmaps (see Figure 1.1) as well as practical missions and reference architectures (see Figure 1.2 for the application-to-application [A2A] integration roadmap).

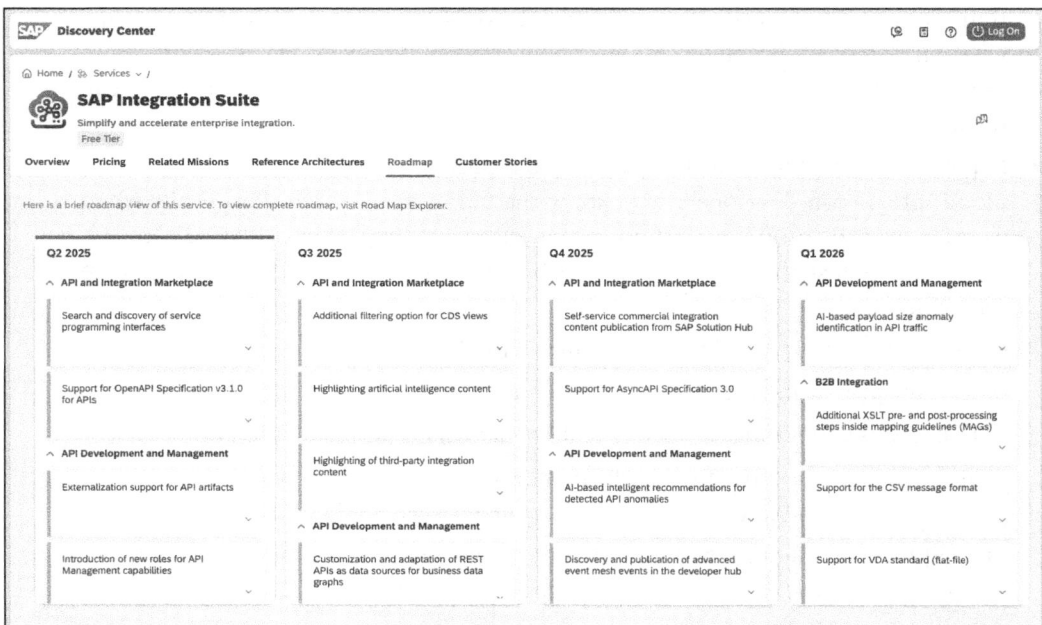

**Figure 1.1** Roadmap

Picture a modern company that wants to manage its data, processes, and user experiences in the cloud. Customer relationship management (CRM) data is stored in Salesforce, enterprise resource planning (ERP) data is stored in SAP S/4HANA, customer interactions are run through a modern e-commerce system, and logistics data comes from Internet of Things (IoT) devices worldwide. But how can we get all these systems to communicate with one another in real time? We need to make sure that the communication is strong, secure, and will work in the future.

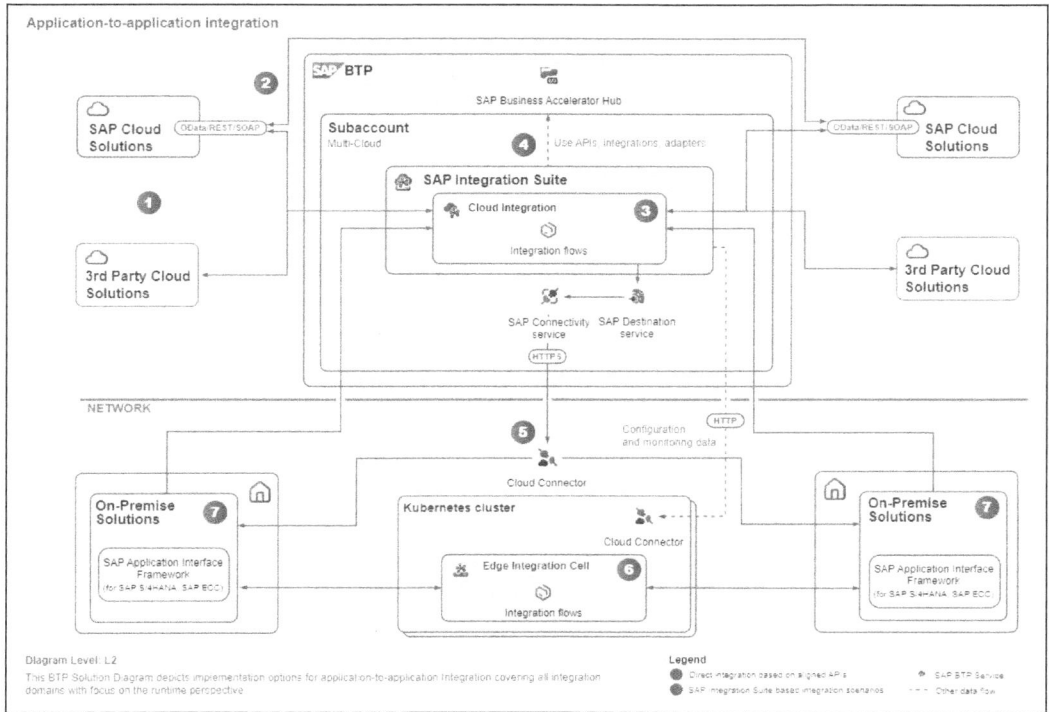

**Figure 1.2** Reference Architecture for A2A Integration

The potential applications of cloud integration capability are highly diverse. These systems range from basic system links to complex, rule-based process orchestrations. The following scenarios, drawn from actual projects, demonstrate the wide range of possibilities:

- **SAP SuccessFactors meets Microsoft Entra ID (formerly known as Active Directory)**
  An international industrial group uses SAP SuccessFactors as its leading HR system. As soon as it hires a new employee, user accounts must be automatically created in *Microsoft Entra ID*, access cards must be requested, and SAP authorizations must be assigned. Cloud Integration orchestrates this onboarding process by querying SAP SuccessFactors via OData, transforming the data, communicating with the Microsoft Graph API by using REST, and forwarding the feedback to a central ticket system.

- **IoT-based machine integration**
  A machine manufacturer sends sensor data from plants to SAP Business Technology Platform (SAP BTP) in real time. Cloud Integration processes these events, converts

them into IDocs, and integrates them into the central SAP S/4HANA system. At the same time, the system sends a warning to the technical field service if certain thresholds are exceeded.

- **International e-invoicing**
  A company operates in several countries and must comply with country-specific requirements for electronic invoices. Cloud Integration processes local billing data, transforms it according to the country (e.g., via XRechnung in Germany, FatturaPA in Italy, or Comprobante Fiscal Digital por Internet [CFDI] in Mexico), and transfers it via web service to the respective government platform. Monitoring, error handling, and legally compliant logging are included.

These examples clearly demonstrate that cloud integration isn't a rigid tool. Rather, it's a modular system for mapping end-to-end processes, and it can also elegantly implement country-specific and industry-specific requirements.

In recent years, the way companies use software and process data has changed fundamentally. Monolithic on-premise systems used to dominate, with all business processes taking place within a single, central ERP system. Today, however, the reality is decentralized, heterogeneous, and highly dynamic: SAP S/4HANA runs in the cloud, HR processes are implemented with SAP SuccessFactors, CRM takes place in Salesforce, e-commerce runs on Shopify or Adobe Commerce, and business intelligence analyses are performed with SAP Analytics Cloud or Microsoft Power BI. This hybrid nature brings enormous advantages in terms of flexibility, speed of innovation, and scalability— but it also presents companies with a key challenge: achieving the smooth, secure, and maintainable integration of these systems. This is exactly where the Cloud Integration capability comes in.

SAP Integration Suite is SAP's central integration platform, and it's delivered as a cloud-based integration platform as a service (iPaaS). It helps companies connect different types of IT systems. The suite isn't a single tool but a set of specialized parts that let called capabilities work together. Detailed information (i.e., service descriptions) and valuable resources on SAP Integration Suite and all other services that are available on SAP BTP can be found in *SAP Discovery Center* (*https://discovery-center.cloud.sap/index.html*). Figure 1.3 presents the service description for SAP Integration Suite.

While *middleware* was often seen as technical "glue" between systems in the past, integration is now becoming a strategic component of the IT portfolio. It's no longer just a matter of sending data from A to B but of modeling, monitoring, and controlling end-to-end processes across system boundaries. SAP Integration Suite brings this process to the cloud—with all the advantages of a modern platform solution—as follows:

- **Scalability on demand**
  It has no size limits, no hardware issues, and no traditional maintenance cycles.

- **Availability from the cloud**
  It provides access worldwide, anytime, and anywhere—with a high level of security

- **Seamless integration into SAP BTP**
  It provides close integration of different capabilities like Cloud Integration, Event Mesh, and API Management, as well as integration with other SAP BTP services like SAP Build Process Automation and SAP Cloud Identity Services.

- **Fast development and deployment**
  It reduces project runtimes through modeling tools and standard content.

- **Clean core-enabled**
  It provides integration as a separate layer outside the core systems in a way that's maintainable, upgrade-friendly, and future-proof.

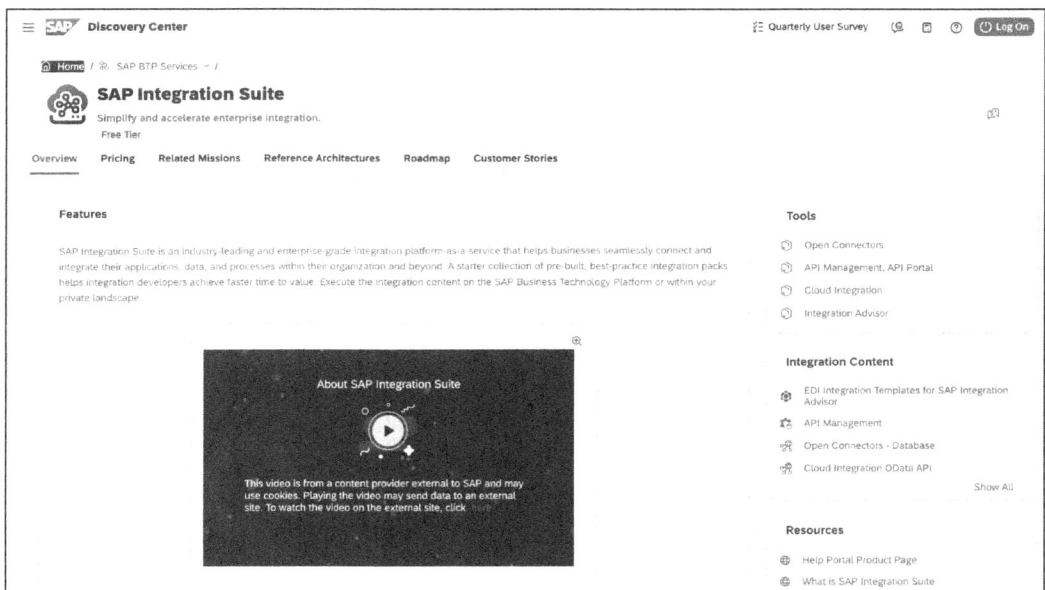

**Figure 1.3**  SAP Integration Suite

SAP Integration Suite isn't just another SAP tool; it's a central element of SAP BTP and therefore closely linked to SAP's overarching strategy. This means it supports the *clean core principle*, which states that no more individual integrations should happen directly in the ERP core. It promotes *API-first architectures* to facilitate integrations via defined interfaces. These interfaces are loosely coupled and reusable. Integration is the foundation for RISE with SAP and is, without a doubt, an enabler of event-based architectures and AI-supported processes.

---

**What is RISE with SAP?**

*RISE with SAP* is a strategic-business-transformation-as-a-service offering from SAP that supports companies on their journey to the cloud—technologically, organizationally, and commercially. Its objective is to expedite digital transformation and furnish companies with a modular, predictable, and low-risk migration to an intelligent, cloud-based ERP.

Core components of RISE with SAP are as follows:

- SAP S/4HANA Cloud (which is available in both public and private editions)
- SAP BTP, which is ideal for extensions, integration, analytics, and AI
- *Business process intelligence* (*SAP Signavio*), which facilitates analysis and optimization of business processes
- The *SAP Business Network* starter pack for SAP S/4HANA, which provides a foundation for becoming a connected enterprise
- Technical migration tools and services for tasks such as custom code checks and readiness assessments

RISE with SAP customers generally have access to SAP Integration Suite as part of the SAP BTP services. That makes cloud-based integration a central building block of RISE projects for interfaces with third-party providers, on-premise systems, and SAP solutions such as SAP SuccessFactors, SAP Ariba, and SAP Concur.

### 1.2.2   Communication Types

Modern IT and integration architectures are defined by a multitude of connections among applications, business processes, devices, people, and institutions. To make this complexity tangible, distinctions among different interface types have become established in integration practice. This classification is based on the communication partners and their roles in the business or system context. SAP Integration Suite is the clear solution here. Before continuing with our discussion of Cloud Integration's capabilities, let's briefly review the relevant interface types.

#### Application-to-Anything

Application-to-anything (A2X) communication is the modern, expanded form of A2A. It describes all integration scenarios in which a software system—such as SAP S/4HANA or a database—communicates with another digital system or object. The *X* in A2X symbolizes any target components: other applications, platforms, devices, services, or even user interfaces (UIs).

Here are some typical A2X scenarios:

- Integrating SAP S/4HANA with a CRM system (e.g., Salesforce)
- Connecting cloud applications (e.g., SAP Analytics Cloud, SAP SuccessFactors)
- Synchronizing production control systems and ERP systems
- Automating data flows among internal backend systems

A2X integrations are the lifeblood of a system landscape. They connect a company's central applications and ensure the continuous, smooth running of business processes.

### Business-to-Business

*Business-to-business* (B2B) communication is structured, automated communication between two legally independent companies. This communication is typically between a manufacturer and its supplier or a retailer and its logistics service provider.

Such interfaces are based on electronic data interchange (EDI) standards such as EDI-FACT and ANSI X12 or on specialized APIs. The objective of B2B communication is straightforward: to seamlessly transfer business-critical information (such as orders, delivery notifications, invoices, and payment notifications) electronically, without manual intervention.

Some typical B2B scenarios include the following:

- Automatic exchange of invoice and order messages between SAP systems
- Integration with SAP Ariba for digital processing of purchasing processes
- Connection of major customers via Applicability Statement 2 (AS2) or Applicability Statement 4 (AS4), both of which are secure and reliable communication protocols for data exchange between business partners. AS4 is the more modern, web services–based evolution of AS2.

B2B interfaces have strict requirements in terms of security, traceability, and standardization because they involve cross-company transactions.

### Business-to-Government

A company uses *business-to-government* (B2G) integrations when it communicates electronically with government agencies or public institutions. These scenarios are typically mandated by law and must adhere to specific criteria regarding format, deadlines, security, and traceability.

Some typical B2G scenarios include the following:

- Transmission of tax data via ELSTER (e.g., advance sales tax returns)
- Electronic invoicing (e.g., via XRechnung or ZUGFeRD) to public contractors
- Reports for Intrastat statistics or customs clearance

B2G interfaces are regulated, subject to audit, and highly formalized. They usually require the use of specific protocols, certificates, or integration platforms.

### Business-to-Consumer

*Business-to-consumer* (B2C) communication is integration with end customer–oriented systems, such as online shops, mobile apps, and customer portals. Its goal is straightforward: to provide information such as order status, customer data, and product availability in real time.

Typical B2C technologies include the following:

- REST APIs (e.g., the product catalog via OData)
- Webhooks (e.g., shipping notifications)
- OAuth-secured interfaces for self-service portals

B2C integrations place special demands on scalability, availability, user experience, and data security, and these demands are directly linked to customer expectations.

### Machine-to-Machine and Internet of Things

*Machine-to-machine* (M2M) interfaces are part of the IoT world and are the key to direct communication among machines, sensors, and devices. These scenarios are typically event driven and use lightweight protocols such as Message Queuing Telemetry Transport (MQTT), Advanced Message Queuing Protocol (AMQP), and Kafka.

Examples of some typical M2M scenarios are as follows:

- Machines sending production data to SAP S/4HANA via Event Mesh
- IoT sensors reporting temperature exceedances in real time
- Vehicles transmitting GPS data to logistics systems.

M2M integrations are a core component of Industry 4.0 and *edge-to-cloud scenarios*, in which huge amounts of data must be processed in real time and integrated into business processes.

### 1.2.3   SAP Integration Suite Capabilities

We'll now examine the individual capabilities of SAP Integration Suite in more detail. We'll also present you with a typical real-world application scenario for each capability.

### Cloud Integration

Cloud Integration is SAP Integration Suite's most renowned capability because it allows for the seamless integration of business processes across cloud and on-premise systems. It serves as the process engine for your landscape.

Imagine that a medium-sized machine manufacturer uses SAP S/4HANA Cloud for its central enterprise resource planning (ERP) but continues to operate an old warehouse management system on-premise. When a customer places an order, the inventory must be queried, a delivery must be planned, and the order status must be synchronized. Cloud Integration connects systems via standard and user-defined interfaces in real time.

Here are some typical use cases for Cloud Integration:

- Streamlining order-to-cash processes
- Integrating employee onboarding between SuccessFactors and Microsoft Entra ID
- Synchronizing master data
- Implementing event-based integrations via Event Mesh

### API Management

The API Management capability enables the management, security, and publication of APIs to open up data to external and internal users in a controlled manner.

Imagine a retail company that wants to enable its suppliers to query inventory levels directly via an API. At the same time, internal developers must be able to quickly build new applications on top of existing APIs. *API Management* provides governance, security, and a developer portal for this purpose.

Here are some typical use cases for API Management:

- Exposing OData APIs from S/4HANA.
- Building an API facade for mobile apps
- Establishing partner connections with access control

### Open Connectors

*Open Connectors* abstracts the complexity of third-party APIs (such as Salesforce, Microsoft 365, and HubSpot) and provides harmonized REST interfaces.

Imagine the marketing department is set on synchronizing contact data from HubSpot with SAP S/4HANA. Open Connectors automates authentication and speeds up integration, and it doesn't require in-depth API expertise.

Here are some typical use cases for Open Connectors:

- Integrating software as a service (SaaS) platforms such as Slack, Twilio, and Shopify
- Centralizing user management
- Synchronizing CRM and ERP data

### Integration Advisor

The Integration Advisor capability provides automated mapping suggestions based on machine learning to support B2B and EDI scenarios.

Imagine an automotive supplier receiving customer data in EDIFACT format and needing to convert it into internal extensible markup language (XML) format. Integration Advisor thoroughly analyzes the structures and automatically suggests a mapping based on best practices and metadata.

Here are some typical use cases for Integration Advisor:

- Creating mappings for EDIFACT, ANSI X12, IDoc, etc.
- Harmonizing data structures
- Reducing manual mapping efforts

---

**EDIFACT, X12, and IDocs**

Structured exchange formats play a central role in electronic business data communication. In the business-to-business context, three of the most important formats are EDIFACT, ANSI X12, and IDoc. They facilitate the automated exchange among systems of standardized business documents, including orders, delivery notes, invoices, and payment notifications. They do it as follows:

- *Electronic Data Interchange for Administration, Commerce and Transport* (EDIFACT) is an international EDI standard that was developed under the auspices of the United Nations (UN). EDIFACT is a widely used standard in Europe, and it covers a range of business processes including orders (ORDERS), invoices (INVOIC), and delivery notifications (DESADV). Its structure is segment based and text based, and it's machine readable. Some of the main areas of application of EDIFACT are in wholesaling, the automotive sector, logistics, and the public sector (e.g., Pan-European Public Procurement Online [PEPPOL]).

- ANSI X12 is a set of EDI standards that were developed by the American National Standards Institute. It's the North American counterpart to EDIFACT, and it's primarily used in the United States, Canada, and parts of Latin America. It's structure is comparable to that of EDIFACT, yet it varies in syntax, segmentation, and numbering. For examples the codes it uses are 850 for a purchase order, 810 for an invoice, and 856 for an advance shipping notice. Typical areas of application are in companies that in the retail, pharmaceutical, and health care sectors in North America.

- Intermediate Document (IDoc) is a proprietary SAP format that's used for data exchange among SAP systems or with external partners. An IDoc comprises a control record, one or more data records, and a status record. These systems are highly structured and closely integrated with SAP modules such as SD, MM, and FI. Types include ORDERS05, DELVRY03, and INVOIC02. Typical areas of application are SAP-to-SAP integration and hybrid EDI scenarios with conversion via middleware.

---

**Trading Partner Management**

*Trading Partner Management* is a relatively new module that centralizes the management of partners, agreements, and communication protocols.

Imagine a trading company that utilizes AS2 and Secure File Transfer Protocol (SFTP) to communicate with over a hundred suppliers. Trading Partner Management oversees the technical and contractual framework of this, including certificate management and compliance.

**1**

> **Applicability Statement 2**
>
> *Applicability Statement 2* (AS2) is a widely used communication protocol for secure and reliable electronic data exchange between business partners. It's used in particular in the EDI environment.

Here are some typical use cases for Trading Partner Management:

- Management of business-to-business partnerships and protocols
- Management of contracts and certificates
- Governance of communication relationships

**Event Mesh**

*Event Mesh* facilitates event-based architectures, whereby systems respond to changes in real time rather than periodically polling for data.

Imagine that a production machine detects a maintenance requirement and sends an event to the central system. This automatically initiates an order with the designated spare parts supplier. This approach eliminates the need for polling jobs and unnecessary interface loads, and it ensures that only targeted responses to business events are executed.

> **Event-Based Architecture**
>
> *Event-based architecture* is a system that organizes data in a way that allows it to be processed in an efficient and timely manner. In this paradigm, systems don't interact with each other through direct API calls. Instead, they respond to events that are generated by other systems. The basic principle is simple. Instead of synchronous point-to-point communication (as in "System A calls system B"), a producer sends an event (e.g., "New customer created," "Invoice created") to a central event infrastructure (e.g., Event Mesh, Kafka, the AMQP broker). Other systems acting as consumers subscribe to these events and respond to them asynchronously.

Here are some typical use cases for Event Mesh:

- Integrating the IoT
- Decoupling systems
- Implementing reactive microservices architectures

**OData Provisioning**

*OData Provisioning* enables the fast and secure provision of OData services based on existing SAP systems, in particular SAP Gateway.

Imagine that a field service employee needs access to customer master data from the on-premise SAP system in a mobile app. OData Provisioning provides an OData service that the app can access securely and efficiently and that includes authentication and data filtering.

Typical use cases for OData Provisioning are as follows:

- Exposing SAP data for SAP Fiori apps
- Integration of mobile applications
- Accessing SAP Business Suite or SAP S/4HANA via OData

### Data Space Integration

The *Data Space Integration* capability enables controlled, sovereign integration of different members of a data space.

Imagine an energy supplier wants to share sensor data with research institutions without opening up its proprietary systems. Data Space Integration allows it to publish metadata, define access rights, and control data usage—with full transparency.

Typical use cases for Data Space Integration are as follows:

- Data exchange across company boundaries
- Establishing sovereign data infrastructures (e.g., in the health care or energy sector)
- Standardized metadata integration

### Integration Assessment

The Integration Assessment capability assists you in strategically planning and evaluating integration projects while ensuring compliance with governance rules.

Imagine a multinational corporation is interested in implementing a group-wide integration policy. It employs Integration Assessment to evaluate ongoing projects according to criteria such as reusability, security level, and strategic relevance, and it systematically prioritizes initiatives.

Here are some typical use cases for Integration Assessment:

- Strategic integration planning
- Portfolio management for interfaces
- Governance definitions for large SAP landscapes

### Migration Assessment

*Migration Assessment* allows companies to assess the suitability of their existing interfaces for migration to the cloud and identify any potential risks.

Imagine that a company with hundreds of SAP Process Integration/SAP Process Orchestration (SAP PI/SAP PO) interfaces is planning to switch to the SAP Integration

Suite. The Migration Assessment automatically analyzes which iFlows use complex mappings or proprietary adapters, and it then provides a recommendation for the conversion.

Here are some typical use cases for Migration Assessment:

- Preparation of PI/PO replacements
- Analysis of legacy system interfaces
- Automated migration roadmaps

It's also important to note that all the capabilities presented in this section are included as standard in the SAP Integration Suite license. This means that every customer who has licensed the Integration Suite—whether via subscription, SAP Business Technology Enterprise Agreement (SAP BTPEA), or the now replaced Cloud Platform Enterprise Agreement (CPEA) and Pay-As-You-Go (PAYG)—has access to all capabilities of the suite. These include Cloud Integration, API Management, Open Connectors, Event Mesh, Integration Assessment, and Data Space Integration.

SAP is deliberately pursuing a modular activation approach. It doesn't make sense to activate all capabilities immediately in every case, so companies should activate and configure only those modules that are required for their specific system landscape and integration scenarios. This keeps the integration environment lean, clear, and maintainable—in line with the concept of a "lean and secure integration landscape."

This is also a recommended approach from a governance and compliance perspective: capabilities that aren't activated don't incur additional costs but can lead to unnecessary complexity, increased maintenance effort, and security risks if used carelessly. A needs-based, step-by-step expansion, on the other hand, allows the advantages of the platform to be exploited in a targeted and controlled manner.

### 1.2.4   SAP Cloud Transport Management

*SAP Cloud Transport Management* on SAP BTP plays a central role in SAP Integration Suite, particularly in governance, compliance, and lifecycle management. With this service, software artifacts such as iFlows, APIs, and UI components can be transported in a structured and controlled manner between different subaccounts and environments—for example, from a development environment to a test or production environment.

Especially in regulated or security-critical industries, it's essential to make changes traceable and audit-proof. SAP Cloud Transport Management does just that, as follows:

- It makes transports—including logging, user context, and versioning—rule-based.
- It supports the separation of development, test and production—even across different SAP BTP subaccounts.
- It eliminates manual exports and imports and thus significantly reduces the likelihood of errors.

For SAP Integration Suite, this means that iFlows, APIs, and Integration Advisor specifications can be packaged, validated, and transferred to other environments as standalone transport packages—without detours or file downloads.

The use of SAP Cloud Transport Management not only contributes to technical quality assurance but also provides the basis for professional change and release management within SAP BTP. Organizations that use it benefit from increased security, improved traceability, and consistent deployments while meeting internal control system and external audit requirements.

---

**Interaction between SAP Cloud Transport Management and SAP Cloud ALM**

SAP Cloud Transport Management is the standard for managing and transferring development artifacts and application-specific content in many SAP cloud products. You can connect SAP Cloud Transport Management to SAP Cloud ALM to gain full visibility into the transport landscape, including the associated transport paths (nodes) and deployments. This allows you to control, monitor, and orchestrate transports across system boundaries directly in SAP Cloud ALM. You can establish the connection via an SAP Cloud Transport Management instance that you subscribe to in SAP BTP.

---

## 1.3   Important Concepts and Technologies

The Cloud Integration capability (sometimes referred to as *process integration* or *CI*) is the core component of SAP Integration Suite when it comes to integrating applications, systems, and services in hybrid and cloud-centric IT landscapes. Cloud Integration enables companies to connect data and business processes across system boundaries—whether they are SAP or non-SAP systems, in the cloud or on-premises. Cloud Integration is more than just a cloud-based successor to SAP PI/SAP PO. It has a clear focus on modularity, extensibility, security, and seamless integration into SAP BTP, particularly for hybrid landscapes.

---

**What Is a Hybrid System Landscape?**

A *hybrid system landscape* is an IT architecture that integrates cloud and on-premise systems. In the SAP environment, it's a common model. For example, SAP S/4HANA (on-premise) is connected to cloud solutions such as SAP SuccessFactors, SAP Ariba, or SAP Analytics Cloud. The two worlds work closely together, with data and business processes linked via integration solutions such as SAP Integration Suite.

A typical hybrid scenario is one with human resources management in the cloud with SAP SuccessFactors while payroll accounting takes place in the local SAP Human Capital Management for SAP S/4HANA. Another is central cloud ERP that communicates with on-premise systems for production or logistics.

---

Having a hybrid landscape is the clear choice for companies that want to keep leveraging their existing investments in on-premise systems while gradually integrating modern cloud technologies. This ensures flexibility, innovation, and investment protection in the digital transformation.

The Cloud Integration capability of SAP Integration Suite is a middleware solution that is fully hosted on SAP BTP and that combines a model-based approach with deep technical capabilities such as mapping, routing, security, and transformation. Users work in a browser-based development tool where they can graphically model, configure, test, and deploy iFlows.

Cloud Integration supports the transition from traditional middleware systems such as SAP PI/SAP PO or third-party solutions to a future-proof, flexible, and low-maintenance cloud-based architecture. It plays a key role in the clean core approach because it facilitates outsourcing integrations from the ERP system and centrally managing them via SAP BTP.

---

**Clean Core**

*Clean core* is a central architectural principle in the SAP environment that aims to leave the core of an SAP system—especially in SAP S/4HANA—as unchanged as possible. Therefore, you should avoid making direct modifications to the SAP standard. Instead, you can implement individual enhancements, adjustments, and integrations via modern, standardized interfaces outside the actual system—for example, on SAP BTP.

The objective of the clean core approach is to enhance system maintainability, ensure upgradeability, and prevent the accumulation of technical debt. In a cloud-based SAP landscape, such as the RISE with SAP model, having a clean core is paramount because automatic updates are installed regularly and can be jeopardized by extensive modifications to the core. With regard to clean core, you can make extensions via key user tools, in-app extensions, and side-by-side developments on the BTP, and you can use official APIs, events, and CDS views. However, we recommend that you don't make any direct modifications to the SAP standard, such as writing customer-specific ABAP code in standard transactions or making changes to SAP database tables. Having a clean core is therefore essential for future-proof, cloud-enabled, and continuously expandable SAP system landscapes.

---

In Section 1.3.1, we'll explain the key components of SAP Integration Suite. In Section 1.3.2, we'll introduce the Cloud Integration capabilities. Finally, in Section 1.3.3, you'll learn about prepackaged content.

### 1.3.1  Key Components

A Cloud Integration tenant consists of several *key components* that work together to efficiently model, execute, monitor, and manage iFlows. We'll discuss the following components in this section: the design time, the runtime engine, the edge integration cell, the keystore/trust store, monitoring and tracing capabilities, and the partner directory.

**Design Time**

The *design time* is the browser-based development environment for Cloud Integration. Using this web-based development tool, integration developers can do the following:

- Use drag-and-drop components, adapters, routing elements, scripts, and mappings to graphically model iFlows.
- Set configuration elements for recipient URLs, headers, tokens, and data conversions.
- Perform test runs to validate the flows locally with sample payloads.
- Log version artifacts and retrieve previous versions.
- Organize integration content: manage iFlows, value mappings, scripts, and resources on a package basis.

The design time is essential to development. It enables visually guided, intuitive modeling while providing precise control over technical details through script integration and fine-grained configuration options (see Figure 1.4).

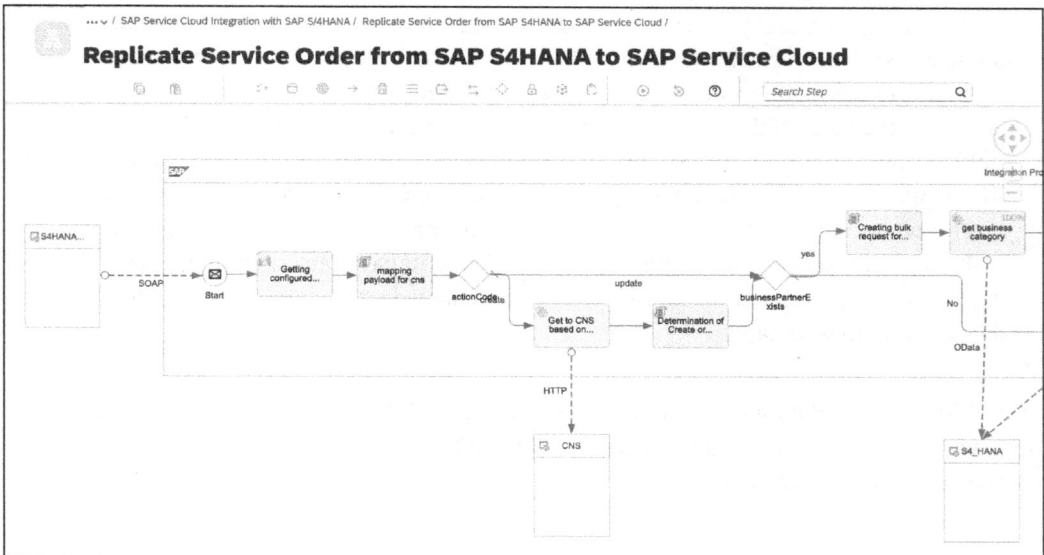

**Figure 1.4**  Cloud Integration Design Time

**SAP Process Orchestration Design Time versus Cloud Integration Design Time**

Design time in SAP PO is based on a highly structured, centralized approach within the enterprise services repository (ESR) and the integration directory. Developers model integration artifacts such as message mappings, service interfaces, and operation mappings based on a strictly typed metadata model—often in conjunction with SAP NetWeaver Developer Studio. The focus is on reusability, strong typing (e.g., through XML Schema Definition (XSD) or Web Services Description Language (WSDL)–based objects), and deep integration into the classic SAP ecosystem.

Design time in Cloud Integration, on the other hand, follows a much more flexible, cloud-native low-code/pro-code approach. The iFlows are modeled directly in a browser-based web UI, where you can seamlessly combine graphical editors, integrated scripting languages (Groovy and JavaScript), converters, routers, and adapters. Unlike SAP PO, you don't need a central object library (such as ESR)—you manage and version content on a package basis. Cloud Integration focuses on agility, extensibility, and continuous delivery, while SAP PO typically follows a classic transport and deployment process. Both environments thus reflect the technological and methodological evolution of the SAP integration strategy.

The design time is the central storage location for all content you create in the Cloud Integration tenant. You can group the content in packages, version it, and prepare it for transport (e.g., via SAP Cloud Transport Management). The repository forms the basis for the lifecycle of integration artifacts and enables structured teamwork with review and approval processes. Integration Packages can contain the following artifacts:

- iFlows that utilize graphical logic and metadata
- Script collections for Groovy and JavaScript
- Value mapping tables that show context-based code assignments (see Figure 1.5)

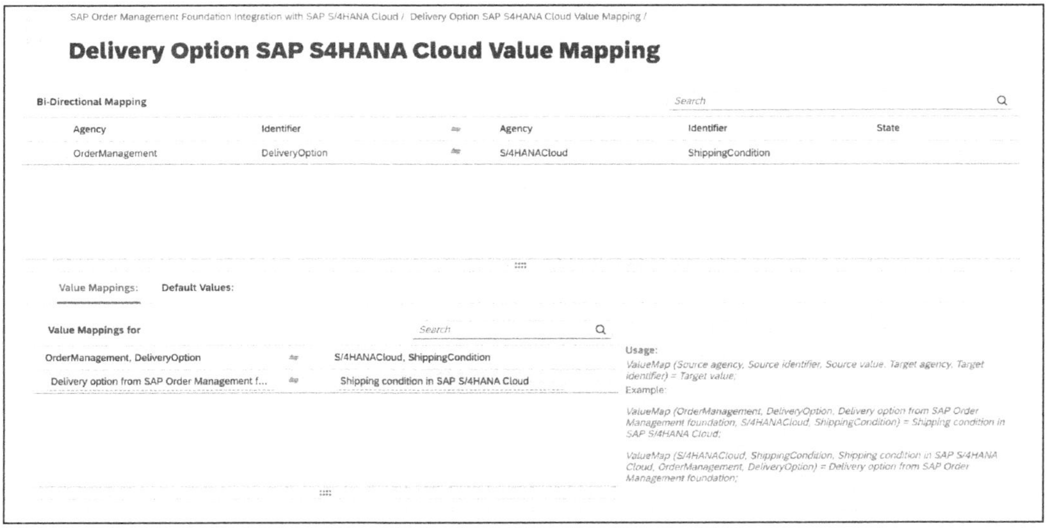

**Figure 1.5** Value Mapping

- Message mapping definitions in graphical models or XML/Extensible Stylesheet Language Transformations (XSLT)–based formats
- Resources such as XSD schemas, JavaScript Object Notation (JSON) schemas, XSLT files, and external archives

### Runtime Engine

The Cloud Integration *runtime*, which is a core component of SAP Integration Suite, is engineered for high scalability, resilience, and adaptability to diverse integration scenarios across cloud, on-premise, and hybrid landscapes. You can design iFlows by using the web-based graphical editor and deploy them to the runtime environment for execution. The runtime supports a wide array of integration patterns, including message transformation, routing, and protocol adaptation. Its modular pipeline architecture allows for flexible composition of processing steps that are tailored to specific integration requirements.

The Cloud Integration runtime runs within the customer's subaccount and offers the following:

- **Multitenant processing**
  Each tenant is logically isolated, multitenant capable, and operates independently.
- **High availability**
  The runtime is guaranteed to provide load balancing, redundancy, and automatic recovery in case of failures.
- **Transaction support**
  Supported protocols ensure seamless database access, which facilitates transactional processes.
- **Parallel processing**
  You can process multiple messages at once to ensure high throughput rates.
- **Scalability**
  Vertical and horizontal scaling is fully SAP-managed, so customers don't need to manually provision runtimes.

A key aspect of message processing during runtime is the *quality of service* (QoS)—in other words, how reliably a message is processed. By default, Cloud Integration guarantees that every incoming message is processed at least once in the tenant, regardless of the adapter technology used. In addition, certain adapters offer advanced options for controlling processing behavior, particularly regarding multiple transmissions and retries (see Table 1.1).

These QoS levels allow differentiated control of the integration logic, depending on business criticality, system behavior, and specific requirements for consistency and fault tolerance.

| Quality of Service | Description |
| --- | --- |
| At Least Once | This setting guarantees that an incoming message will be processed at least once by the system. If the same message arrives multiple times (e.g., due to a technical retry), each instance will be processed regardless of whether the system recognizes it as a duplicate. This setting is available by default for all sender adapter types. |
| Best Effort | In this mode, processing is synchronous. The integration platform responds immediately to the incoming request—without additional retry logic or delivery guarantees. This behavior is particularly suitable for noncritical scenarios with real-time characteristics. |
| Exactly Once | This option ensures that each message is processed exactly once—even if it's transmitted multiple times by the sending system. Cloud Integration checks whether the message has already been processed, for example, by using a message ID (such as an SAP Exchange Infrastructure message ID). Any duplicates are detected and automatically discarded. |

**Table 1.1** Message Processing QoS

You can also monitor the runtime engine with native monitoring and alerting functions, including performance statistics, error analysis, and trace logs.

### Edge Integration Cell

The *SAP Integration Suite, edge integration cell* allows for data processing within a private landscape. This allows sender and receiver systems to exchange data without passing through the internet, as the data is hosted exclusively in an on-premise environment. SAP Integration Suite's cloud-based environment is the tool of choice for designing integration content, and this content is then deployed within the organization's firewall at a private runtime location. The runtime environment is realized as a Kubernetes container, which facilitates secure, internal data exchange. If you want to process data within your private landscape, the edge integration cell is the solution. For example, consider a case where you want to have sender and receiver systems exchange data and be exclusively hosted in an on-premise environment. In this case, passing the data through the Internet isn't the best option.

You can use the cloud-based environment of SAP Integration Suite to design your integration content. Instead of deploying the integration content on your cloud-based Cloud Integration runtime, activate a private runtime location (for example, within the firewall of your organization) and have this runtime process the messages.

At runtime, the messages exchanged between sender and receiver systems are exclusively passed through your private landscape, as Figure 1.6 shows.

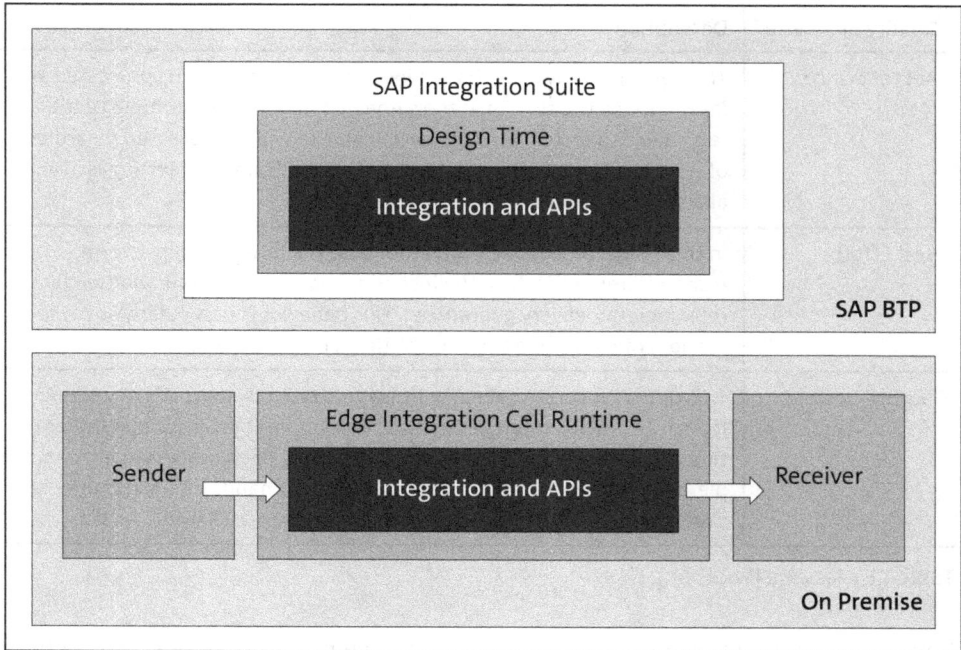

**Figure 1.6** Edge Integration Cell

You must use the Cloud Integration Monitor app to manage your integration artifacts and security content and to monitor message processing. As Figure 1.6 clearly shows, the edge integration cell runtime processes the integration scenarios and connects to the cloud at regular intervals to synchronize data, such as deployed artifacts that are required to reliably operate your integration scenarios.

**Keystore and Trust Store**

The *keystore* is a security-critical component of the cloud integration environment. It contains the following:

- **Private keys and certificates**
  These are for hypertext transfer protocol secure (HTTPS) and SFTP communication, client certificate authentication, and AS2 encryption

- **OAuth 2.0 client credentials**
  These are used for token-based authentication, such as with Microsoft Graph API and SAP SuccessFactors (see Figure 1.7)

- **Trust anchors**
  These are essential for validating server certificates when establishing HTTPS connections (see Figure 1.8)

Administrators can add, update, and delete keystore entries via the Cloud Integration UI or via REST/OData APIs.

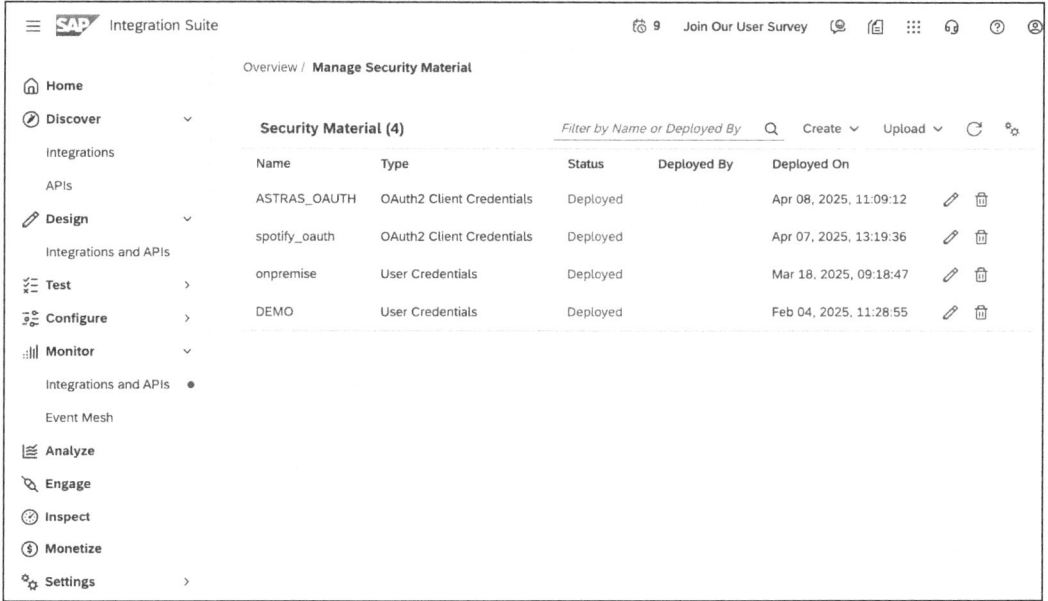

**Figure 1.7**  Security Materials

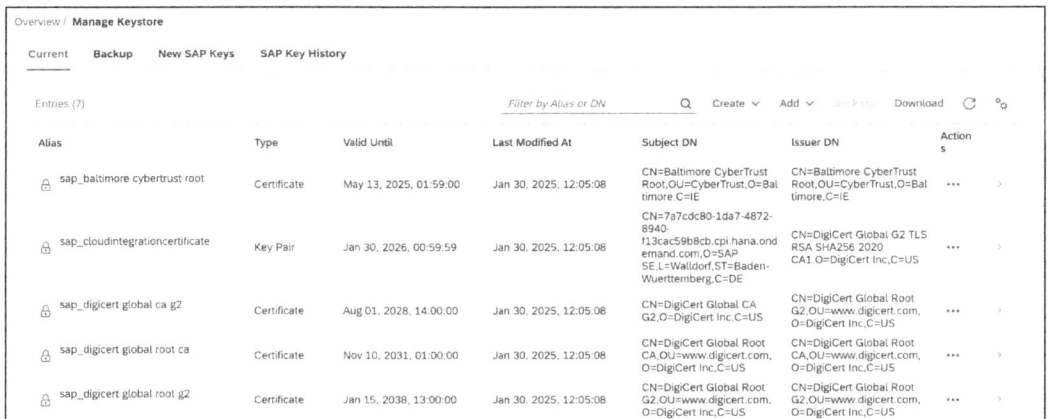

**Figure 1.8**  Keystore

## Monitoring and Tracing

*Monitoring* is a key discipline in the operation of iFlows. Cloud Integration provides several tools for monitoring and tracing, as follows:

- **Message processing log**
  This is a tool that allows users to efficiently manage and process messages (see Figure 1.9). It displays all messages passing through, along with their status (success, failed, or in progress), payloads, headers, and processing steps.

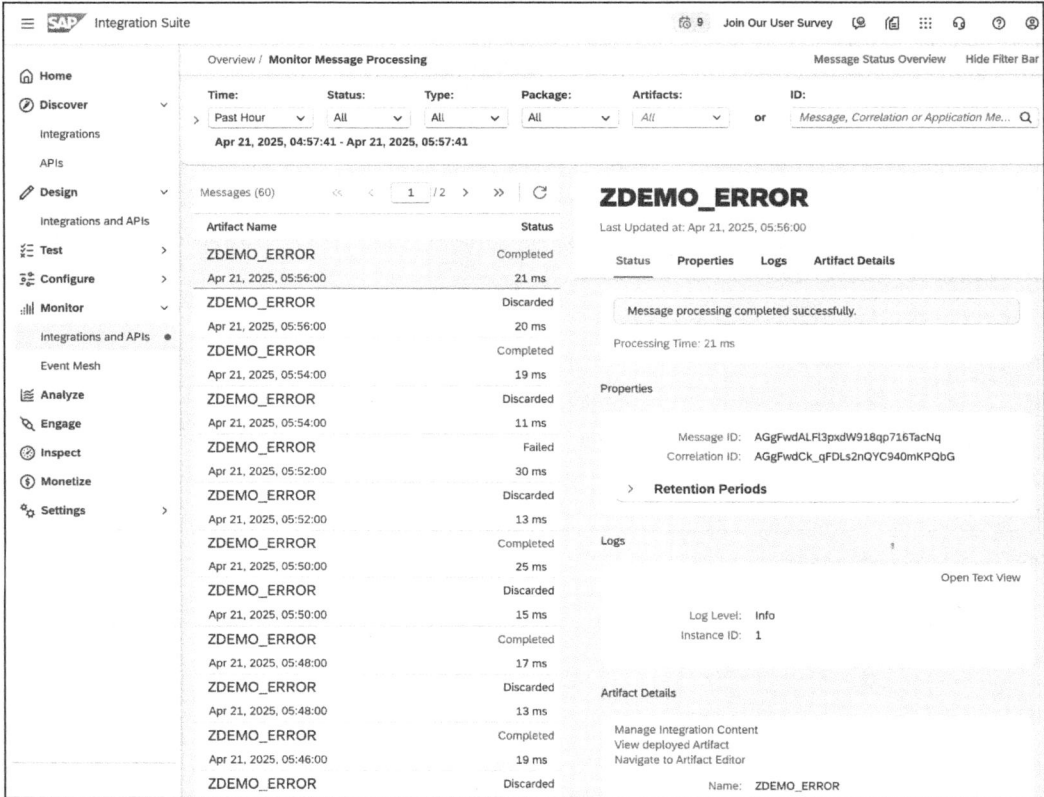

**Figure 1.9**  Cloud Integration Message Processing Log

- **System logs**
  You'll get detailed runtime information in the form of a trace log. For security-related information, the HTTP access log is also accessible (see Figure 1.10).

- **Trace mode**
  You can activate trace mode for individual flows or global sessions. It displays all values that are manipulated during runtime, in detail.

- **Custom logging (via Groovy)**
  Developers can set specific logs (e.g., for API responses, exception details).

Monitoring is also accessible via API, which allows seamless integration with leading log management systems like Splunk and SAP Cloud ALM.

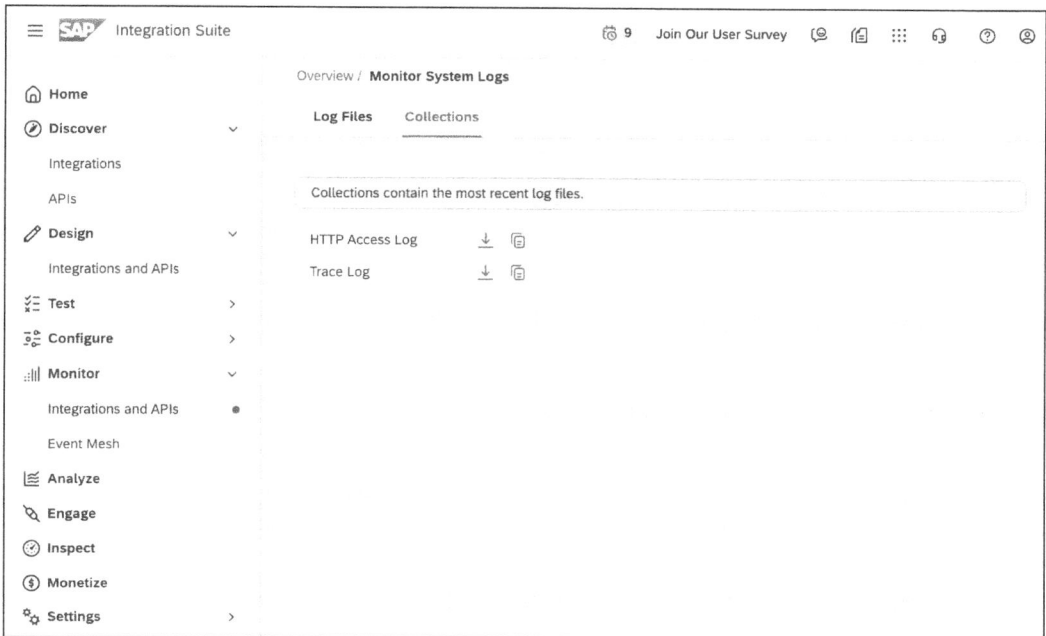

**Figure 1.10** System Logs

## Partner Directory

The *partner directory* is a tenant-specific storage component within Cloud Integration. It's specially designed for managing business partners within complex integration networks. It also serves as the central repository for partner-related *metadata*, which includes identifiers, communication endpoints, certificates, and routing information. It also enables highly flexible control of iFlows in the context of dynamic B2B scenarios.

The central concept here is configurable partner parameters. iFlows can be modeled with generic parameters, especially in scenarios with many communication partners, such as those found in classic EDI, supply chain, or payment transaction environments. The specific values—such as target URLs, partner IDs, and authentication data—aren't stored in iFlow itself. Instead, they are loaded dynamically from the partner directory at runtime. This is based on information from the incoming message context, such as headers, properties, and payload fields.

The partner directory approach has the following clear advantages:

- It has centralized maintenance, meaning partner data is maintained in one place and can be versioned.
- It boasts a high degree of reusability. A single generic iFlow can serve several hundred or several thousand partners.
- This guarantees zero downtime because new partners can be integrated into the existing architecture without you having to adapt or redeploy existing iFlows.

■ It allows for fast scaling. The establishment of new partner connections is significantly accelerated because there are no development cycles.

You can access the partner directory programmatically via an OData API (e.g., for maintenance via self-service portals or backend processes) and at runtime in the iFlow itself.
For example, you can access it through parameterized steps such as content modifiers,
routers, or request-reply calls with dynamic target values.

This architecture makes the partner directory a strategic enabler for scalable, maintainable, and reliable integrations in a B2B context. It offers maximum flexibility and minimal maintenance effort.

### 1.3.2   Cloud Integration Capabilities

Cloud Integration provides a rich set of capabilities that are designed to support
diverse message processing patterns between senders and receivers. These capabilities
are fundamental to building robust, scalable, and maintainable iFlows across heterogeneous system landscapes.

In this section, we explore the wide range of integration capabilities that are available
in Cloud Integration. These capabilities are categorized by function and are designed to
handle different scenarios such as message transformation, routing, content enrichment, persistence, security, and extensibility.

**Message Transformation**

*Message transformation* is one of the core functionalities in any integration scenario. It
ensures that the structure and semantics of data sent by a source system can be understood and correctly interpreted by the target system. Cloud Integration offers multiple
tools to support this need, and each one serves a specific purpose in data format trans-
lation and manipulation. The main subcapabilities under this category include the following:

■ **Mapping**
*Mapping* transforms the data structure and format used by the sender into a structure and format that the receiver can interpret. Cloud Integration supports the following:

  – Graphical message mapping with a visual editor, which is compatible with XSD
    and Entity Data Model XML (EDMX) (see Figure 1.11)

  – Script-based mapping with Groovy or JavaScript (see Figure 1.12)

  – XSLT-based transformations with support for XSLT 2.0 (see Figure 1.13)

  Mapping is essential in cross-application or cross-enterprise scenarios where structural differences in message formats must be harmonized.

■ **ID mapping**
*ID mapping* maps a source message ID to a target message ID, and it plays a critical
role in implementing exactly-once processing and deduplication logic.

- **Content modifier**
  The *content modifier* allows developers to read from and write to the message body, header, or property container (in a process called *message exchange*). This enables dynamic manipulation of message content to prepare it for target systems or downstream steps.

- **XML modifier**
  The *XML modifier* removes external document type definitions (DTDs) and XML declarations from inbound XML messages.

- **Converter**
  A *converter* is used to switch between formats like XML, JSON, comma-separated values (CSV), and EDI.

- **Decoder/encoder**
  A *decoder* is used to decode incoming encoded data or encoding outgoing messages.

- **Filter**
  A *filter* is used to extract specific nodes from a payload.

- **Message digest**
  A *message digest* is used to generate hashes from payloads. It's a compact, unique checksum (hash value) that's calculated from the content of a message. It serves as a digital fingerprint of the message and is used to ensure integrity and authenticity.

- **Script**
  A *script* is used to implement advanced, custom transformation logic with scripting languages.

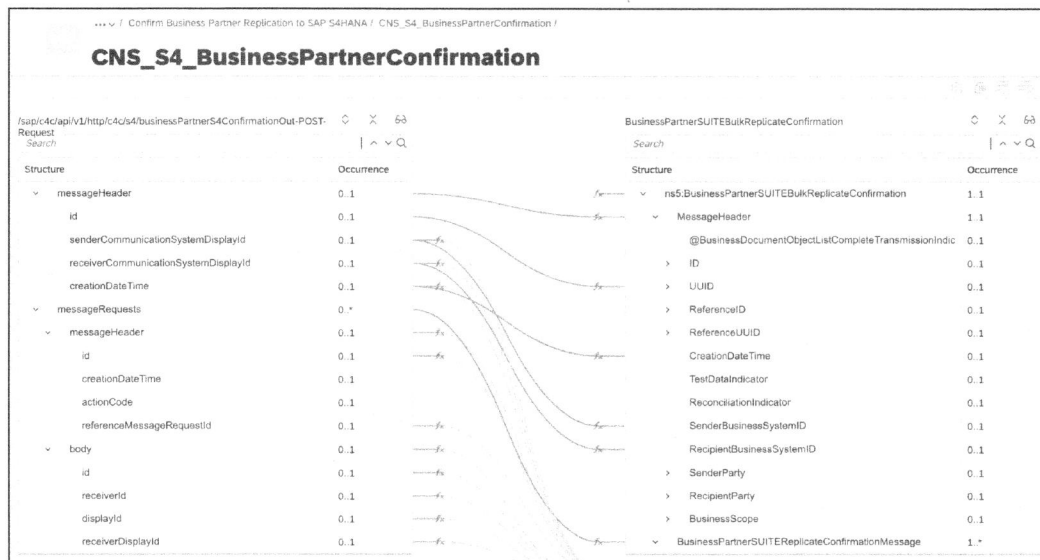

Figure 1.11  Graphical Message Mapping

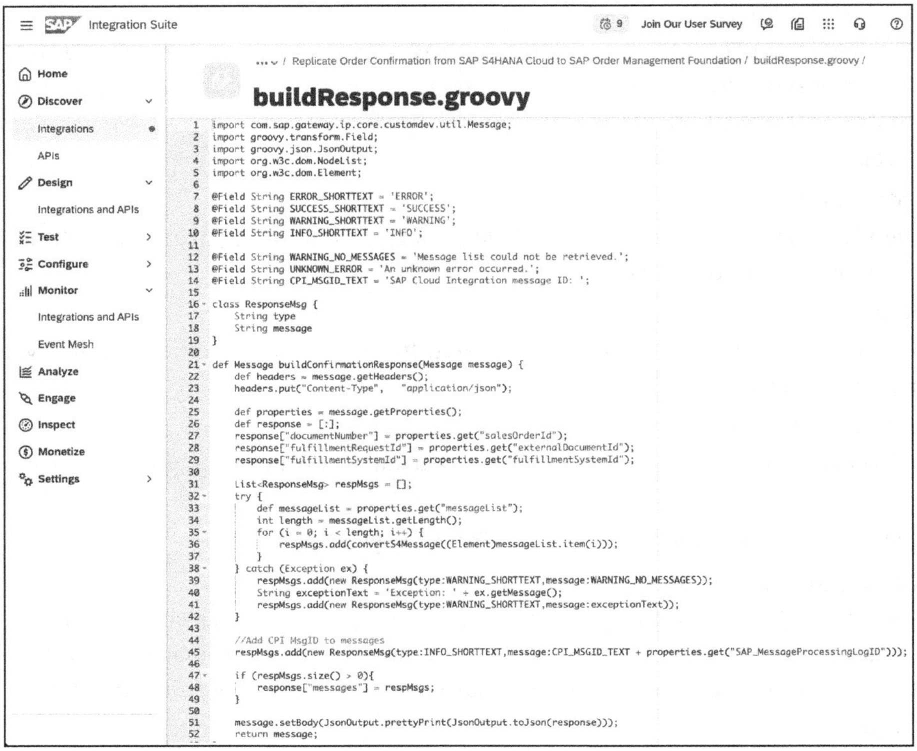

**Figure 1.12** Groovy Script

**Figure 1.13** XSLT-Based Transformation

### External System Calls and Process Modularization

In many integration scenarios, you need to communicate with external systems during message processing or to modularize complex logic into reusable components. Cloud Integration provides several capabilities to help you handle such cases efficiently, and they include the following:

- **Request-reply and send**
  This is used for synchronous and asynchronous communication with external services.

- **Content enricher and poll enricher**
  These are used to retrieve and enrich data from other systems.

- **Process call, looping process call, and idempotent process call**
  These are used to define and invoke reusable subprocesses and control duplicate execution.

---

**What Does Idempotent Mean?**

An operation is *idempotent* if it makes no difference whether it's called once or multiple times. The result and the state of the system don't change after it has been successfully executed once. In Cloud Integration, idempotency is crucial for avoiding duplicate executions, especially in the event of technical retries, network problems, or incorrect repetitions by source systems. The idempotent process call is the solution for this purpose. It reliably checks whether a message with the same ID has already been processed. If so, it isn't processed again; it's discarded or marked, for example.

---

### Routing

*Routing* is essential for directing messages to the correct recipient based on business logic or message content. Cloud Integration provides a flexible set of routing options and thus allows messages to be distributed, split, and recombined. Key capabilities under routing include the following:

- **Router**
  This implements conditional routing logic, and it supports content-based routing with header values, payload content, or properties.

- **Multicast**
  This is used to send the same message to multiple receivers simultaneously or sequentially.

- **Splitter**
  This breaks composite messages into individual parts for separate processing.

- **Join** and gather
  This recombines messages or collects responses from multiple branches.

**Data Persistence and Temporary Storage**

There are scenarios in which messages must be stored temporarily or persisted for later inspection or replay. Cloud Integration supports this with specialized capabilities for durable storage and variable management, including the following:

- **Persist message**
  This stores the full message payload for future inspection. The persist step uses the message store as the logical storage location, and the message store and the data store use the same tenant database.

- **Data store operations**
  These are temporary storage and retrieval operations (SELECT, GET, WRITE, and DELETE) for scenarios that require message buffering or manual retry logic. You can use them to implement the push-pull pattern where the message received from a sender is stored in the data store and a receiver polls (reads) the message from there during another message processing run. You can also use it to store the message temporarily for future reference. The data store is an ideal solution for storing messages so that you can use them across different iFlows. This component stores data on your tenant by using SAP Adaptive Server Enterprise (SAP ASE), platform edition. Please be advised that there is an overall database space limit of 32 GB.

- **Write variables**
  You can use this to set and use dynamic variables during runtime and to define variables to share data across different iFlows (that are deployed on the same tenant).

**Message Protection and Security**

*Data privacy* and *secure communication* are nonnegotiable in enterprise integration. Cloud Integration provides strong encryption, signature, and verification mechanisms to protect messages during transmission and processing. These include the following:

- **Encryptor/decryptor**
  You can use this to protect message content using Pretty Good Privacy (PGP) or PKCS #7 standards. *PGP* is used to asymmetrically encrypt and digitally sign message content, especially in B2B scenarios with high security requirements. *PKCS #7/ CMS* is a cryptographic standard that's used to encrypt and sign structured message formats, such as those found in EDI or AS2 contexts. It facilitates the secure transmission of data by employing standard-compliant cryptography.

- **Signer/verifier**
  You can use this to ensure message authenticity and integrity with digital signatures. When you sign a document, a digital signature is generated to verify the message's origin and integrity. The verify function inspects this signature to ensure its authenticity and prevent manipulation or forgery.

### 1.3.3   Prepackaged Content

SAP Integration Suite offers a suite of preconfigured integration content to provide companies with a streamlined approach to common integration scenarios. These integration packages contain ready-to-use iFlows, configurations, mappings, and documentation, specifically for frequently required SAP-to-SAP integrations such as SAP S/4HANA to SAP SuccessFactors, SAP Ariba, SAP Fieldglass, or SAP Analytics Cloud.

You can access this content directly via the **Discover** section of the SAP Integration Suite (see Figure 1.14).

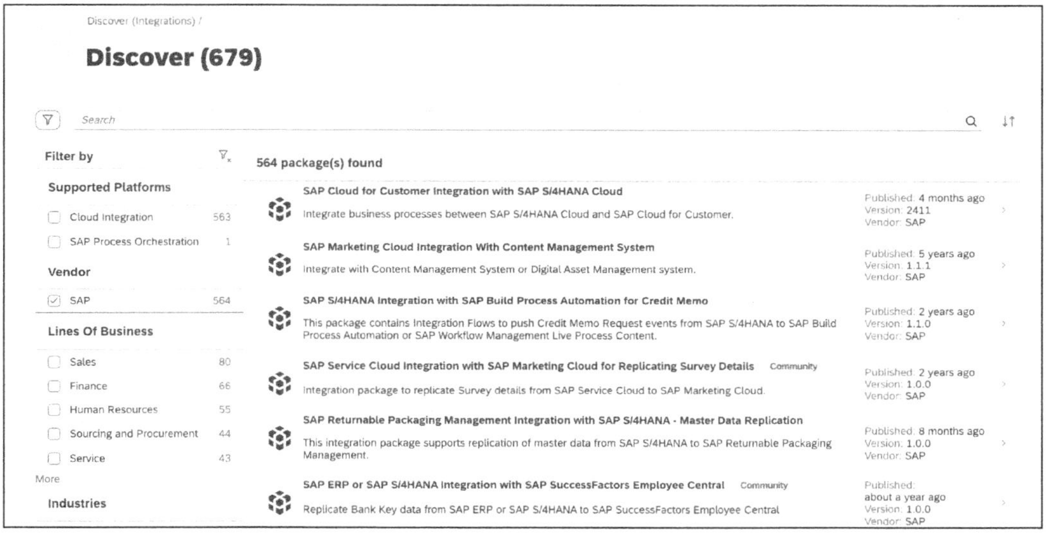

**Figure 1.14** Prepackaged Content in SAP Integration Suite

You can also discover prepackaged content via the *SAP Business Accelerator Hub* (formerly SAP API Business Hub; see *https://api.sap.com*), which contains not only SAP's proprietary software packages but also a wide range of SAP partner content, including industry-specific connectors, add-ons, and integration accelerators for third-party applications. Figure 1.15 shows the prepackaged content you can use to integrate SAP S/4HANA Cloud Public Edition with SAP SuccessFactors.

A particular highlight of the platform is the commitment of the SAP community. *Community packages* are also available via SAP Business Accelerator Hub, and they are openly licensed, freely editable integration packages under the Apache 2.0 license. These packages are often developed through direct exchanges among integration experts, customers, and partners while working to address real-world requirements. You can customize, expand, and use these resources in customer projects as needed— whether in a consulting environment or in productive cloud landscapes. See Figure 1.16 for an example of prepackaged community content.

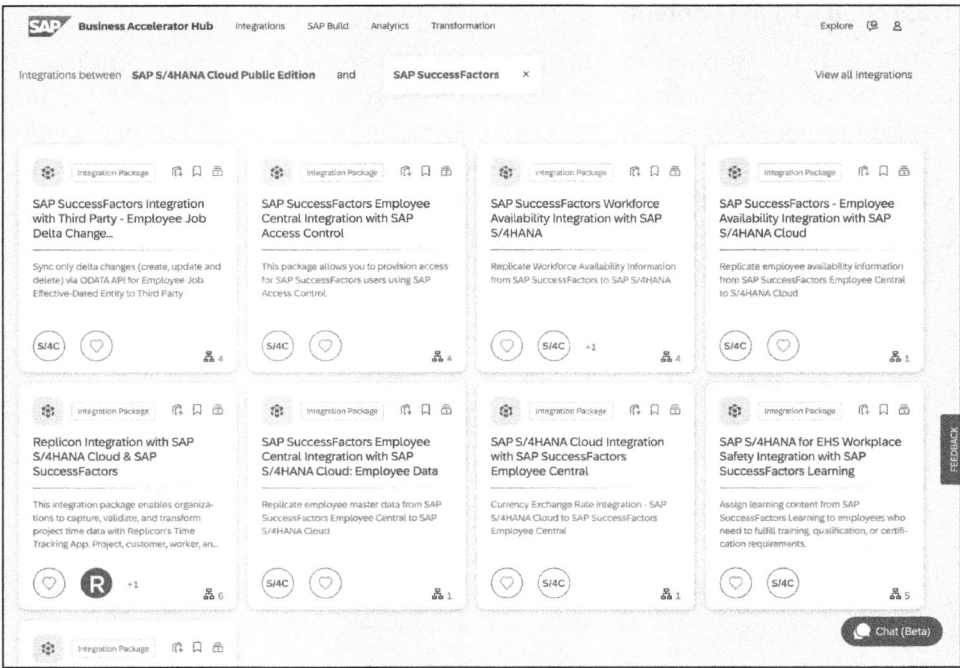

**Figure 1.15** SAP Business Accelerator Hub

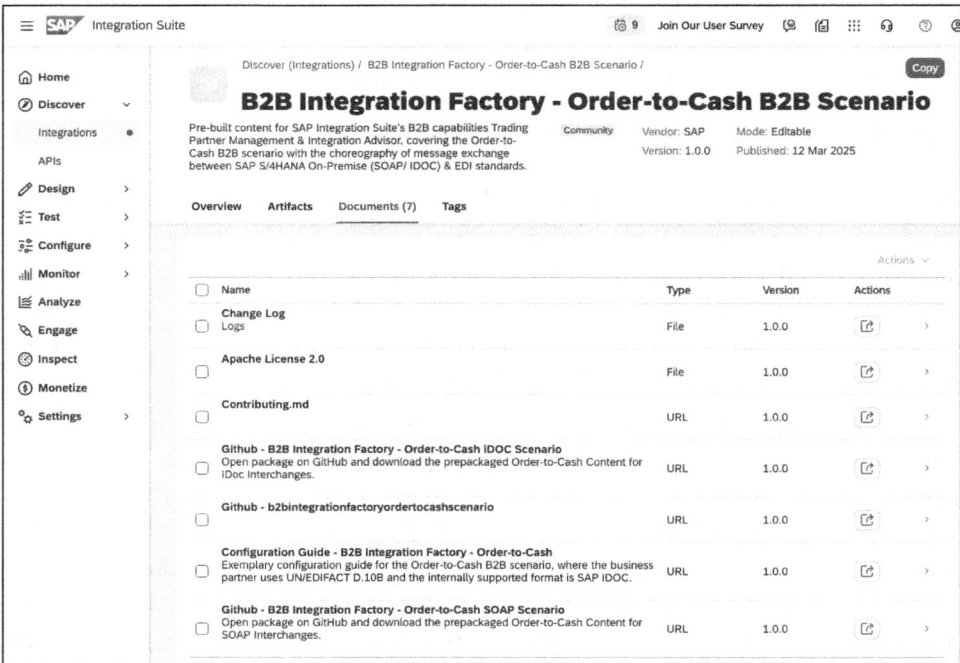

**Figure 1.16** Prepackaged Community Content

When you're using prebuilt content, SAP distinguishes between two types of usage:

- **Unmodified use**
  If you use an iFlow directly from an official package without modification, the processed messages are metered but not billed.

- **Modified use**
  Once you've edited the iFlow—for instance, to include your own process logic, transformations, or security mechanisms—billing is based on the number of messages processed (metered and billed).

For cross-tenant use cases, such as global rollouts and multiclient setups, SAP advises adhering to the documented best practices. This approach ensures that unmodified content doesn't incur unnecessary costs, even in distributed scenarios.

A *message* is defined as an electronic communication that's exchanged via the capabilities of the Cloud Integration. If a message exceeds 250 KB, the excess will be charged at a rate of one additional message for each 250 KB or portion thereof. This is the standard metering metric that's used to meter the usage of all SAP Integration Suite capabilities. In Cloud Integration, messages are counted for measurement and billing purposes, and the number of outgoing messages sent via receiver adapters—not the number of iFlow calls—is used for this calculation. (For further details, see SAP Note 2942344.)

The fundamental principles are as follows:

- Each time an iFlow is executed, the total size of all outgoing messages across all receiver adapters is tallied. Please note that a single message is counted for every 250 kilobytes of aggregated size.

- In instances where the total size exceeds 250 kilobytes, the figure is rounded up. For example, if the measurement is 270 kilobytes, two messages are generated; if the measurement is 520 kilobytes, three messages are generated.

- If a response is also received (e.g., via a request-reply step), the calculation is based on the sum of the request and the response.

Now, let's check this with some examples:

- 1 recipient, 70 KB → 1 message (less than 250 KB)
- 2 recipients, 70 KB + 40 KB = 110 KB → 1 message
- 2 recipients, 150 KB + 110 KB = 260 KB → 2 messages
- 120 KB request + 160 KB response = 280 KB → 2 messages
- 3 messages of 70 KB each (after splitter) → 3 messages

As you might expect, there are some special cases and exceptions, as you can see in the following list. Please note that only the input size at the receiver adapter is relevant. The final output quantity "over the wire" isn't considered.

- **Splitter**
  Each split generates its own messages, which are counted individually. If multiple splitters are present in the same iFlow, the splitter with the highest number of split messages will be used for metering.

- **Retries**
  Each repetition by retry mechanisms is counted separately.
- **JMS and process direct adapters**
  Messages via these adapters aren't counted because they are processed internally within the SAP Integration Suite tenant.
- **Subprocesses**
  The same rules apply here—outbound messages via receiver adapters are counted.

It's important to note that at present (August 2025), the actual measurement of message size isn't yet available in runtime. Until it is, the following guidelines will be in effect: each iFlow run with at least one receiver adapter will be counted as one message, regardless of the actual size of the messages.

The current metering statistics can either be found in the SAP BTP cockpit on the subaccount level (see Figure 1.17) or in the SAP BTP cockpit on the global account level.

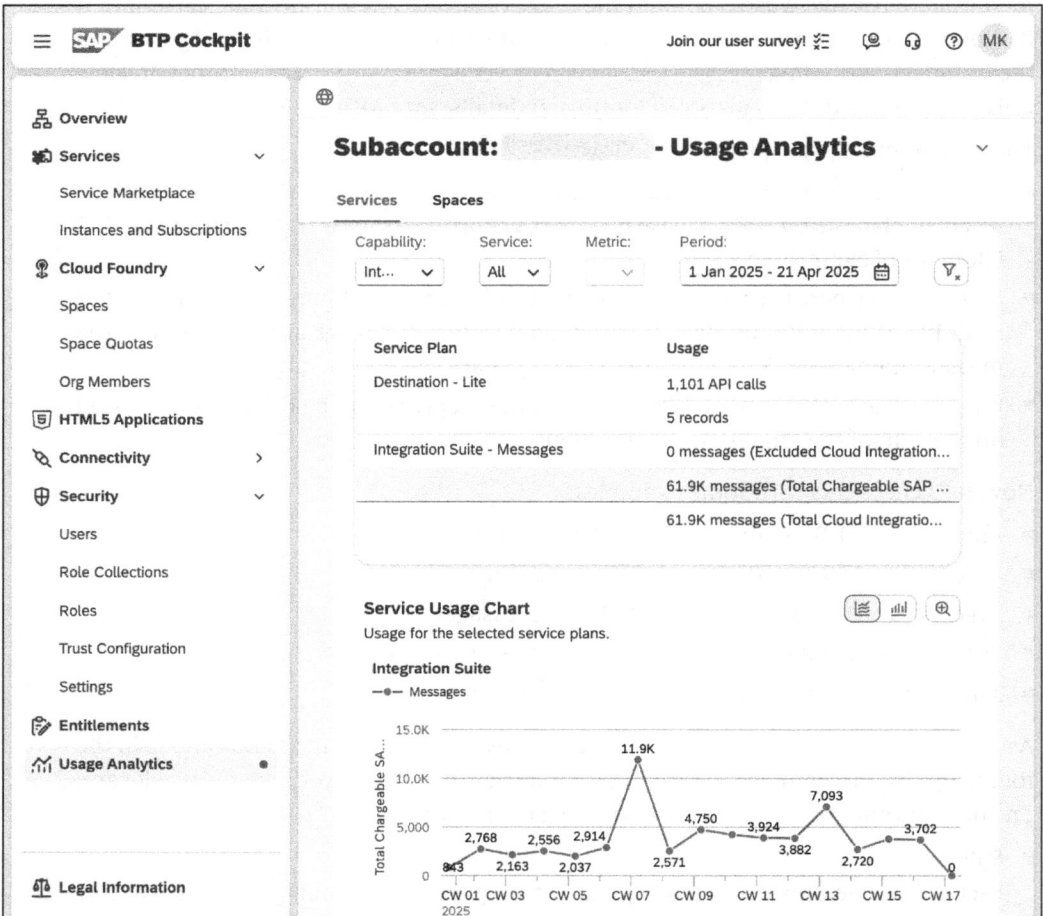

**Figure 1.17**  Message Metering Usage Analytics

## 1.4   Overview of Supported Adapters

*Connectivity adapters* are fundamental building blocks of Cloud Integration. They provide the technical bridge between SAP Integration Suite and the diverse ecosystem of external systems, applications, and protocols. Adapters determine how messages are received and sent across the integration landscape, whether you're connecting to SAP S/4HANA, exchanging data with third-party SaaS platforms, integrating legacy databases, or supporting complex B2B transactions.

Adapters abstract protocol-specific details and encapsulate connectivity logic, such as security, authentication, error handling, and data formatting. Cloud Integration clearly differentiates between *sender adapters*, which receive data into an iFlow, and *receiver adapters*, which send data to an external system. Other adapters are bidirectional and have both sender and receiver variants.

Adapters support a wide range of *protocols* and *integration patterns*, such as REST, Simple Object Access Protocol (SOAP), OData, IDoc, FTP/SFTP, JMS, AS2/AS4, Java Database Connectivity (JDBC), Lightweight Directory Access Protocol (LDAP), social media, and fiscal authority connections such as ELSTER. Cloud Integration supports both enterprise-scale and edge-level scenarios, including hybrid and event-driven architectures.

In the following sections, we'll categorize and explain each adapter in detail, providing both functional and technical context to help architects and developers make informed design decisions. Adapters are categorized based on common technical and functional patterns, and each category includes an introductory explanation, followed by detailed descriptions of the adapters belonging to that group.

### 1.4.1   HTTP–Based Adapters

Adapters in this category use the HTTP protocol to communicate, and they play a critical role in API-based integrations. These adapters are the go-to solution for integrating cloud applications, SAP backend services, and third-party APIs that use REST, SOAP, and OData protocols. They support various message formats, including XML, JSON, and *AtomPub*, as well as a range of authentication mechanisms, such as *Basic Auth*, *OAuth 2.0*, and *mutual transport layer security* (mTLS). HTTP-based adapters enable both synchronous and asynchronous communication patterns and are highly configurable to match different service endpoint requirements.

From a protocol perspective, HTTP-based adapters allow you to define headers, query parameters, URL paths, request/response payloads, and error-handling behavior. These adapters are the foundation of service exposure, consumption, and orchestration in API-centric architectures.

The HTTP receiver adapter reliably performs REST-style HTTP calls, and it offers full support for GET, POST, PUT, and DELETE methods. It offers granular control over HTTP headers and URL parameters, supports both JSON and XML payloads, and allows for asynchronous or synchronous processing. It's the go-to solution for connecting to REST APIs from external cloud systems, including customer portals, web shops, and mobile apps.

---

**REST and IETF RFCs**

Representational State Transfer (REST) isn't a protocol; it's an architectural style for web services. REST defines how systems should communicate via the HTTP protocol: stateless, resource-oriented, or standards-compliant. REST uses the mechanisms described in the official HTTP specifications (remote function calls [RFCs]) in a targeted way.

The technical foundations for REST-based APIs are the HTTP RFCs. In particular, they are as follows:

- RFC 7230–7235 (HTTP/1.1) defines methods such as GET, POST, PUT, DELETE, headers, status codes, and caching mechanisms.
- RFC 9110–9112 (HTTP/1.1 & HTTP/2) clearly specifies the current successors, which further standardize and optimize HTTP.

REST doesn't say how to implement HTTP, but rather, it builds on the HTTP standards. For example, the following rules govern the use of GET, POST, PUT, and DELETE:

- GET is used to retrieve resources.
- POST is used to create new content.
- PUT is used to replace existing content.
- DELETE is used to delete content.

These rules are based on the definitions and rules of the HTTP RFCs.

---

Let's look at the key HTTP-based adapters:

- **HTTPS sender adapter**
  This is the key to integrating secure HTTP endpoints because it enables seamless consumption from external systems via HTTPS. It supports OAuth 2.0 for token-based authentication, mTLS for certificate-based trust, and custom header manipulation. The adapter is perfect for webhook-driven scenarios and when third-party systems need to push data into Cloud Integration.

- **SOAP adapter (sender/receiver)**
  This provides full support for SOAP 1.1 and 1.2 protocols. The adapter integrates seamlessly with WSDL-based web services and supports advanced features like Web Services Security (WS-Security), SOAP headers, and attachments (Multipurpose Internet Mail Extensions [MIME] and the Message Transmission Optimization Mechanism [MTOM]). It's frequently used to connect to enterprise service buses and legacy SOAP-based systems.

- **OData adapter (sender/receiver)**
  This is the tool you need to communicate with OData v2 and v4 services. The adapter consumes or exposes entity sets and supports operations such as CREATE, READ, UPDATE, and DELETE (CRUD). It's used in SAP S/4HANA, SAP SuccessFactors, and SAP Cloud for Customer integrations. It boasts advanced features such as cross-site request forgery (CSRF) token handling, key-based filtering, and pagination.

- **ODC receiver adapter**
  This is the perfect solution for SAP Gateway's OData channel (ODC). It directly executes operations in ABAP backends by invoking OData services that are defined in the SAP Gateway framework. It's perfect for fine-grained service invocation, and it works well with the broader OData adapter when tighter ABAP integration is needed.

- **OpenConnectors receiver adapter**
  This allows you to unite your business with a single integration layer that's accessible to over 150 non-SAP SaaS applications through normalized REST APIs. It abstracts away the complexity of dealing with different authentication schemes, object models, and pagination mechanisms. Examples include integrations with Salesforce, Zendesk, HubSpot, ServiceNow, and Google Workspace.

### 1.4.2   Messaging and Event-Based Adapters

*Messaging and event-based adapters* are essential for decoupled, asynchronous, and event-driven integration architectures. They rely on established messaging protocols like AMQP, JMS, and Kafka to ensure reliable communication patterns among distributed components. Their use unquestionably promotes scalability, resiliency, and the ability to react to real-time events without introducing tight coupling between systems.

In Cloud Integration, these adapters are used in microservice architectures, IoT scenarios, and when integrating with message brokers like Event Mesh, Apache Kafka, and RabbitMQ. They also support features such as guaranteed delivery, message persistence, retry logic, and message correlation, which are essential for robust enterprise integration.

Let's look at the following key messaging and event-based adapters:

- **AMQP adapter (sender/receiver)**
  This integrates with AMQP brokers such as Event Mesh and open-source platforms like RabbitMQ. It supports the publication of events as a receiver adapter and the consumption of messages from queues or topics as a sender adapter. This adapter is perfect for loosely coupled communication in event-driven systems.

- **JMS adapter (sender/receiver)**
  This implements the JMS standard for message queuing. It provides capabilities for reading from and writing to JMS queues and topics. The adapter supports delayed

message processing, retries, and decoupling of producers and consumers. JMS is the go-to solution for hybrid SAP landscapes and for buffering messages between iFlows.

- **Kafka adapter**
  This enables Cloud Integration to seamlessly communicate with external Kafka brokers by using the Kafka protocol. The adapter supports publishing messages to Kafka topics and consuming records from them, and it's perfect for high-throughput, low-latency streaming scenarios, such as real-time analytics, telemetry, and data lake ingestion.

- **ProcessDirect adapter (sender/receiver)**
  This ensures seamless synchronous communication between iFlows within the same tenant. It provides a lightweight and performant mechanism for composing integration logic from reusable components. This adapter is the go-to solution for modular design, and it's used extensively for encapsulating shared logic such as authentication, lookups, and error handling.

### 1.4.3  B2B and EDI Adapters

The adapters in this category ensure standardized, secure, and interoperable exchange of structured business documents. These capabilities are essential for supply chain management, finance, logistics, and procurement, especially when you're working with external partners, regulatory bodies, or legacy B2B infrastructures. The supported protocols—such as AS2, AS4, IDoc, and XI—ensure reliable transmission of business-critical data (e.g., purchase orders, invoices, shipping notices) with end-to-end compliance and traceability.

B2B adapters in Cloud Integration meet enterprise integration requirements, which include nonrepudiation, message-level security, acknowledgments, and long-running asynchronous messaging. These adapters are highly configurable and support both file- and payload-based message bodies in various structured formats like XML and EDIFACT.

Let's look at the following key B2B and EDI adapters:

- **AS2 adapter (sender/receiver)**
  This implements the AS2 protocol, which is widely used for EDI document exchange. The adapter supports encryption, digital signatures, and message disposition notifications. AS2 is the go-to solution for partner integration, especially in retail and manufacturing, where secure document transfer and acknowledgment are essential.

- **AS4 adapter (sender/receiver)**
  This is based on the ebMS 3.0 messaging standard, and it provides advanced features for secure, reliable document exchange over web services. It's especially crucial for the public sector and utilities, where policy-based messaging, audit trails,

and compliance are paramount. The adapter supports payload encryption, signing, and compression.

- **IDoc adapter (sender/receiver)**
  This reliably processes IDoc (Intermediate Document) messages via SOAP communication, it supports both synchronous and asynchronous messaging, and it retains control record structures as defined in the SAP S/4HANA system. It's perfect for tightly integrated SAP-to-SAP scenarios and when transitioning from SAP PI/SAP PO to Cloud Integration.

- **XI adapter (sender/receiver)**
  This supports the XI 3.0 messaging protocol, and it ensures compatibility with SAP PI and SAP PO. It enables reuse of existing SAP PI/SAP PO artifacts and facilitates hybrid integration landscapes, thus allowing staged migrations from on-premise to cloud-based architectures.

---

**XI3.0 Protocol**

*XI 3.0* is a proprietary SAP protocol for cross-system message exchange within SAP PI or SAP PO. It's based on SOAP and was developed specifically for asynchronous, XML-based message transport in the SAP environment. The following are the typical features:

- **Transport envelope**
  This is SOAP 1.1.

- **Communication**
  This is asynchronous with optional QoS.

- **Header information**
  This is SAP's own, including message ID, sender/receiver service, and interface name.

- **Use**
  This is to connect SAP and non-SAP systems via central middleware.

In modern cloud scenarios, XI 3.0 is usually replaced by standard APIs (OData/REST), but it's still required for migration or connection of classic SAP PI/SAP PO landscapes.

---

### 1.4.4   File-Based Adapters

File-based adapters are the solution for integration scenarios where file exchange is the primary communication mechanism, especially in legacy or hybrid IT landscapes. These adapters are essential for integrating with systems that produce or consume structured or unstructured data in file format, such as ERP systems, bank file interfaces, and legacy middleware. Despite the rise of API-based communication, file exchange remains common due to its simplicity, batch-processing nature, and widespread tooling support.

File-based adapters in Cloud Integration support scheduled polling, file pattern matching, directory traversal, and postprocessing actions such as archiving and deletion. They also offer different authentication models and transport encryption options, depending on the protocol used.

Let's look at the following key file-based adapters:

- **SFTP adapter (sender/receiver)**
  This is the solution for secure file transfer over the Secure File Transfer Protocol (SFTP). The adapter supports key-based authentication, polling intervals for file availability, filename pattern matching, and options for archiving or deleting processed files. It's the preferred choice for secure and compliant file exchange, particularly in regulated industries.

- **FTP(s) adapter (sender/receiver)**
  This implements the traditional File Transfer Protocol (FTP) to enable file upload and download from FTP servers. FTP is less secure than SFTP due to its lack of encryption, yet it remains a popular choice in older environments where encryption isn't enforced. The adapter unquestionably supports authentication, polling, and basic postprocessing operations.

### 1.4.5 Mail and Communication Adapters

Mail adapters enable integration scenarios based on email communication, and they are essential in situations where systems can't expose APIs or messaging endpoints and instead rely on email as the primary data exchange mechanism. Use cases include automated report delivery, alerting, document-based communication (e.g., invoices, order confirmations), and interaction with legacy systems and external partners.

These adapters support both polling-based inbound communication (using the Internet Message Access Protocol [IMAP] or Post Office Protocol version 3 [POP3]) and outbound delivery (via the Simple Main Transfer Protocol [SMTP]). Advanced features include filtering by subject, sender, or attachments, and secure transport via Secure Sockets Layer (SSL)/Transport Layer Security (TLS).

A mail adapter is the tool you need to send and receive emails. On the sender side, it uses SMTP to dispatch emails with configurable headers, subjects, and body content. You can add attachments dynamically from integration payloads. On the receiver side, it supports IMAP and POP3 to poll mailbox folders, filter incoming messages, extract attachments, and trigger further processing. Our system uses secure authentication methods and encrypted channels to comply with modern security standards.

### 1.4.6 SAP-Specific Application Adapters

These adapters are purpose built to integrate directly with SAP core systems, cloud solutions, and compliance frameworks. They offer optimized communication with SAP applications like SAP SuccessFactors, SAP Ariba, SAP S/4HANA, and SAP Gateway, and they are designed to make full use of SAP's native data models, APIs, and security frameworks. These adapters reduce implementation effort, ensure compatibility with SAP-specific formats (such as IDoc, commerce eXtensible Markup Language [cXML], and Business Application Programming Interface [BAPI]), and support SAP's best-practice integration content.

These adapters are essential in scenarios involving HR data exchange, procurement automation, or tax reporting. They offer robust and prealigned mechanisms that ensure high data quality, process orchestration, and SAP-specific error handling.

Let's look at the following key SAP-specific adapters:

- **SuccessFactors adapter (sender/receiver)**
  This integrates with SAP SuccessFactors by using REST, SOAP, and OData v2/v4 protocols. It can be used to query or manipulate entities such as employees, positions, job applications, and learning activities. The adapter supports a variety of authentication mechanisms, including OAuth 2.0 and basic auth. It's versatile and essential for modern HR integrations, and it supports operations such as QUERY, INSERT, UPDATE, UPSERT, and DELETE.

- **Ariba adapter (sender/receiver)**
  This is the key to seamlessly exchanging business documents with SAP Business Network using cXML. Common use cases include purchase orders, order confirmations, invoices, and shipping notices. It ensures secure communication and adheres to SAP Business Network integration standards to guaranteeing seamless interoperability in procurement and supplier collaboration scenarios.

- **RFC receiver adapter**
  This reliably connects to on-premise ABAP systems using the RFC protocol. It allows you to invoke RFC-enabled function modules (RFMs) from the cloud. Configuration is required via the cloud connector and trusted RFC destinations. It's well-suited for synchronous calls into business logic implemented in SAP ERP or SAP S/4HANA.

- **IDoc adapter (sender/receiver)**
  This adapter is responsible for transmitting and receiving IDocs via SOAP. It preserves the structure of control and data records as defined in the SAP system, and it supports both synchronous and asynchronous IDoc processing. It's a staple in logistics, finance, and master data replication between SAP systems.

- **ELSTER receiver adapter**
  This is designed for one purpose: to submit tax-related data to German tax authorities via the ELSTER platform. It ensures secure, encrypted communication and compliance with legal requirements for data formats and transfer protocols. Use cases include VAT declarations and wage tax reporting.

### 1.4.7  Social Media and Public APIs

These adapters enable Cloud Integration to interact with publicly accessible platforms and social media APIs, thus facilitating use cases such as sentiment analysis, brand monitoring, user engagement, and real-time communication. These adapters are used in marketing automation, customer feedback loops, and analytical dashboards.

Integrating external public data into SAP systems gives businesses visibility into how their brands, products, or services are perceived in open channels. Furthermore, these adapters respond dynamically to user-generated content and enable data-driven engagement strategies.

Let's look at the following key social media/public API adapters:

- **Facebook receiver adapter**
  This connects to the Facebook Graph API to retrieve public data, such as posts and comments. It supports keyword- and profile-based filtering, pagination of large result sets, and structured parsing of metadata. This adapter is used in social listening platforms, sentiment analysis tools, and marketing analytics.

- **Twitter receiver adapter**
  This allows you to both read and post tweets by using the Twitter API. It supports filtering by hashtags, user mentions, geolocation, and time windows, and it's ideal for use in real-time dashboards, customer engagement workflows, and public event monitoring. Authentication and rate-limiting compliance are built in to align with Twitter's developer guidelines.

### 1.4.8   Directory and Database Integration

These adapters enable Cloud Integration to interact directly with structured data repositories, such as relational databases and directory services. This category of adapters is essential for scenarios that involve master data synchronization, identity and access provisioning, reporting, or backend integration with custom or legacy systems.

These adapters are essential for connecting Cloud Integration with critical IT infrastructure, such as HR systems, ERP databases, and LDAP directories. They facilitate real-time queries, updates, and user management to ensure seamless integration and optimal performance.

Let's look at the following key directory and database integration adapters:

- **JDBC receiver adapter**
  This facilitates structured query language (SQL)–based interaction with relational database systems. The adapter supports execution of dynamic SQL statements (`SELECT`, `INSERT`, `UPDATE`, and `DELETE`), stored procedures, and batch operations. Payloads can be passed with structured XML or direct SQL command strings, and result sets must be mapped back into message payloads for downstream processing. The supported databases are as follows:
  - SAP HANA Cloud and SAP HANA, platform edition
  - SAP ASE

- IBM Db2
- Microsoft SQL Server
- Oracle
- PostgreSQL

The adapter's got options for transactional control, error handling, and configurable connection pooling. It's frequently used in hybrid integrations where cloud-based iFlows need to access or synchronize data with on-premise databases.

- **LDAP receiver adapter**
This provides unmitigated read/write access to Lightweight Directory Access Protocol (LDAP) systems. This adapter is used for identity management tasks, such as creating, modifying, or deleting user entries in corporate directories. Common use cases include role provisioning, group assignment, and replication of user data from cloud HR systems to on-premise directories. It supports standard LDAP operations and attributes, and it integrates with Active Directory and other LDAP-compliant services.

## 1.5    iFlow Design Guidelines

Designing iFlows is at the heart of working with SAP Integration Suite. As an integration developer, you're responsible for building flows that are functional, robust, secure, and maintainable. These flows often support mission-critical business processes, and poorly designed flows inevitably lead to performance bottlenecks, security vulnerabilities, and increased operational costs—risks that enterprises simply can't afford.

SAP presents a set of practical guidelines to help you design high-quality iFlows. They cover three main areas:

- **Learning the basics**
Before moving on to advanced scenarios, you need to master the foundational modeling capabilities of Cloud Integration. This includes understanding how to structure iFlows, work with adapters, configure routing, and use core building blocks such as content modifiers and message mappings. You should master these basics so you can model iFlows efficiently and consistently. For more information, see *https:// help.sap.com/docs/cloud-integration/sap-cloud-integration/learn-basics*.

- **Guidelines for designing enterprise-grade iFlows**
Enterprise-grade iFlows must meet higher standards of security, reliability, and maintainability. This section of the guidelines will teach you the design principles that will make your flows ready for production. You must apply the highest security standards (encryption, authentication, and authorization), and you must keep flows readable and well-structured to ensure that they remain understandable even years later. You should also implement robust error handling and monitoring to ensure

smooth operations. These guidelines will help you safeguard business-critical processes while keeping complexity under control. For more information, see *https://help.sap.com/docs/cloud-integration/sap-cloud-integration/guidelines-to-design-enterprise-grade-integration-flows*.

- **Guidelines for implementing specific integration patterns**
  You can address many integration challenges by applying well-established enterprise integration patterns. This of the guidelines provides you with clear instructions on implementing patterns, such as the following:
  - Content-based routing is the process of directing messages dynamically, based on payload content.
  - In content enrichment, messages are augmented with additional data from external systems.
  - A splitter breaks a single message into multiple parts for parallel processing.

  For more information, see *https://help.sap.com/docs/cloud-integration/sap-cloud-integration/guidelines-to-implement-specific-integration-patterns*.

Master these patterns and you'll design flexible, scalable solutions.

## 1.6 Summary

In Section 1.1, we clearly stated this chapter's mission: to provide practical, hands-on insights into (not just theoretical knowledge of) SAP Integration Suite and especially Cloud Integration. In Section 1.2, we provided an overview of Cloud Integration as the central capability of SAP Integration Suite and the digital backbone of hybrid landscapes. In Section 1.3, we delved into the principles that shape modern integration, and we explained the concept of hybrid landscapes, the role of Cloud Integration as cloud-based middleware, and the significance of the clean core approach. We also introduced key technical components such as design time, runtime, monitoring, security stores, and partner directories, alongside advanced message-processing features, prepackaged integration content, and metering rules.

In Section 1.4, we shifted the focus to connectivity and provided a structured overview of the wide range of adapters available in SAP Integration Suite. These include HTTP, OData, SOAP, event-driven, B2B/EDI, file-based, mail, SAP-specific, database, and social media adapters. We also positioned each adapter family within typical enterprise scenarios to show how they extend Cloud Integration into every corner of a heterogeneous IT landscape.

Finally, in Section 1.5, we introduced the foundational best practices for modeling iFlows. We emphasized modular design, reusability, and scalability to preparing you for the in-depth, scenario-driven guidance in later chapters. At this point, you've gotten a strategic and technical overview of SAP Integration Suite, and you're ready to tackle practical project-level challenges.

# Chapter 2

# Infrastructure and Configuration Best Practices

*The process of successful integration doesn't commence with the initial line of code; rather, you initiate it by establishing the appropriate foundation.*

In the context of an evolving IT landscape that's characterized by the integration of cloud and on-premise systems, APIs, event streams, and data sources from diverse domains, the establishment of a clean architecture and configuration is a pivotal factor in ensuring success. It's imperative for organizations to meticulously plan their integration platform from the outset to ensure the establishment of a foundation that's conducive to stability, scalability, and security.

This chapter presents the optimal methodology for preparing and configuring SAP Integration Suite. It'll teach you how to establish the platform within SAP Business Technology Platform (SAP BTP) through a methodical approach, emphasizing the adherence to structured architectural principles and security concepts. Also, it'll give you a comprehensive overview of typical deployment scenarios, ranging from cloud-to-cloud to hybrid scenarios, and it emphasizes both security and efficiency in implementation. Then, it will methodically introduce you to best practices that can take you from having a functioning integration solution to an excellent one.

But before you can dive into detailed best practices, you need to understand the foundational elements that make up the infrastructure and configuration landscape of SAP Integration Suite. Each core section of this chapter will introduce you to a specific building block that contributes to establishing a robust, secure, and future-ready integration environment. Together, they form a logical sequence—from conceptual design to technical implementation—that you can use to makes sure your integration projects begin on solid ground.

Section 2.1 outlines the modular structure of SAP Integration Suite within SAP BTP and explains how its component-based design supports diverse integration requirements. From integration flow (iFlow) orchestration to API Management, Open Connectors, Event Mesh, and the cloud connector, this section introduces the technological backbone that underpins scalable and maintainable integration architectures.

Section 2.2 guides you through the initial provisioning of SAP Integration Suite. From Cloud Foundry activation in the SAP BTP subaccount to capability selection and activation, this part of the chapter provides hands-on guidance on how to initialize the integration platform properly. It emphasizes the importance of environment separation, structured configuration, and role-aware provisioning.

Section 2.3 shifts the focus to access control. It details how the role-based access control model in SAP BTP ensures secure, transparent, and governed use of the platform. It covers both manual assignment in the cockpit and automated provisioning via Identity Authentication, and it demonstrates how to clearly define and maintain access and responsibilities across project lifecycles.

Section 2.4 introduces a crucial piece of hybrid scenarios that provides secure access to on-premise systems without exposing internal infrastructure to the internet. You'll learn how to install, configure, and connect the cloud connector to SAP BTP and thus enable trusted communication with backend systems like SAP S/4HANA—while maintaining strict network and identity controls.

Finally, Section 2.5 addresses identity federation and secure user login via centralized authentication. This includes establishing trust relationships between SAP BTP and Identity Authentication, configuring domains and origin keys, and enabling seamless single sign-on (SSO). The chapter further contextualizes this process by showing how users can be managed, grouped, and granted access based on attributes and group membership to ensure consistency and compliance across the system landscape.

## 2.1   Architecture

SAP Integration Suite functions as a centralized integration platform within SAP BTP, thereby empowering companies to implement comprehensive, flexible, and future-proof integration of cloud and on-premise systems. In the contemporary context of increasingly heterogeneous IT landscapes, there is a growing imperative for the effective integration of data sources, applications, and processes, irrespective of their physical or logical location. A meticulously conceived architecture isn't merely a prerequisite for technical stability; it's also of paramount importance for operational agility, scalability, and security.

SAP Integration Suite is distinguished by its modular and service-oriented architecture, which is itself distinguished by a substantial number of components, each of which is designed to fulfill distinct integration requirements while exhibiting seamless interlocking capabilities. Its objective is to provide companies with a malleable array of functionalities that encompass synchronous and asynchronous communication, API management, event-driven architecture (EDA), and hybrid scenarios.

This section provides a comprehensive view of the architectural building blocks that make up SAP Integration Suite and their role in shaping a reliable and enterprise-ready integration platform. It begins with an overview of the suite's key functional components, including the iFlow engine, API Management, and event-based messaging capabilities. It then explores common deployment models, ranging from cloud-only to hybrid scenarios, that illustrate how the suite adapts to diverse system landscapes. It then shifts focus to the platform's embedded security mechanisms, including identity propagation, data protection, and access control. Finally, the section outlines design best practices that help ensure long-term scalability, maintainability, and operational resilience in real-world integration projects.

### 2.1.1    Overview of Key Components

The iFlow engine, which is based on Apache Camel, occupies a central position within the architectural framework. It's responsible for the execution and orchestration of integration processes: the iFlows. These iFlows are modeled with a web-based graphical tool, and they contain the entire logic for message processing: from receipt to validation, transformation, mapping, and forwarding to target systems. Standard protocols such as HTTP(S), SOAP, OData, SFTP, and numerous others can be utilized. The iFlow engine's forte is its adaptability, which enables the implementation of a diverse array of integration patterns. These patterns range from rudimentary point-to-point connections to intricate, multistage process chains that incorporate conditional branching, error handling, and dynamic routing.

In addition to the iFlow engine, SAP Integration Suite offers *API Management*, which is a comprehensive tool for managing, securing, and publishing APIs. APIs are becoming increasingly important in modern IT landscapes—not only as a technical interface, but also as a strategic channel for digital transformation. You can use API Management to facilitate the provision of REST or SOAP-based interfaces, the versioning of these interfaces, the assignment of policies (e.g., for rate limiting, quotas, or token checks), and the documentation of these interfaces via a developer portal. Authentication and authorization are also controlled centrally—either via OAuth 2.0, API Keys, or SAML.

Another essential component is the Open Connectors service, which offers standardized connections to over 160 popular SaaS applications and web services, including Salesforce, HubSpot, Google Drive, Facebook, and Zendesk. SAP offers standardized REST APIs with a converged data model. This indicates that a "Contact" object from Salesforce possesses the same structure as the corresponding object from Microsoft Dynamics, and this development has the potential to streamline a significant portion of the customization work for developers when integrating such systems. Open Connectors provide a multifaceted array of functionalities, including access methods, integrated authentication mechanisms, dynamic schema mappings, and event handling.

The increasing importance of asynchronous communication is reflected in the Event Mesh component. This is based on the principle of event-driven architecture, and it thus enables decoupled communication between systems, services, and applications.

Rather than making direct API calls, publishers have the option of sending messages or events to defined topics. Subscribing components, such as other applications or iFlows, can then retrieve and process these as required. This development has been shown to enhance resilience and scalability by eliminating the need for producers and consumers to be concurrently active. Event Mesh is compatible with SAP's proprietary event standards, including those from SAP S/4HANA, as well as open standards such as CloudEvents and AMQP.

A critical yet often overlooked component of the cloud connector is its role in facilitating seamless integration with external systems. This configuration establishes a secure, bidirectional connection between on-premise systems and SAP BTP, thereby obviating the need for local infrastructure to be directly exposed to the internet. The connector functions as a lightweight agent within the customer network that selectively enables outgoing connections to the cloud while permitting the precise delineation of services or resources that may be exposed. To illustrate, OData services from a local SAP gateway system or RFC functions from an ERP system can be securely accessed via the cloud. Access to this system is granted through whitelists, user roles, and audit functions, all of which operate with complete transparency within the company's security architecture.

### 2.1.2  Typical Deployment Scenarios

SAP Integration Suite has been demonstrated to be applicable in a variety of scenarios, ranging from purely cloud-based architectures and hybrid models to complex multi-cloud and multitenant environments. Let's look at the following key deployment scenarios:

- **Cloud-to-cloud scenarios**
  A distinguishing feature of these scenarios is the utilization of cloud-based systems, which are characterized by their remote, internet-based operation. A paradigmatic illustration of this integration can be observed in the convergence of SAP S/4HANA Cloud with SAP SuccessFactors or SAP Ariba. The transmission of information is typically facilitated through REST APIs or OData services. The advantages of this approach include high speed, simple authentication via OAuth 2.0, and the direct availability of standard integrations via SAP Business Accelerator Hub.

- **Cloud-to-on-premise scenarios**
  A significant number of companies continue to be heavily reliant on conventional ERP systems and databases that are managed within their local networks. The integration of these on-premise systems with modern cloud applications necessitates technical connectors, as well as security solutions and network architecture concepts. In such scenarios, the cloud connector plays a pivotal role. The challenges associated with this approach include latency, authentication across different domains, and the secure handling of sensitive data.

- **Hybrid scenarios**
  The majority of real integration projects fall into this category, and in this context, cloud-based systems and on-premise systems coexist in a state of parallel operation. This may encompass data synchronization, event-driven communication, or the integration of legacy systems into contemporary business processes. Hybrid scenarios necessitate not only technical integration but also governance and compliance strategies, particularly for internationally operating companies with divergent data protection requirements.

### 2.1.3   Security Architecture

The architecture of SAP Integration Suite has been demonstrated to meet the highest standards in terms of security, data protection, and regulatory requirements. Security isn't conceptualized as a technical add-on but rather as an integral component of the platform itself.

Authentication is typically facilitated by the Identity Authentication service, which can be integrated into existing identity landscapes, such as Microsoft Entra ID and other Security Assertion Markup Language (SAML)–compatible services. Token-based processes, such as OAuth 2.0, are utilized for APIs, which generate dynamic, short-lived access tokens and thus prevent misuse. Authorization is governed by roles and groups, which can be managed centrally within SAP BTP.

At the network level, TLS encryption, IP whitelisting, certificate verification, and logic separation are employed. The client separation is a particularly salient feature, and each tenant of SAP Integration Suite is logically and physically separated from the next, including data storage, process runtime, and administration. This approach not only ensures adherence to standards such as the General Data Protection Regulation (GDPR) but also facilitates secure DevOps processes by ensuring the separation of development, testing, and production environments.

Additionally, SAP provides comprehensive logging and audit capabilities. It's imperative to note that each and every access, data transfer, and administrative change is meticulously logged and can be viewed centrally via the SAP BTP cockpit or external tools such as SAP Cloud ALM. Additional encryption solutions, including SAP Data Custodian and proprietary key management systems (KMS), are also available for companies with elevated security requirements, such as those in the banking, health care, and public sectors.

### 2.1.4   Best Practices and Design Suggestions

The design and implementation of a successful integration architecture with SAP Integration Suite requires a combination of technical expertise, strategic thinking, a comprehensive understanding of operational processes, and a consistent application of best practices. This section lays out the key best practices that have been empirically validated to bring success in projects of varying sizes and from diverse industries.

A fundamental tenet of contemporary integration solutions is the principle of *loose coupling*, which holds that systems must be integrated in a way that ensures their operation is as autonomous as possible. This approach enhances flexibility while concurrently fortifying the overall solution's resilience to errors. Event-driven architectures (EDAs), such as those supported by Event Mesh, promote this decoupling by making system interactions asynchronous and nonblocking. The ability to temporarily store messages and process them as required also ensures greater system stability and reliability.

Another factor that contributes to the success of such a system is the consistent separation of development, testing, and production environments. In the context of SAP BTP, the establishment of distinct subaccounts facilitates the implementation of specialized role and authorization concepts. When utilized in conjunction with SAP Cloud Transport Management, this approach establishes a regulated lifecycle for integration artifacts. This approach mitigates the likelihood of adverse outcomes such as regressions and unintended side effects, particularly in instances where substantial modifications are made to the process.

Monitoring and logging are essential elements of any productive integration architecture, and SAP Integration Suite is equipped with a centralized monitoring dashboard that offers a visual representation of the status of active iFlows, error messages, response times, and other critical performance indicators in real time. If you need such a comprehensive solution, we recommend that you establish a connection to SAP Cloud ALM, which is a system that integrates project management, test and change data, and operational data. This approach has been demonstrated to enhance the efficiency of error analyses, root cause analyses, and performance optimizations.

The significance of standardization and reusability is frequently underestimated, but these aspects are of paramount importance. The development of integration projects is often marked by a rapid escalation in complexity, which often results from the implementation of disparate solutions for each distinct use case. Conversely, central building blocks—including iFlow fragments, mapping templates, naming conventions, and validation rules—ought to be designed for reusability and maintenance in a shared repository. The utilization of predefined content packages from SAP Business Accelerator Hub provides tested integrations for SAP and third-party systems, so it significantly reduces development time while minimizing maintenance effort.

You shouldn't think of documentation a downstream task; rather, think of it as an essential part of the integration process. We can't overstate the importance of well-structured and well-maintained documentation in the context of training new team members. Such documentation isn't merely a convenience; it's an essential component of various processes, including audits, recertifications, and long-term maintenance. The scope of the document should encompass both the technical level (e.g., interface descriptions, data formats, iFlow logic) and the functional meaning (e.g., process context, business rules, responsibilities).

In industries with strict regulatory requirements, such as pharmaceuticals, finance, and public administration, regular security checks and compliance audits are imperative. This isn't merely an issue of technical vulnerabilities; it's also a concern for organizational processes. For instance, it's essential to ascertain whether roles have been assigned correctly, data encryption is applied consistently, and interface access is logged properly. SAP offers its own proprietary tools for this purpose, including SAP Data Custodian and integrated audit logs within SAP BTP.

Another design principle pertains to service governance. In numerous organizations, the integration landscape evolves organically, often over the course of years, in tandem with shifting responsibilities. In the absence of clearly defined responsibilities, naming conventions, and rules for versioning and publishing, an *integration proliferation* rapidly emerges and is challenging to maintain. In this context, the establishment of an integration center of excellence is a pivotal strategy because it fosters the establishment of architectural standards, templates, and release processes.

In terms of performance and scalability, you must exercise caution during the initial stages of development to ensure that integrations don't impede the efficiency of the overall system. SAP Integration Suite offers two options for automatic scaling: horizontal, through parallel processing; and vertical, through the allocation of additional computing resources. Nevertheless, we recommended that you implement regular performance tuning, for example, by buffering large amounts of data, batch processing, prioritizing iFlows, or load balancing across different endpoints.

Having an up-to-date understanding of architecture means not only understanding current circumstances but also constantly evaluating and integrating innovations. SAP's approach to innovation involves the implementation of brief innovation cycles, which is a strategy that underscores the company's commitment to ongoing development and refinement of its platform. You'll need to regularly test new functions—for example, in the areas of artificial intelligence (AI), machine learning, self-healing mechanisms, and low-code integrations—for their benefits and integrate them if necessary. This approach guarantees that the architecture will remain not only robust but also future-proof.

## 2.2    Set Up

Before you can initiate integration scenarios, you need to ensure that SAP Integration Suite is configured and set up correctly. The entry point for all activities is SAP BTP, or more precisely a subaccount that serves as a logical environment for the provision, administration, and execution of services. The subaccount will be the foundation for the subsequent segregation of development, testing, and production landscapes that will facilitate the systematic organization of all associated resources.

This section walks you through the steps involved in the activation, configuration, and preparation of SAP Integration Suite within the context of a subaccount. It encompasses not only the purely technical steps but also important configuration decisions, role and authorization concepts, and requirements you must meet before actual activation. The objective of this section is to help you establish a ready-to-use, secure, and scalable integration environment that will serve as a foundation for all your subsequent development and operational activities.

You'll use the Cloud Foundry environment to run services in SAP BTP, and to start with, you'll need to activate it in the subaccount. To enable Cloud Foundry, click the **Enable Cloud Foundry** button in the subaccount dashboard in the SAP BTP cockpit, as shown in Figure 2.1. If you don't enable Cloud Foundry first, central services—including SAP Integration Suite—won't be available to you.

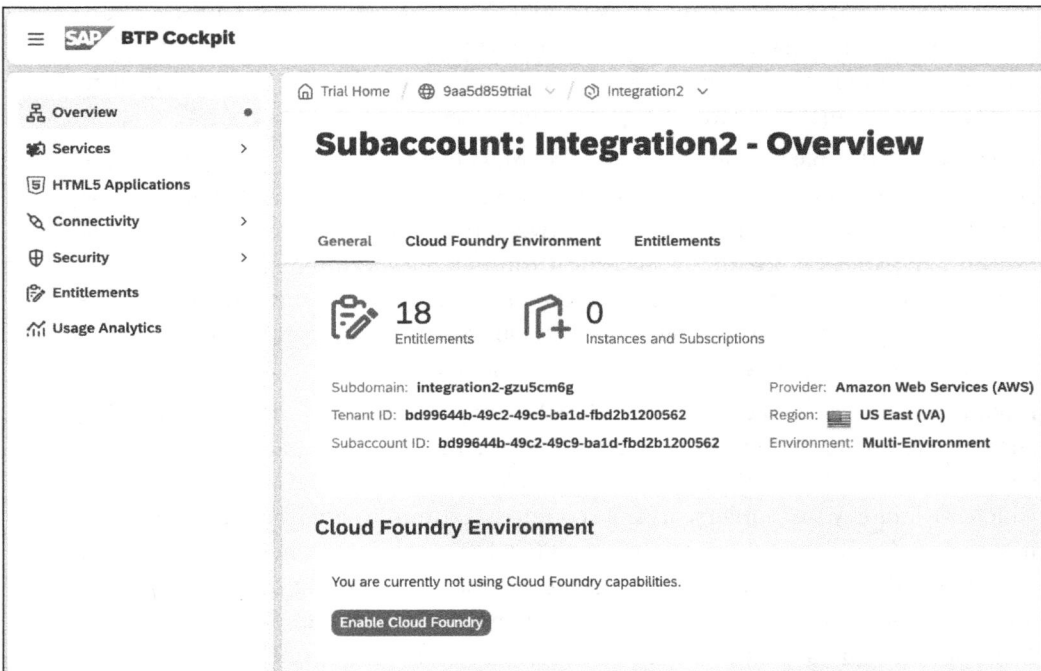

**Figure 2.1** Enabling Cloud Foundry

After you clicking the button, a window will open where you can set up the environment. As shown in Figure 2.2, you can enter the **Instance Name** and the **Org Name**. The selected **Plan** in this example is **trial**, which is intended for evaluation and testing. To finish setting up the environment, click **Create**.

After you activate the environment, it will appear in the dashboard. You'll see that an API URL has been generated, and the current status, organization ID, and name will be displayed in the middle of the subaccount overview. You can then create a space by clicking the **Create Space** button (see Figure 2.3).

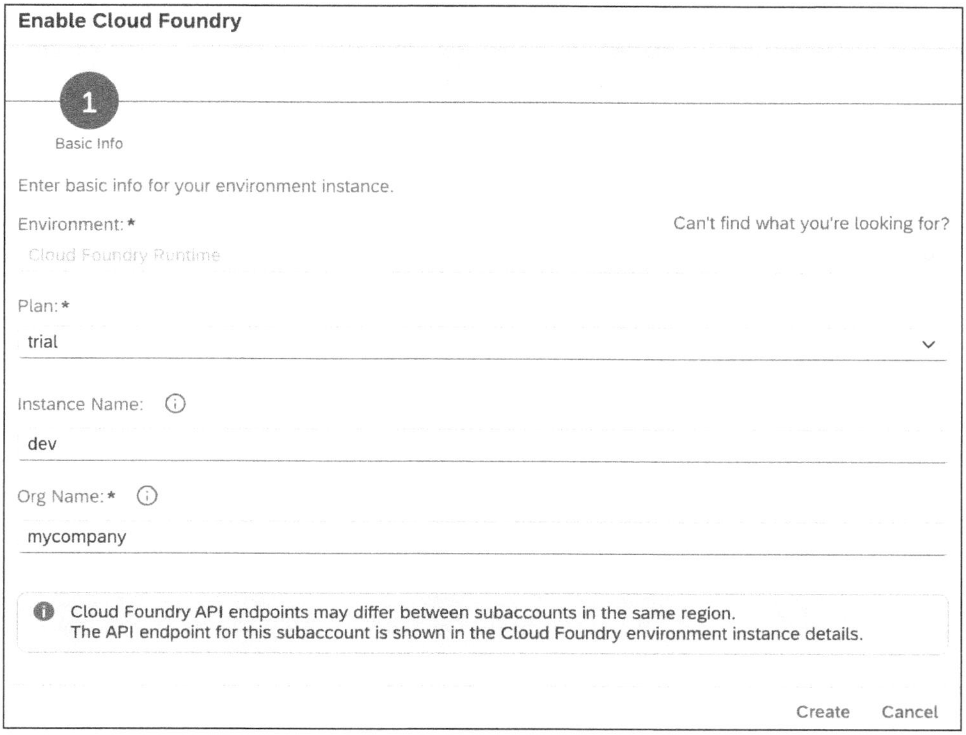

**Enable Cloud Foundry**

**1**

Basic Info

Enter basic info for your environment instance.

Environment: *                                                        Can't find what you're looking for?

Cloud Foundry Runtime

Plan: *

trial

Instance Name:   ⓘ

dev

Org Name: *   ⓘ

mycompany

ⓘ   Cloud Foundry API endpoints may differ between subaccounts in the same region.
The API endpoint for this subaccount is shown in the Cloud Foundry environment instance details.

Create     Cancel

**Figure 2.2**  Inserting Cloud Foundry Details

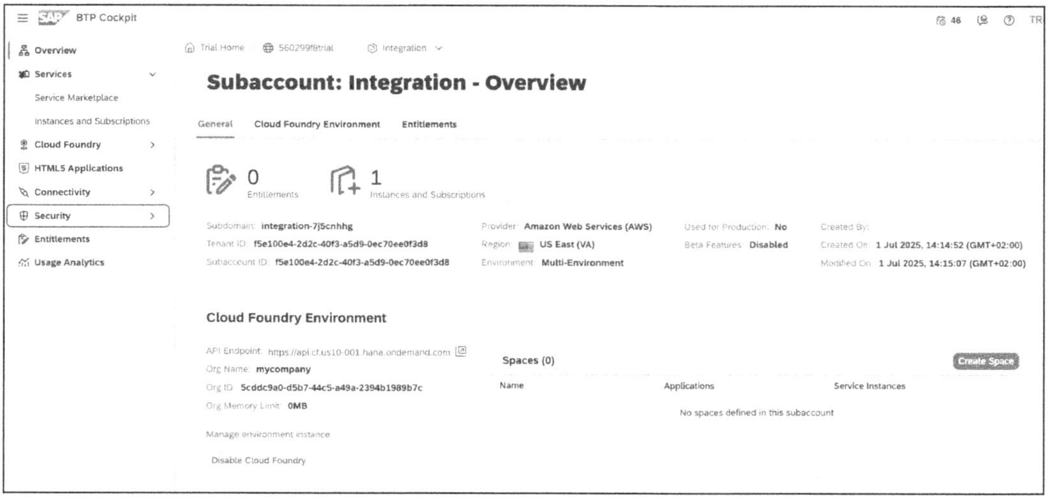

**Figure 2.3**  Creating Space

A *space* is a logical subunit of the Cloud Foundry organization. In the example shown in Figure 2.4, an admin has created a space called **dev.** They've also assigned a user the

**Space Developer** and **Space Manager** roles by checking the appropriate boxes, so that the user can deploy and manage services later. The admin would then click **Create** to move on to the next step.

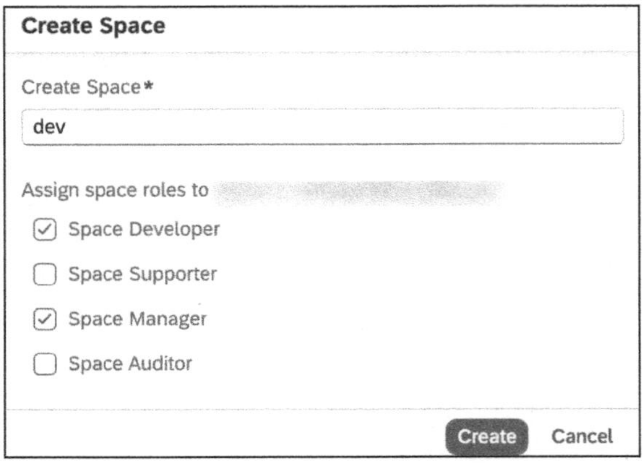

**Figure 2.4** Inserting Space Details

Once you've set up Cloud Foundry, go to the **Service Marketplace** section and search for the **Integration Suite** service. You'll see several integration services, one of which is **Integration Suite**. Select it and then click the **Create** button on the right to set up a new subscription (see Figure 2.5).

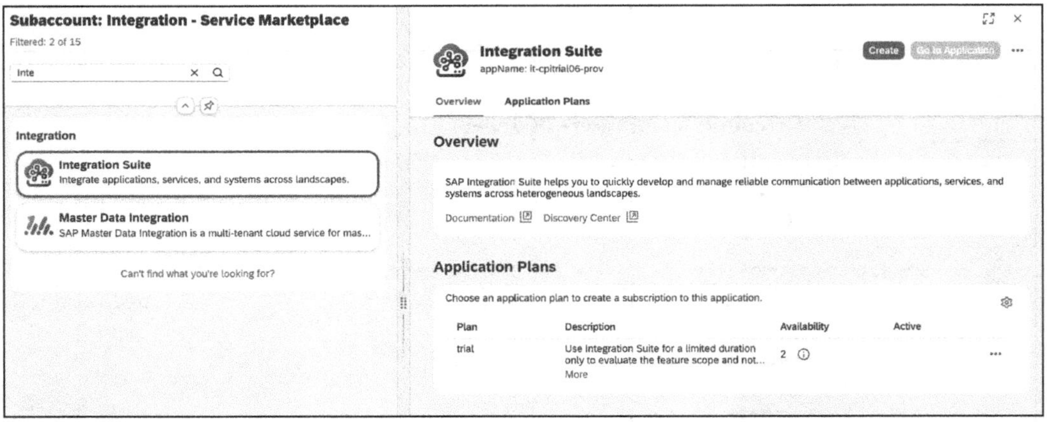

**Figure 2.5** Choosing SAP Integration Suite

On the screen shown in Figure 2.6, you'll see the subscription that you created, and all you'll need to do is confirm the **Service** (**Integration Suite**) and **Plan** (**trial**) and then start the instance by clicking **Create**. After you successfully complete all of these steps, SAP Integration Suite will be available to you.

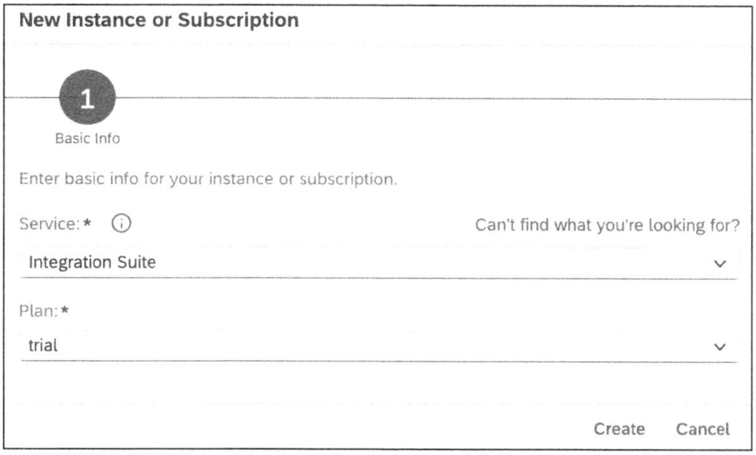

**New Instance or Subscription**

(1) Basic Info

Enter basic info for your instance or subscription.

Service: *  (i)                              Can't find what you're looking for?

Integration Suite                                                              ⌄

Plan: *

trial                                                                          ⌄

                                                              Create    Cancel

**Figure 2.6** Inserting SAP Integration Suite Details

In the **Security · Users** section of the example shown in Figure 2.7, you can see a user who has already been granted basic access rights to the subaccount. In this case, the user has the **Subaccount Administrator** role, and you can add more authorizations by clicking the **Assign Role Collection** button.

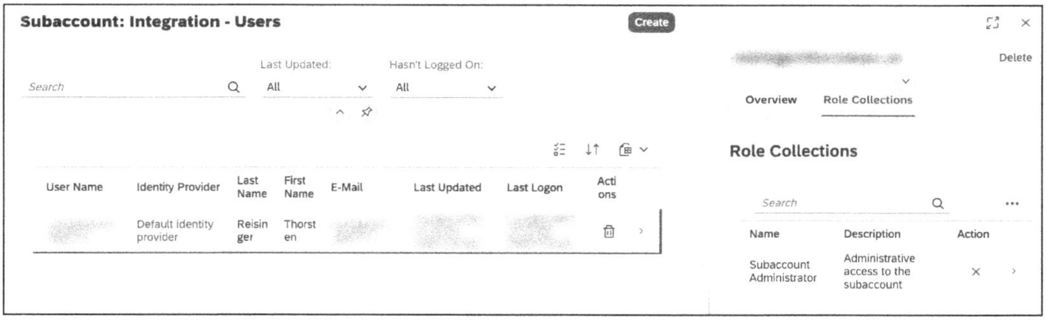

**Subaccount: Integration - Users**          Create                                    ⌗  ✕

                                                                               Delete
Search          Q   Last Updated:   Hasn't Logged On:
                    All         ⌄   All          ⌄        Overview   Role Collections
                              ∧  ✄

                                        ☰  ↕  ⊞ ⌄        **Role Collections**

| User Name | Identity Provider | Last Name | First Name | E-Mail | Last Updated | Last Logon | Actions |
|---|---|---|---|---|---|---|---|
|  | Default identity provider | Reisinger | Thorsten |  |  |  | 🗑 › |

Search                    Q      ...

| Name | Description | Action |
|---|---|---|
| Subaccount Administrator | Administrative access to the subaccount | ✕  › |

**Figure 2.7** Authorizing User

In Figure 2.8, the **Integration_Provisioner** role is explicitly assigned to the user, who needs to have this role to set up SAP Integration Suite and provide integration services. It includes basic authorizations for setting up capabilities, assigning them, and managing subscriptions. Other roles, such as **Integration Developer** and **Integration Viewer**, may be involved later in the project lifecycle, depending on the team's area of responsibility.

You can open SAP Integration Suite through the **Instances and Subscriptions** navigation menu that's displayed in Figure 2.9. The service will appear with the **Subscribed** status, and clicking the icon on the right will take you to the web frontend of SAP Integration Suite.

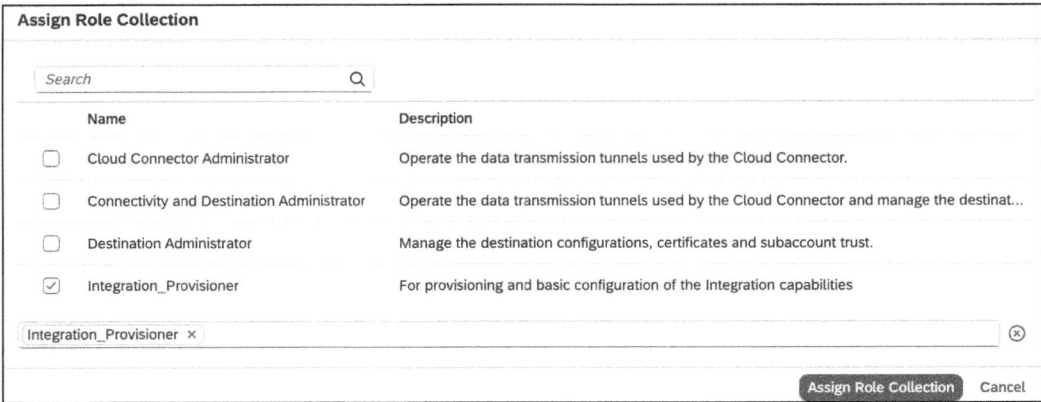

**Figure 2.8** Choosing Integration_Provisioner Role

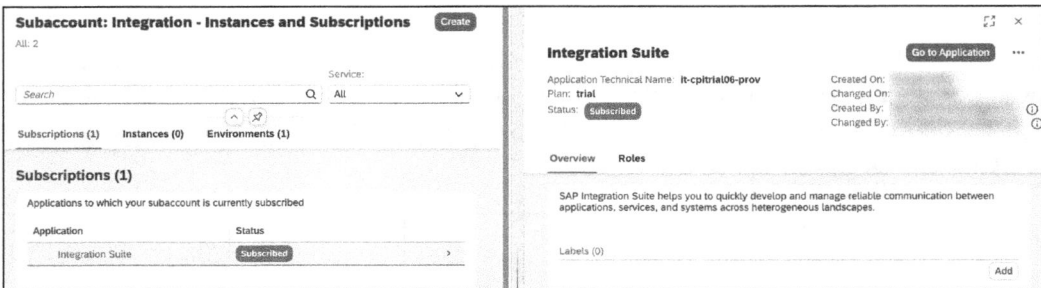

**Figure 2.9** Opening SAP Integration Suite

After you start SAP Integration Suite, the main home screen of the application will appear. It displays information tiles on news, documentation, community, and learning paths. We'll focus on the **Capabilities** module, which includes the specific functions and areas of use that you can activate within the suite. In the example in Figure 2.10, the **You haven't added any capabilities** message appears because none have been activated yet. You can click the **Add Capabilities** button to set the specific functions of SAP Integration Suite.

You can find more details about SAP Integration Suite's functional scope on this screen. SAP offers a modular structure, which means that individual functions (capabilities) can be activated depending on the project's requirements. All available modules are highlighted in Figure 2.11, and selecting such modules lets you use both classic system integration and modern API and event-driven approaches on the same platform.

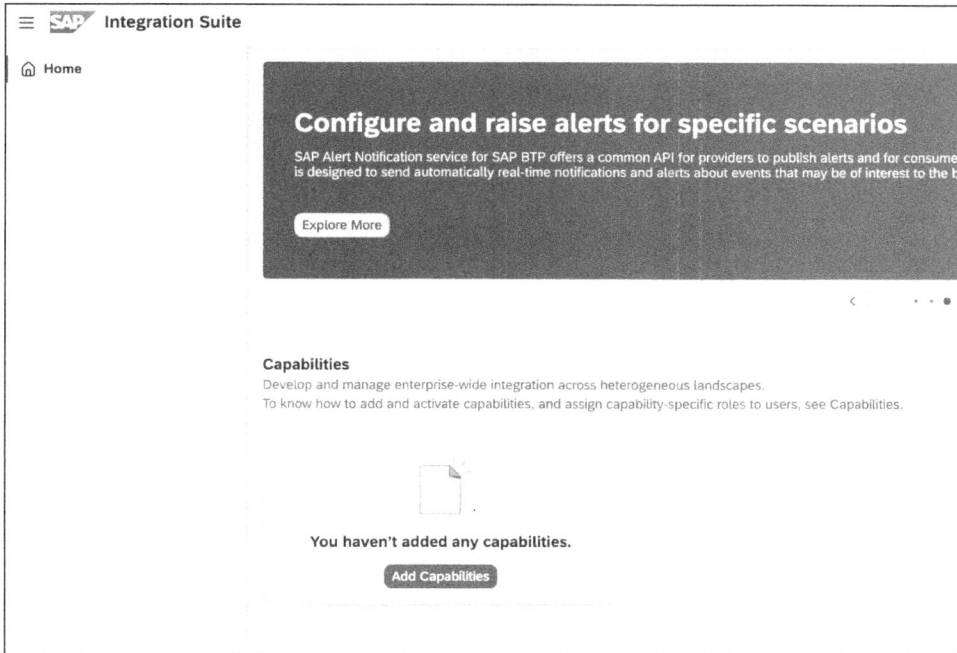

**Figure 2.10** Adding Capabilities

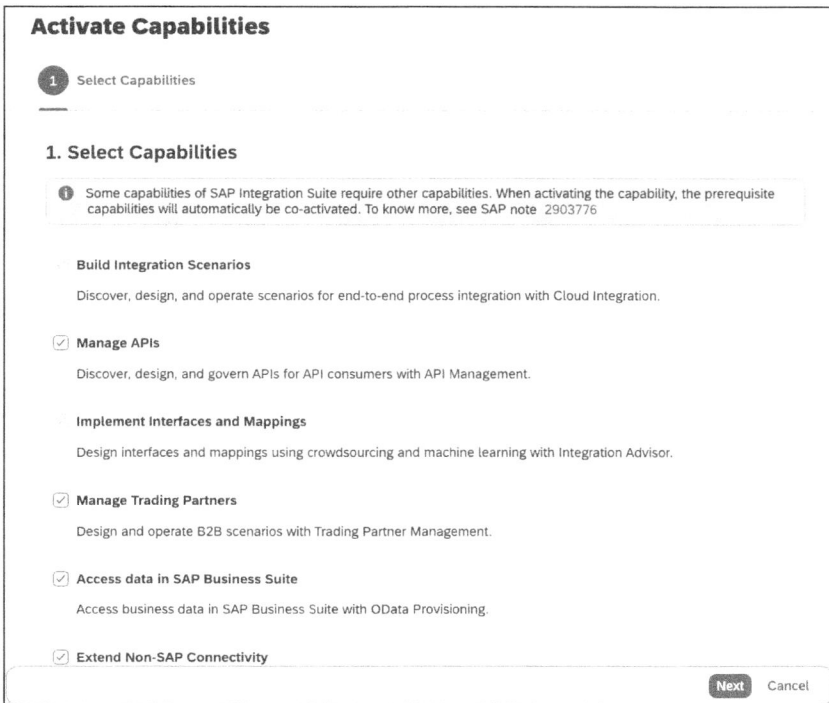

**Figure 2.11** Selecting Capabilities

Before you activate these capabilities, a summary of all the ones you've selected will appear (see Figure 2.12). You'll see extra options, like turning on message queues in Cloud Integration and getting the developer hub in API Management. After checking these options, you can activate the functions you selected by clicking the **Activate** button. This process starts the necessary back-end components and can take a few minutes, and after that, SAP Integration Suite will be ready to use.

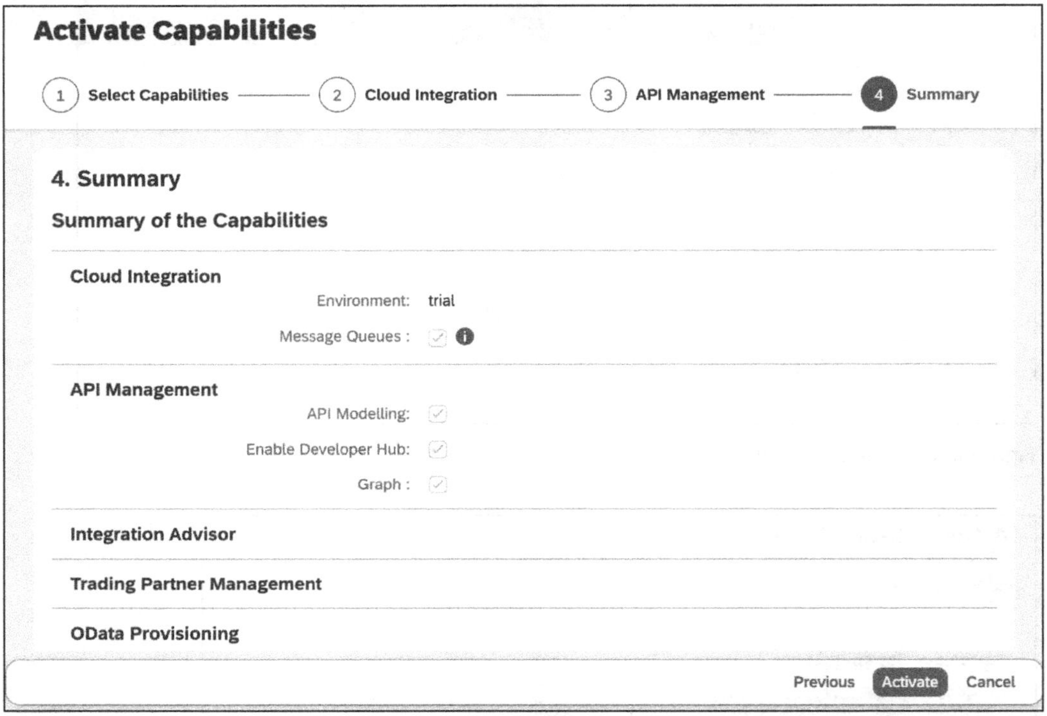

**Activate Capabilities**

①  Select Capabilities ———— ②  Cloud Integration ———— ③  API Management ———— ④  Summary

### 4. Summary

**Summary of the Capabilities**

**Cloud Integration**

Environment:   trial

Message Queues :   ☑ ⓘ

**API Management**

API Modelling:   ☑

Enable Developer Hub:   ☑

Graph :   ☑

**Integration Advisor**

**Trading Partner Management**

**OData Provisioning**

Previous   [ Activate ]   Cancel

**Figure 2.12**  Summary of Capabilities

Once you've activated the capabilities you selected, you'll see a summary of all the functions in SAP Integration Suite that are currently active (see Figure 2.13). In the **Cloud Integration** section, you'll find information about message queues, including how many queues there can be, the memory limit, and the transaction volume.

The subcomponents of API Management—the API portal, developer hub, and Graph—will be fully activated and available for defining, publishing, and managing APIs. The Integration Advisor, Trading Partner Management, and OData Provisioning components will also be activated and ready for use. These functions let you do the following:

- Create integration content (via Integration Advisor)
- Map B2B scenarios with partners (via Trading Partner Management)
- Provision all the SAP landscape's data (via OData Provisioning)

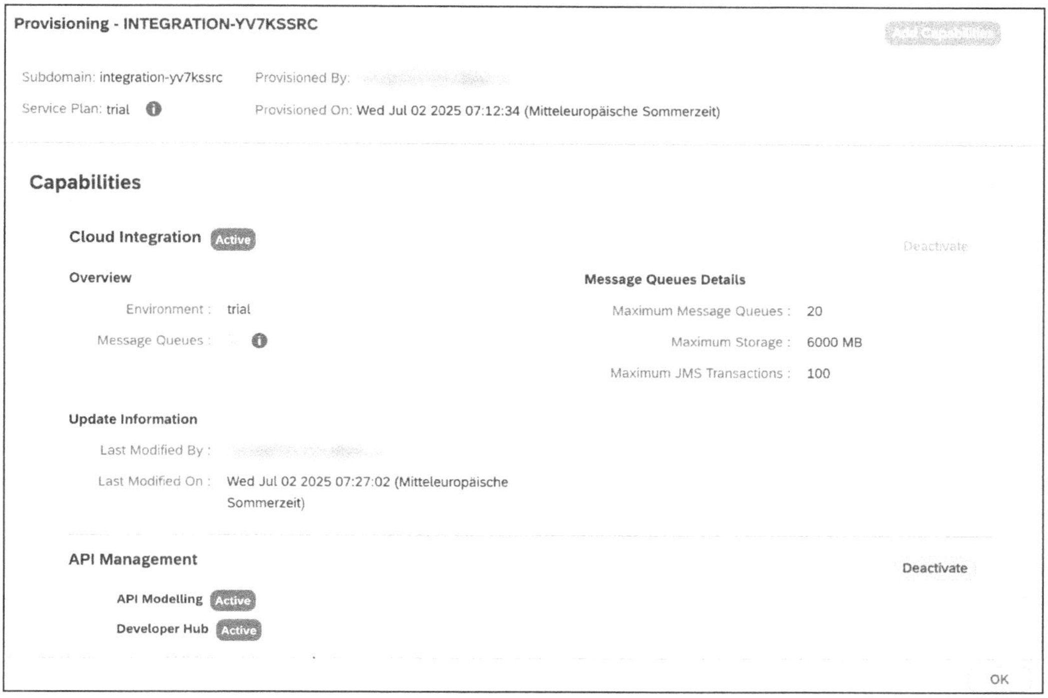

**Provisioning - INTEGRATION-YV7KSSRC**

Subdomain: integration-yv7kssrc       Provisioned By:

Service Plan: trial   ℹ               Provisioned On: Wed Jul 02 2025 07:12:34 (Mitteleuropäische Sommerzeit)

**Capabilities**

**Cloud Integration** `Active`                                                              Deactivate

**Overview**                                          **Message Queues Details**

    Environment :   trial                     Maximum Message Queues :   20

    Message Queues :   ℹ                       Maximum Storage :   6000 MB

                                                       Maximum JMS Transactions :   100

**Update Information**

    Last Modified By :

    Last Modified On :   Wed Jul 02 2025 07:27:02 (Mitteleuropäische
                            Sommerzeit)

**API Management**                                                                   Deactivate

    **API Modelling** `Active`

    **Developer Hub** `Active`

                                                                                     OK

**Figure 2.13** Active Capabilities

## 2.3   Roles and Authorizations

To create a modern IT landscape, you need to achieve technical excellence in the integration of systems, processes, and data sources, and you also need a well thought-out authorization concept. Particularly in a platform such as SAP BTP, which offers users and developers extensive options for developing, integrating, and extending applications, you must assign the correct roles and authorizations to ensure both security and smooth operations. In such a versatile and open system, you need to precisely control who is allowed to do what.

A well thought-out authorization concept serves the following goals:

- **Security**
  It prevents unauthorized access and protects critical actions (e.g. deleting instances, changing iFlows).
- **Transparency and traceability**
  It gives each user a clearly defined role and responsibilities.
- **Efficiency**
  It authorizes role-specific access so users only see what they actually need to see.
- **Governance and compliance**
  It helps companies adhere to guidelines that stipulate the separation of functions, auditability, and audit security, for example.

Especially in larger organizations and regulated industries, these aspects are not only useful but also legally binding.

At the heart of the SAP BTP authorization concept is the role-based access control model, which stipulates that a user's authorizations should not be assigned directly but via roles. A *role* bundles certain technical authorizations (which are called *scopes*) or groups of authorizations (which are called *role templates*) and combines them into *role collections*, which are then assigned to users or groups. The basic principle follows this four-level model:

1. **Scope**
   This includes individual authorizations (e.g., "edit destination").
2. **Role template**
   This is a group of scopes for a specific use case.
3. **Role (business role)**
   This is a concrete, instantiated role.
4. **Role collection**
   This is a summary of several roles that are assigned to a user.

In practice, you'll assign authorizations exclusively via role collections. SAP BTP organizes environments via global accounts, subaccounts and the environments they contain, such as Cloud Foundry. Roles and authorizations are context dependent—in other words, a user can act as an administrator in one subaccount while having no rights in another subaccount. You must therefore configure roles individually for each SAP BTP environment. The subaccount structure offers a clear separation of responsibilities and also allows multitenant approaches—for example, for separate development, test and production environments.

In the following sections, we'll walk you through assigning roles via the SAP BTP cockpit and using Identity Authentication.

### 2.3.1   Assigning Roles in the SAP BTP Cockpit

You'll usually assign roles via the SAP BTP cockpit, via the **Security • Users** menu path. There, you as an administrator can specifically assign one or more role collections to a user. In a typical configuration, you may need to follow these steps:

1. Select the relevant subaccount.
2. Create a new role collection or select an existing one via **Security • Role Collections**.
3. Assign this role to the user (via their e-mail address or user ID)
4. Optionally, you can use identity providers to control which users are logged in via which authentication system. We'll see exactly how this works in Section 2.5.

Now, let's take a look at it in the system. You'll want to assign your user the authorization to create, edit, and monitor iFlows in SAP Integration Suite, so as already described, you need to assign the corresponding role collections for this. These role collections, which you'll assign in a moment, have been created automatically due to the addition of the **Build Integration Scenarios** capability to SAP Integration Suite. First, you switch to SAP BTP and to the subaccount where SAP Integration Suite was created. Then, you navigate to **Security · Users,** as shown in Figure 2.14. Next, you navigate to the bottom right-hand corner of the screen under the **Role Collections** heading and click **Assign Role Collection**.

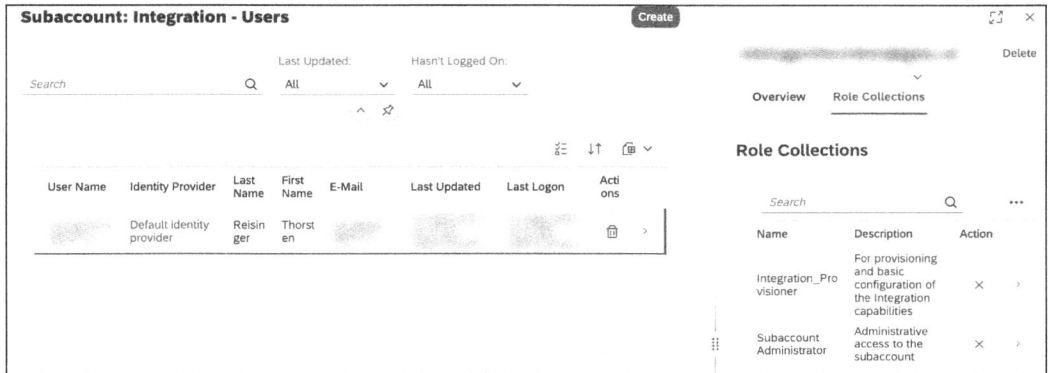

**Figure 2.14** Assigning Role Collection to User

That will bring up a new window where all available roles that haven't yet been assigned to the user are displayed. You should select the **PI_Administrator**, **PI_Business_Expert**, and **PI_Integration_Developer** role collections, as in Figure 2.15. Then, click **Assign Role Collection** to assign the role collections to the user.

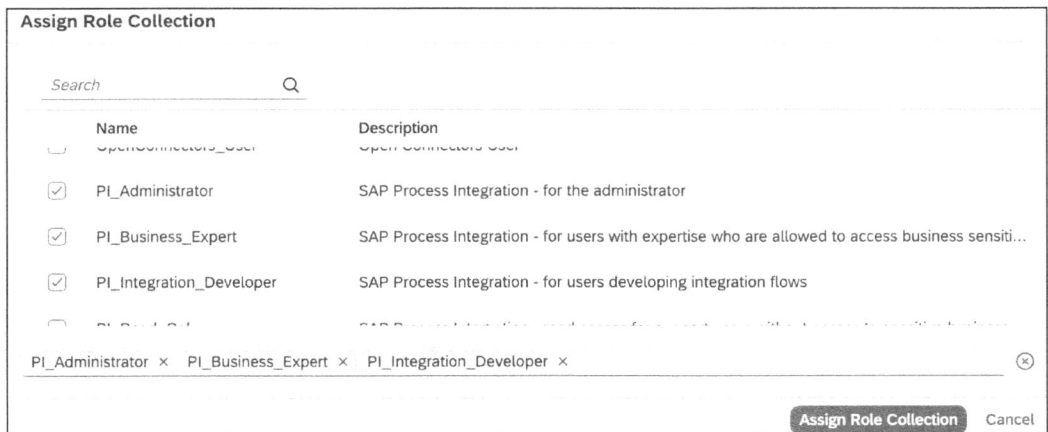

**Figure 2.15** Selecting PI Role Collections

After you've made these assignments, the authorizations may not yet be effective, so to resolve the problem, restart SAP Integration Suite in an incognito tab. When the role collections are properly assigned, it should look like the screen shown in Figure 2.16.

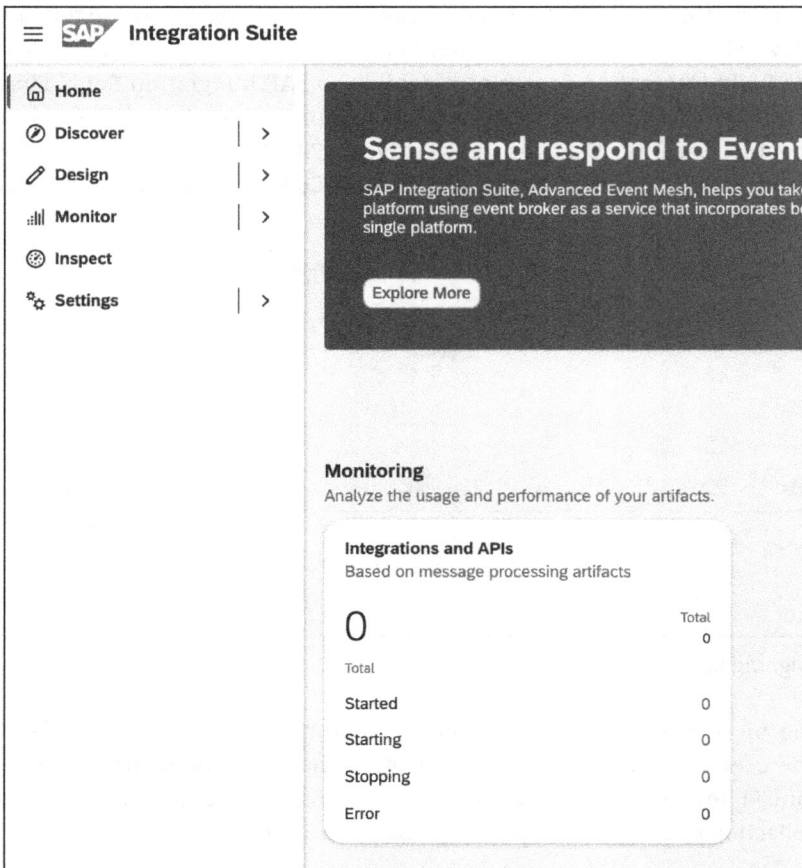

**Figure 2.16**  SAP Integration Suite Build Integration Scenarios

### 2.3.2   Assigning Roles with Identity Authentication

You can connect SAP BTP directly to Identity Authentication to make it serve as a central identity provider through which users can log in. One of the strengths of Identity Authentication is its ability to configure automated role assignments through mapping tables, group rules, or attributes (e.g., departmental affiliation).

For example, you can define a rule so that all users in the **integration_developers** group automatically receive the **PI_Integration_Developer** role collection. This will reduce manual effort and improve the consistency of role assignment.

Before you can set up these authorizations, you must create the **integration_developers** group in Identity Authentication and assign it to the user. To do this, navigate to the Identity Authentication service that has a trust to the subaccount where SAP Integration Suite has been created. How exactly this works is explained in Section 2.5. Then, click the **Users & Authorizations** tab and then click **Groups**, as shown in Figure 2.17.

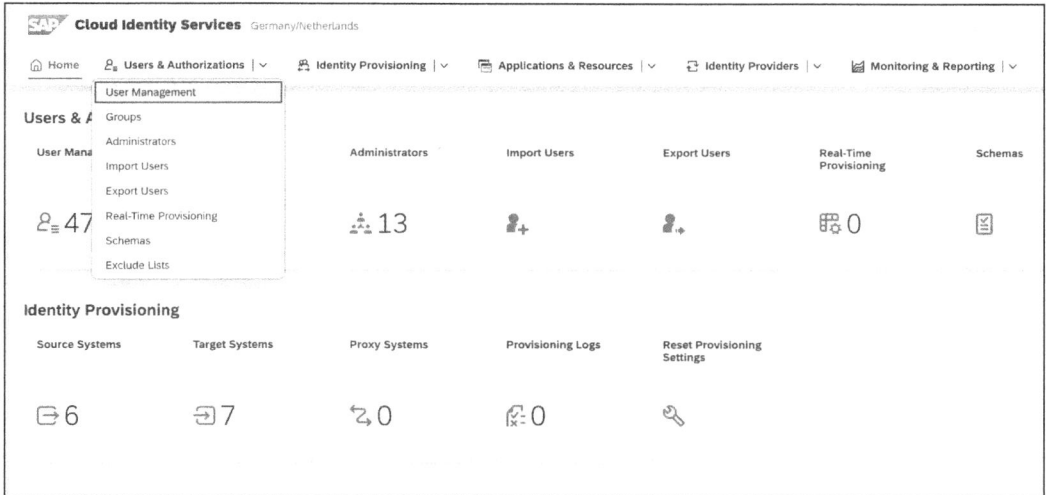

**Figure 2.17**  Identity Authentication Overview

Once there, click **Create** at the top right and fill in the technical name and the display name. You can ignore the remaining steps for this scenario, and as soon as you've filled in all the required fields (**Group Name** and **Display Name**), click **Finish** to finalize the group creation as shown in Figure 2.18.

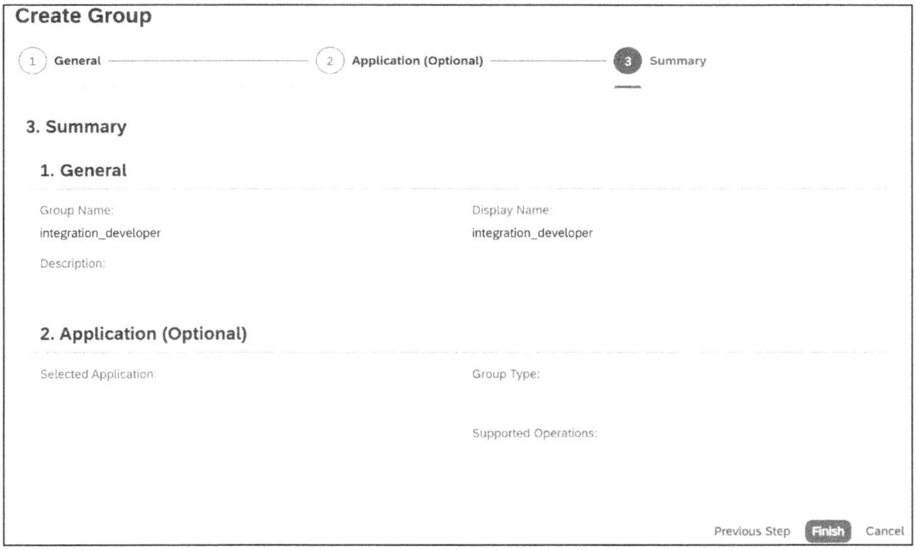

**Figure 2.18**  Creating Identity Authentication Group

Now, you need to assign the Identity Authentication user to the Identity Authentication group. To do this, click **Add** in the top right-hand corner of the group and select the user you want (see Figure 2.19).

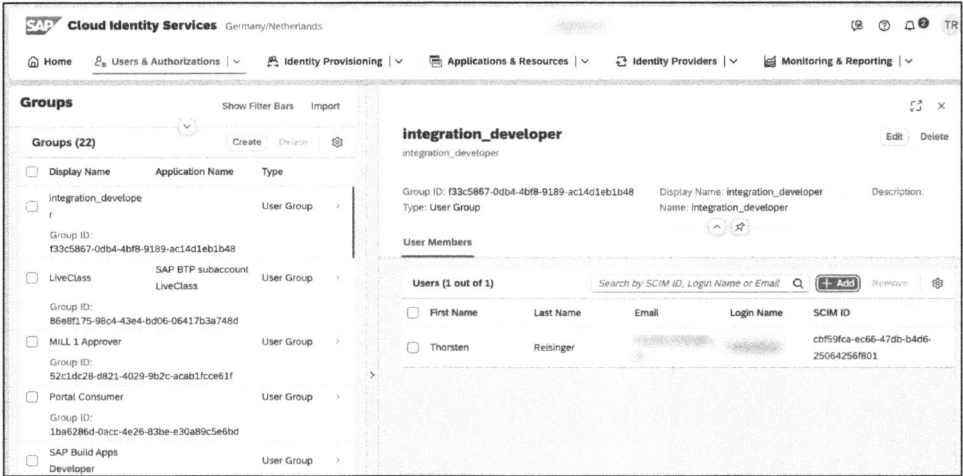

**Figure 2.19** Assigning Identity Authentication User to Identity Authentication Group

Finally, you need to tell the role collection in SAP BTP which Identity Authentication group should automatically receive this role collection. To do this, switch to the SAP BTP **Role Collection** page, select the role collection you want, and click **Edit**. Then, as shown in Figure 2.20, scroll down until you reach the **User Groups** area, select the Identity Authentication service, enter the technical name of the Identity Authentication group in the **Name** field, and click **Save** to finalize the change.

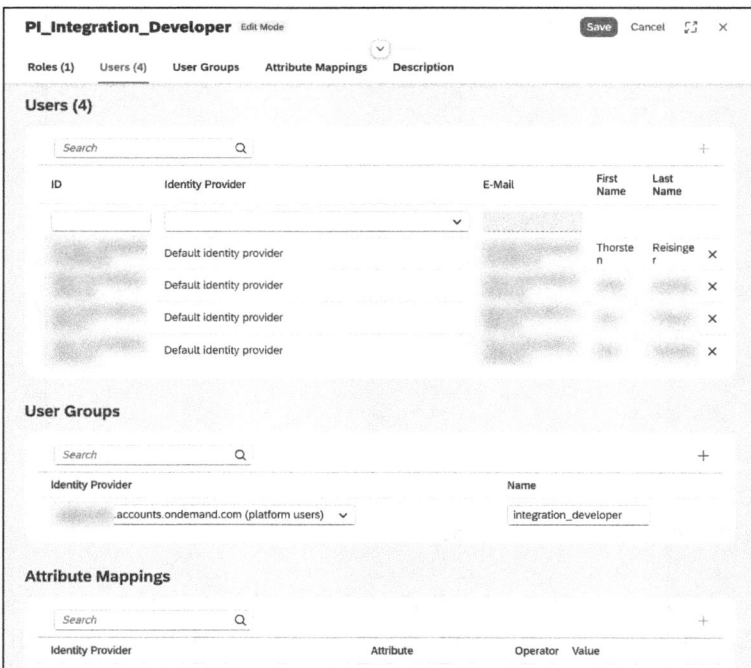

**Figure 2.20** Assigning Role Collection Identity Authentication Group

After that, when you log in to SAP Integration Suite with the Identity Authentication user, the user will have Role Collection authorization.

## 2.4 Integrating the Cloud Connector

The *cloud connector* is a central tool in hybrid SAP landscapes. It acts as a secure bridge between SAP BTP in the cloud and a company's local (on-premise) systems. Nowadays, when many organizations are gradually migrating to the cloud, the cloud connector is essential: it makes it possible to connect cloud applications with existing backend systems—such as SAP ERP, S/4HANA, databases, and non-SAP systems—without compromising security or network architectures.

You'll install the cloud connector locally in the on-premise infrastructure—typically, on a dedicated server or in a secure network segment. From there, it will establish an outbound tunnel to SAP BTP. It's important to note that the connection isn't initiated from the outside but established exclusively from the inside to the outside, and that makes the cloud connector particularly secure. No ports are opened on internal firewalls, and no VPNs are set up—access is controlled, encrypted, and based on defined approvals.

Within the cloud connector configuration, you can define very granularly which internal resources are to be shared: individual systems, services, RFC endpoints, OData APIs, and even specific directories. You can make these releases per subaccount so that you can operate several cloud projects in parallel with one or more cloud connectors.

---

**Disclaimer**

Because this book is limited to the basic configuration of the cloud connector, to make it work for a standard configuration, it doesn't cover the installation and monitoring of the cloud connector. For that, we recommend that you read *Configuring SAP Business Technology Platform*, which you can find at *https://www.sap-press.com/configuring-sap-business-technology-platform_6080/*.

---

When you start the cloud connector for the first time, you must enter the administrator credentials. Enter "Administrator" as the user name and "manage" as the password. After you successfully authenticate, you must change the password, and then, you'll be redirected to the **Define Subaccount** area (see Figure 2.21). There, you'll have the option of connecting to a subaccount and defining an HTTPS proxy.

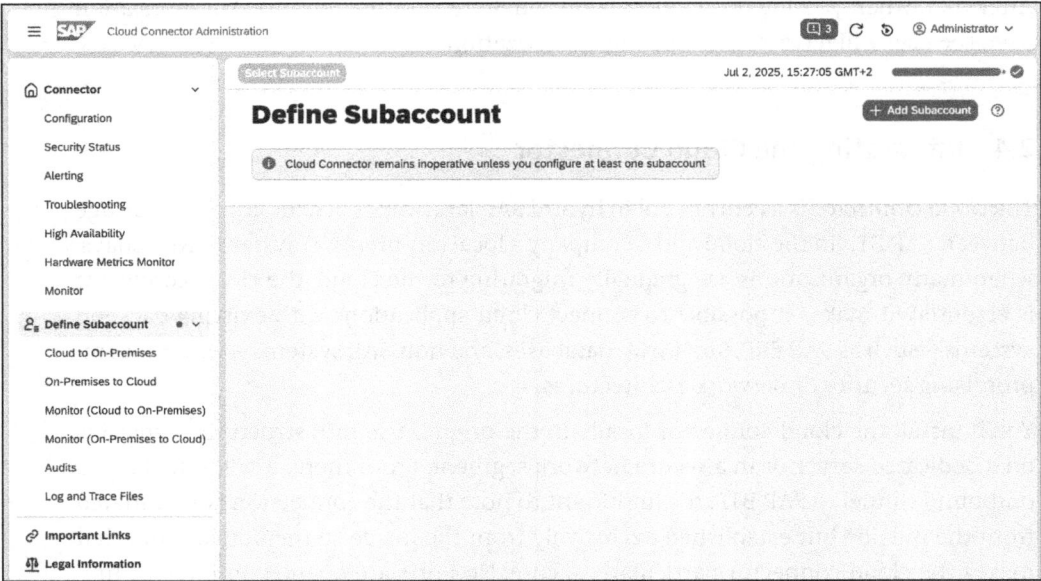

**Figure 2.21**  Defining Subaccount in Cloud Connector

Next, you'll need to connect the cloud connector to your subaccount. To do this, click **+ Add Subaccount** at the top right. You'll then be asked whether you want to configure a proxy (see Figure 2.22), but for now, skip that by clicking **Next** and leave all fields empty since your cloud connector can access the internet without a proxy.

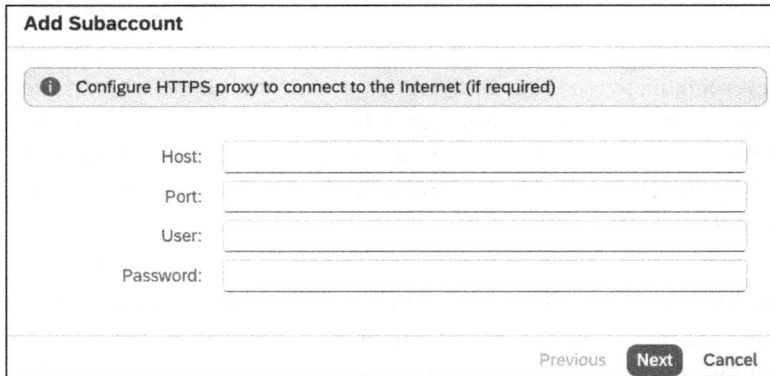

**Figure 2.22**  Defining Proxy

Next, you'll be asked whether you want to create the subaccount manually or via authentication data. Select **Configure using authentication data** as shown in Figure 2.23.

**Add Subaccount**

Select:    ◯  Configure manually

⦿  Configure using authentication data

Previous    **Next**    Cancel

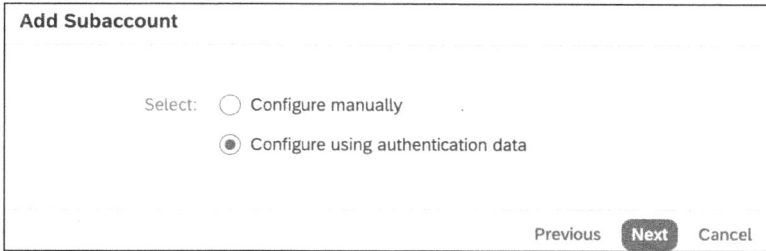

**Figure 2.23**  Adding Subaccount via Cloud Connector

Now, you need to access the authentication data of your subaccount. To do this, navigate back to SAP BTP, select the subaccount you want, navigate within it to **Connectivity** on the left-hand side, select **Cloud Connectors**, and click **Download Authentication Data** (see Figure 2.24). You'll then receive a *.data* file.

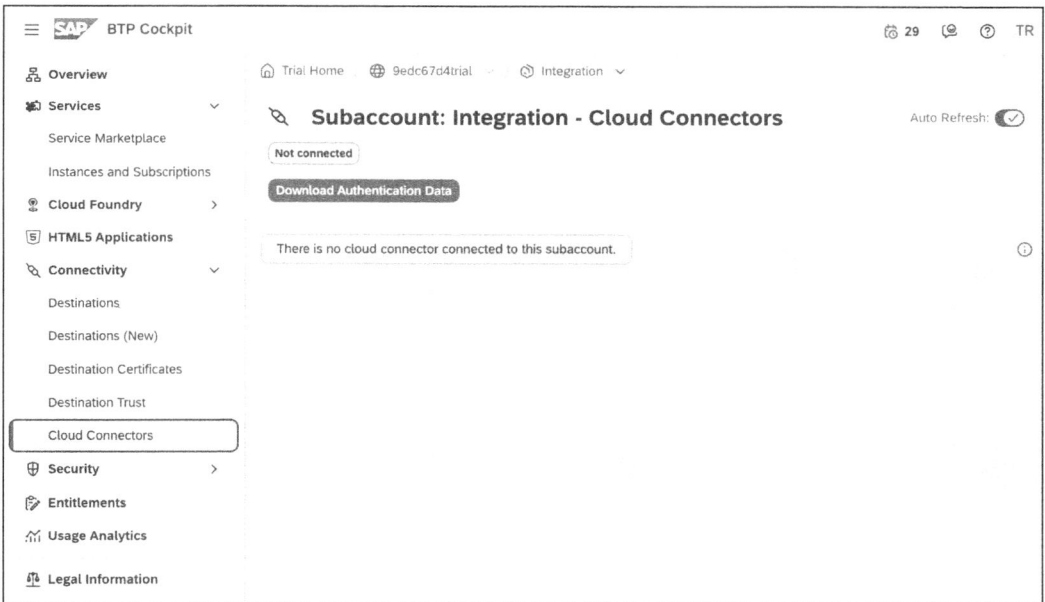

≡  SAP  BTP Cockpit                                                                 🕙 29   (2)   ⑦   TR

⛊ **Overview**

🔌 **Services**                    ⌄        ⌂ Trial Home      ⊕ 9edc67d4trial  ⌄     ⊘ Integration  ⌄

   Service Marketplace                     ✎    **Subaccount: Integration - Cloud Connectors**          Auto Refresh:  ◉

   Instances and Subscriptions            [ Not connected ]

⚐ **Cloud Foundry**               >        [ Download Authentication Data ]

⑤ **HTML5 Applications**                   ┌─────────────────────────────────────────────────────┐
                                           │  There is no cloud connector connected to this subaccount. │    ①
🔌 **Connectivity**               ⌄        └─────────────────────────────────────────────────────┘

   Destinations

   Destinations (New)

   Destination Certificates

   Destination Trust

   [ Cloud Connectors ]

⊕ **Security**                    >

🗂 **Entitlements**

ᚆ **Usage Analytics**

⚖ **Legal Information**

**Figure 2.24**  Downloading Authentication Data

After that, switch back to your cloud connector, where you'll have to select how you want to configure the subaccount. As described earlier, select **Configure using authentication data**, select the **Add Subaccount authentication data from file** option, click the **Browse** button to find the *.data* file you've just downloaded, and select it. The end result should look like the one shown in Figure 2.25.

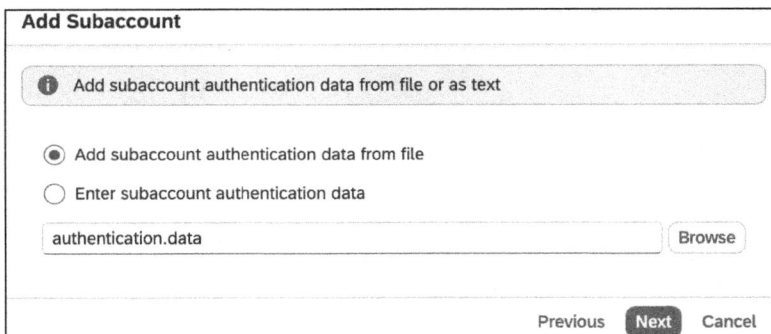

**Figure 2.25** Selecting Authentication Data File

If the import is successful, the subaccount data will be displayed (see Figure 2.26). **UNKNOWN** is always displayed as the **Region**. However, this is retrieved from the subaccount when the connection is completed. The initial **Display Name** is the subdomain of the subaccount. You can change the **Display Name** after successfully establishing the connection, as shown in Figure 2.27. Click **Finish** to create the connection between the subaccount and the cloud connector.

**Figure 2.26** Summary Add Subaccount

If you perform all these steps correctly and completely, the connection will be established, the cloud connector should look like Figure 2.27, and the subaccount should look like Figure 2.28. Then, you can change the display name in the **Subaccount Overview** section off the screen shown in Figure 2.27 by using the pencil icon in the upper right-hand corner. If the authentication data has expired during import, you'll need to download it again from the subaccount, as previously shown in Figure 2.24.

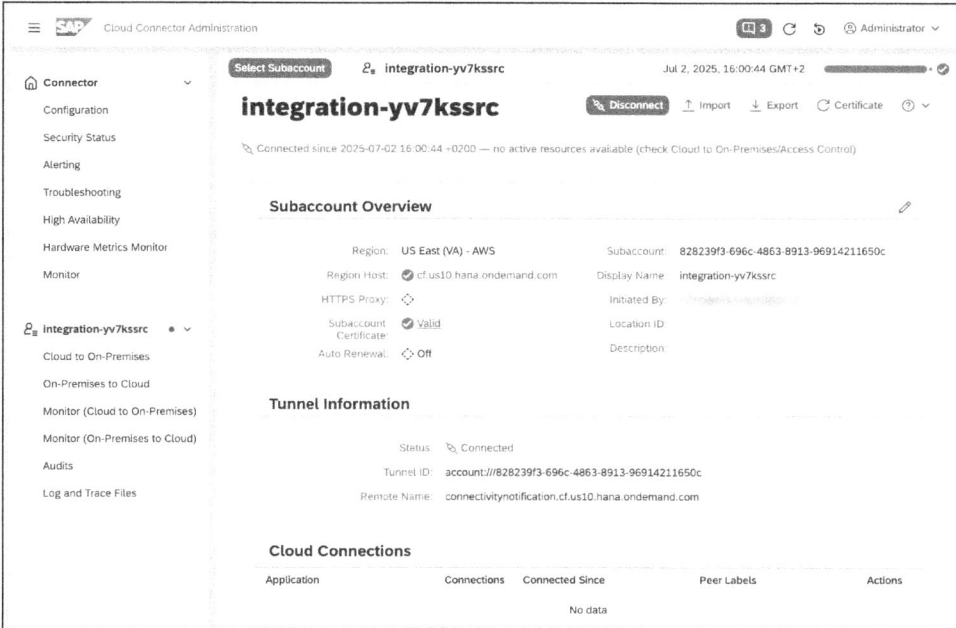

**Figure 2.27** Completed Subaccount

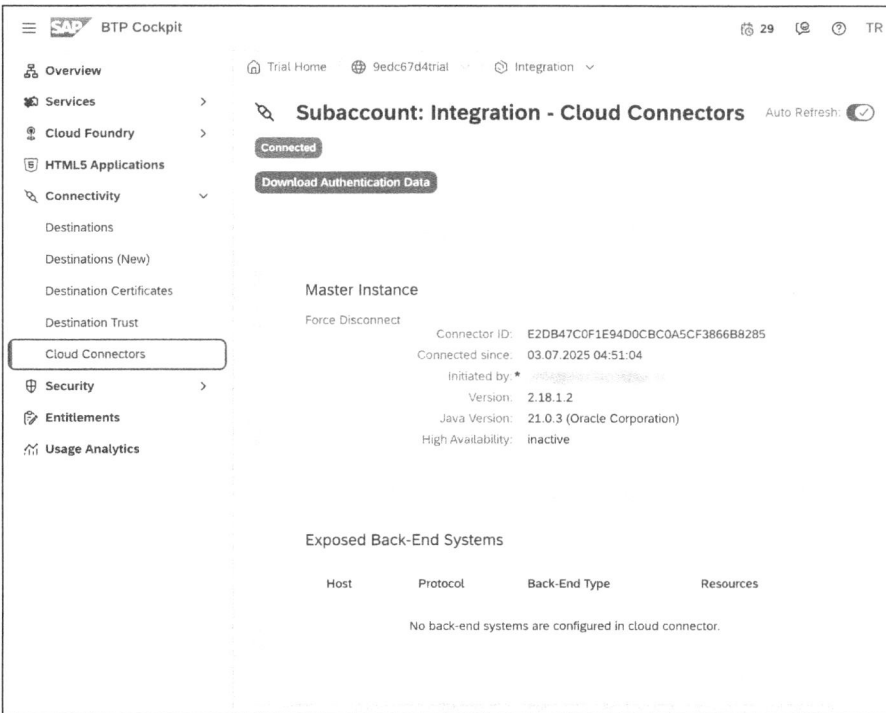

**Figure 2.28** Subaccount Connected to Cloud Connector

The next step is to establish the connection to your SAP S/4HANA system. To do this, navigate to the section with the name of your subaccount and then click on **Cloud to On-Premises**. Once there, navigate to the **Access Control** tab (see Figure 2.29).

To connect to your SAP S/4HANA system, you need to create a mapping. To do this, click on the **+** symbol on the right-hand side of **Mapping Virtual to Internal System**.

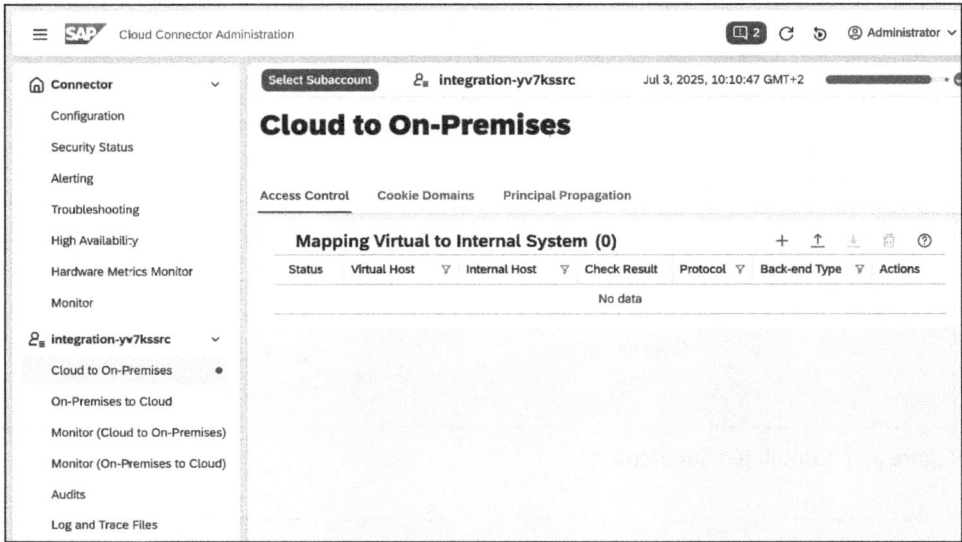

**Figure 2.29**  Cloud to On-Premises Screen

That will open a popup where you'll need to determine the **Back-end Type**. Since you want to connect to an on-premise SAP S/4HANA system, select **ABAP System** from the dropdown list and click **Next** (see Figure 2.30).

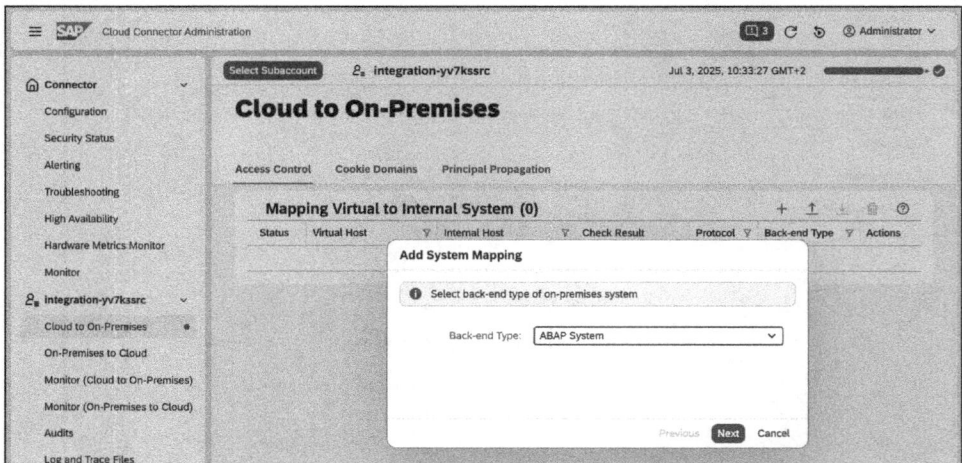

**Figure 2.30**  Adding System Mapping and Choosing Back-End Type

On the next page, you have to select the protocol with which you want to connect to your SAP S/4HANA system. Select **HTTPS** as the **Protocol** and click **Next** (see Figure 2.31).

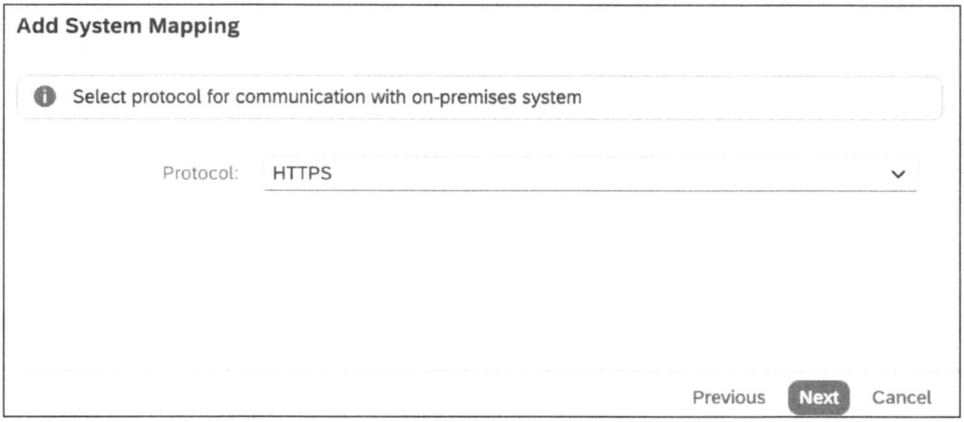

**Add System Mapping**

ⓘ  Select protocol for communication with on-premises system

Protocol:      HTTPS                                                              ⌄

Previous    **Next**    Cancel

**Figure 2.31**  Selecting Protocol

Now, you have to enter the **Internal Host** and **Internal Port** of the HTTPS endpoint of your SAP S/4HANA system, as shown in Figure 2.32. If you don't know it, look it up in the SAP S/4HANA system via Transaction SMICM, where you can select **GoTo · Services** to find the necessary information.

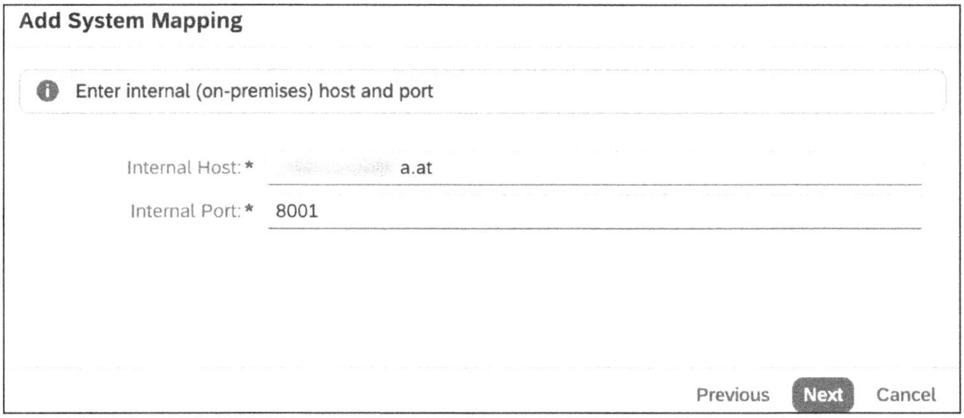

**Add System Mapping**

ⓘ  Enter internal (on-premises) host and port

Internal Host: *          a.at

Internal Port: *     8001

Previous    **Next**    Cancel

**Figure 2.32**  Configuring Internal Host and Port

The next step is to define a **Virtual Host** and **Virtual Port**. You can use the **Virtual Host** to access the cloud connector in SAP BTP, and it can (but doesn't have to) use the same host and port as the internal host. The host and port are different in this scenario (see Figure 2.33).

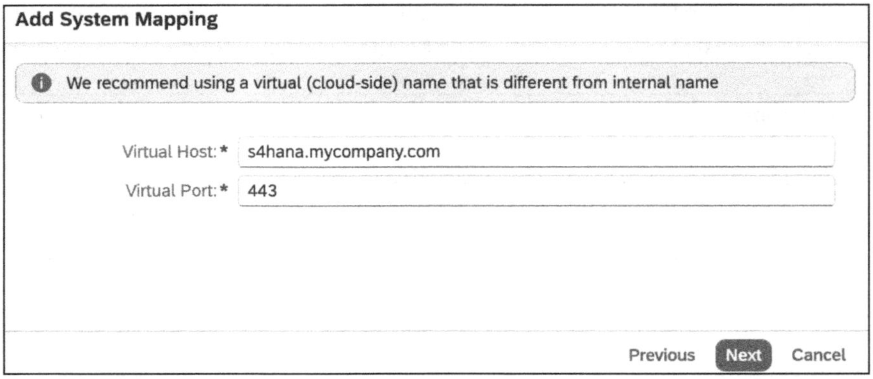

**Add System Mapping**

ⓘ We recommend using a virtual (cloud-side) name that is different from internal name

Virtual Host: *    s4hana.mycompany.com

Virtual Port: *    443

Previous    **Next**    Cancel

**Figure 2.33**  Configuring Virtual Host and Port

On the screen shown in Figure 2.34, you need to decide whether or not to **Allow Principal Propagation** for this connection. *Principal propagation* is based on the concept of identity propagation along a technical communication chain, where the aim is to pass the user identity of the calling user from a cloud application—for example, from SAP Integration Suite—through to the connected backend system in order to execute authorized actions on behalf of the user. In classic scenarios, communication between systems often takes place with a technical user, which means that the actual end user identity is lost. Principal propagation raises this model to a new level of security and transparency: the identity of the calling user is retained and is transmitted to the on-premise system via tokens, certificates, or SAML assertions. This can check authorizations, perform logging, or grant differentiated access based on the known identity.

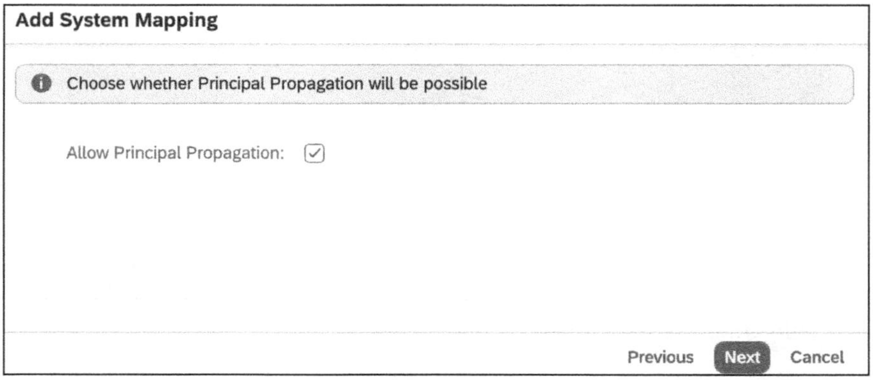

**Add System Mapping**

ⓘ Choose whether Principal Propagation will be possible

Allow Principal Propagation:    ☑

Previous    **Next**    Cancel

**Figure 2.34**  Allowing Principal Propagation

Next, you choose **X.509 Certificate** as the **Principal Type**, check the **System Certificate** for **Logon**, and for **Host in Request Header**, choose **Use Virtual Host**. After this, you type in the **System ID** of your SAP S/4HANA system (as shown in Figure 2.35), click **Next** to see the summary of your configuration, check the **Check Internal Host** button, and finalize it by clicking **Finish**.

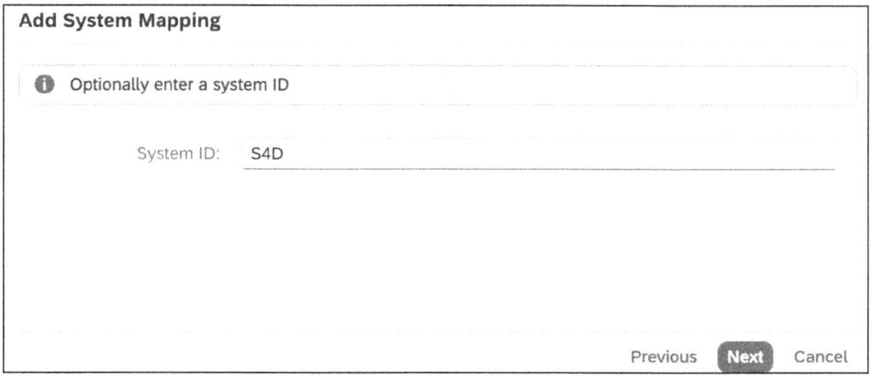

**Add System Mapping**

ⓘ  Optionally enter a system ID

System ID:   S4D

Previous   **Next**   Cancel

**Figure 2.35** Specifying System ID

Then, you'll see that the **Check Result** says **Not Reachable** (as in Figure 2.36). That's because your cloud connector needs to trust your SAP S/4HANA system, and to make it do that, you need to upload the X.509 certificate of the SAP S/4HANA system.

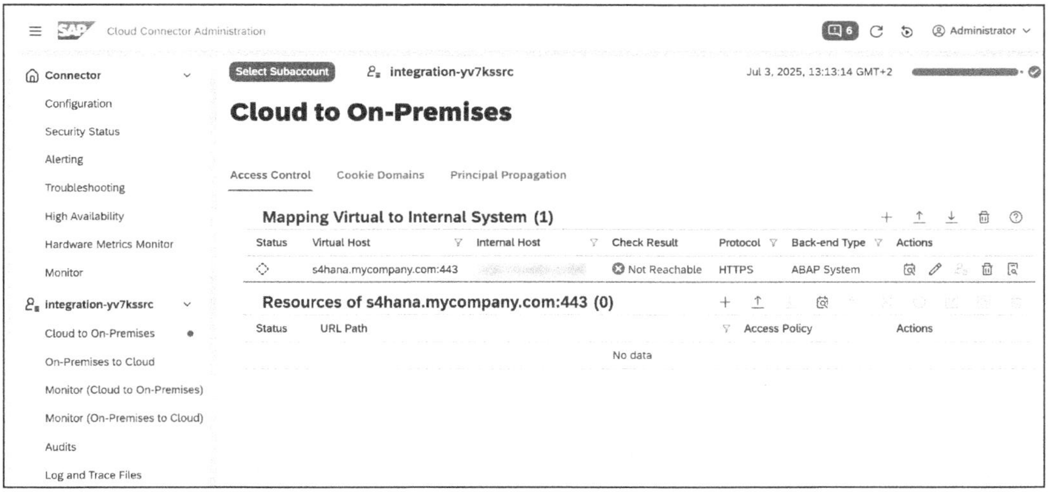

**Figure 2.36** Check Result Is Not Reachable

To do that, navigate to **Connector** on the left-hand side and then to **Configuration**. Then, in the **On-Premises** tab, scroll down until you reach **Backend Trust Store**. Here, you'll have to upload the X.509 certificate of the S/4HANA system by clicking the **+** symbol (see Figure 2.37)—but only if **Determining Trust Through Allowlist** is set to **ON**. If it isn't, all backend systems will be automatically trusted.

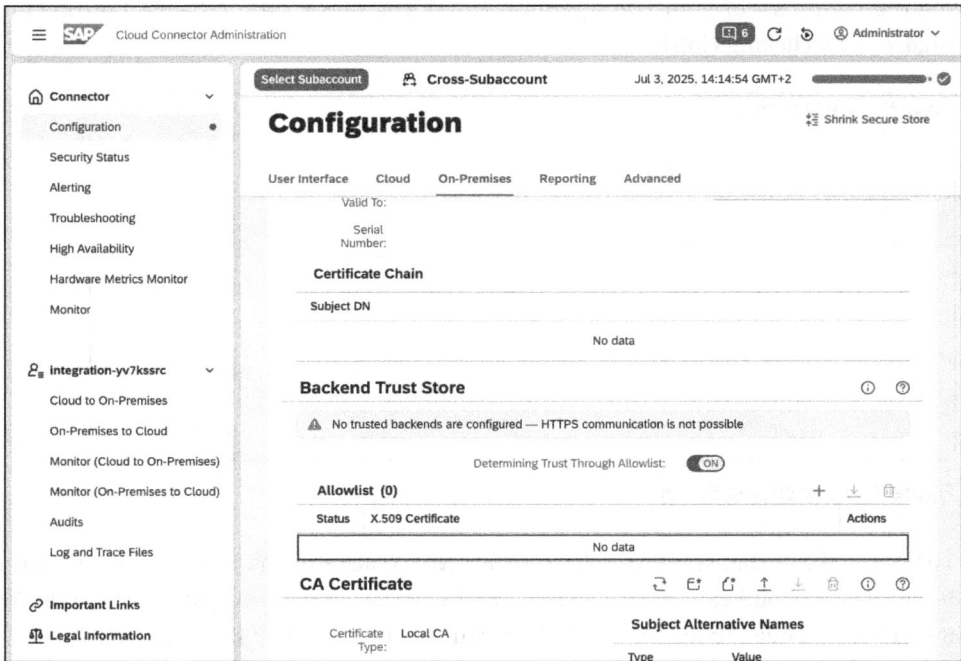

**Figure 2.37**  Uploading X.509 Certificate

When you go back to your mapping, select the **Actions** menu, and click the **Check Availability** button, the **Check Result** will be shown as **Reachable** (see Figure 2.38).

Once your SAP S/4HANA system is reachable, you'll only need to release the resources that the cloud connector can access, which are ICF paths. To do this, click on the **+** symbol to the right of **Resources of s4hana.mycompany.com:443**.

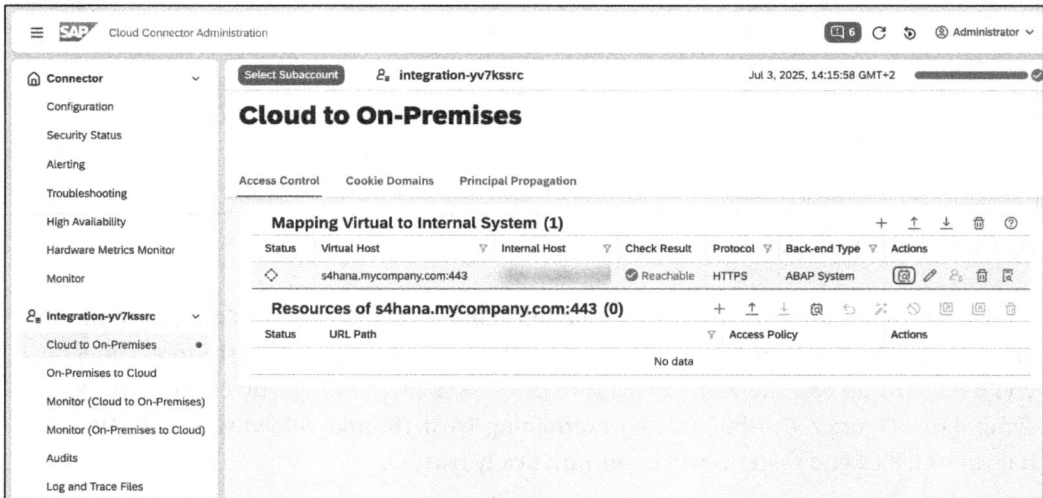

**Figure 2.38**  Check Result Is Reachable

That will open a window where you have to enter the URL. In this example, you enter "/" as the **URL Path,** enable it by activating the **Path and All Sub-Paths** radio button (to let the cloud connector reach all ICF paths), and finalize it by clicking **Save** (see Figure 2.39).

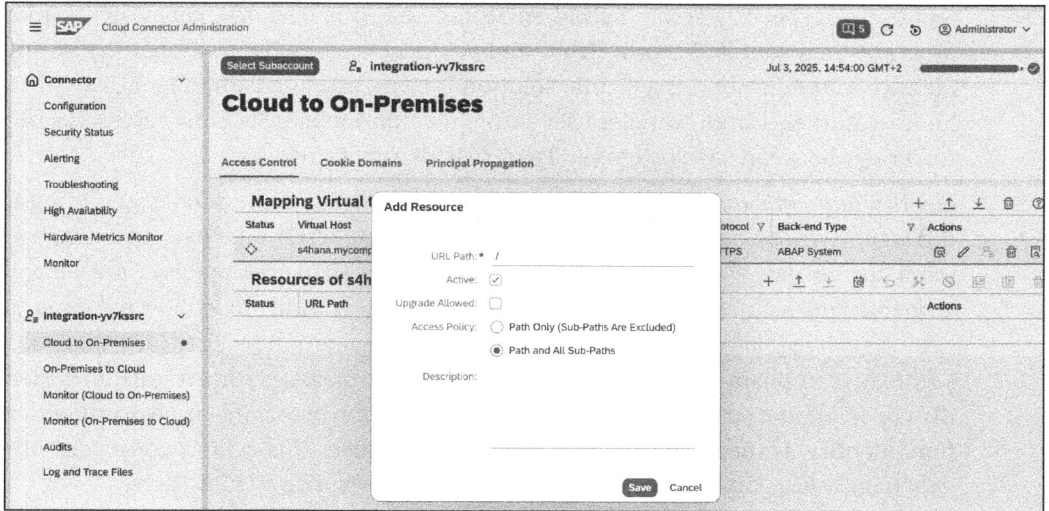

**Figure 2.39** Defining Available Resources

If you switch back to your SAP BTP, you'll see that the virtual host is now displayed and ready for use (see Figure 2.40).

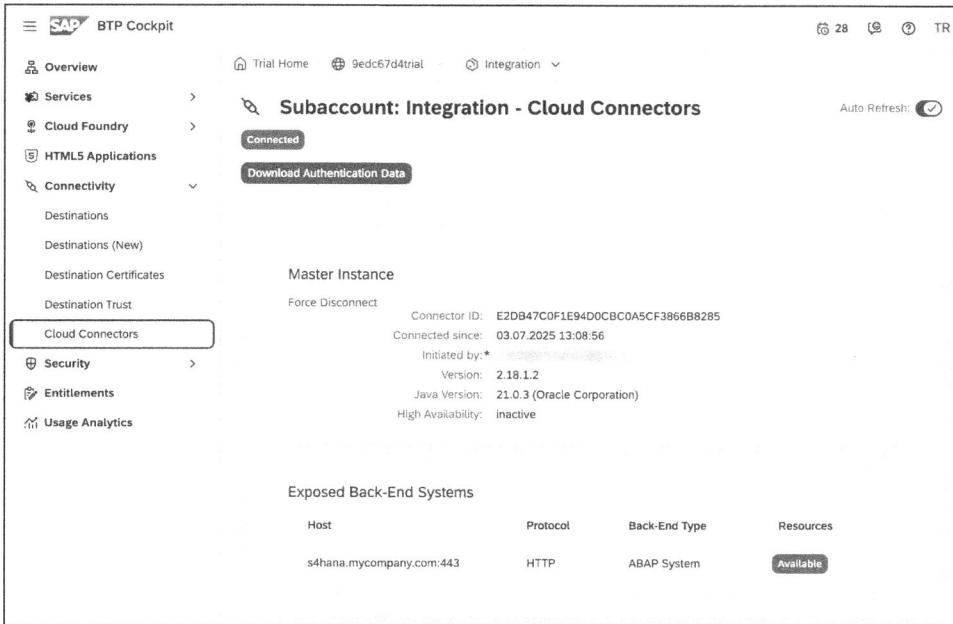

**Figure 2.40** Successfully Connected Cloud Connector

## 2.5   Connecting to the Identity Authentication Service

The secure management of user identities is one of the key challenges in modern, cloud-based corporate landscapes. With the increasing networking of services, systems and users—especially across hybrid environments—the need to map authentication processes in a standardized, scalable, and compliance-compliant manner is growing. Identity Authentication is a central solution for this, and as a component of SAP BTP, Identity Authentication handles user authentication, enables SSO, and serves as a central bridge between SAP systems and external identity providers.

*Identity Authentication* is a cloud-based identity provider that operates according to the *identity federation approach*. This means that it doesn't take over user management in the narrower sense, but rather, it ensures that user requests are reliably authenticated—either against its own user directory or against a connected external system such as Microsoft Entra ID, Active Directory Federation Services (ADFS), Okta, or SAML 2.0–compliant sources. Identity Authentication plays a central role in SAP's identity consolidation strategy: instead of each cloud system managing its own authentication, Identity Authentication acts as a central instance. This creates consistent user experiences (e.g., through SSO) and clearly traceable security architectures.

Integration into SAP BTP takes place via trust configurations. Identity Authentication is registered as a trusted identity provider in the SAP BTP subaccount, and applications that are operated in this subaccount—such as SAP Integration Suite and SAP Build Work Zone—then rely on Identity Authentication as an authentication source.

Identity Authentication can be used in numerous scenarios, including the following:

- Single sign-on (SSO) between SAP cloud applications (e.g. SAP SuccessFactors, SAP Integration Suite, SAP Build) and company directories
- Federated authentication via SAML 2.0 or OpenID Connect
- Multifactor authentication (MFA) for increased security
- User login portals (e.g., with self-service functions)
- Branding and UI customization of the login process
- Central authentication for SAP BTP subaccounts

A common use case is an employee logging into an SAP Fiori app that's running in SAP BTP by using the same credentials they use internally for Microsoft 365—without having to reenter their access data. This is made possible by a trust relationship between Identity Authentication and the company's own identity provider.

In the following sections, we'll walk through configuring the connection between the SAP BTP subaccount and Identity Authentication and then show you how to log in to an application by using the Identity Authentication user.

### 2.5.1   Configuring the Connection Between Subaccount and Identity Authentication

To configure the connection, you must first make Identity Authentication available. You can obtain the Identity Authentication tenant from the person from whom you obtained SAP BTP access, and that person will provide you with a productive Identity Authentication service and a test Identity Authentication service and then assign them to your SAP BTP global account.

Next, to connect your SAP BTP subaccount to your Identity Authentication, navigate to the subaccount you want. Once you're there, navigate to the **Security** menu item on the left-hand side and then to **Trust Configuration**. By default, only the **Default identity provider** is stored there, as you can see in Figure 2.41. The default identity is the standard SAP identity provider, and it enables login with the global ID.

Then, click the **Establish Trust** button in the top right-hand corner to start the connection between the SAP BTP subaccount and Identity Authentication.

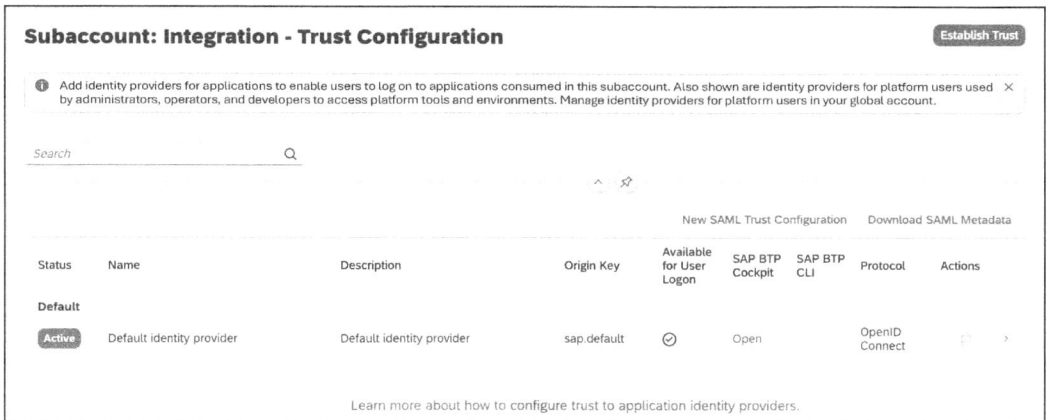

**Figure 2.41** Trust Configuration Overview

That will open a new window where all Identity Authentication services assigned to the global account are displayed. Select the Identity Authentication service you want and click on **Next** at the bottom right (see Figure 2.42).

Next, you need to select the Identity Authentication domain that you want to link to your SAP BTP subaccount. To do this, click on the dropdown menu in the **Domain** field in the **Choose Domain** step in the configuration screen. There, select the domain you want that's assigned to the Identity Authentication tenant. This domain will later form the basis for a uniform single sign-on experience, and you should therefore use it consistently for all relevant SAP BTP and non-SAP BTP applications. Once you've selected the correct domain, click on **Next** to continue with the configuration (see Figure 2.43).

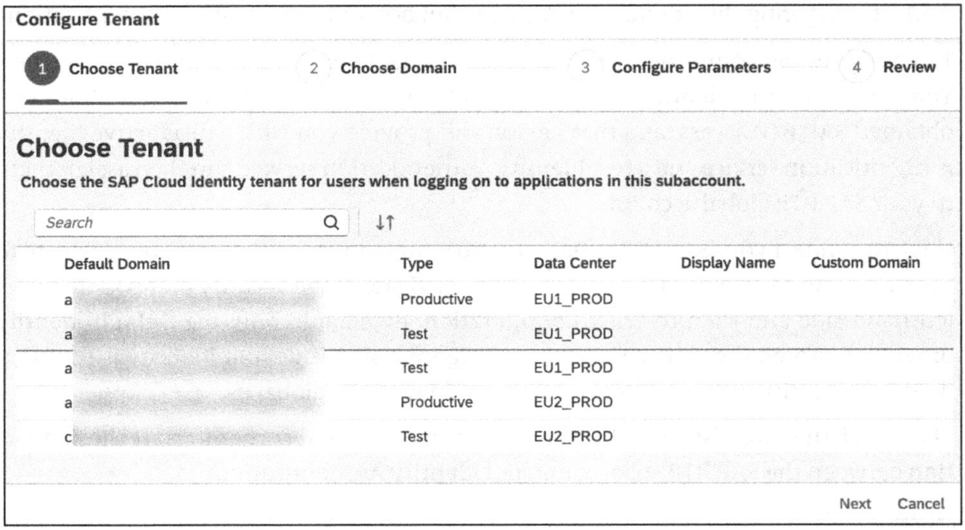

**Figure 2.42**  Selecting Identity Authentication Tenant

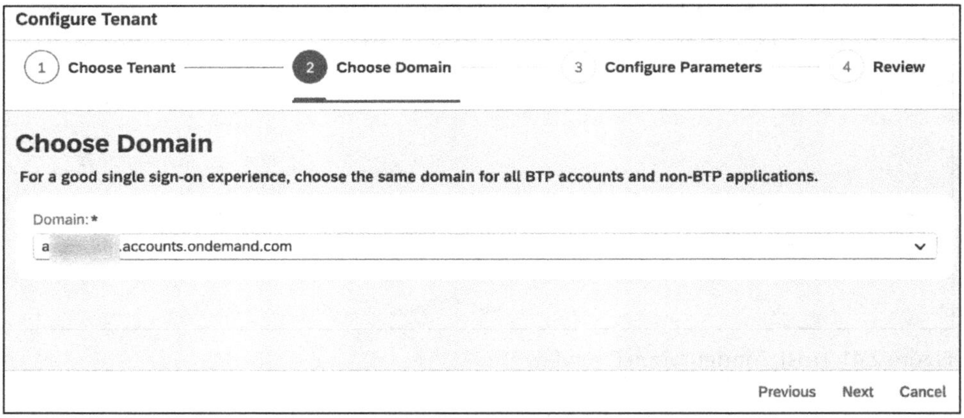

**Figure 2.43**  Selecting Identity Authentication Domain

In the next step, you'll be on the **Configure Parameters** screen, where you'll define the essential parameters of the trust relationship between the subaccount and the Identity Authentication service. In the **Name** field, enter a descriptive name for the configuration—typically in the form of the domain with an explanatory addition—for example, "(business users)" is a name you could enter. Optionally, in the **Description** field, you can enter a description that explains the purpose of the connection. The **Origin Key** is particularly important: you'll use this technical identifier later to assign the user origin context—for example, in role or access assignments. Under **Link Text for User Logon**, you specify what users should see as a selection option in the login window, such as the name of their company or their email domain. Optionally, you can also check the **Create Shadow Users During Logon** box to automatically create users in SAP BTP when logging in. Finally, click on **Next** to proceed to the review (see Figure 2.44).

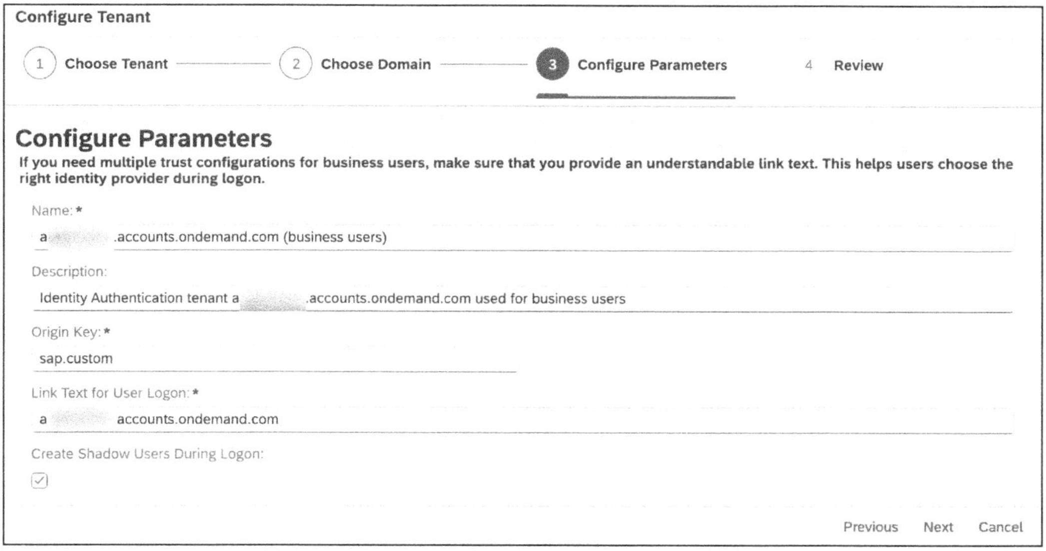

**Figure 2.44** Configuring Trust Parameter

The last step of the configuration takes you to the **Review** view, in which all the settings you've made are summarized once again. Here, you can see the selected tenant URL, the assigned domain, the origin key, and whether the automatic creation of shadow users has been activated. This overview serves as a final check before you activate the trust configuration. If all the information is correct, you complete the setup by clicking the **Finish** button (see Figure 2.45).

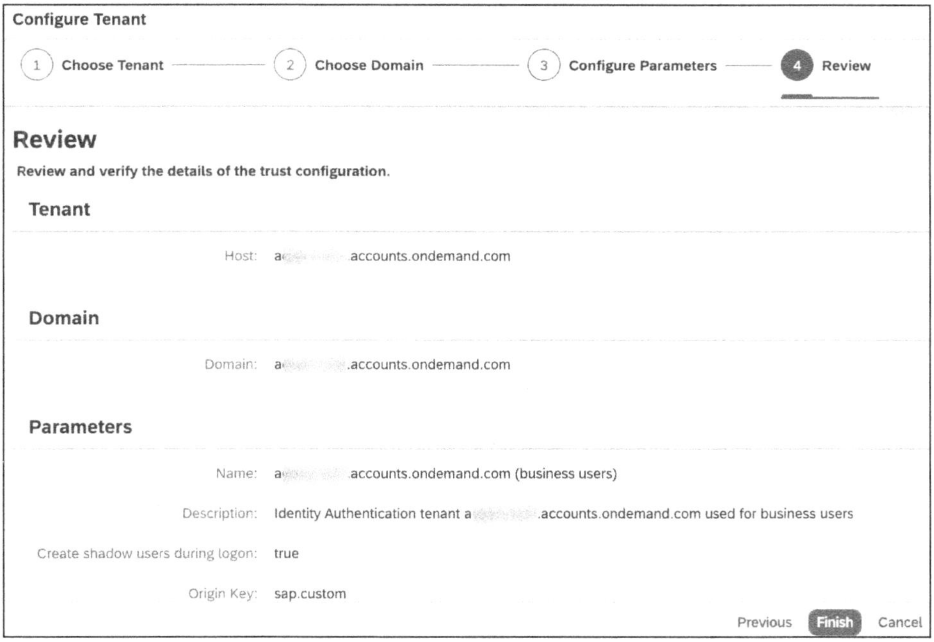

**Figure 2.45** Reviewing Identity Authentication Trust Configuration

Once you've successfully completed the setup, the new trust configuration will appear in the overview in the **Trust Configuration** section of the SAP BTP subaccount. Here, you'll see the status of the connection (e.g., **Active**), the assigned name, and the authentication protocol variant used (e.g., OpenID Connect). At that point, the configuration will be ready for use (see Figure 2.46).

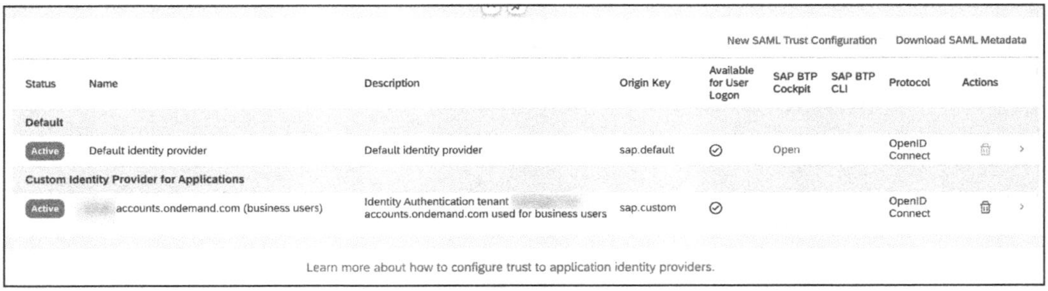

| Status | Name | Description | Origin Key | Available for User Logon | SAP BTP Cockpit | SAP BTP CLI | Protocol | Actions | |
|---|---|---|---|---|---|---|---|---|---|
| New SAML Trust Configuration | | | | | | | Download SAML Metadata | | |
| **Default** | | | | | | | | | |
| Active | Default identity provider | Default identity provider | sap.default | ⊘ | Open | | OpenID Connect | 🗑 | › |
| **Custom Identity Provider for Applications** | | | | | | | | | |
| Active | accounts.ondemand.com (business users) | Identity Authentication tenant accounts.ondemand.com used for business users | sap.custom | ⊘ | | | OpenID Connect | 🗑 | › |

Learn more about how to configure trust to application identity providers.

**Figure 2.46** Completed Identity Authentication Trust Configuration

### 2.5.2   Logging in with the Identity Authentication User

Once you've successfully configured Identity Authentication as a trusted identity provider, a user can log in to a connected application with their Identity Authentication identity. When accessing the application, the user will automatically be redirected to the Identity Authentication login page, provided the application uses the previously configured origin key. There, the user can enter their login details (e.g., email address and password) or authenticate themselves with an alternative method such as two-factor authentication (2FA), if it's activated. Identity Authentication handles the authentication and issues a corresponding token (e.g., SAML, OpenID Connect), which is then transferred to the target application, which in turn grants access based on this information and forwards it on a role-based basis.

Once it's set up, this login enables a seamless and secure single sign-on experience across different cloud and on-premise applications—without the user having to authenticate themselves multiple times. It's also important to assign the corresponding role collections to the Identity Authentication user in the subaccount to ensure functional authorization within the application.

The first step is to create the user who will later log in via Identity Authentication in the SAP BTP subaccount. To do this, you navigate to **Security**, click on **Users** in the left-hand menu in the SAP BTP cockpit, and click **Create** in the top right-hand corner. In the dialog box that opens, enter the **User Name**, **Identity Provider**, and **E-Mail** address you want. You should also select the previously configured Identity Authentication tenant so you can log in later via Identity Authentication, and then, you confirm the entries by clicking the **Create** button (see Figure 2.47).

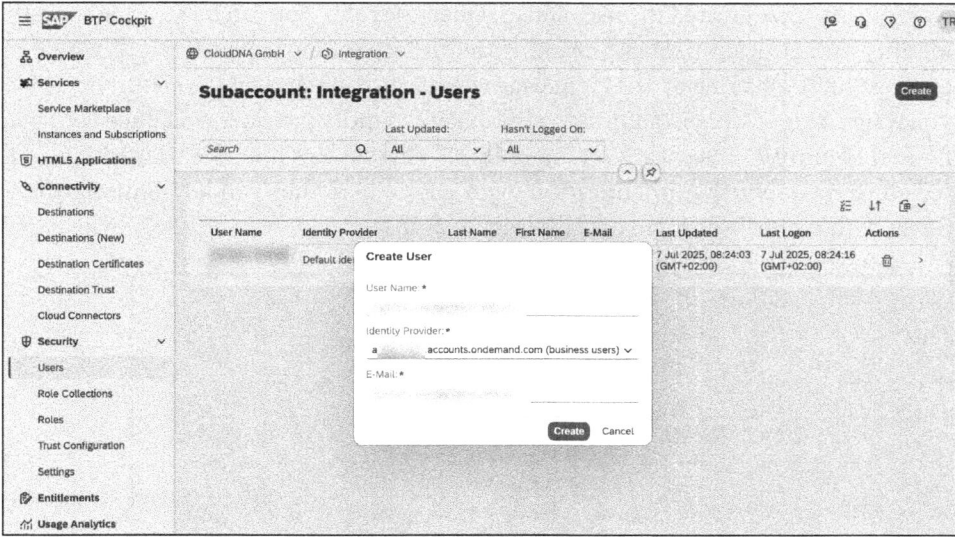

**Figure 2.47**  Creating Identity Authentication User in BTP

In the next step, you assign the necessary authorizations to the user. To do this, open the user in the subaccount and click on **Assign Role Collection** in the **Role Collections** area. That will open a dialog box where you can search specifically for relevant authorization roles—in this case, for the roles for SAP Integration Suite (see Figure 2.48).

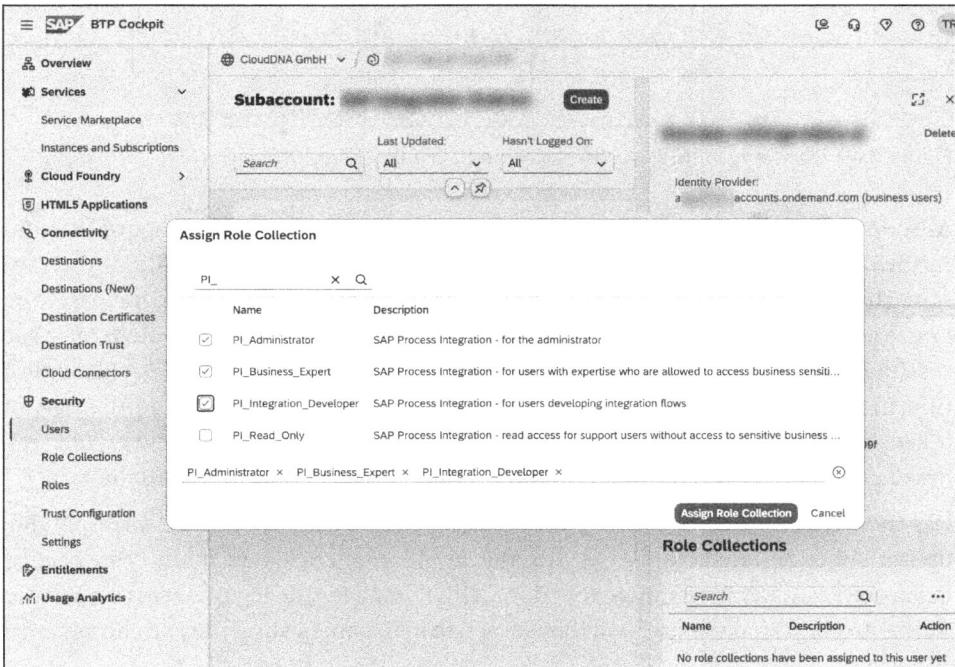

**Figure 2.48**  Assigning Users Role Collections

As soon as you've created the user and assigned the roles, you can test the login via the Identity Authentication tenant. Opening the corresponding application (e.g., SAP Integration Suite) will bring up the login page where you can select which identity provider you want to log in with. In addition to the default identity provider, you'll see your configured Identity Authentication tenant. By clicking on this link, the user will be redirected to the Identity Authentication login page where they can authenticate themselves with their stored login data (see Figure 2.49).

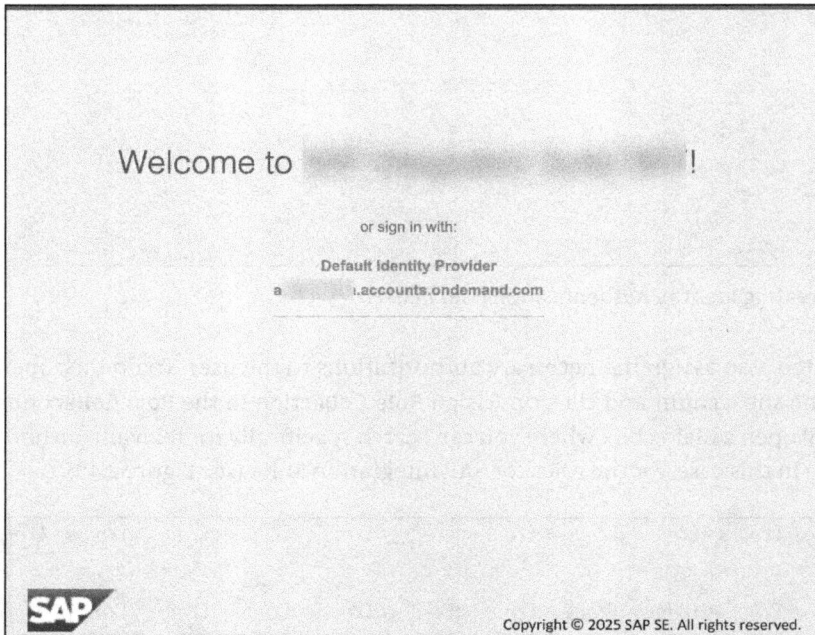

**Figure 2.49** Choosing Identity Authentication to Log In

As soon as the user selects their identity provider—in this case, the configured Identity Authentication tenant—they'll be redirected to the login page of Identity Authentication. There, they can enter their login details, such as their email address and password, or use an extended authentication procedure such 2FA, if it's been activated. After successful authentication, the user will be automatically redirected back to the application, in our case to SAP Integration Suite. The application will check the authentication token issued by Identity Authentication (e.g., SAML, OpenID Connect) and grant access based on the roles and authorizations you assigned in the SAP BTP subaccount.

Depending on the role collections you've assigned, the user will see the corresponding user interface or functionality—such as the development of iFlows, monitoring of existing processes, or administrative functions. This completes the login process via Identity Authentication, and the user will then be active in the application they've chosen.

## 2.6 Summary

In this chapter, we've established the technological and organizational basis for the operation of SAP Integration Suite within SAP BTP. Our emphasis wasn't exclusively on the technical provision, but rather on the deliberate, strategically conceived configuration of a secure, scalable, and maintainable integration platform. The chapter is divided into several sections that build on each other and describe the configuration of a complete environment as a whole.

First, we had you create a structure within SAP BTP, and you carried out the configuration at the level of a subaccount, which is a logical unit for the provision and management of services. After Cloud Foundry activation, you selected SAP Integration Suite and subscribed to it in the subaccount's Service Marketplace.

After the technical provisioning, you set up the user administration in the subaccount. An important step in the overall configuration was the connection of Identity Authentication to the subaccount, and by configuring a trust configuration, you determined that users would be authenticated via Identity Authentication in the future. Parallel to the cloud configuration, we described the integration of the cloud connector, which you installed locally and connected to the previously set up subaccount.

The configurations we've described in this chapter are not an end in themselves. Rather, they serve to build a well-founded, maintainable, and expandable integration ecosystem that meets the following requirements:

- Technical security through TLS, SAML, role models, and Identity Authentication connection
- Organizational separation of responsibilities through role-based access
- Scalability and maintainability through modular capabilities and standardized architecture
- Compliance capability through client separation, the shadow user principle, and auditability
- Future-proofing through the integration of modern services such as Event Mesh, API management, and identity federation

This ecosystem therefore forms the basis for all subsequent integration activities. Productive integration scenarios can only be implemented efficiently and with low risk once the system is securely configured, cleanly structured, and set up in an identity-conscious manner.

# Chapter 3

# Mappings: Groovy, XSLT, and Message Mappings

*The integration of heterogeneous systems has long been one of the central challenges in enterprise IT, and in the age of digital transformation, the complexity of this task has only intensified. Companies no longer operate in isolated, monolithic environments; instead, they rely on diverse landscapes made up of on-premise systems, cloud applications, and third-party services. These components must exchange information continuously and reliably to support modern business processes, which are increasingly distributed and event driven.*

The shift toward cloud-based architectures—and the hybrid environments that result—has introduced new technical and organizational demands on integration platforms. Data must not only be moved between systems but also transformed, validated, enriched, and routed in a way that ensures semantic consistency across different domains. Formats vary widely, from XML-based business documents to lightweight JSON payloads from REST APIs, and even to unstructured or semistructured data from legacy interfaces. Without a robust integration strategy, mismatches among these formats, protocols, and data models can create friction, errors, and costly delays.

Cloud Integration has been designed to address precisely this challenge. It provides a flexible, cloud-native framework for modeling, orchestrating, and executing cross-system data flows. Built on a scalable, multitenant architecture and powered by a mature message processing engine, Cloud Integration supports a wide range of adapters, transformation tools, and processing patterns. This makes it a versatile choice for organizations seeking to streamline communication between SAP and non-SAP systems.

Among the many capabilities of Cloud Integration, two functional domains stand out as fundamental to successful integration scenarios: scripting and message mapping. These are not competing approaches but complementary disciplines, and each serves a distinct role in the transformation and control of messages as they traverse the integration layer.

This chapter introduces the three primary transformation techniques that are available within Cloud Integration: graphical message mapping, Extensible Stylesheet Language Transformations (XSLT), and Groovy scripting. Each of these methods supports a different class of transformation problems. *Graphical message mappings* enable developers to define field-level relationships visually and thus provide an accessible and maintainable approach to XML-based scenarios. *XSLT*, on the other hand, offers a standards-based solution that's ideal for rules-driven XML transformations with fine-grained control. *Groovy scripting* introduces full programmatic flexibility and thus allows integration developers to implement complex logic that can't be easily expressed through declarative tools. The introductory section of this chapter will guide you through the fundamentals of what scripting and mapping really means, the last part of the chapter includes a real-world and guided example of it.

You must understand when and how to use each technique to build robust and maintainable iFlows. While you may be able to fully address simple mappings through graphical tools, more advanced use cases—such as dynamic message construction, runtime decision-making, and interaction with external context—may require scripting or a hybrid approach.

## 3.1   Introduction to Mapping and Scripting

In the evolving landscape of enterprise integration, it's more and more important for organizations to transform data reliably and efficiently. As they increasingly adopt hybrid system architectures that comprise a mix of cloud-based services, on-premise systems, and external applications, the integration layer must do more than simply connect systems. It must mediate between differing data formats, interface definitions, and semantic models to ensure that business processes function seamlessly across technological boundaries.

SAP Integration Suite, running on SAP BTP, provides the foundation for addressing this challenge. At the core of Cloud Integration, which is a component of the suite, lies a powerful and flexible transformation engine that's capable of converting, enriching, and routing messages among disparate systems. The effectiveness of these transformations largely depends on the use of mappings and scripting.

Modern enterprise IT architectures are increasingly shaped by hybrid system landscapes in that they combine legacy on-premise systems with cutting-edge cloud services and third-party SaaS applications. As companies digitalize and expand, their ability to ensure seamless interoperability across these heterogeneous components becomes paramount. In such a dynamic environment, integration isn't merely a technical layer—it's a business enabler.

A key challenge in these scenarios is the diversity of data formats, communication protocols, and data models. Systems often "speak different languages" because they're shaped by their respective vendors, standards, and business requirements. Effective integration must reconcile these discrepancies, and in Cloud Integration, this reconciliation is achieved through a range of transformation and scripting capabilities that are collectively referred to as *mappings*.

At the heart of every integration project lies the transformation of data from one structure to another. A *mapping* defines the logical relationship between the source data model (e.g., a CRM system's output) and the target model (e.g., the input format that's expected by an ERP system). This may involve renaming fields, reformatting values, nesting structures, or aggregating elements.

Cloud Integration provides three principal approaches to handling these requirements:

- **Groovy scripting (see Section 3.2)**
  This approach provides full programmatic control over message content, format, and processing logic. For more advanced requirements, procedural logic becomes necessary. Using Groovy empowers integration developers to perform complex manipulations that go beyond what declarative mappings can achieve. This includes conditional branching, aggregation across message structures, content enrichment using external properties, and dynamic message creation.

  Cloud Integration supports Groovy natively within the iFlow designer. Scripts are executed as processing steps, and they interact with the `Message` object, including its body, headers, properties, and attachments. The syntax is concise, and the language is easy to adopt for developers with Java experience. You can reuse scripts across projects and maintain them in centralized script collections.

  Scripting, such as in Groovy or JavaScript, allows you to write transformation logic as programs. Scripting is the most flexible approach, and it enables integration developers to address extremely complex, non-XML input/output (I/O) formats or to apply business logic that is unfeasible with declarative mapping. Scripting is rarely the first choice for standard XML-to-XML transformations, but it's invaluable for JSON, CSV, and nonstandard protocols. Its advantage is "no limits" expressiveness, but it comes at the cost of losing mapping-specific tooling and readability.

- **Message mapping (see Section 3.3)**
  This is a visual, model-driven approach to straightforward XML-to-XML transformations. Message mapping is widely used in SAP integration scenarios, especially by consultants who are familiar with SAP Process Integration (PI/PO). It allows users to drag and drop connections between source and target elements by using a visual editor. Predefined functions—such as `substring`, `exists`, and `ifWithoutElse`—make common operations accessible without coding. You can also inject custom logic with user-defined functions (UDFs) that you can write in Java.

  Message mappings are ideal for structured, schema-based XML transformations with predictable patterns. However, they're limited to XML inputs and outputs and can become difficult to maintain for deeply nested or highly conditional logic.

  Message mapping is the most visual and user-friendly tool. It features a graphical interface where the source and target message structures are rendered as trees, and you can perform mapping by drawing connections between them. You can apply built-in functions, value mappings, and conditional expressions via the interface. Message mapping is well-suited to standard transformation tasks where the mapping logic is relatively straightforward, and it's accessible to users who have a minimal programming background. Key benefits are ease of use, high productivity, and maintainability for simple to moderately complex mappings

- **XSLT mapping (see Section 3.4)**
  This approach addresses scenarios that require more structured or rule-based logic and that often involve advanced looping, template reuse, and conditional formatting. As a World Wide Web Consortium (W3C)–standardized transformation language, XSLT is particularly well suited to transforming hierarchical XML content, it provides clear separation of logic and data, it's ideal when existing XSLT libraries or governance rules are already in place, and it enables structured, rule-based XML transformations by using standardized stylesheets.

  XSLT mapping is a code-centric approach that offers significantly greater flexibility and power than message mapping. It allows direct authoring of XSLT stylesheets, which means developers can leverage the full capabilities of the XSLT language and XML Path Language (XPath). XSLT Mapping is ideal for complex transformations, recursive structures, advanced filtering, and scenarios where the visual mapping editor falls short. Its strengths are precision, conciseness for complex logic, support for dynamic decision-making, and the ability to handle edge cases cleanly—though it requires knowledge of XSLT and experience with XML data structures.

Both message mapping and XSLT mapping share the advantage of transparency and reusability, but they remain inherently declarative, so they're limited in their ability to compute dynamic values, interact with message headers, and fetch external resources.

Each mapping method has its strengths and trade-offs, and in practice, iFlows often benefit from a hybrid approach. Declarative mappings offer clarity, speed, and visual transparency—so they're ideal for standard transformations. Scripting, on the other hand, delivers the flexibility you need to address edge cases and evolving business logic.

Section 3.5 demonstrates this hybrid approach by presenting a real-world integration scenario. It illustrates how you can use message mapping for structural transformations, while Groovy handles validation and enrichment and XSLT wraps the payload in a standards-compliant envelope. This combination leverages the strengths of each technique to create solutions that are robust, maintainable, and adaptable to future changes.

Table 3.1 is a quick comparison chart for the three approaches. In practice, these mapping options are complementary. In straightforward scenarios, message mapping suffices; as complexity grows and specialized transformation logic is needed, XSLT mapping and groovy scripting become essential tools in the integrator's toolkit.

| Mapping Approaches | Suitable Scenarios | Required Skills | Advantages | Limitations |
|---|---|---|---|---|
| Groovy scripting | Very complex or non-XML formats | Programming (Groovy/Java-Script) | Unlimited flexibility and the ability to use any logic | Code maintenance and testing complexity |
| Message mapping | Standard XML-to-XML and simple logic | Integration developer | Easy, graphical, built-in functions | Limited for advanced or recursive tasks |
| XSLT mapping | Advanced XML transformations | XSLT and XML proficiency | Powerful, flexible, reusable, standards | Requires XSLT skills and is less visual |

**Table 3.1** Overview of Mapping and Scripting Options

## 3.2   Scripting with Groovy

In the syntactic sense, Groovy is of course based on Java, but there are various adaptations and optimizations. These are possible due to the partially more dynamic typing, and they also help with readability, among other things. A few helpful things Groovy has that Java doesn't are as follows:

- Variable definition using the def keyword or explicit type
- Simpler null checking
- Support for lists and maps
- Closures

The following sections introduce the key points of Groovy. They start by covering technical details and basic syntax fundamentals, and then, they moves into helpful concepts that are especially useful in Cloud Integration. They also cover closures, String manipulation, pattern matching using a regex, and working with XML and JSON constructs. The finish by covering logging within the scripts for better debugging and tracing.

### 3.2.1   Technical Overview of Groovy and Its Internal Architecture

Groovy is a powerful, optionally typed, and dynamically compiled programming language that runs on the Java Virtual Machine (JVM). Designed to complement Java by reducing boilerplate and increasing flexibility, Groovy is particularly well-suited to scripting scenarios where agility, brevity, and seamless Java interoperability are priorities. Within Cloud Integration, Groovy serves as the primary scripting language for customizing message flows, transforming content, and controlling runtime behavior.

Groovy is both dynamically typed and statically compilable. In its dynamic mode—which is used by default in Cloud Integration—it defers type resolution to runtime and thus allows developers to write flexible, expressive code with minimal syntax. Despite its dynamic nature, Groovy is built entirely on top of the Java platform. It compiles source code into Java bytecode, which is executed by the standard JVM.

The Groovy runtime consists of the following:

- **Groovy compiler (groovyc)**
  This translates *.groovy* source files into *.class* files (bytecode).
- **GroovyClassLoader**
  This loads Groovy classes at runtime to extend Java's ClassLoader mechanism.
- **Metaobject protocol (MOP)**
  This provides dynamic method dispatching and property resolution at runtime.
- **Groovy runtime libraries**
  This is a collection of classes that extend Java's standard library (e.g., enhanced String, Collection, Closure, and XML handling classes).

Groovy's tight integration with Java means that every Groovy class is a Java class under the hood. This ensures that Groovy scripts can do the following:

- Directly instantiate and manipulate Java objects.
- Use existing Java libraries.
- Be called from Java code (and vice versa).

This interoperability is a key reason why SAP chose Groovy as the scripting standard in Cloud Integration—it allows the runtime to leverage existing Java infrastructure while exposing a simplified scripting interface to developers.

In Cloud Integration, Groovy scripts are executed within the Apache Camel–based runtime of the iFlow engine. When a message passes through a script step, the engine invokes a Groovy interpreter, which does the following:

1. It loads the script with a secure GroovyClassLoader.
2. It injects the current `Message` object and context properties into the script.
3. It executes the `processData()` method synchronously.
4. It replaces the message body and/or headers, based on the returned result.

Cloud Integration uses a sandboxed Groovy environment to ensure security and runtime stability. This environment enforces the following:

- **Class whitelisting**
  This makes sure only approved packages (e.g., `java.util`, `groovy.xml`, `java.text`) are accessible.
- **No file system or network access**
  This prevents external calls, unless they're explicitly enabled via connectivity adapters.
- **Limited memory and execution time**
  This avoids runaway scripts that could degrade tenant performance.

All scripts run within the same JVM that hosts the iFlow runtime, so they share memory with the main integration engine but are isolated in terms of scope.

Groovy scripts interact with messages via the Message API, which is part of the `com.sap.gateway.ip.core.customdev.util` package. This API provides access to the following:

- **Body**
  This is the actual message payload, which is convertible to String, InputStream, or a Java object.

- **Headers**
  These are the key-value pairs that are used for routing, tracking, and metadata.

- **Properties**
  These are transient variables that are set during message processing (e.g., content modifier, router).

- **Exchange context**
  This is limited access to runtime state for dynamic behavior.

While Groovy scripting enables fine-grained control over message processing, it operates within a broader orchestration framework that governs the flow of messages across systems and components. At the heart of this framework lies *Apache Camel*—a lightweight, open-source integration framework that provides the backbone for message routing and mediation in Cloud Integration.

Understanding how Apache Camel works internally—and how it's used in Cloud Integration—provides valuable insight into the behavior of iFlows, including message routing, transformation, exception handling, and step execution.

Apache Camel is a rule-based routing and mediation engine that allows developers to define enterprise integration patterns (EIPs) by using a simple and expressive domain-specific language (DSL). It provides a standardized and extensible architecture for handling messaging systems, protocols, and message transformations.

Camel is built around the concept of a *route*, which defines the path that a message takes from a source (the producer) to one or more destinations (consumers). A route is composed of steps such as the following:

- `from()`
  This defines the input endpoint (e.g., HTTP, JMS, SFTP).

- `to()`
  This defines the target endpoint.

- `process()`
  This allows custom processing logic.

- `choice()`, `filter()`, `split()`
  This enables conditional routing and message manipulation.

Camel supports over 300 components for connectivity, including file systems, databases, cloud services, and APIs.

In Cloud Integration, each iFlow is essentially a Camel route that has been modeled with a graphical editor instead of traditional code. SAP abstracts Camel's DSL behind a visual representation and a set of predefined building blocks. Under the hood, however, the iFlow is translated into a series of Camel route definitions and processors.

Key concepts from Camel in Cloud Integration are as follows:

- **Processors**
  Each step in the iFlow (e.g., content modifier, router, Groovy scripting) is a Camel processor that transforms or handles the message.

- **Exchange objects**
  The Camel Exchange represents the message and its associated metadata. In Cloud Integration, this is exposed as the Message object that's used in Groovy scripts.

- **Route definitions**
  The iFlow XML is converted into Camel route configurations that are executed at runtime.

- **Message headers and properties**
  These correspond directly to Camel's headers and exchange properties so that they're accessible within Groovy and mapping steps.

SAP maintains its own runtime implementation based on Camel, but it includes additional layers of abstraction, security, multitenancy, and cloud-readiness. This ensures compatibility with SAP's managed BTP infrastructure and tenant isolation requirements.

When you deploy and trigger an iFlow (e.g., by receiving an HTTP message), Apache Camel initializes a route and begins processing the message through each defined step. The execution order follows the route definition in the iFlow, with each step treated as a Camel processor:

1. The message arrives at a sender adapter (e.g., HTTPS, IDoc, SFTP), which acts as a Camel consumer.

2. Steps such as content modifiers, mappings, and scripts modify the Exchange as it passes through.

3. Routers, filters, and choices implement conditional logic using Camel's routing engine.

4. Exception subprocesses are implemented as Camel error handlers using onException, doCatch, and doFinally constructs.

This flow is executed in memory and synchronously, unless it's otherwise configured with parallel or asynchronous steps.

Some benefits of Camel in Cloud Integration are as follows:

- **Standardized message flow**
  Camel provides a reliable, battle-tested engine for processing and transforming messages.

- **Modularity**
  Each step is a reusable, composable component that's aligned with the integration patterns outlined in Gregor Hohpe's *Enterprise Integration Patterns*.

- **Scalability**
  Routes can be parallelized and load-balanced to supporting scale-out in SAP's cloud runtime.

- **Observability**
  Camel supports interceptors and logging, thus enabling SAP to offer tools like message monitor, trace, and logging in the Cloud Integration UI.

To simplify the development experience, SAP abstracts much of Apache Camel's native configuration and hides the complexity of its routing DSL. As a consequence, developers working within Cloud Integration don't have direct access to define raw Camel routes or leverage the full range of Camel components. The creation of custom connectors or processors is restricted to predefined extensibility mechanisms that are offered by the platform. Furthermore, the execution order of processing steps is strictly governed by the graphical iFlow model, and this order isn't exposed or configurable through Camel's domain-specific language.

Despite these abstractions, understanding Camel's internal role will help you understand runtime behaviors, such as the following:

- Why certain steps can be placed in parallel subprocesses
- How exceptions are caught and handled
- How messages can be split, aggregated, or multicast across branches

### 3.2.2 Variables and Data Types

In Groovy, variables are fundamental building blocks that store references to values—whether they be simple data types like strings and numbers or complex objects like XML trees or JSON maps. One of Groovy's strengths lies in its *dynamic typing*, which allows developers to write concise, readable code without declaring variable types explicitly. This flexibility is especially useful in the context of Cloud Integration, where scripts are typically lightweight and task oriented.

Since Groovy is based on dynamic typing, we'll start off this section by highlighting the differences between dynamic and static typing. Then, we'll follow up with some useful groups of types and touch on the field of type inference and how strictly a variable must hew to its initial type. Then, we'll review constant values and null values. Finally, we'll cover type considerations within Cloud Integration.

**Dynamic Versus Static Typing**

Groovy supports both dynamically and statically typed variables. In practice, most scripts in Cloud Integration use dynamic typing, which is declared with the def keyword, as shown in Listing 3.1.

```
def name = "Customer"
def count = 5
def active = true
```

**Listing 3.1** Groovy Scripting: Loosely Typed Variables

Developers may also use *explicit typing* when clarity or type-specific operations are required, as shown in Listing 3.2.

```
String name = "Customer"
Integer count = 5
Boolean active = true
```

**Listing 3.2** Groovy Scripting: Explicitly Typed Variables

As you can see in Listing 3.1 and Listing 3.2, the name is a String, the count is an Integer, and the active is a Boolean. Groovy infers the types from Listing 3.1 at runtime, based on the assigned value, and this is manually defined and overridden in Listing 3.2. For most definitions—especially for simple temporary values—implicit typing may be sufficient. The explicit types, on the other hand, can come in handy for autocompletion and when types are expected—as an explicitly typed variable will throw an error on mismatching assignment.

**Primitive and Reference Types**

Groovy supports the same *primitive types* as Java, including the following:

- int, long, float, and double
- boolean and char

However, Groovy automatically wraps them in their *object counterparts* (e.g., Integer, Boolean, Character) when necessary so they can be treated like objects. Some of the common data types in groovy are listed in Table 3.2.

| Types | Descriptions |
|---|---|
| String | Textual data; it supports interpolation. |
| Integer | Whole numbers. |
| Boolean | True/false logic. |

**Table 3.2** Groovy Scripting: Vanilla Data Types

| Types  | Descriptions                                    |
|--------|-------------------------------------------------|
| List   | Ordered collection (like an array).             |
| Map    | Key-value collection (like a dictionary).       |
| Range  | Number or character sequences (1..10).          |
| Object | Base class of all types.                         |

**Table 3.2**  Groovy Scripting: Vanilla Data Types (Cont.)

### Type Inference and Reassignment

Because Groovy is dynamically typed, you can reassign variables you declared with def to values of different types. The importance of type reassignment lies on the implicit typing, as mentioned, by using the def keyword. On the other hand, you can't alter the type of explicitly typed variables after you initially define them (see Listing 3.3).

```
def data = "123"
data = 123   // Now an Integer and possible due to "def"

String data2 = "123"
data2 = 123 // Results in an error due to explicit type of String
```

**Listing 3.3**  Groovy Scripting: Type Reassignment Possibilities

While this flexibility can be powerful, it may reduce code clarity. In practice, it's best to avoid reassigning types within the same scope unless you're intentionally working with polymorphic logic.

### Immutability and Constants

Groovy doesn't have a built-in final keyword for def declarations, but you can declare constants as final or static final when using explicit typing.

In Cloud Integration scripts, you'll often declare constants at the top of the script to improve readability and maintainability, as shown in Listing 3.4.

```
final String COUNTRY = "DE"
```

**Listing 3.4**  Groovy Scripting: Defining Constant by Using Explicit Type

### Working with Null Values

Groovy provides several language features that help you handle null values safely. These enhance the standard Boolean comparison operations via null-safe access. You can shorten the inline ternary operator by using the Elvis operator (?:). To access class-level functions, you can use a JavaScript like a preceding question mark (?.).

These operators, which are shown in Listing 3.5, reduce the need for verbose null checks and are particularly useful in message processing, where optional fields are common:

```
def name = null
def result = name ?: "Unknown"        // Elvis operator
def length = name?.length()           // Safe navigation
```

**Listing 3.5**  Groovy Scripting: Null-Safe Variable Access Operations

### Groovy in Cloud Integration

Within Cloud Integration, the most common use of variables involves the following:

- Extracting or modifying message bodies (typically in XML or JSON as String)
- Reading message headers and properties (e.g., Map<String, Object>)
- Building collections for iteration
- Preparing payloads for external systems (e.g., REST API calls)

As depicted in Listing 3.6, body is accessed and returned as a String, customerId is likely a String, and enrichedData is a Map.

```
def body = message.getBody(String)
def customerId = message.getHeaders().get("CustomerID")
def enrichedData = [id: customerId, payload: body]
```

**Listing 3.6**  Groovy Scripting: Cloud Integration Message Content Extraction

### 3.2.3   Operators and Expressions

Groovy offers a rich set of operators that make scripting expressive and concise. In many cases, these operators are *syntactic sugar* over Java method calls, but Groovy extends and enhances them with additional capabilities that are particularly useful in integration scenarios—such as null handling, conditional assignment, and safe property access. This section introduces the most commonly used operators in Groovy, along with examples that are relevant to Cloud Integration.

### Assignment and Arithmetic Operators

Groovy supports the standard assignment and arithmetic operators found in Java. Arithmetic operations work on numbers and also on strings (concatenation), lists (combining), and even ranges, as shown in Table 3.3 and Listing 3.7.

```
def total = 4 + 3          // 7
def message = "Hello, " + "world!"  // "Hello, world!"
def range = 1..3 + 4..5    // [1, 2, 3, 4, 5]
```

**Listing 3.7**  Groovy Scripting: Arithmetic and Assignment

| Operators | Descriptions | Examples | Results |
|-----------|--------------|----------|---------|
| = | Assignment | x = 5 | Assigns 5 to x |
| + | Addition/concatenation | a + b | Adds or concatenates |
| - | Subtraction | a - b | Subtracts b from a |
| * | Multiplication | a * b | Multiplies a and b |
| / | Division | a / b | Divides a by b |
| % | Modulo | a % b | Remainder of a ÷ b |

**Table 3.3**  Groovy Scripting: Arithmetic and Assignment Operators

## Comparison Operators

Groovy allows comparison between values with intuitive operators, much like Java does. However, unlike Java, Groovy's == operator performs *value comparison* (via .equals()), not reference comparison, as seen in Listing 3.8.

```
def a = "123"
def b = "123"
assert a == b      // true
assert !a.is(b)    // false (different references)
```

**Listing 3.8**  Groovy Scripting: Operators and Comparison

To overcome the issue of not being able to perform reference checks, Groovy introduced the is operation (.is()) for true reference comparison. The most common comparison operators can be found in Table 3.4.

| Operators | Descriptions | Examples | Results |
|-----------|--------------|----------|---------|
| == | Equality | a == b | True if value is equal |
| != | Inequality | a != b | True if value isn't equal |
| <, >, <=, >= | Relational comparisons | a > b | Compares values |
| .is(…) | Identity | a.is(b) | Compares reference instead of value |

**Table 3.4**  Groovy Scripting: Operator Overview

## Logical Operators

Logical operations in Groovy follow standard Java semantics. As shown in Listing 3.9, these logical operators are particularly useful in Groovy conditions and filters:

```
if (customerId && amount > 1000) {
    // high-value transaction
}
```

**Listing 3.9**  Groovy Scripting: Use of Logical Operators in Control Structures

They're listed in Table 3.5.

| Operators | Descriptions | Examples | Results |
|-----------|--------------|----------|---------|
| && | Logical AND | a && b | True if both are true |
| \|\| | Logical OR | a \|\| a | True if either one is true |
| ! | Logical NOT | !a | Negates a Boolean value |

**Table 3.5**  Groovy Scripting: Logical Operators

### Safe Navigation and Null Handling

One of Groovy's standout features is its support for *null-safe operations*, which help avoid `NullPointerException`—which is particularly useful in message transformations where optional fields are common. Let's look at the following three key operators in this space:

- **Safe navigation operator**
  The safe navigation operator (`?.`) prevents dereferencing a method or property call on an undefined object, which would end up as runtime error and instead returns with `null` again. As shown in Listing 3.10, accessing `city` will safely return `null` if customer or `address` isn't defined, instead of throwing an exception.

  ```
  def name = customer?.address?.city
  ```

  **Listing 3.10**  Groovy Scripting: Safe Navigation Operator

- **Ternary operator**
  Groovy supports a full ternary conditional expression (`_ ? _ : _`) as an inline alternative to using a complete conditional structure block. Listing 3.11 demonstrates the use of conditionally assigning a flagged value based on an input condition. This enhances code readability and reduces cluttered and bloated control structures for a simple condition-based assignment.

  ```
  def status = (amount > 1000) ? "VIP" : "Standard"
  ```

  **Listing 3.11**  Groovy Scripting: Ternary Operator for Shortened Conditional Assignment

■ **Elvis operator**
The Elvis operator (?:) provides a fallback value when the left-hand side is null or appears false (with an empty string, a zero, etc.). This shortens the ternary operator even further by making it need only two inputs instead of three. Worth mentioning is the automatically returned value of the first input, in case it results in a true condition. This eliminates the possibility of alternating the first input. It would be possible by using the ternary operator, but in the case of Listing 3.12, it's unnecessary.

```
def result = input ?: "default"
// same, but shortened as
def sameResult = input ? input : "default"
```

**Listing 3.12** Groovy Scripting: Comparison of Elvis and Ternary Operators

**Other Useful Operators**

Let's quickly look at a few other useful operators:

■ **Spread operator (*.)**
You can use this to apply a method/property to all items in a collection. This is equivalent to using a closure in combination with the .each() loop of said collection.

■ **As operator** (x as String)
You can use this for type casting or converting values. Compared to Java, using the as operator doesn't only type-cast the value into the given type but can also parse its value when converting a textual number into an actual integer, for example.

■ **In operator** ("value" in list)
This is a shortened and more readable version of Java's .contains() function. It will test for membership in a given collection or range

In real-world Cloud Integration scenarios, operators are used to do the following:

■ Check whether a message header exists and assign default values.

■ Transform or enrich payloads.

■ Control the flow within Groovy scripts.

The script in Listing 3.13 uses the Elvis operator for a default value of the priority, type coercion with as, and logical operators to evaluate business logic.

```
def priority = message.getHeaders().get("Priority") ?: "Normal"
def amount = message.getProperties().get("Amount") as Double

if (priority == "High" && amount > 1000) {
    messageLog.setStringProperty("Flagged", "true")
}
```

**Listing 3.13** Groovy Scripting: SAP-Based Message Access

### 3.2.4  Control Flow Statements

*Control flow statements* allow Groovy scripts to make decisions, execute logic conditionally, and repeat actions. In the context of Cloud Integration, control flow is commonly used to inspect message content, apply conditional transformations, handle validations, and control processing steps based on runtime metadata.

Groovy inherits most of its control flow constructs from Java but enhances them with more concise syntax and flexibility. This section covers conditional statements (if, switch), loops (for, while, and each), and introduces closures as functional control elements.

#### Conditional Logic with if, else if, and else

The if statement is the most common conditional structure in Groovy. Its syntax is intuitive and similar to Java but doesn't require parentheses around the condition (though they're allowed).

Conditions can include comparison operators and logical operators, as mentioned in Section 3.2.3. As shown in Listing 3.14, in Groovy's truthiness model, non-zero numbers, nonempty strings, and non-null values all evaluate as true in conditions.

```
def amount = 750

if (amount > 1000) {
    status = "High"
} else if (amount > 500) {
    status = "Medium"
} else {
    status = "Low"
}

def name = "John"
if (name) {
    // Executes because name is not null or empty
}
```

**Listing 3.14** Groovy Scripting: Conditional Control Statements

#### The switch Statement

Groovy's switch statement is more powerful than Java's. In addition to classic constant cases, it supports the following:

- Type checking
- Pattern matching
- Collection membership
- Closures as conditions

As shown in Listing 3.15, the switch statement's functionality makes it particularly useful when evaluating multiple types of conditions in a single block (e.g., string values, numeric ranges, custom logic).

```
def value = 42

switch (value) {
    case 0:
        result = "Zero"
        break
    case 1..10:
        result = "Low"
        break
    case Integer:
        result = "Number"
        break
    case { it % 2 == 0 }:
        result = "Even number"
        break
    default:
        result = "Other"
}
```

**Listing 3.15** Groovy Scripting: Conditional switch Statement

### Looping Constructs

Groovy supports traditional loop structures (for and while), which also support control-altering keywords. The following keywords are used in the same way as they're used in Java:

- break
  This exits the current loop or switch block.
- continue
  This skips to the next iteration of a loop.
- return
  This returns a value from a method or script.

Compared to Java, Groovy does support looping over map entries natively, without the need to call .entries() that exists in Java. This can be seen in the second example of Listing 3.16. Since Groovy is oriented in a more functional coding style, the still supported do-while loop is very uncommon.

```
def names = ["Anna", "Ben", "Cara"]
for (name in names) {
    println name
}
```

```
def userMap = [id: 101, name: "Tom"]
for (entry in userMap) {
    println "${entry.key} = ${entry.value}"
}

def i = 0
while (i < 3) {
    println i
    i++
}
```

**Listing 3.16** Groovy Scripting: Loops Using for and while

### 3.2.5   Closures

In addition to traditional looping control statements, Groovy adapts the concept of Java's lambda expressions, which are called closures. A *closure* is an anonymous block of executable code that can be assigned to a variable, passed as a parameter, or executed inline. Closures are one of Groovy's most powerful features, and you'll frequently use them in collection iteration, filtering, transformation, and aggregation—especially when working with XML or JSON payloads in Cloud Integration.

You define a closure by using curly braces { }, and you can declare optional parameters before the -> symbol. If you don't define parameters, Groovy will use an implicit nullable it variable.

Closures are *first-class objects*, meaning they can be assigned to variables, stored in data structures, or passed to methods. This makes them highly flexible for declarative-style programming. As seen in Listing 3.17, you can call the assigned variables as functions when containing a closure.

```
def greet = { name -> println "Hello, $name!" }
greet("SAP Developer")  // Output: Hello, SAP Developer!

def printer = { println "Called!" }
printer()

def doubler = { it * 2 }
println doubler(4)  // Output: 8
```

**Listing 3.17** Groovy Scripting: Simple Closure Definition as Inline Functions

### 3.2.6   Working with Strings

In Cloud Integration, string manipulation is one of the most common tasks you'll perform in Groovy scripts. Whether you're parsing message content, building payloads, formatting values, or working with message headers, you'll need to have a clear understanding of how Groovy handles strings. Groovy enhances Java's native String class with additional operators and methods to offer powerful and concise ways to manipulate textual data.

Groovy supports the following multiple styles for declaring strings, depending on whether you need interpolation, multiline support, or plain text:

- **Single quoted** (`'text'`)
  This is immutable vanilla Java String.

- **Double quoted** (`"text"`)
  This is interpolatable Groovy String (GString).

- **Triple occurrence quoted** (`'''text'''` or `"""text"""`)
  This is multiline enhanced String or GString (depending on the quotes).

Double-quoted strings in Groovy are GString objects, which allow *interpolation*—the embedding of variables and expressions directly into the string using the $ syntax—while expressions need additional curly braces (`${}`). Interpolation allows you to inject values or expressions into a string, and Groovy evaluates them at runtime. Be aware that GStrings are not strictly equal to plain String objects. This rarely causes issues in Cloud Integration, but it's worth noting when you're interacting with external libraries that require strict Java types. To accommodate plain String objects when you need them, you can convert a GString into a String by explicitly calling `.toString()`.

```groovy
def plain = 'Hello World'  // plain String

def name = "Anna"
def message = "Hello, $name"

def price = 10
def tax = 0.19
def total = "Total: €${price * (1 + tax)}"

def xml = '''<note>
  <to>User</to>
  <message>Hello!</message>
</note>'''

def plainAgain = "$name".toString()
```

**Listing 3.18** Groovy Scripting: Defining Text with Strings and GStrings

Groovy provides an extensive set of utility methods that were inherited from Java and are extended by Groovy's `StringGroovyMethods`. Compared to Java, Groovy extends the String operators (see Table 3.6) by two more features. This enhances the concatenation (a + b) by stripping off text at the end ("file.txt" - ".txt" = "file") as well as repetition ("*" * 5 = "*****").

| Methods | Descriptions | Examples |
|---|---|---|
| toUpperCase() | It converts to uppercase. | "abc".toUpperCase() → "ABC" |
| toLowerCase() | It converts to lowercase. | "ABC".toLowerCase() → "abc" |
| trim() | It removes surrounding whitespace. | " text ".trim() → "text" |
| replace() | It replaces part of a string. | "2025".replace("5", "4") → "2024" |
| substring() | It extracts part of a string. | "abcdef".substring(2, 5) → "cde" |
| split() | It splits a string with a delimiter. | "A,B,C".split(",") → ["A", "B", "C"] |
| contains() | It checks whether a string contains a value. | "Hello".contains("lo") → true |
| startsWith() | It checks a prefix. | "SAP".startsWith("S") → true |

**Table 3.6** Groovy Scripting: Common String Methods

### 3.2.7   Pattern Matching and Regular Expressions

Groovy integrates regular expressions (regexes) natively into the language to make pattern matching far more expressive and concise than in Java. This is particularly valuable in Cloud Integration, where you'll frequently need to use string validation, field extraction, and conditional logic based on message formats.

To define a regular expression in Groovy, you use the regex literal syntax: ~/pattern/. This creates a java.util.regex.Pattern object at runtime, as shown in Listing 3.19.

```
def pattern = ~/^\d{4}-\d{2}-\d{2}$/  // Matches a date like 2025-07-13
```

**Listing 3.19** Groovy Scripting: Regex Pattern Object Definition

Groovy provides two special operators for regex matching:

- ==~ (the exact match operator)
  This operator tests whether the entire string matches the pattern, which is useful for validating formats (e.g., checking whether a message ID or timestamp is valid).
- =~ (the partial match operator, which creates a matcher)
  This operator returns a matcher object that you can use to inspect or extract parts of

a string that contain a match. The result of =~ is an instance of `java.util.regex.Matcher`. You can then interact with this matcher in multiple ways:

– **Boolean check**
  The matcher object resolves into a Boolean value that represents at least one found match.

– **Matched groups**
  Each found element mentioned in the regular expression is included in the matcher, and you can be accessed it via an indexed lookup by using square brackets.

– **Iterable**
  The matcher is both iterable and indexable, and you can looped over it and call it with `.each` and a closure.

Each matched pattern occurrence contains its capturing groups as child elements. This is shown in Listing 3.20.

```
def text = "User: John, ID: 456"
def matcher = text =~ /ID:\s*(\d+)/
if (matcher) {
    println matcher[0][1]  // Output: 456
}
```

**Listing 3.20** Groovy Scripting: Regex Pattern and Group Access

While Groovy offers many convenient string methods like `.contains()` and `.startsWith()`, regexes provide much more powerful ways to perform:

- Pattern validation
- Dynamic replacements
- Complex extraction logic (with groups)
- Multipattern scanning

Groovy's integration with regexes via =~, ==~, and ~/.../ syntax provides a concise and powerful toolset for pattern matching and text analysis. The =~ operator enables rich interaction with match groups and supports iteration and extraction, while the ==~ operator is ideal for validation use cases.

In Cloud Integration, these capabilities are indispensable when you're processing unstructured payloads, validating external IDs or formats, or extracting specific data elements from mixed content. When you use it carefully, Groovy's regex support greatly simplifies logic that would otherwise require verbose procedural code.

### 3.2.8   Collections and Iteration

*Collections* are a core part of Groovy's data model. They allow developers to work with groups of values—such as lists of items, maps of key-value pairs, and ranges of numbers or characters—in a concise, expressive, and functional style. In Cloud Integration, you'll frequently use collections when processing structured data like line items in XML, parsing JSON arrays, and building grouped values dynamically.

Groovy simplifies working with collections through enhanced syntax and a rich set of methods that support filtering, transformation, aggregation, and iteration. In this section, we'll investigate types of collections and their commonly used versions in Groovy (and Java as well). Furthermore, we'll tackle interacting and working with collections.

**Common Collection Types in Groovy**

Groovy primarily works with three types of collections:

- **Lists**
  These are simple, ordered collections backed by Java's `java.util.List`, with the possibility of mixing datatypes arbitrarily.

- **Maps**
  These are unordered collections of key-value-pairs, and they're often used for structured data or property sets.

- **Ranges**
  These are ordered, sequential collections of numbers or characters.

Whereas lists and maps share the same bracket notation (the use of square brackets), the key difference lies in the colon notation that separates the key and value of a map. Additionally, the range notation—which is indicated by two dots between the starting and ending values—creates a list of numbers or characters that span the boundaries while including both end values. Listing 3.21's first section, which is about lists, shows how multiple datatypes are possible. You'll typically access the elements of given collections by using square brackets and including the *name* of the element, which in the case of a list is the index. When you're dealing with a map lookup, Groovy also supports the dot notation.

```
// List
def list = [10, 20, 30, 'string element', true]
list[0]  // Accesses 10
list << 40  // Adds 40 to the list

// Map
def map = [id: 15, name: 'SAP Specialist']
map['name']  // SAP Specialist
map.id  // 15 (dot notation is supported)
```

```
// Range
def nums = 1..5       // [1, 2, 3, 4, 5]
def chars = 'a'..'e' // ['a', 'b', 'c', 'd', 'e']
```

**Listing 3.21** Groovy Scripting: Collection Types

### Interacting with Collections

Groovy provides an elegant and highly expressive model for iterating over collections. Unlike traditional Java, which often relies on verbose for loops, Groovy embraces a more functional style that allows developers to operate on lists, maps, and ranges by using closures and concise method calls. This approach makes code more readable, reduces boilerplate, and fits naturally with transformation tasks in Cloud Integration.

Whether your goal is to read every element, transform the structure, filter out unwanted data, or aggregate values, Groovy offers a range of methods tailored to your use case. These iteration mechanisms are particularly powerful when applied to JSON arrays, XML node lists, and dynamically constructed maps during message processing.

At its core, collection iteration in Groovy is based on the idea that each element in a collection can be passed into a closure (an anonymous code block) where custom logic is applied.

Some of the most common collection methods include the following:

- **each and** `eachWithIndex`
  Both of these methods iterate over every element, and the `eachWithIndex` variant also contains the index as second parameter.

- `collect`
  You'll use this method as a transformation function like `map` in the Java context when you're working with Java Streams.

- **find and** `findAll`
  When searching through lists, you'll often need to exclude certain elements or find a specific one. Whereas the `findAll` method works like a filter and returns a list of conditionally matching elements, the `find` version only looks for the first occurrence and returns a single element (or `null` if it finds nothing).

- **Any and every**
  In certain cases, collections might be searched only to resolve into a logical condition. For this purpose, the `any` and `every` methods provide a lookup feature to indicate whether at least one (`any`) or all (`every`) elements of the collection match the provided function.

- `inject`
  This method serves as a collector—like Java's `reduce` version (not to be confused with Groovy's `collect` method)—to join all elements into a single value. You need to call `inject` with an initial value and a closure to combine the elements together.

- **join**
  In contrast to the `inject` variant, `join` will collect all items into a single string that separates each value with the provided string from the method call.

While all methods provide very useful functionality, the real power comes when you're chaining the methods together in a functional programming style. With this style, you can filter for certain items, adjust them accordingly, and reduce them into a combined value.

Groovy collections are mutable by default, so you can add, remove, or update elements easily. Unlike in plain Java, you insert elements at the end via the << operator, instead of using a method. On the other hand, removing items by hand in Groovy works somewhat similarly to how it works in Java: you use the `.remove(idx)` method and provide the element's index (instead of the element itself, as in Java). You can update the value at any given position (both index and key based) in the same way as you assign a value to a normal variable, by using =.

In Listing 3.22, we define an incoming flow of item prices. These would typically come from a message payload, but for simplicity, we omit that. We then check whether the list contains higher-priced values that need to be normalized to only 30% of the initial value. While you can skip the conditional section and only keep the first `mainList` access, it can be more performant to check for potential items first. If you don't need to normalize any items need, you can skip the entire pipeline and you don't need to work through an empty list.

After collecting the inbound list from Listing 3.22 into a resulting collection—which is needed, due to two possible insertion points—we reduce all items into a single sum of values.

```
def mainList = [1, 5, 17, 4, 12, 44, 9, 3, 10]
def highPriceLimit = 10
def normalized = []

if (mainList.any(it >= highPriceLimit) {
  // find all exceeding elements and normalize them by a certain factor
  mainList.findAll { it >= highPriceLimit }.collect { it * 0.3 }.each { normalized << it }
}

mainList.findAll { it < highPriceLimit }.each { normalized << it }
def total = normalized.inject(0) { accumulated, item -> accumulated + item }
```

**Listing 3.22** Groovy Scripting: Collection Methods with Method Chaining

Groovy collections simplify working with grouped data, which is essential in message-based systems like Cloud Integration. With its support for lists, maps, and ranges—

combined with expressive methods like each, collect, and inject—Groovy makes it easy to traverse, filter, and transform data structures in a concise and readable way.

These collection capabilities are foundational when you're processing JSON arrays, looping over XML nodes, or building outbound payloads programmatically. Once you've mastered these patterns, you as a developer will be able to handle even the most complex transformation and enrichment scenarios with confidence.

### 3.2.9 XML Handling

Groovy scripting in SAP BTP offers powerful capabilities for processing XML data efficiently within iFlows. XML remains a dominant data format in enterprise integration, and Groovy's built-in XML libraries provide flexible APIs for parsing, manipulating, and serializing XML documents with minimal code and excellent readability.

On SAP BTP and Cloud Integration, the most common classes to parse XML are groovy.util.XmlSlurper and groovy.util.XmlParser:

- XmlSlurper
  This provides a lazy, efficient approach in which the XML is parsed into a tree of lightweight objects. It's best suited for reading and navigating XML, where you don't need to modify it extensively.

- XmlParser
  This parses the entire XML into a mutable DOM-like structure, and it's ideal when you need to add, update, or remove nodes.

An example of using XmlSlurper to parse an input XML string from the message body is as follows:

```
import groovy.util.XmlSlurper
def inputXml = new XmlSlurper().parseText(message.getBody(String))
```

This allows you to navigate easily by using Groovy's GPath expressions (e.g., inputXml.Order.Item.each { item -> ... }).

Once parsing is done, you can access elements simply by their names, as follows:

```
def customerName = inputXml.Order.customer.name.text()
def items = inputXml.Order.Item
```

You can also loop over repeating elements, as shown in Listing 3.23.

```
items.each { item ->
    def product = item.product.text()
    def price = item.price.text()
    // Process each item as needed
}
```

Listing 3.23 Groovy Scripting: Simple Closure Definition Working Like for-each

While XmlSlurper is read-only, XmlParser lets you modify the XML DOM, as shown in Listing 3.24.

```
import groovy.util.XmlParser
def xmlParser = new XmlParser()
def xmlDoc = xmlParser.parseText(message.getBody(String))

// Add a new element
def newElem = new Node(xmlDoc.Order[0], 'discount', '10%')

// Update an element value
xmlDoc.Order.Item[0].price[0].value = '29.99'

// Remove an element
xmlDoc.Order.remove(xmlDoc.Order.DebugInfo[0])
```

**Listing 3.24** Groovy Scripting: XML Traversal and Content Modification

After making any modifications, you can serialize back to String by using XmlUtil. serialize, as follows:

```
import groovy.xml.XmlUtil
String resultXml = XmlUtil.serialize(xmlDoc)
message.setBody(resultXml)
```

Now, let's move on to building new XML structures. Groovy's StreamingMarkupBuilder is ideal for constructing new XML documents dynamically, as shown in Listing 3.25.

```
import groovy.xml.StreamingMarkupBuilder
def builder = new StreamingMarkupBuilder()
def outputXml = builder.bind {
    root {
        Order {
            customer {
                name(inputXml.Order.customer.name.text())
            }
            items {
                inputXml.Order.Item.each { item ->
                    itemNode {
                        product(item.product.text())
                        price(item.price.text())
                    }
                }
            }
        }
    }
}
message.setBody(outputXml.toString())
```

**Listing 3.25** Groovy Scription: Creating Payload with Markup Builder

This approach cleanly separates reading from writing and is well-suited for transform-ing one XML structure into another.

Finally, let's look at the following practical patterns you can use in Cloud Integration:

- `XmlSlurper`
  Use this for lightweight parsing and data extraction when source XML doesn't need modifications.

- `XmlParser`
  Use this when you must alter the XML tree or add or remove nodes.

- `XmlUtil.serialize`
  Use this to serialize modified XML before setting it back to the message.

- `StreamingMarkupBuilder`
  Use this to create completely new XML outputs from scratch or combine values cre-atively.

You can also log intermediate XML states during scripting to help with troubleshooting large or complicated XML transformations.

### 3.2.10   JSON Handling

JSON has become one of the most ubiquitous data formats in modern integration sce-narios, APIs, and web services, including those on SAP BTP. With Groovy scripting inside Cloud Integration, you need to work efficiently with JSON payloads to dynami-cally transform, validate, and enrich message data.

Groovy offers powerful and straightforward APIs to parse, manipulate, and generate JSON, so it's a natural choice for JSON processing within Cloud Integration Groovy scripts. We describe these provided techniques for working with JSON in the following sections. We start with getting JSON data into Groovy object structures, and then we cover modi-fying the entities. After that, we deal with serializing structured data back into JSON string representations, and we conclude with some use cases and good practices.

**Parsing JSON Strings into Groovy Objects**

Groovy's `JsonSlurper` is the primary class you'll use to parse JSON text. It reads a JSON string and converts it into nested Groovy data structures, which are typically maps, lists, and primitives—all of which are accessible with intuitive Groovy syntax. Listing 3.26 shows an example of how to parse the message body into a JSON object. The `json-Object` becomes a dynamic Groovy map-like representation of your JSON content that you can use to access fields directly by property names.

```
import groovy.json.JsonSlurper

def jsonInput = message.getBody(String)
```

```
def jsonParser = new JsonSlurper()
def jsonObject = jsonParser.parseText(jsonInput)
```

**Listing 3.26** Groovy Scripting: JSON Parsing

For instance, take the JSON input shown in Listing 3.27.

```
{
  "orderId": "12345",
  "customer": {
    "name": "Jane Doe",
    "email": "jane.doe@example.com"
  },
  "items": [
    {"product": "Widget A", "quantity": 5},
    {"product": "Widget B", "quantity": 2}
  ]
}
```

**Listing 3.27** Resulting JSON Payload Structure

Using this, you can read values like the following:

```
def orderId = jsonObject.orderId
def customerName = jsonObject.customer.name
def firstItemProduct = jsonObject.items[0].product
```

This makes navigating even complex, nested JSON straightforward.

### Modifying JSON Data Structures

Since `JsonSlurper` returns standard Groovy collections, you can modify those data structures with Groovy's rich collection API, as shown in Listing 3.28.

```
// Increase quantity of first item
jsonObject.items[0].quantity += 1

// Add a new field
jsonObject.customer.vip = true

// Remove a field
jsonObject.remove('obsoleteField')
```

**Listing 3.28** Groovy Scripting: Manipulating JSON Objects

You can manipulate nested lists, add or remove elements, and perform any data transformation you need.

### Creating and Serializing JSON

To let you convert Groovy objects back into JSON strings, Groovy provides `JsonOutput`, as shown in Listing 3.29.

```
import groovy.json.JsonOutput

def outputJson = JsonOutput.toJson(jsonObject)
message.setBody(outputJson)
```

**Listing 3.29**  Groovy Scripting: JSON Serialization

This serializes the Groovy map/list structure back into a JSON-formatted string that's suitable for passing downstream in the Cloud Integraiton flow.

For more human-readable JSON (in pretty print), use the following:

```
String prettyJson = JsonOutput.prettyPrint(outputJson)
message.setBody(prettyJson)
```

### Building JSON From Scratch

You can build JSON documents programmatically with Groovy maps and lists, which convert seamlessly into JSON, as shown in Listing 3.30.

```
def newOrder = [
    orderId: "98765",
    customer: [
        name: "John Smith",
        email: "john.smith@example.com"
    ],
    items: [
        [product: "Gadget", quantity: 3],
        [product: "Thingy", quantity: 1]
    ]
]

def jsonString = JsonOutput.toJson(newOrder)
message.setBody(jsonString)
```

**Listing 3.30**  Groovy Scripting: JSON Payload Creation

This is especially useful when you're transforming XML or other formats into custom JSON structures.

### Handling Nested JSON and Arrays

Groovy makes it easy to navigate and manipulate arrays (lists) within JSON.

To iterate over items, use the following code:

```
jsonObject.items.each { item ->
  println "Product: ${item.product}, Quantity: ${item.quantity}"
}
```

You can filter or transform lists by using Groovy's collection methods, as follows:

```
def filteredItems = jsonObject.items.findAll { it.quantity > 2 }
```

You can also transform items, as follows:

```
def productNames = jsonObject.items.collect { it.product }
```

### Use Cases and Best Practices

Common use cases in Cloud Integration are as follows:

- **Validation**
  Checking mandatory fields, value ranges, and data formats
- **Enrichment**
  Adding fields based on logic or lookup data
- **Conversion**
  Transforming the JSON structure from source schema to target schema
- **Splitting**
  Breaking large JSON arrays into smaller chunks for downstream processing
- **Merging**
  Combining multiple JSON message sources into one

Beyond simple parsing and manipulation, you can integrate JSON schema validation in Groovy via third-party libraries (though this requires bundling dependencies) to ensure that the incoming JSON is well-formed and matches the expected structure.

You can also implement conversion between XML and JSON by using Groovy's combined XML and JSON APIs, which enable seamless format bridging within the same script.

Some best practices are as follows:

- Always parse JSON input inside try-catch blocks to handle malformed JSON gracefully.
- Use `JsonSlurper`'s `parseText` for strings; it also supports other input streams if needed.
- Manipulate data by using native Groovy collections for efficiency.
- When serializing, prefer `JsonOutput.prettyPrint` during development; switch to compact form for production to reduce message size.
- Use logging to output JSON snippets during processing for easier troubleshooting.
- Avoid deeply nested or overly large JSON outputs, due to possible performance bottlenecks.

### 3.2.11   Logging Features

`Logging` is a critical capability in SAP BTP integrations, especially when you're using Groovy scripting in Cloud Integration. Effective logging facilitates troubleshooting, auditing, and understanding the runtime behavior of complex message flows. Groovy scripts that run inside Cloud Integration have direct access to Cloud Integration's logging framework, and that enables development teams to produce meaningful runtime insights without disrupting performance or cluttering logs unnecessarily.

This section outlines how Groovy integrates with Cloud Integration's logging architecture, demonstrates practical logging techniques, and offers best practices to help you maintain traceable, maintainable, and performant iFlows.

#### Accessing Cloud Integration Logging from Groovy Scripts

In Cloud Integration, each message that's processed through an iFlow is represented by a `Message` object that's accessible in Groovy scripts. This object exposes methods for interacting with the message payload, headers, properties, and importantly, the message log.

To work with logging in Groovy, you obtain a message log object from the message context via the `messageLogFactory`, as follows:

```
def messageLog = messageLogFactory.getMessageLog(message)
```

If the message log isn't available (which is common in production environments when logging is disabled), `messageLog` will be null. Therefore, always check for null before adding logs to avoid runtime errors. You use the message log API primarily to add attachments (payload snapshots) or string properties (key-value pairs) that represent logs.

#### Logging Payloads as Attachments

You can log the current message body (payload) as an attachment for post-execution review, as shown in Listing 3.31.

```
if (messageLog != null) {
    def body = message.getBody(String)
    messageLog.addAttachmentAsString("Payload Snapshot", body, "text/plain")
}
```

**Listing 3.31** Groovy Scripting: Conditionally Adding Message Content to Logs

You can add multiple attachments to each message, and Cloud Integration will display them in the monitoring UI under the **Attachments** tab.

#### Logging Headers and Properties

You can record all headers and properties to the message log for visibility, as shown in Listing 3.32.

```
if (messageLog != null) {
    def headers = message.getHeaders()
    headers.each { key, value ->
        messageLog.setStringProperty("Header:" + key, value.toString())
    }
    def properties = message.getProperties()
    properties.each { key, value ->
        messageLog.setStringProperty("Property:" + key, value.toString())
    }
}
```

**Listing 3.32** Groovy Scripting: Adding Log Properties

This helps surface metadata that influences processing decisions.

### Custom Debug and Info Messages

While Groovy's `println` statements print to the tenant logs (which are less visible and less organized), logging that you perform via Cloud Integration's message log attaches contextually to the message and is easier for operators to trace.

In the absence of a direct logging API for arbitrary text messages, developers commonly add custom string properties or attachments that contain debug info. For example, to log a custom note, you can use the code in Listing 3.33.

```
if (messageLog != null) {
    messageLog.setStringProperty("Debug", "Order processing started for " +
orderId)
}
```

**Listing 3.33** Groovy Scripting: Adding Debug Logs

### Controlling Log Levels

Cloud Integration controls logging verbosity through the following log levels: TRACE, DEBUG, INFO, WARN, and ERROR. Your integrations will benefit from logging only the necessary amount of information in each environment, so you should do the following:

- Use TRACE or DEBUG levels for detailed insight during development or support troubleshooting.
- Limit logging in production environments to WARN or ERROR levels to avoid performance impact or log bloat.

While Groovy doesn't directly expose the Cloud Integration log level, you can retrieve it from message properties. Unlike other logging facilities, Groovy doesn't support supplying the logging methods with the according log level. Instead, it does it the other way around by making you retrieve the current log level and wrap your logging statements in corresponding conditional blocks. You can do this as shown in Listing 3.34.

```
def logConfig = message.getProperty("SAP_MessageProcessingLogConfiguration")
def logLevel = logConfig?.logLevel as String

if (logLevel == "DEBUG" || logLevel == "TRACE") {
    // Detailed logging logic here
}
```

**Listing 3.34** Groovy Scripting: Executing Trace Logs on Property Log Level

This conditional logging strategy improves efficiency.

### Best Practices and Advanced Techniques

Let's look at some best practices and advanced techniques for logging:

- Log only what's necessary. Avoid logging full payloads or large binary attachments in production, unless it's required for audits or troubleshooting.
- Use descriptive names. Name attachments and properties clearly to indicate their purpose and stage (e.g., "Before Mapping Payload," "After Enrichment Headers").
- Modularize logging logic. Encapsulate common logging routines in reusable Groovy classes or functions to simplify maintenance.
- Externalize control flags. Use iFlow/global parameters or tenant-level properties to enable or disable logging dynamically, without code changes.
- Handle null message log gracefully. Always check for message log availability to avoid errors in different deployment stages.
- Avoid sensitive data leaks. Be cautious about logging personally identifiable or confidential information and apply masking or omission as you need to.
- Use timestamps and correlation IDs. Include contextual data to help correlate logs across distributed systems.
- Leverage attachments for complex data. When logging structured or large data fragments, use attachments rather than string properties.
- Test logging output regularly. Before deploying to production, verify log visibility and volume in development or test tenants.
- Reusable logger classes. Implement POJO-like Groovy classes with standard logging methods (info, warn, error, and debug) that internally call Cloud Integration message log APIs with proper level checks.
- Log as attachments with metadata. Log intermediate serialized XML or JSON payload fragments as attachments with specific MIME types (e.g., application/xml).
- Conditional and contextual logging. Use script logic to selectively log based on message content, (e.g., only log error payloads or large transactions).
- Correlation with external systems. Include IDs or tokens in logs to facilitate end-to-end traceability across integrated applications.

## 3.3   Message Mapping

*Message mapping* is one of the core transformation capabilities that are available in Cloud Integration. It allows integration designers to visually transform data from one XML structure to another by using a graphical mapping editor. Message mapping lets designers avoid having to do manual coding, and it still supports powerful data manipulation features. This section introduces the concept, purpose, and working principles of message mapping to set the stage for later deep-dives into advanced functions and best practices.

In integration scenarios, source and target systems often represent business data differently—one system might send an order in a simple XML schema, while another might expect a nested, enriched version. Message mapping bridges this gap by providing a visual tool you can use to define transformation rules between the source message structure and the target message structure. Instead of writing transformation code (as in XSLT or Groovy), you work within a mapping canvas:

- The left pane shows the source structure.
- The right pane shows the target structure.
- You draw lines (mapping connections) between fields that correspond, or you connect them through transformation functions from the function palette.

Message mapping in Cloud Integration supports hierarchical XML formats and handles everything from simple one-to-one mappings to highly complex multistep transformations.

There are several compelling reasons to use graphical mapping:

- **Ease of use**
  Drag-and-drop connections make it accessible, even to those without deep coding skills.

- **Clarity**
  The visual representation makes relationships between fields explicit and easy to explain to business stakeholders.

- **Built-in functions**
  Many standard transformations (string manipulation, math, date formatting, and conditional logic) are available without custom scripting.

- **Reusability**
  Once you've built mappings, you can reuse them in other iFlows with the same structures.

- **Testing support**
  You can test mappings within the Cloud Integration tooling by using sample XML payloads before deployment.

A typical message mapping process follows these steps:

1. Define source and target structures: upload or reference schemas (XML Schema Definitions [XSDs]), Web Services Description Languages (WSDLs), or message types. These form the basis for the mapping canvas.

2. Map fields.

3. Perform direct mapping: directly link fields with identical meanings.

4. Perform transformative mapping: apply functions between fields to adjust content.

5. Use built-in functions: chain functions for conversions, aggregations, or conditional value setting.

6. Handle repeating elements: map repeating structures to target lists to controlling occurrence and order.

7. Perform test mapping: run the mapping in the Cloud Integration editor with sample input and inspect the generated output.

Cloud Integration's graphical mapping tool supports the following types of mapping:

- **One-to-one field mapping**
  This is the simplest form.

- **One-to-many/many-to-one mapping**
  This involves using functions to combine or split field values.

- **Value mapping**
  This involves replacing source values with corresponding target values via lookup tables.

- **Conditional mapping**
  This involves using IfWithoutElse, IfThenElse, or Boolean functions to select values based on criteria.

- **UDFs**
  This involves extending mappings with custom Groovy/JavaScript code for specialized transformations.

Also, to perform advanced mappings, you must understand Cloud Integration's queue and context concepts (which are important for repeated elements and grouping).

### 3.3.1   Fundamentals of Graphical Mapping

*Graphical mapping* in Cloud Integration is the practical environment in which you visually define how data fields in one XML structure map to one another. While the concept seems straightforward—connecting a source field to its corresponding target field—the underlying framework offers a wide range of tools and concepts that can support everything from a direct data copy to intricate conditional transformations. The graphical mapping editor in Cloud Integration provides a two-pane view, as shown in Figure 3.1. The key components of this view are as follows:

**❶ Schema definition resources**

This is an overview panel for XSD files that are locally and globally available to the mapping editor. You can use these resource files as source and target structures.

**❷ Left pane: source structure**

This represents the incoming XML message structure. It's usually based on an XSD (schema) or WSDL definition that's imported into your package. The tree view allows you to drill down into nested elements and attributes.

**❸ Right pane: target structure**

This shows the required XML structure that the mapping must produce. Like the source, it's drawn from a schema or service definition.

**❹ Middle mapping space: canvas**

This is the interactive area between the panes where you draw connections between source and target fields, and it's also where you insert function nodes from the function palette to transform data before passing it to the target.

**❺ Function palette**

This is typically located above or to the side of the mapping canvas, and it contains predefined functions that are grouped into categories: **String, Boolean, Arithmetic, Node Functions, Date/Time handling, Context Functions**, and more.

**❻ Mapping toolbar**

This provides options to validate mapping, test with sample messages, align lines, and manage connections.

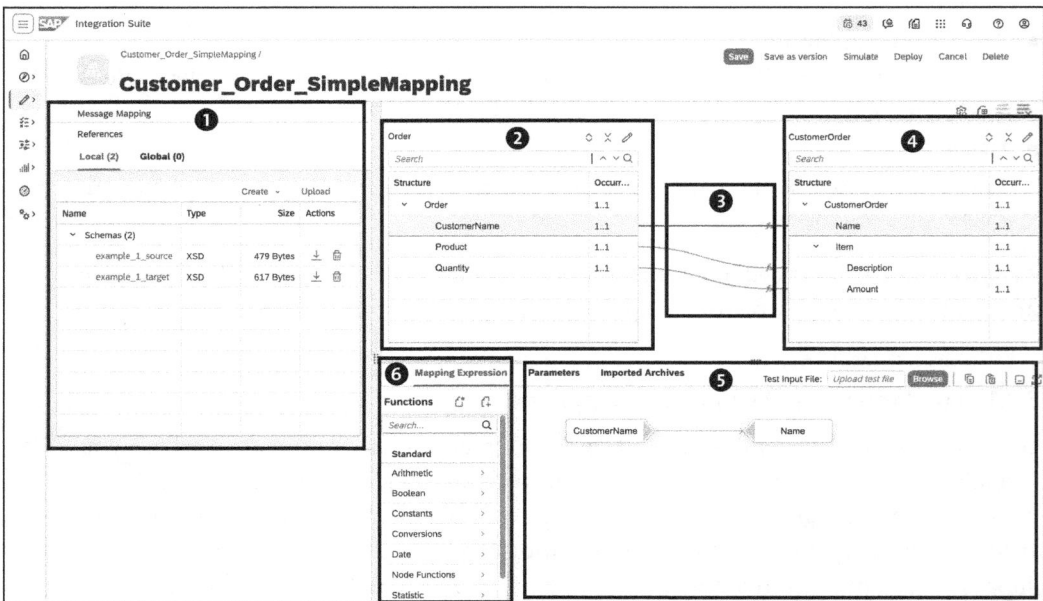

**Figure 3.1** Message Mapping: Overview of Graphical Mapping Editor Layout

Before you can create a mapping, you need to define both source and target message types:

- You can upload XSD files directly. Cloud Integration reads their element definitions to display the hierarchy.

- If your interface is based on a WSDL, Cloud Integration can extract the relevant message part.

- You can also use message types that you define directly in an IFlow's data store or reference types from an integration package.

---

**Tip**

Keep schema definitions as precise and minimal as possible. Overly broad schemas make navigation harder and can impact mapping performance.

---

The simplest mapping is a one-to-one connection. Here's how you create one:

1. Expand source and target trees until the desired elements are visible.
2. Drag from the source element to the target element. This draws a connector line showing the mapping path.

Once connected, the mapping engine will automatically pull the source value into the target field during runtime. Not all target fields have a direct source equivalent. In such cases, you can use constants by following these steps:

1. Drag a **Constant** function from the palette.
2. Double-click it to set the constant value.
3. Connect it to the target element or attribute.

For example, you can do this to set a fixed "USD" currency code in a target's currency attribute.

The palette is at the heart of transformation capabilities, and you can find the following there:

- **String functions**
  These include `concat`, `substring`, `toUpper`, `replace`, and `padString`.
- **Number functions**
  These include `add`, `subtract`, `multiply`, `divide`, and `abs`.
- **Boolean functions**
  These include `ifThenElse`, `and`, `or`, and `equals`.
- **Date/time functions**
  These include `formatDate`, `addDays`, `currentDate`.
- **Node functions**
  These include `count`, `exists`, `removeContexts`.

Functions are represented as nodes. You connect inputs (from source fields or other functions) to the function node and connect its output to the target. For example, you can concatenate a first name and a last name into a single target full name by following these steps:

1. Connect the `FirstName` and `LastName` source fields to a `concat` node.

2. Set the delimiter as a space.

3. Connect the concat's output to the target's `FullName` element.

When you're mapping repeating structures (like multiple `<Item>` elements), Cloud Integration works in terms of context and queue:

- Each repeating node appears in a queue of values.

- Functions respect contexts, which define how values are grouped.

- Correct handling ensures that when you map `Item/Name` and `Item/Price`, the nth name is matched with the nth price.

You rarely need to manipulate contexts for simple one-to-one repeats, but advanced scenarios (like grouping or flattening multiple collections) may require you to use context functions such as these:

- `splitByValue`

- `collapseContexts`

There are also times when you should only populate the target field under certain conditions. The following Boolean or conditional functions help with this:

- `ifThenElse(condition, valueIfTrue, valueIfFalse)`
  This lets you route data accordingly.

- `ifWithoutElse(condition, valueIfTrue)`
  This makes the target field populate only if the condition meets certain criteria.

For example, you can only set `DiscountFlag` to "Y" if `Amount` > 1,000.

Cloud Integration supports value mapping tables to replace source values with designated target values. Follow these steps to implement this:

1. Configure value mappings in the value mapping artifact.

2. In the graphical mapper, use the **Value mapping** function to specify the mapping group and the relevant agency/schema.

This is essential for cross-system code conversion (e.g., country codes, unit of measure).

You don't need to deploy an iFlow every time you wish to test a mapping. Instead, you can do this:

1. Open the mapping artifact.

2. Switch to the **Test** tab.

3. Provide a sample XML payload that matches the source schema.

4. Execute the test to see the transformed output XML instantly.

Testing here is schema driven; if your test XML doesn't match the source schema, errors will occur.

Common issues in mapping include the following:

- **XPath/path mismatches**
  Element may not be found because of schema mismatches or incorrect hierarchies.

- **Context mismatches**
  Outputs may not align correctly with repeating inputs.

- **Data type formatting errors**
  Use explicit conversion functions if you need to.

Validate your mappings regularly during development to avoid runtime surprises.

Best practices for foundational mappings are as follows:

- Start simple. Build direct mappings before adding transformations.

- Use comments. Document the intent of nonobvious mappings by using the **Documentation** field in functions or mapping connections.

- Name function nodes. Replace default names with meaningful ones (e.g., "CalcNet-Price").

- Clean connections. Avoid unnecessary crossings; group related functions visually for clarity.

- Test incrementally. After each logical section, run a test to ensure correctness.

### 3.3.2 Core Mapping Operations and Patterns

In graphical message mapping, the *core operations* go beyond simply connecting a source field to a target field. Transformation in Cloud Integration often demands combining, reshaping, filtering, splitting, and recontextualizing source data to deliver the structure and semantics the target system needs.

### Understanding Queues and Contexts

The Cloud Integration mapping engine processes XML data in *queues* (lists of values) in each field.

A *context* defines how values in a queue are grouped or reset as the mapping executes, as follows:

- **One value per queue entry**
  Each occurrence of an element in the source becomes a queue entry.

- **Context change**
  This creates new subgroups inside the queue when a higher-level element changes.

Why does this matter? If two fields come from parallel repeating structures, you must ensure that their queues have matching contexts so that the nth value from one queue aligns with the nth value from the other.

Let's look at an example problem in Listing 3.35.

```
<Lines>
  <Line><Product>P001</Product></Line>
  <Line><Product>P002</Product></Line>
</Lines>
<Prices>
  <Price>100</Price>
  <Price>200</Price>
</Prices>
```

**Listing 3.35** Example XML Payload

If Product and Price structures are separate in the schema, then their contexts may not align automatically and mapping them together without adjustment could mismatch values. The solution is to use context functions like removeContexts or changeContext to align them.

**Field-to-Field Mapping Types**

Even basic mappings fall into patterns:

- **Direct mapping**
  This is single source to single target, with no transformations.
- **One-to-many**
  A single-source value populates several target fields.
- **Many-to-one**
  This combines multiple source fields (using concat, add, etc.) into one target.
- **Conditional mapping**
  The target field will be populated only if the source meets certain criteria.

Common functions of many-to-one mappings are as follows:

- **concat**
  This chains strings with an optional separator.
  Example: FirstName + ' ' + LastName.
- **Add and subtract**
  These are for numeric aggregation or adjustments.

For example, let's say you have the following attributes:

- Price: 1000
- Discount: 200
- Target: NetPrice = Price - Discount = 800

For the mapping, you'll connect `Price` and `Discount` to the `subtract` function, which will lead to the output of `NetPrice`.

Occasionally, a single composite value must populate multiple target fields. To make this happen, you need to split fields (one-to-many). For example, let's say you have a source field called `FullName="John Smith"` and target fields called `FirstName="John"` and `LastName="Smith"`. You can use `splitBySeparator` or `substringBefore/substringAfter` functions to extract parts from the source.

---

**Testing Core Patterns**

You should test each complex mapping with the following:

- **Minimum data**
  An edge case is one record that's missing optional fields.
- **Maximum data**
  Use this to handle stress context for performance.
- **Typical data**
  Use this for common operational payloads.

The graphical test tool will immediately reveal whether contexts or functions yield unexpected empty or misaligned target nodes.

---

**Key Functions**

Now, let's look at some of the key functions that are used in mapping operations and patterns. We've split them into categories based on their use, as follows:

- **Applying conditional logic**
  The mapping palette provides the following Boolean and control functions:
  - `ifThenElse`
    This conditions related to this function are `valueTrue` and `valueFalse`.
  - `ifWithoutElse`
    This function makes the target populate only when the condition is true.
  - `equals`, `not`, `and`, and `or`
    These functions are for comparisons and Boolean logic.
- **Date/time transformation patterns**
  Dates are often format-specific, so use functions like the following:
  - `currentDate` and `currentDateTime`
    These are self-explanatory.
  - `formatDate`
    This reformats the date between patterns (e.g., `yyyy-MM-dd` → `dd.MM.yyyy`).
  - `addDays` and `subtractDays`
    These calculate delivery dates.

- **Context changes for control**
  Key context functions in Cloud Integration message mapping are as follows:
  - `removeContexts`
    This flattens queues so all values are in one context.
  - `createContexts`
    This introduces breaks at specific points (e.g., before each header element).
  - `splitByValue`
    This creates new context breaks whenever the selected value changes.
  - `collapseContexts`
    This removes consecutive duplicate values.

  These functions are crucial for grouping and aligning repeating data between source and target.

- **Node functions and aggregation**
  Node-level operations process entire queues, as follows:
  - `count`
    This counts the number of values in a queue.
  - `exists`
    This returns a Boolean if the queue has any values.
  - `sort`
    This reorders values.
  - `min` / `max`
    This performs a numeric or string comparison.

- Value mapping lookups
  To translate codes between systems (e.g., DE → Germany), do the following:
  - Maintain *value mapping* tables in Cloud Integration artifacts.
  - Use the `ValueMapping` function with the agency/schema/mapping group configured.
  - If no match is found, set the default value (for resilience).

You can connect functions in a series to achieve complex results without scripting. For example, let's say you want to transform a name into uppercase and append "(VIP)" if a customer spends more than $1,000. To do so, use the following code:

```
Name → toUpper → ifThenElse(Spend>1000, concat(Name," (VIP)"), Name)
```

**Best Practice**

Break into multiple small function nodes for clarity, rather than one large complex chain.

**Handling Attributes**

XML attributes are mapped just like elements but may require careful navigation:

- Attribute values appear in schema below their parent element with a preceding @ in XPath.
- Constants are often used to set attribute values (e.g., currency).

**Best Practices and Pitfalls in Core Operations**

For core operations, you should do the following:

- Use clear naming for complex function nodes.
- Group related logic spatially on the canvas.
- Use constants and lookups for system codes, instead of hard-coding into schemas.
- Validate context alignment when mapping multiple repeating structures.
- Avoid creating long unbroken function chains—they're harder to debug.
- Don't rely on implicit context behavior if it's not understood; explicit is safer.
- Avoid overly broad schemas that will increase navigation complexity and runtime.

### 3.3.3  Advanced Mapping Functions

*Advanced mapping functions* in Cloud Integration allow integration developers to extend the capabilities of graphical message mapping well beyond direct field transfers and basic function chains. These capabilities will come into play when you're dealing with complex XML structures, multiple data sources, and advanced conditional requirements, or when you're restructuring data in ways that basic mapping logic can't achieve easily. You should reach for advanced patterns whenever you encounter any of the following:

- Source and target structures differ significantly in hierarchy or grouping.
- You need to merge or cross-reference multiple repeating structures.
- Business rules require you to use calculated or derived data with multiple conditions.
- You need to reuse complex logic across multiple mappings without duplication.
- Performance and maintainability demand consolidated logic, rather than scattered chains.

Let's start by diving a little deeper into context manipulation. Key functions for context manipulation and their uses are as follows:

- splitByValue
  This creates a new context whenever a value in the input field changes. One possible scenario is to group line items by product category.

- changeContext
  This forces a context break at a specified higher-level element for alignment.

- collapseContexts
  This removes successive empty contexts or redundant groupings while still preserving breakpoints where values actually change.

- removeContexts
  This flattens data completely, and you can implement it before aggregation to make totals consider all items across contexts.

Sometimes, you need to pick specific positions from a repeating queue, and you can use the following:

- **index function (or position filtering)**
  This helps retrieve the nth occurrence.

- exists
  This, combined with conditional logic, ensures that you only map when expected data exists.

Beyond simple concatenation, you have the following operations:

- substringBefore/substringAfter
  This extracts a portion, based on a delimiter.

- tokenize
  This splits a string into multiple values (in a new queue), based on a regex or separator.

- normalize
  This removes redundant whitespace and special characters.

- replaceAll
  This is a regex-based replacement for complex cleaning tasks.

One practical case is parsing a combined City, ZipCode string into separate City and PostalCode elements.

More complex date/time scenarios require the following combinations:

- Using convertToDateTime (to parse a textual timestamp into an actual Date object) and then formatDateTime (to stringify the actual Date object into textual form)

- Using addDays/subtractDays as date manipulation to add/subtract a given number of days to or from the date while respecting the boundaries of months and years

- Multi–time zone conversions if Cloud Integration adapter properties carry UTC/ local offsets

A common advanced requirement is mapping from two or more source messages in a single mapping step. This happens when an iFlow merges content (e.g., a main order XML + a lookup XML from a datastore). You'll need to know about the following features to handle such a requirement:

- Cloud Integration's graphical mapping can load multiple source structures.
- Functions like `join`, `lookup`, or context alignment are used to merge data.
- You might want to use key-based matching by isolating keys from one structure and then indexing/selecting the corresponding values from another.

Advanced mappings often involve subtotaling or summarizing data, as follows:

- Using `sum`, `avg`, `min`, and `max` in context after flattening to the required level
- Combining with `splitByValue` to subtotal per category or per parent node

For example, if you want to calculate total quantity per `WarehouseID`, you'd follow these steps:

1. Use `splitByValue` on `WarehouseID` (grouping).
2. Put `removeContexts` in the quantity field.
3. Sum within each group.

Deep nesting sometimes leads to similar substructures at different levels. Cloud Integration supports applying functions repeatedly down the tree, although without a code loop—meaning you replicate mapping patterns for each level.

For very deep recursion, consider complementing mapping with an XSLT or UDF approach. When built-in functions are insufficient, you can embed custom Groovy or JavaScript logic as a UDF inside the graphical mapping. We'll take a detailed dive into UDFs in Section 3.3.4.

Several standard functions become "advanced" tools when combined with user-defined function:

- `mapWithDefault`
  This supplies a fallback when input is missing.
- `sort`
  This orders values before mapping.
- `distinctValues`
  This removes duplicates prior to output.
- `concat` + `collapseContexts`
  This aggregates unique values into a single string list.

> **Performance and Optimization**
>
> Complex mappings can impact runtime, particularly with large payloads and multiple nested loops, excessive context changes and reintroductions, and unnecessary complex UDF logic inside high-frequency mappings. Therefore, we recommend that you do the following:
>
> - Flatten data early if aggregating and avoid repeated context reshaping.
> - Consolidate related rules into fewer function nodes or UDF calls.
> - Test with large payloads to spot bottlenecks.

Best practices in advanced mapping are as follows:

- Prototype incrementally so that you build and test each transformation segment to avoid troubleshooting a huge monolith.
- Encapsulate repeated logic in UDFs or cloneable function segments.
- Document complex nodes and mapping paths for team clarity.
- Prefer built-in functions where performance is a concern; reserve UDFs for logic that truly needs it.
- Guard UDF code with null checks and data type handling to prevent runtime errors when inputs are missing or empty.

### 3.3.4   User-Defined Functions

While Cloud Integration's graphical mapping tool offers a rich set of built-in functions, there are situations in which no combination of these functions can implement a business rule cleanly or efficiently. *User-defined functions* (UDFs) solve this problem by allowing you to embed custom Groovy or JavaScript code directly into the mapping environment. They effectively extend the tool's capabilities with logic that's tailored to your specific integration needs.

UDFs combine the readability and structural advantages of graphical mapping with the power and flexibility of programming languages—without forcing you into a full script-mapping step. You might choose a UDF in any of the following situations:

- Built-in functions can achieve the result, but only through overly complex chains that hurt readability and maintainability.
- A rule depends on complex conditions (multiple inputs and nested logic) that aren't easily represented graphically.
- You need to apply a custom algorithm (a mathematical calculation, text parsing, encryption, hashing, etc.).
- You want to centralize a rule that will be reused in many mappings to ensure that changes happen in one place only.

Let's walk through the steps you'll use to create a UDF in the graphical mapping tool:

1. Open your mapping. In the Cloud Integration web UI or Eclipse-based tool, open the graphical mapping you want to extend.

2. Select **Functions • User-Defined.** This inserts a placeholder function node into your mapping canvas.

3. Choose a language. Groovy is recommended for Cloud Integration because it integrates natively with the Cloud Integration runtime and has rich Java interoperability. JavaScript is the alternative option, and it's sometimes used when porting code from non-SAP environments.

4. Define parameters. Give meaningful names to each input parameter (these correspond to the mapping connections coming in from source fields or other functions).

5. Define the return type. In Cloud Integration mapping, this is generally a String, String[] (multiple values), or number.

6. Write your code. The editor lets you write your transformation logic by using the passed-in arguments. Listing 3.36 shows an example of this in Groovy.

```
// Groovy UDF example: Calculate net amount
def calcNetAmount(String amount, String taxRate) {
    if (!amount || !taxRate) return ""
    def amt = new BigDecimal(amount)
    def tax = new BigDecimal(taxRate)
    return (amt - (amt * tax / 100)).toPlainString()
}
```

**Listing 3.36** User-Defined Function Preview Showing Groovy-Based Syntax

7. Save and test. Connect your UDF inputs to source fields and its output to target fields. Then, use the mapping test tool with sample data to confirm results.

Inside a UDF, there are some things to note when it comes to handling input and output:

- Parameters are received as strings (or string arrays), even when the source is numeric or date data. You must explicitly convert them to other types for arithmetic or date operations (new BigDecimal(str) or Date.parse(...)).
- The return type determines how the mapping engine processes the output:
  - A single string goes directly to the mapped target field.
  - A string array has multiple values output, one per queue entry/context.
- Always handle null or empty strings gracefully to avoid runtime exceptions.
- If your input field has multiple occurrences in the current context, Cloud Integration delivers them as an array in your UDF.

UDFs are stored inside the mapping artifact. This means that to reuse them in multiple mappings, you can copy them between mappings or design a template mapping that contains commonly used UDFs for cloning.

**Best Practice**

Keep a version-controlled library of UDF snippets outside Cloud Integration for reference and quicker migration.

Some advanced patterns with UDFs are as follows:

- **Data enrichment**
  This calls out to static data inside the function. Since using static data in the mapping itself requires constant blocks—which are especially impractical when there are multiple ones for one mapped field—it can come in handy to create a simple UDF that contains the static content (see Listing 3.37).

```
def countryFullName(String code) {
def map = [DE:"Germany", US:"United States", FR:"France"]
  return map[code] ?: "Unknown"
}
```

**Listing 3.37** UDF Example Snippet for Data Enrichment

- **Complex conditionals**
  These handle multicriteria branching more elegantly than nested if-then-else blocks in the mapper. Even though you can manage a complex condition directly in the mapping, it gets quite unreadable quickly, and it's very hard to debug and maintain when you need to make changes. A much more readable and expandable solution is to move it into a designated UDF, as presented in Listing 3.38.

```
def shippingMethod(String country, String amount) {
    def amt = amount.toBigDecimal()
    if (country == 'US' && amt > 1000) return "Express"
    if (country == 'US') return "Standard"
    return "International"
}
```

**Listing 3.38** UDF Example Snippet for Complex Conditional Conversion

- **Text parsing and regex**
  These leverage full Java regex APIs for advanced matching, as demonstrated in Listing 3.39.

```
def extractDigits(String input) {
    return input?.replaceAll("\\D+", "")
}
```

**Listing 3.39** UDF Example Snippet for Regex-Based String Manipulation

- **Mathematical computations**
  These are complex formulas, roundings, and percentage calculations that are not directly supported by built-ins.

- **Reusable subfunctions**
  Within a UDF, you can define private helper methods for internal clarity, though Cloud Integration will only expose the main function as a mapping node.

Unlike standalone Groovy scripts in Cloud Integration, debugging inside mapping UDFs is limited to the following:

- Using `println()` statements (which are visible in tenant logs but not the mapping test UI)
- Returning intermediate values temporarily to inspect output
- Creating a dedicated "test" mapping for the UDF with simplified input structures to isolate logic

If an error occurs at runtime, Cloud Integration usually throws a mapping error with the UDF name and exception message, so you should inspect it to locate the issue.

Some best practices for UDFs are as follows:

- Keep UDF logic simple where possible; complex loops over large queues can slow mappings.
- Consider preprocessing data before mapping if heavy computation is needed.
- Avoid unnecessary object creation in tight loops.
- Built-in functions are generally faster than equivalent UDF logic, due to engine optimization.
- Sanitize and validate all inputs, especially if they'll be used in concatenations or lookups.
- Avoid performing external network calls from UDFs. Cloud Integration mapping UDFs are designed for data transformation within the mapping step, not integration I/O.

Let's look at a scenario that involves UDF-driven aggregation and labeling. You'll classify customers as Gold, Silver, or Bronze, based on total spend in an XML order. Follow these steps:

1. Use a separate mapping connection and sum function to total all Order/Amount values per customer context.

2. Pass this total into a UDF classifyCustomer, as shown in Listing 3.40.

```
def classifyCustomer(String total) {
def val = total.toBigDecimal()
if (val >= 10000) return "Gold"
```

```
if (val >= 5000) return "Silver"
return "Bronze"
}
```

**Listing 3.40** Classification UDF for Handling Simple Conditions

3. Output the classification to the `LoyaltyTier` target field.

This is far cleaner than trying to represent multithreshold logic via nested `ifThenElse` functions.

### 3.3.5   Best Practices and Maintainability

Graphical message mapping in Cloud Integration provides powerful, code-free data transformation capabilities. But as mappings grow in complexity and integration landscapes evolve, the same features that made the tool accessible can lead to sprawling, hard-to-manage logic. A mapping that works today might become problematic tomorrow in any of the following situations:

- Source or target schemas change.
- Business logic grows more complex.
- Multiple integrators work on the same artifacts.
- The number of iFlows that reuse a mapping increases.

Without best practices, you risk any or all of the following:

- **Technical debt**
  This comes from overly complex canvases with tangled lines and redundant logic.
- **Performance bottlenecks**
  These come from excessive function nodes and unnecessary context reshaping.
- **Maintenance nightmares**
  These come from unclear transformation rules and duplicated logic.

It's important to remember that readable design is maintainable design. The mapping canvas should tell its "story" visually, so you should do the following things:

- Implement logical grouping. Arrange nodes and connections so that related rules are visually close. For complex scenarios, separate sections for header mapping, line-item mapping, and constants.
- Minimizing crossovers. Don't make lines cross unnecessarily; use function nodes to reroute them cleanly.
- Use descriptive labels. Rename function nodes (especially UDFs) so they have clear, business-relevant names (`CalcNetPrice`, `FormatOrderDate`, etc.).
- Use colors and markers. If you're working in a tool version that supports node highlighting, use it to distinguish key transformation steps.

You should break big transformations into modular, reusable segments by doing these things:

- Use value mapping artifacts for standard code conversions so they're maintained centrally.
- Encapsulate multistep conversions into UDFs rather than long, chained built-in functions.
- If a mapping does two or more unrelated transformations (e.g., customer data and invoice items), consider splitting into separate mapping artifacts or applying multi-mapping patterns.

Documentation inside a mapping is just as important as in code, so here are some best practices for it:

- The function node names should reflect a purpose, so for example, use `JoinFirst-LastName` instead of `concat1`.
- For parameter names in UDFs, use explicit terms like `amount` and `taxRate` instead of `arg1` and `arg2`.
- In inline documentation, most mapping tools in Cloud Integration allow you to add notes or comments on nodes. Use them to explain complex rules.
- Maintain a mapping specification document externally. It should include field-mapping definitions, business rules, and references to artifacts.

Reusability reduces duplication and increases consistency, so you should follow these best practices as well:

- Store common value mappings, constants, and UDFs in shared libraries or template mappings.
- Use global variables or iFlow parameters (if applicable) to inject constants that vary by environment instead of hard-coding in mappings.

Source and target schemas evolve, so to protect your mappings, do the following:

- Avoid overly tight coupling. Don't map based on assumed fixed positions in a queue; rely on element names.
- When possible, design mappings defensively by using `ifWithoutElse` or `mapWithDefault` for potentially missing elements.
- After performing schema imports, verify that newly added fields don't change existing field paths.

Performance matters, especially for large payloads or high-volume flows, so you should do the following:

- Flatten early. For aggregations, remove contexts before summing to avoid repeated operations.
- Limit UDF use in high-frequency loops; built-in functions execute faster.

- Avoid redundant functions. For example, don't `removeContexts` twice in a chain unless you need to.
- Test with large payloads. Cloud Integration's test tool accepts big sample XMLs, so you should use them to spot slow segments.

For mappings with many functions, do the following:

- Use *"staging" outputs*, which are intermediate nodes whose output goes to both testing and final targets.
- Build and test the mapping area by area, rather than all at once.
- Consider *multimapping* (splitting message processing into multiple smaller mapping steps) if a single mapping becomes unmanageable.

Treat mappings like source code, as follows:

- Export and store them in a Git repository or content versioning system.
- Tag and document notable changes—especially those driven by schema updates or new business rules.
- Maintain a test suite of I/O pairs to validate mapping after changes.

To make mappings easier to debug, do these things:

- Temporarily insert diagnostic functions to output intermediate values during testing.
- For UDFs, insert `println()` statements (which are visible in Cloud Integration logs) to verify parameter values.
- In production, control logging via Cloud Integration log levels and conditional logic to avoid overhead.

Some best practices in action are as follows:

- **Context-safe aggregation**
  When summing values from multiple contexts, put `removeContexts` in the numeric field, pass to `sum`, and (optionally) reintroduce a context break with `createContexts`. This prevents misaligned sums and ensures the right totals per group.
- **Defensive conditional mapping**
  When mapping an optional field, use `ifWithoutElse(exists(FieldX), FieldX)` to skip creating an empty tag if data is missing.
- **Controlled UDF use**
  When a rule is repeated more than two times in a mapping, it's a candidate for a UDF.

To future-proof your mappings, keep the following in mind:

- Isolate business logic. Keep change-prone rules (discount rates and code lists) outside the mapping if possible—for example, in parameter tables or lookups.

- Follow SAP standards. Align with Cloud Integration's recommended naming and folder conventions.
- Perform regular reviews. Schedule mapping reviews to detect obsolete functions, unused UDFs, and outdated value mappings.

### 3.3.6   Troubleshooting and Common Issues

Even with the best design practices, you can run into issues when performing graphical message mappings in Cloud Integration. This can be due to mismatched schemas, unexpected payloads, or subtle context and queue misalignments. So, you need to be able to efficiently diagnose and resolve these problems to keep integrations running smoothly.

When a mapping fails or produces incorrect output, you'll want to follow these steps:

1. Reproduce the issue by using the same payload and runtime context that triggered it.
2. Isolate the scope by narrowing the problem to a particular segment or field.
3. Test incrementally by validating sections of the mapping in isolation before reintegrating.
4. Document your findings by recording problematic inputs, observed outputs, and the fixes you applied.

Some useful tools and features for troubleshooting are as follows:

- **Mapping test tool**
  In the mapping artifact editor, you can switch to the **Test** tab and run transformations with sample XML input to view output instantly without deploying the full iFlow.
- **Tracing the log level**
  In the **Manage Integration Content** section, you can temporarily set the iFlow's log level to Trace to capture payload snapshots before and after the mapping step.
- **Message processing logs**
  After executing a flow, you can open the processing log, check step details, and download payloads for comparison.
- **Attachments in logging**
  Within Groovy scripts before/after the mapping, you can attach payloads for deeper inspection.

Some common issues and how to address them are as follows:

- **Schema mismatches**
  - **Symptoms**
    The mapping test fails with parsing errors, or target output is missing expected fields.

- **Causes**

  The XSD structure in the mapping artifact doesn't match the actual runtime XML shape, often after service changes.

- **Fix**

  Reimport the correct schema, then validate element names and hierarchy against incoming payload.

- **Path or context misalignment**

  - **Symptoms**

    Values don't appear where expected; repeating elements are mismatched or merged incorrectly.

  - **Causes**

    Different parent structures cause queues to have unrelated contexts.

  - **Fix**

    Either adjust with `changeContext` or `removeContexts` or remap within correct parent loop.

- **Empty output elements**

  - **Symptoms**

    Target fields are created but empty.

  - **Causes**

    The source element is missing or the conditional prevents population.

  - **Fix**

    Verify the source element path; use `mapWithDefault` or constants for fallback values.

- **Wrongly aggregated data**

  - **Symptoms**

    Sums, concatenations, or aggregations produce wrong totals.

  - **Causes**

    The context is not flattened before aggregation, there are duplicate values, or there's incorrect grouping logic.

  - **Fix**

    Before aggregating, use `removeContexts` and check grouping with `splitByValue` as needed.

- **Performance degradation**

  - **Symptoms**

    The mapping step takes excessive time in large payloads.

  - **Causes**

    There are unnecessary repeated functions, large UDF loops, or excessive context shifts.

  - **Fix**

    Optimize function paths, move heavy lifting to preprocessing, and simplify UDF logic.

Let's walk step-by-step through a diagnostic workflow, as follows:

1.  Verify schemas and sample data. Check that mapping's source and target XSDs reflect reality, then, validate the sample XML against XSD.

2.  Run test mapping with minimal payload. Keep one occurrence per repeating element and then verify transformation logic in its simplest form.

3.  Add a problematic payload segment. Gradually introduce real data complexity until the issue reappears—this will narrow your search.

4.  Inspect contexts in the test tool. Many mapping editors let you hover over lines to see queue/context details; use them to confirm alignment.

5.  Isolate functions. Temporarily bypass function nodes to pinpoint whether the problem lies in raw mapping or transformation logic.

6.  Check UDF behavior. Pass known values into UDFs in a test mapping, then print outputs to logs if they're running in iFlow.

When debugging user-defined functions, you should wrap code in `try/catch` blocks and log errors to Cloud Integration logs, as shown in Listing 3.41.

```
try {
    // logic
} catch (Exception e) {
    println "UDF Error: ${e.message}"
    return ""
}
```

**Listing 3.41** Groovy Syntax for Handling Exceptions

Finally, you should add intermediate returns to see which transformation step is breaking. Avoid depending on global state; always pass required context into parameters.

## 3.4    XSLT Mappings

Extensible Stylesheet Language Transformations (XSLT) is a specialized language that's designed to transform XML data. In the context of Cloud Integration, XSLT serves as a powerful mapping tool that facilitates conversions and rearrangements of XML-based messages as they traverse integration processes among diverse software systems. The fundamental purpose of using XSLT in data integration is to enable seamless communication and interoperability among disparate systems that often use varying data formats and structures.

To understand the role of XSLT in SAP BTP, it's helpful to approach it step by step. We'll begin with a general introduction before moving into the technical fundamentals of XSLT, which provide the foundation for practical usage. Since XPath expressions are central to navigating XML structures, we'll also explore their fundamentals. Building on this, we'll then illustrate how XSLT enables filtering, grouping, and sorting of data, as well as performing calculations and string operations. Then, we'll discuss the handling of repeated or grouped data—which is a common scenario in enterprise integrations—and we'll conclude with how XSLT can dynamically assign headers and properties within Cloud Integration flows. Together, these subsections aim to give you a holistic view of XSLT, from its basics to advanced applications, to ensure that you have a strong foundation for practical implementation.

### 3.4.1   Introduction

Data integration projects frequently involve exchanging data among business applications that were designed independently and that therefore represent information differently. For example, a customer order message might have one XML schema when it originates from a CRM system and a different schema when it heads into an ERP back-end. XSLT provides the declarative means to define how a source XML structure should be mapped to a target XML structure, including the logic for renaming elements, transforming values, filtering nodes, and much more.

In Cloud Integration, the XSLT mapping step is incorporated into the iFlow as a standard mapping approach alongside graphical and scripting-based mappings. The use of XSLT is particularly relevant when the mapping logic is intricate or when requirements extend beyond the capabilities of the Message Mapping tool—such as when advanced XML navigation, dynamic value computation, or conditional data transformations are needed.

The basic process of XSLT mapping in data integration involves three core artifacts:

- **Source XML document**
  This represents the input data.
- **XSLT stylesheet**
  This defines transformation rules by using XSLT syntax, including templates and XPath expressions.
- **Result XML**
  You generate this by applying the XSLT stylesheet to the source XML and using an XSLT processor (which is embedded in Cloud Integration).

With XSLT in Cloud Integration, you can do the following:

- Transform XML data from one format to another, to support complex mapping logic.
- Filter and aggregate data to produce exactly the payload that's required by the target system.
- Implement conditional mapping rules and value transformations, such as formatting dates and recalculating currencies.
- Enable modular, reusable transformation logic that can be applied across multiple iFlows and scenarios.

SAP's implementation supports XSLT 3.0 and XPath 3.1, so it gives developers access to the latest features of the language and ensures compatibility with modern XML processing requirements.

Whether you should choose XSLT mapping over message mapping or scripting depends on the specific requirements of your integration scenario and your team's skills. Situations in which you should use XSLT mapping are as follows:

- You have complex source-target structures with many nested, repeating, or recursive elements, where visual mapping becomes unwieldy.

- You want to do advanced filtering, grouping, or conditional transformations that would require convoluted expressions in message mapping or large scripting blocks.

- Your business logic requires standard XML processing tools, such as XPath and XSLT extension functions.

- You want to reuse existing XSLT assets, for example, when transitioning from SAP PI/SAP PO or other middleware.

- You're performing performance-critical integrations, since XSLT processors are optimized for XML transformation.

On the other hand, the advantages of using XSLT in Cloud Integration are as follows:

- XSLT mappings are deterministic and well-structured, and they enable precise, reliable transformations.

- Code reuse and template-based design are intrinsic to XSLT, so they improve maintainability.

- Powerful constructs like keys, templates, and XPath functions allow transformations that can't be represented graphically.

- XSLT mappings are standard and portable, so they ease migration between platforms and tools.

- Separation of mapping logic from process logic improves testability and reusability.

However, teams must weigh these benefits against the steeper learning curve: XSLT is less approachable than drag-and-drop mapping, and debugging can be more challenging without specialized tools. Therefore, you should avoid XSLT in the following situations:

- **Doing very simple, direct field mappings**
  Message mapping is faster and easier to maintain.

- **Working with non-XML payloads**
  Scripting is necessary.

- **Team expertise in XSLT is lacking**
  The graphical message mapping tool may be preferable to ensure maintainability and knowledge transfer.

In summary, you should use XSLT mapping in Cloud Integration for complex XML transformation needs that go beyond the capabilities of message mapping or where advanced, precise XML handling is required—provided that the necessary skills are available on the team.

### 3.4.2  Technical Fundamentals of XSLT

To master XSLT mapping in Cloud Integration, you need to develop a solid understanding of XSLT's building blocks, syntax, and execution model. XSLT is a W3C standard XML-based language that's designed to transform XML documents into other XML documents and alternative formats such as HTML, plain text, and others by leveraging XPath expressions.

At its core, every XSLT transformation is governed by an *XSLT stylesheet*, which is a specialized XML document that defines the rules (which are also known as *templates*) for transforming input XML into the desired output. An XSLT stylesheet typically begins with the structure shown in Listing 3.42.

```
<xsl:stylesheet version="3.0"
    xmlns:xsl="http://www.w3.org/1999/XSL/Transform">
    <!-- Template definitions go here -->
</xsl:stylesheet>
```

**Listing 3.42** XSLT Stylesheet Definition Boilerplate

The root element in Listing 3.42 declares the stylesheet's version (Cloud Integration supports up to XSLT 3.0) and the necessary XSL namespace.

Templates, which you define with `<xsl:template>`, represent the fundamental unit of logic in XSLT. Each template matches nodes in the source XML by using patterns—primarily, XPath expressions—that are specified by the match attribute. When a node in the input XML matches the pattern, the corresponding template fires and dictates how to process or transform the matched node.

In this template (see Listing 3.43), whenever the root `<Order>` element is encountered in the source, the template starts an `<Invoice>` element in the output and instructs XSLT to process each child `<Item>` by applying other appropriate templates.

```
<xsl:template match="/Order">
  <Invoice>
    <xsl:apply-templates select="Item"/>
  </Invoice>
</xsl:template>
```

**Listing 3.43** XSLT Template Snippet

Templates can match any part of the source document, including the root (/), specific elements, attributes, and even text nodes (text()). The modularity of templates makes XSLT particularly adept at handling large and variably structured XML payloads.

Most XSLT operations rely on XPath, which is a concise language for referencing nodes, attributes, and values in XML. XPath expressions are used in template matches, in data select expressions, and wherever the transformation logic needs to pinpoint information in the source XML.

In the following code, the select attribute employs XPath to extract the value of the nested <name> element within <customer>:

```
<xsl:value-of select="customer/name"/>
```

XPath supports navigation by the following:

- Child axis (e.g., customer/name)
- Attributes (e.g., @id)
- Predicates to filter nodes (e.g., Item[price > 100])
- Functions to compute or format data (e.g., string(), format-date())

Cloud Integration's XSLT engine supports XPath 3.1 and thus allows advanced queries and functions. We'll take a deeper dive into XPath in the following chapter.

XSLT processors work in a pattern-matching, rule-driven manner. When an XSLT transformation runs, the following sequence of events occurs:

1. The processor starts with the root node.
2. It searches for the template whose match pattern aligns with the current node.
3. The body of the matching template determines the output structure.
4. The process continues recursively, guided by instructions such as <xsl:apply-templates> (which triggers template matching for children) or <xsl:for-each> (which iterates over node sets).

The default template mechanism ensures that every node is processed, even if it's not explicitly handled—although production stylesheets should specify templates for all meaningful nodes.

The templates produce output in two ways:

- **Literal result elements**
  Standard XML elements that are written directly in the template become part of the output.
- **XSLT instructions**
  The XSLT built-in elements, which are prefixed by xsl:, fetch, manipulate, or create content dynamically.

Common XSLT instructions include the following:

- `<xsl:value-of select="..."/>`
  This writes the text value of the selected node.

- `<xsl:for-each select="...">`
  This loops over nodes selected by the XPath expression.

- `<xsl:choose>`, `<xsl:when>`, `<xsl:otherwise>`
  This implements conditional logic.

- `<xsl:attribute name="...">`
  This dynamically creates attributes.

- `<xsl:copy>`, `<xsl:copy-of>`
  This copies nodes or node sets wholesale.

XSLT encourages modularity through named templates (which are invoked by `<xsl:call-template>`), global parameters, and the use of modes for template specialization. Such modular design is vital for large Cloud Integration projects and encourages the reuse of mapping logic across different integration scenarios.

Now, let's look at an example: suppose you have the input XML shown in Listing 3.44.

```
<Order>
  <customer>
    <name>Jane Doe</name>
  </customer>
  <Item>
    <product>Widget</product>
    <price>29.99</price>
  </Item>
</Order>
```

**Listing 3.44** XSLT Example: XML Source Structure Preview

An XSLT template might transform this to the code shown in Listing 3.45.

```
<Invoice>
  <CustomerName>Jane Doe</CustomerName>
  <ProductName>Widget</ProductName>
  <Total>29.99</Total>
</Invoice>
```

**Listing 3.45** XSLT Example: Simplified XML Target Structure

By creating appropriate templates for each node and using XPath to extract and insert values, XSLT facilitates a clear separation between input structure and output requirements.

### 3.4.3 XPath Language Fundamentals: Navigating XML in XSLT

Mastering *XML Path Language* (XPath)—the query language that allows you to locate, filter, and manipulate information inside XML documents—is a cornerstone of writing effective XSLT stylesheets. In Cloud Integration, XPath expressions are woven throughout XSLT mappings, where they power template matches, data extraction, and sophisticated transformation rules.

XPath is a W3C standard that's designed to address and operate on the nodes in an XML document. Every segment of a path in XPath corresponds to a relationship in the XML tree—be it an element, an attribute, or even a text value. In practice, XPath serves a function akin to SQL for relational data: it enables you to pinpoint exactly which "rows" (nodes) you want to work with and in what order.

Cloud Integration supports XPath 3.1, which is one of the most advanced iterations. It brings powerful features, function libraries, and data type support to XSLT developers.

In XPath, XML documents are abstracted as a tree of nodes. There are several types of nodes:

- Element nodes (e.g., `<Order>`)
- Attribute nodes (e.g., `@id`)
- Text nodes (the text within elements)
- Namespace nodes
- Processing instruction and comment nodes (which are used infrequently in mapping contexts)

XPath uses the concept of *axes* to specify relationships and navigation paths. The most common axes are as follows:

- **Child axis (default)**
  This selects child elements. For example, `Order/customer` picks the `<customer>` element directly beneath `<Order>`.
- **Attribute axis**
  This accesses attributes with @ (e.g., `Order/@id`).
- **Descendant axis**
  This traverses all levels below the current node. For example, `//Item` selects all `<Item>` elements anywhere in the document.
- **Parent and ancestor axes**
  These navigate upward in the tree, and they're useful for context-aware mappings.

Some of the most important XPath expressions and syntax are as follows:

- **Basic paths**
  XPath uses these familiar file path-like notations:
  - `/Order/Item`
    This is the direct path from the root `<Order>` to its `<Item>` child.

- `//price`

    This selects all `<price>` elements, wherever they occur.

- **Predicates for filtering**

    XPath predicates are designated with square brackets (`[  ]`), and they provide filter logic for selecting nodes based on their properties or position. Some different predicates are as follows:

    - `Item[price > 100]`

        This selects `<Item>` elements with a `<price>` subelement that's greater than 100.

    - `customer[name='Jane Doe']`

        This finds `<customer>` elements whose `<name>` child exactly matches `"Jane Doe"`.

    - `Item`

        This is the first `<Item>`. (Indexes in XPath are 1-based.)

- **Attributes and functions**

    You can reference attributes and use built-in functions for extraction, computation, or formatting, as follows:

    - `@status`

        This accesses the "status" attribute in the current context.

    - `count(Item)`

        This computes the number of `<Item>` elements.

    - `string(name)`

        This converts the `<name>` child element into a string.

    XPath 3.1 dramatically expands the available functions, including `format-date`, mathematical operations, string processing (`replace` and `substring`), and many more.

- **Combining paths**

    You can combine paths with logical and, or, and union (`|`):

    - `Item[price > 100 or price < 10]`

        This selects items outside a specified price range.

    - `/Order/customer | /Order/vendor`

        This selects both `<customer>` and `<vendor>` under `<Order>`.

XPath also appears throughout an XSLT stylesheet:

- `<xsl:template match="Order/Item[price > 100]">`

    This appears in the `match` attribute of `<xsl:template>`.

- `<xsl:value-of select="customer/name"/>`

    This appears in `select` attributes for extracting values.

- `<xsl:for-each select="Item[@status='urgent']"> ... </xsl:for-each>`

    This appears in iteration.

Because of XPath's expressive power, you rarely need to use cumbersome code to traverse the source document. The right expression delivers exactly the nodes you need, directly to your template logic.

With support for XPath 3.1, Cloud Integration enables the following:

- **Sequences and higher-order functions**
  These are for complex value aggregations and transformations.

- **String and date/time manipulation**
  This is essential for business document processing.

- **Grouping and distinct values**
  For example, `distinct-values(Item/category)` provides unique categories.

Now, let's look at a practical example. Suppose your XML contains multiple customers and you only want to create output for those who are in a specific country, as shown in Listing 3.46.

```
<Customers>
  <Customer>
    <name>Jane Doe</name>
    <country>DE</country>
  </Customer>
  <Customer>
    <name>John Smith</name>
    <country>US</country>
  </Customer>
</Customers>
```

**Listing 3.46** XSLT Example: XML Tree

In your XSLT, the XPath in Listing 3.47 selects just the German customers.

```
<xsl:for-each select="Customers/Customer[country='DE']">
  <GermanCustomer>
    <xsl:value-of select="name"/>
  </GermanCustomer>
</xsl:for-each>
```

**Listing 3.47** XSLT Example: XSLT Path Selection Loop

When you're given a list of customers with different country attribute values (see Listing 3.46), you can filter for these values directly in the XPath selection, as shown in Listing 3.47.

### 3.4.4   Filtering Data with XSLT

Filtering data is one of the fundamental tasks in XSLT mapping because it enables you to include only relevant portions of the source XML in the output payload, based on specific conditions. In Cloud Integration, where messages often contain large or complex XML structures, you must precisely filter data to optimize downstream processing and ensure that the target systems receive exactly what they need.

This section explores key concepts, patterns, and practical techniques for filtering XML data by using XSLT within Cloud Integration. We'll cover XPath predicate filtering, conditional processing, exclusion strategies, and the use of built-in or custom logic to handle common filtering scenarios.

In typical integration scenarios, source messages may carry rich datasets, including multiple records, optional fields, and nested elements that are not always required by the receiving system. Sending the full source data can be inefficient or even problematic—for example, it can cause performance bottlenecks, bloat network traffic, or lead to errors if unexpected nodes appear. Filtering lets you do the following:

- Reduce message size by dropping unnecessary elements.
- Enforce business rules (e.g., only process orders above a minimum value).
- Customize payloads dynamically, based on runtime data.
- Ensure compliance with target system interfaces by removing invalid or unsupported nodes.

XSLT excels at filtering because it inherently processes XML data node by node, thereby offering you the ability to exert fine-grained control by using XPath conditions combined with template logic.

The fundamental filtering techniques are as follows:

- **Predicate filtering with XPath**
  The most direct and common way to filter nodes is to use XPath predicates (e.g., [condition]) within template matches or selection expressions. For example, you may want to select only <Item> elements where the <price> is greater than 100. Here (see Listing 3.48), only <Item> nodes that meet the predicate appear in the output or further processing. A predicate can be any valid XPath expression that resolves to a Boolean.

```
<xsl:template match="Item[price > 100]">
  <!-- Process expensive items only -->
</xsl:template>
```

**Listing 3.48**  XSLT: Predicate Filtering with XPath

This method is efficient because the filtering happens during node selection and irrelevant nodes are never processed.

- **Conditional filtering with `<xsl:if>` and `<xsl:choose>`**
  Sometimes, you want to include or exclude parts of the output conditionally, inside templates. You can do this by using the following conditional concepts:
  - `<xsl:if>`
    This executes content only if the test condition is true.
  - `<xsl:choose>`
    This acts as a switch-like composure with multiple branches.

  Both of these can be seen in Listing 3.49, with the classic `if` in the upper section and a switch-alike in the lower area.

```
<xsl:if test="price > 100">
  <ExpensiveItem>
    <xsl:value-of select="product"/>
  </ExpensiveItem>
</xsl:if>

<xsl:choose>
  <xsl:when test="price > 100">
    <ExpensiveItem>...</ExpensiveItem>
  </xsl:when>
  <xsl:otherwise>
    <RegularItem>...</RegularItem>
  </xsl:otherwise>
</xsl:choose>
```

**Listing 3.49** XSLT Syntax for Conditionals Using if and choose

This approach is useful when filtering depends on complex logic or multiple criteria and when you want to retain template modularity.

- **Excluding nodes using template overrides**
  By default, XSLT applies built-in templates that copy nodes recursively. To explicitly exclude certain elements, you can write empty templates that match those nodes and do nothing. For example, you can use the following to remove all `<DebugInfo>` elements from output:

```
<xsl:template match="DebugInfo"/>
```

  Because this template matches `<DebugInfo>` and contains no processing instructions, those nodes are effectively skipped. This exclusion pattern is especially helpful when consuming large XML payloads that contain diagnostic or nonessential elements that must not appear at the receiver.

- **Combining filters with logical operators**
  Predicates and tests support Boolean operators like `and`, `or`, and `not()`, and they allow very expressive filtering criteria, as follows:

```
<xsl:template match="Item[price > 100 and @status='confirmed']">
  <!-- Process confirmed expensive items -->
</xsl:template>
```

You can also negate conditions, as follows:

```
<xsl:if test="not(discontinued)">
  <!-- Only include items not marked discontinued -->
</xsl:if>
```

This flexibility lets you refine your filtering rules closely to fit business logic.

Some practical considerations when filtering in Cloud Integration are as follows:

- Use predicates to limit processing as early as possible. This reduces unnecessary template invocation and improves performance.
- Beware of data types. XPath comparisons depend on data types and formats. For example, comparing text elements that represent numbers requires conversion or explicit casting (number(price)).
- Handle optional nodes carefully. Always check for whether optional nodes exist before accessing subelements or attributes, to avoid errors.
- Leverage variables for readability. You can abstract complex filters into named variables to improve maintainability.
- Test filters independently. Use Cloud Integration's mapping test features or stand-alone XSLT processors to verify your XPath expressions and template matches.

Imagine a scenario where you receive an XML with many <Order> elements (see Listing 3.50) but need to forward only those that were confirmed and created within the last 30 days. You can do this with a filtering XSLT mapping. Here, the select condition also includes a type conversion into date as well as two more date functions to calculate the offset of the order's created date until today (see Listing 3.51).

```
<Orders>
  <Order>
    <id>1C01</id>
    <status>confirmed</status>
    <createdDate>2024-07-01</createdDate>
  </Order>
  <Order>
    <id>1002</id>
    <status>pending</status>
    <createdDate>2024-06-01</createdDate>
  </Order>
</Orders>
```

**Listing 3.50** Source XML Snippet

```
<xsl:template match="Orders">
  <FilteredOrders>
    <xsl:for-each select="Order[status='confirmed' and xs:date(createdDate) >=
current-date() - xs:dayTimeDuration('P30D')]">
      <xsl:copy-of select="."/>
    </xsl:for-each>
  </FilteredOrders>
</xsl:template>
```

**Listing 3.51** XSLT Template to Filter Orders

In this example, we and the system do the following:

- We use XPath 3.1 functions like `xs:date()` and `current-date()` for date comparison.
- The predicate filters out nonconfirmed orders and orders created more than 30 days ago.
- The system copies the filtered orders as-is to the output.

### 3.4.5  Grouping and Sorting XML Data with XSLT

In many Cloud Integration scenarios, the source XML contains multiple repeating elements that need to be logically grouped or sorted before integration. Grouping and sorting are common data transformation tasks you'll need to do when generating reports, aggregating records, and preparing structured output for downstream systems.

This section explores how to implement grouping and sorting effectively with XSLT by leveraging features that were introduced in XSLT 2.0 and 3.0—and are supported by Cloud Integration—to organize XML data efficiently.

Raw XML messages often include multiple entries without any inherent order or categorization. For example, an e-commerce order might include many line items, but the target system will expect them grouped by product category or sorted by price. Proper grouping and sorting accomplish the following:

- They enhance data clarity and usability.
- They uphold business rules that require aggregation by key fields.
- They enable downstream systems to process data more efficiently.
- They simplify reporting and analytics.

Without XSLT grouping and sorting, XML transformations become cumbersome and message payloads can become bloated or disorderly.

XSLT 1.0 didn't have native grouping capabilities, so it forced developers to use complex, inefficient "Muenchian grouping" techniques that were based on keys and node sets. However, since the release of XSLT 2.0 and 3.0 (which are used in Cloud Integration), grouping capabilities have been built-in and easier to implement, thanks to the `<xsl:for-each-group>` instruction.

The core element for grouping in modern XSLT is as follows:

```
<xsl:for-each-group select="nodes-to-group" group-by="expression">
  <!-- body for each group -->
</xsl:for-each-group>
```

Here, we have the following:

- **select**
  This is an XPath expression that selects the nodes to group.

- **group-by**
  This is an XPath expression that specifies the grouping key for each node.

Each group's representative is processed in order, and within the group, you can access members by using `current-group()` and `current-grouping-key()`.

You can sort the nodes that are processed inside an XSLT loop with `<xsl:sort>`, and you can apply sorting inside or outside grouping structures as follows:

- **Sorting groups**
  Add `<xsl:sort>` inside `<xsl:for-each-group>` to sort groups by the grouping key, as follows:

  ```
  <xsl:for-each-group select="Item" group-by="category">
    <xsl:sort select="current-grouping-key()" data-type="text" order="ascend-
  ing"/>
    ...
  </xsl:for-each-group>
  ```

- **Sorting nodes within groups**
  Use `<xsl:sort>` inside the child `<xsl:for-each>`, as follows:

  ```
  <xsl:for-each select="current-group()">
    <xsl:sort select="price" data-type="number" order="descending"/>
    ...
  </xsl:for-each>
  ```

  This example sorts items within each category group by descending price. Also, `group-adjacent` groups adjacent nodes that share the same grouping key, as follows:

  ```
  <xsl:for-each-group select="Item" group-adjacent="category">
    ...
  </xsl:for-each-group>
  ```

Use this style when nodes are presorted and you want to group only consecutive nodes with the same key.

Finally, you can use `group-starting-with`/`group-ending-with` to group nodes that start or end with specific patterns.

### 3.4.6   Calculations and String Operations in XSLT

In Cloud Integration, transforming and enriching XML data often requires you to perform calculations and manipulate textual information within the payload. XSLT—especially versions 2.0 and 3.0, supported by Cloud Integration—provides you with a rich set of functions and expressions for arithmetic, date/time, and string operations directly within the stylesheet. Mastering these capabilities significantly enhances your ability to implement dynamic, business-friendly transformations. This section explores how to leverage calculations and string handling in XSLT to build precise, maintainable integration mappings.

Business processes frequently require you to do the following:

- Calculate totals, taxes, discounts, and currency conversions.
- Format dates and times to align with target system expectations.
- Extract, concatenate, or replace parts of strings to clean or rearrange data.
- Apply conditional logic that depends on numeric or textual content.

Integrations that handle invoices, orders, shipments, or master data benefit from native XSLT functions that embed these rules without reverting to external scripting or complex postprocessing.

XSLT expressions can directly perform arithmetic. Here are the expressions:

- +

  This is for addition.

- -

  This is for subtraction.

- *

  This is for multiplication.

- div

  This is for division.

- mod

  This is for modulo.

Listing 3.52 shows an example of how to calculate a total price, including tax.

```
<xsl:variable name="price" select="number(Item/price)"/>
<xsl:variable name="taxRate" select="0.19"/>
<xsl:variable name="total" select="$price + ($price * $taxRate)"/>
<TotalPrice>
  <xsl:value-of select="format-number($total, '#0.00')"/>
</TotalPrice>
```

**Listing 3.52** XSLT Variable Definitions With Referencing in select Clauses

A few things of note in this code are as follows:

- `number()`
  This function converts string values to numeric for calculations.

- `format-number()`
  This formats numeric output (e.g., two decimal places).

You can use variables to store intermediate computation results and thus improve readability and reuse.

XSLT 3.0 with XPath 3.1 introduces strong support for date/time types and functions, which is vital for transformations that involve timestamps, schedules, or reporting periods. Common functions include the following:

- `current-date()`, `current-time()`, and `current-dateTime()`
  You use these to get the current system time.

- `xs:date()` and `xs:dateTime()`
  You use these to cast strings into date values.

- **Arithmetic you perform on dates with durations**
  For example, you can subtract or add days with `xs:dayTimeDuration()`.

Listing 3.53 shows an example of how to calculate whether an order date was within the last 30 days.

```
<xsl:variable name="orderDate" select="xs:date(OrderDate)"/>
<xsl:if test="$orderDate >= current-date() - xs:dayTimeDuration('P30D')">
  <RecentOrder>Yes</RecentOrder>
</xsl:if>
```

**Listing 3.53** XSLT Conditional Expression Evaluation into Variable

This enables dynamic filtering and conditional processing based on date criteria.

You can handle strings in XSLT with a broad suite of functions and operators:

- **Concatenation**
  You can join strings by using `concat()` or the `||` string concatenation operator (this is an XSLT 3.0 feature).
  `<xsl:value-of select="concat(Customer/firstName, ' ', Customer/lastName)"/>`

- **Substring extraction**
  You can use `substring(string, start, length)` to trim or cut out parts.

- **String replacement**
  The `replace()` function allows regular-expression-based text substitution.
  `<xsl:value-of select="replace(ProductName, '\s+', ' ')"/>`

- **String length**
  The `string-length()` function returns character count.

**3**

- **Normalization and trimming**
  Combining `normalize-space()` trims leading and trailing whitespace and collapses internal spaces.
- **Uppercase and lowercase**
  The `upper-case()` and `lower-case()` functions convert text to uniform casing.

For example, you can format a customer name by trimming spaces and capitalizing, as follows:

```
<xsl:variable name="rawName" select="Customer/name"/>
<xsl:value-of select="upper-case(normalize-space($rawName))"/>
```

Often, logic must vary based on data values, and that requires conditions around calculations and string manipulation. You can use `xsl:choose` or XPath's `if-then-else` inline expressions for this. Listing 3.54 shows an example of how to apply a discount if the purchase amount exceeds a threshold.

```
<xsl:variable name="amount" select="number(Order/amount)"/>
<xsl:variable name="discountedAmount" select="
  if ($amount &gt; 100)
  then $amount * 0.9
  else $amount
"/>
<xsl:value-of select="format-number($discountedAmount, '#0.00')"/>
```

**Listing 3.54** XSLT Variable with Choose Block Evaluation

The best practices for calculations and string operations in Cloud Integration XSLT are as follows:

- Always cast values explicitly. Use `number()`, `xs:date()` to avoid type mismatch errors during transformations.
- Use variables liberally to store intermediate values for readability and avoid repeating expensive expressions.
- Test formatting functions across sample inputs to ensure output matches target system expectations.
- Limit complex string processing to what's necessary; when transformations grow too intricate, consider whether scripting might be warranted.
- Combine conditional logic with calculations to keep your mappings flexible and maintainable.

### 3.4.7   Handling Repeating or Grouped Data

In Cloud Integration scenarios, XML often contains repeating structures: multiple `<Item>` nodes within an `<Order>`, lists of `<Customer>` records, or sets of `<Transaction>`

entries. You'll need to process these repeats systematically, sometimes preserve them as-is, reorganize them, or transform them into aggregated/grouped formats. You'll need to understanding how to iterate over, selectively process, and restructure repeating data in XSLT so you can build robust mappings.

In enterprise integrations, payloads are frequently generated from relational datasets with one-to-many relationships. Think orders with many items, customers with multiple addresses, or invoices with multiple tax lines. You'll need to do the following with these repeating structures:

- Preserve them in their entirety for the receiver.
- Conditionally filter them to remove irrelevant repeats.
- Group and transform them to meet reporting or batch-processing needs.
- Flatten them into a different structural layout.

If you handle these repeating structures incorrectly, you can end up with output that contains duplicated, missing, or incorrectly ordered content, and that will lead to downstream processing errors.

The most direct approach to processing repeating nodes is `<xsl:for-each>`, as shown in Listing 3.55.

```
<xsl:for-each select="Order/Item">
  <Product>
    <xsl:value-of select="product"/>
  </Product>
</xsl:for-each>
```

**Listing 3.55** Simple XSLT Snippet for Repetitive Values

You use `select` to target all matching nodes (sequence) in document order. The loop body executes once for each node in sequence, and the context node changes to the current iteration element inside the loop.

When logic is more complex or needs to be modularized, `<xsl:apply-templates>` with specific `<xsl:template>` matches provides cleaner design, as shown in Listing 3.56.

```
<xsl:apply-templates select="Order/Item"/>
<xsl:template match="Item">
  <Product>
    <xsl:value-of select="product"/>
  </Product>
</xsl:template>
```

**Listing 3.56** XSLT Template Repetition Logic

This approach separates the iteration (apply-templates) from its processing definition (the matching template) to enable reuse and clearer structure.

Real-world XML often has nested repeats, such as the one shown in Listing 3.57.

```
<Customer>
  <Orders>
    <Order>
      <Item>...</Item>
      <Item>...</Item>
    </Order>
    <Order>...</Order>
  </Orders>
</Customer>
```

**Listing 3.57** XML Source Structure with Repeating Order/Item Constructs

When looping through nested repeats, ensure you manage context changes carefully, as shown in Listing 3.58.

```
<xsl:for-each select="Customer/Orders/Order">
  <OrderBlock>
    <xsl:for-each select="Item">
      <Product><xsl:value-of select="."/></Product>
    </xsl:for-each>
  </OrderBlock>
</xsl:for-each>
```

**Listing 3.58** XSLT Logic for Nested Item Looping

Each <xsl:for-each> resets the context to the current node in that level's repeat.

Sometimes, you have repeats that you want to aggregate by key values. You can combine <xsl:for-each-group> from Section 3.4.5 with iteration to create meaningful groups, as shown in Listing 3.59.

```
<xsl:for-each-group select="Order/Item" group-by="category">
  <Category name="{current-grouping-key()}">
    <xsl:for-each select="current-group()">
      <Name><xsl:value-of select="product"/></Name>
    </xsl:for-each>
  </Category>
</xsl:for-each-group>
```

**Listing 3.59** XSLT Alternative for Group-Based Nested Loop

By default, node lists iterate in document order. You can use <xsl:sort> inside loops to set custom sorting, as shown in Listing 3.60.

```
<xsl:for-each select="Order/Item">
  <xsl:sort select="price" data-type="number" order="descending"/>
  <Product><xsl:value-of select="product"/></Product>
</xsl:for-each>
```

**Listing 3.60** Listing 1.60: XSLT Snippet with Sorting Subsection

Sorting is applied before the loop body executes for each node.

---

**Avoiding Common Mistakes**

Some common mistakes to avoid are as follows:

- **Not resetting context**
  If you try selecting relative paths without considering current context, you may unintentionally select from the wrong level in the hierarchy.
- **Using absolute paths in loops**
  This can cause the same nodes to be processed multiple times if the path always starts at /.
- **Ignoring** position()
  For tasks like numbering or conditional formatting, position() inside a loop gives the current iteration index.

---

Let's look at an example of flattening nested repeats into a single list. Listing 3.61 shows the source XML that contains the actual useful items that are nested within order blocks, which again are nested in a parent list of orders. Listing 3.62, on the other hand, presents a much simpler structure and also omits the intermediate order grouping.

```
<Orders>
  <Order id="1">
    <Item>A</Item>
    <Item>B</Item>
  </Order>
  <Order id="2">
    <Item>C</Item>
  </Order>
</Orders>
```

**Listing 3.61** Given Nested Source XML Structure Example

```
<AllItems>
  <Item>A</Item>
  <Item>B</Item>
  <Item>C</Item>
</AllItems>
```

**Listing 3.62** Desired Flat Target XML Structure Example

To convert the source into the corresponding target structure by using XSLT, you can define a template and match based on the sources node's XPaths. The necessary transformation logic is presented in Listing 3.63, which uses the match and select operations to extract the content into the designated elements on the target side.

```
<xsl:template match="Orders">
  <AllItems>
    <xsl:for-each select="Order/Item">
      <xsl:copy-of select="."/>
    </xsl:for-each>
  </AllItems>
</xsl:template>
```

**Listing 3.63** XSLT Snippet That Flattens Nested Items

Let's look at a second example where we have repeating data with aggregation. Listing 3.64 shows our source XML structure, including a collection of transactions that all have repeated but the same child attributes.

```
<Sales>
  <Transaction>
    <region>East</region>
    <amount>100</amount>
  </Transaction>
  <Transaction>
    <region>East</region>
    <amount>200</amount>
  </Transaction>
</Sales>
```

**Listing 3.64** XML Parent-Child Snippet

Our goal is to group our data by region and to sum the amount, as shown in Listing 3.65.

```
<xsl:for-each-group select="Transaction" group-by="region">
  <Region name="{current-grouping-key()}">
    <Total>
      <xsl:value-of select="sum(current-group()/amount)"/>
    </Total>
  </Region>
</xsl:for-each-group>
```

**Listing 3.65** LisXSLT Snippet with Looped/Repeated Grouping

Some best practices in this space are as follows:

- Use `<xsl:apply-templates>` for cleaner, reusable logic where possible.
- Keep iterations focused, and filter records early with predicates.

- Combine with grouping and sorting to enrich datasets while iterating.
- Benchmark on large payloads because nested loops can grow expensive quickly.
- Ensure that your XPath selects exactly the repeat level you want, to avoid redundancy.

### 3.4.8   Dynamic Header and Property Assignment Using XSLT

While XSLT's main purpose in Cloud Integration is to transform XML payloads, it can also play a vital role in shaping message headers and exchanging properties dynamically. This allows you to influence routing, logging, and integration logic downstream—all based on values that are extracted or computed from the payload during mapping. In many iFlows, headers and properties are used to do the following:

- Drive conditional routing (e.g., send message to different endpoints based on country or order type).
- Set technical metadata (e.g., correlation IDs, filenames, content types).
- Pass computed business values to later steps without modifying the payload.
- Support decision-making logic in routers, scripts, and further mappings.

When implemented efficiently, dynamic assignment ensures cleaner flow design, as business or routing logic is derived directly from the data at runtime.

In Cloud Integration, the XSLT mapping step can both output the transformed payload and set header or property values via special XML structures. You achieve this by producing an envelope-like output that Cloud Integration interprets.

The specifics are as follows:

- The XSLT produces an XML payload for the main message body.
- Optionally, the XSLT can also generate XML fragments in the message header or exchange property sections when using the Cloud Integration-specific XML format.
- This is most often used in combination with the XSLT mapping step's *message attachments and headers* capability.

As we just mentioned, Cloud Integration supports certain patterns in which you can set dynamic values by using exchange property or header schemas as part of the transformation result. Listing 3.66 shows an example with output structure, with separate sections for headers and body (pseudo-format).

```
<Message>
  <Headers>
    <Header name="TargetSystem">ERP_Germany</Header>
    <Header name="FileName">Order_4711.xml</Header>
  </Headers>
  <Body>
```

```
    <Order> ... transformed order content ... </Order>
  </Body>
</Message>
```

**Listing 3.66** Pseudo Message Payload Structure

You can configure subsequent integration steps (e.g., a content modifier) to extract these header nodes into actual Cloud Integration message headers.

### Dynamic Property Assignment Example

Now, let's look at a dynamic property assignment. Imagine you need to store a computed IsUrgent flag for later decision making, as shown in Listing 3.67.

```
<xsl:variable name="priorityFlag" select="
  if (number(/Order/amount) > 1000) then 'true' else 'false'
"/>
<Property name="IsUrgent">
  <xsl:value-of select="$priorityFlag"/>
</Property>
```

**Listing 3.67** XSLT Snippet for Variable-Based Condition

This property can then be read in a router step to direct urgent messages to a priority processing channel.

In real projects, you often need to transform the payload and set metadata in one step. Let's walk through an example scenario with the following goals:

- Convert an incoming <Order> XML to <SalesOrder> format.
- Set a FileName header based on the order ID and the current timestamp.
- Set a Region property from the customer address.

Listing 3.68 shows the pseudo XSLT that accomplishes these goals.

```
<xsl:variable name="orderId" select="/Order/orderId"/>
<xsl:variable name="timestamp" select="format-dateTime(current-dateTime(), '[Y][
M01][D01][H01][m01][s01]')"/>

<Headers>
  <Header name="FileName">
    <xsl:value-of select="concat('Order_', $orderId, '_', $timestamp, '.xml')"/>
  </Header>
</Headers>

<Properties>
  <Property name="Region">
```

```
    <xsl:value-of select="/Order/customer/region"/>
  </Property>
</Properties>

<Body>
  <SalesOrder>
    <!-- actual payload transformation -->
  </SalesOrder>
</Body>
```

**Listing 3.68** XSLT Example: Payload Transformation and Metadata Assignment

These generated header/property nodes can then be extracted post-XSLT.

The best practices in this space are as follows:

- Keep payload and metadata logic clear. Separate your payload transformation templates from metadata logic to ensure maintainability.

- Avoid hardcoding where possible. Use dynamic XPath selections and variables to adapt header/property values from the running message.

- Test with full Cloud Integration flow. Some Cloud Integration–specific behavior for header/property mapping is only visible when running in a deployed iFlow, not in local XSLT testing.

- Leverage date/time formatting carefully. Ensure that time zones and formats meet the requirements of the target system.

- Validate for null values. Before setting headers/properties from payload data, check whether data exists to avoid empty or misleading values in message metadata.

## 3.5   Practical Design Journey

Let's start with our scenario: you receive a nested XML purchase order document from an e-commerce platform, and for legacy ERP ingestion, you must convert it into a flat XML structure and map nested items and attributes to simple field names.

The transformation must do the following:

- Map from nested list structures to flattened elements.
- Split a complex address into individual fields.
- Calculate a "total price" field for each line item.
- Use a UDF to format customer names consistently (e.g., by converting "JANE DOE" or "jane doe" to "Jane Doe").

Your mapping requirements are as follows:

- Flatten into XML with these elements: `OrderID`, `CustomerName`, `Street`, `City`, `ZIP`, `ProductID`, `Quantity`, `UnitPrice`, and `TotalPrice`.
- Format `CustomerName` by using a UDF to ensure consistent capitalization.
- `TotalPrice` = `Quantity` × `UnitPrice`.

In the following sections, we'll walk through the inbound data format, the targeted outbound data format, and how to create the mapping.

### 3.5.1    Inbound Data Format

The e-commerce platform provides the purchase order data in a structured and nested XML dataset. Given the example input data in Listing 3.69, which is described by the XSD description from Listing 3.70, we can paste the XSD content into a definition file. We'll return to this file later to upload it into the Message Mapping panel to represent the source side.

```
<PurchaseOrder>
    <OrderID>12345</OrderID>
    <Customer>
        <Name>jane DOE</Name>
        <Address>
            <Street>123 Main Street</Street>
            <City>Berlin</City>
            <ZIP>10115</ZIP>
        </Address>
    </Customer>
    <Items>
        <Item>
            <ProductID>P001</ProductID>
            <Quantity>2</Quantity>
            <UnitPrice>50</UnitPrice>
        </Item>
        <Item>
            <ProductID>P002</ProductID>
            <Quantity>1</Quantity>
            <UnitPrice>100</UnitPrice>
        </Item>
    </Items>
</PurchaseOrder>
```

**Listing 3.69**  Design Journey: Graphical Mapping Input Dataset

```
<xs:schema xmlns:xs="http://www.w3.org/2001/XMLSchema">
    <xs:element name="PurchaseOrder">
        <xs:complexType>
            <xs:sequence>
                <xs:element name="OrderID" type="xs:string"/>
                <xs:element name="Customer">
                    <xs:complexType>
                        <xs:sequence>
                            <xs:element name="Name" type="xs:string"/>
                            <xs:element name="Address">
                                <xs:complexType>
                                    <xs:sequence>
                                        <xs:element name="Street" type="xs:string"/>
                                        <xs:element name="City" type="xs:string"/>
                                        <xs:element name="ZIP" type="xs:string"/>
                                    </xs:sequence>
                                </xs:complexType>
                            </xs:element>
                        </xs:sequence>
                    </xs:complexType>
                </xs:element>
                <xs:element name="Items">
                    <xs:complexType>
                        <xs:sequence>
                            <xs:element name="Item" maxOccurs="unbounded">
                                <xs:complexType>
                                    <xs:sequence>
                                        <xs:element name="ProductID" type="xs:string"/>
                                        <xs:element name="Quantity" type="xs:integer"/>
                                        <xs:element name="UnitPrice" type="xs:decimal"/>
                                    </xs:sequence>
                                </xs:complexType>
                            </xs:element>
                        </xs:sequence>
                    </xs:complexType>
                </xs:element>
            </xs:sequence>
        </xs:complexType>
    </xs:element>
</xs:schema>
```

**Listing 3.70** Design Journey: XSD Description of Input Data

### 3.5.2 Targeted Outbound Data Format

The legacy ERP system, on the other hand, needs a slightly adjusted XML structure. While the order positions still occur as a list, the orders head data concerning the customer details need to be flattened into the purchase order itself. You can see this in Listing 3.71, and the corresponding structural XSD description is depicted in Listing 3.72. The XSD is also necessary in file format for uploading to the mapping editor that's to be used on the target side.

```xml
<PurchaseOrder>
    <OrderID>12345</OrderID>
    <CustomerName>Jane Doe</CustomerName>
    <Street>123 Main Street</Street>
    <City>Berlin</City>
    <ZIP>10115</ZIP>
    <OrderLines>
        <Line>
            <ProductID>P001</ProductID>
            <Quantity>2</Quantity>
            <UnitPrice>50</UnitPrice>
            <TotalPrice>100</TotalPrice>
        </Line>
        <Line>
            <ProductID>P002</ProductID>
            <Quantity>1</Quantity>
            <UnitPrice>100</UnitPrice>
            <TotalPrice>100</TotalPrice>
        </Line>
    </OrderLines>
</PurchaseOrder>
```

**Listing 3.71** Design Journey: Desired Outbound Example Data

```xml
<xs:schema xmlns:xs="http://www.w3.org/2001/XMLSchema">
    <xs:element name="PurchaseOrder">
        <xs:complexType>
            <xs:sequence>
                <xs:element name="OrderID" type="xs:string"/>
                <xs:element name="CustomerName" type="xs:string"/>
                <xs:element name="Street" type="xs:string"/>
                <xs:element name="City" type="xs:string"/>
                <xs:element name="ZIP" type="xs:string"/>
                <xs:element name="OrderLines">
                    <xs:complexType>
                        <xs:sequence>
```

```
                    <xs:element name="Line" maxOccurs="unbounded">
                        <xs:complexType>
                            <xs:sequence>
                                <xs:element name="ProductID" type="xs:string"/>
                                <xs:element name="Quantity" type="xs:integer"/>
                                <xs:element name="UnitPrice" type="xs:decimal"/>
                                <xs:element name="TotalPrice" type="xs:decimal"/>
                            </xs:sequence>
                        </xs:complexType>
                    </xs:element>
                </xs:sequence>
            </xs:complexType>
        </xs:element>
    </xs:sequence>
</xs:complexType>
</xs:element>
</xs:schema>
```

**Listing 3.72** Design Journey: Target Data Format XSD

### 3.5.3   Creating the Mapping

To start modeling the mapping, you need to add a message mapping into your integration package. To begin with, you need to switch to edit mode and click the **Add** button in the **Artifacts** tab, as shown in Figure 3.2.

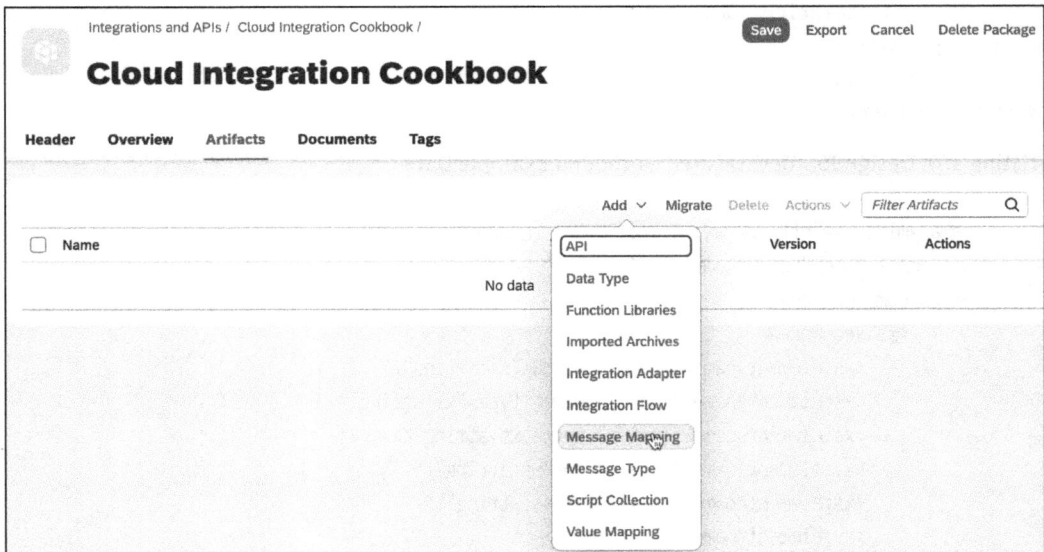

**Figure 3.2** Design Journey: Adding Message Mapping in Integration Package

When you select the **Message Mapping** entry, you'll be prompted with a dialog (see Figure 3.3) in which you need to add a **Name** and **ID** for your mapping. In this case, you input "PurchaseOrder_eCommerce_ERP-legacy" as a name and leave the **ID** as prefilled by syncing to the name. Finally, you submit the dialog by clicking on the **Add and Open in Editor** button. Alternatively, you could use the plain **Add** button, but that would require you to open the mapping manually because you want to modify it anyway.

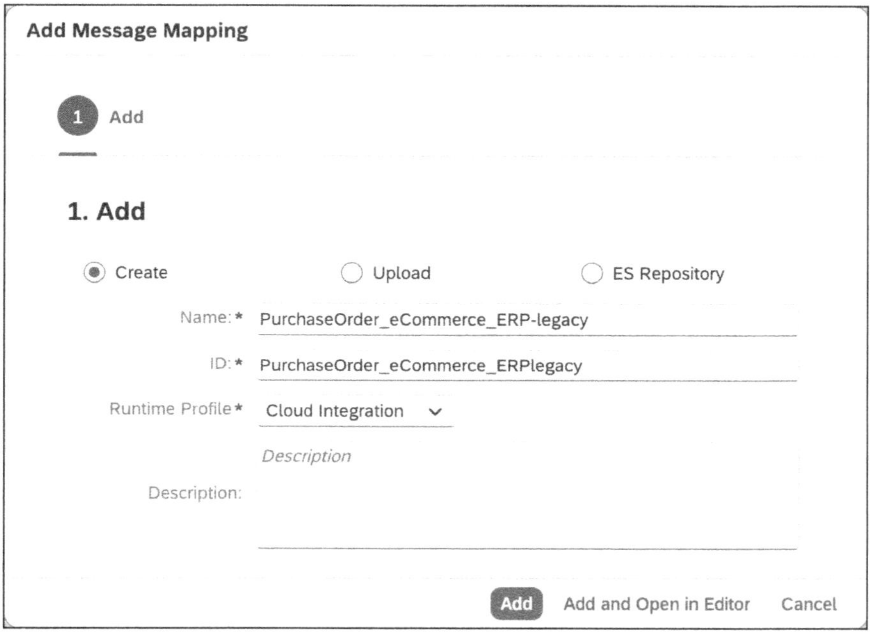

**Figure 3.3** Design Journey: Creating Message Mapping

After the dialog closed itself, you'll be navigated to the mapping editor with your newly created mapping already opened (see Figure 3.4). The editor will be in **View Only** mode and won't allow changes, so you'll need to switch to edit mode by clicking the **Edit** button in the top row. Notice that there aren't any mapping resources (see the **References** section at the left of the screen). You need these resources for the source and target panes to display your message structure for you to connect the input and output fields to each other.

To mitigate the lack of resources, you can upload them into the references section, which will make them available locally for this specific message mapping. To do so, click the **Upload** button (see Figure 3.5) to open a selection/upload dialog.

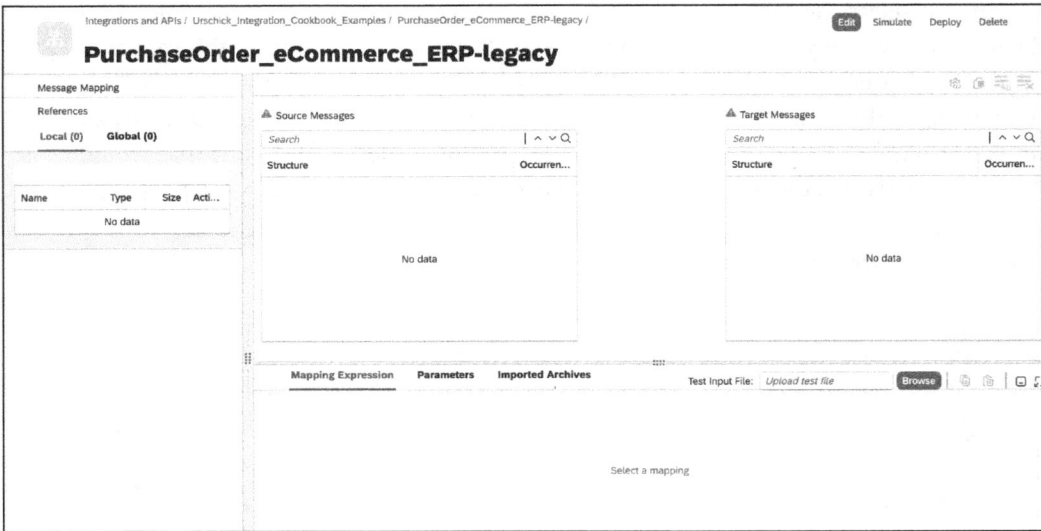

**Figure 3.4** Design Journey: Empty Message Mapping

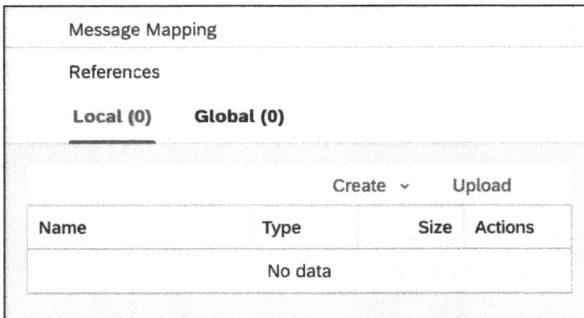

**Figure 3.5** Design Journey: XSD Resources Section

This dialog (see Figure 3.6) lets you upload different file formats. In your case, you want to add the XSD definitions from Listing 3.70 and Listing 3.72. Because it prompts you to upload files instead of the content itself, you'll need the previously mentioned XSD files. You perform the upload by clicking **Add**, and because the upload dialog only supports single files, you need to click it twice to upload the source file first and the target file second.

Once you've done all this, you'll be presented with the two definition files that are listed in the resource section (see Figure 3.7).

The next step toward completing your mapping is to select the source and target structures inside the editor. Because there are no compositions selected on either side, you need to specify them, and you can do this by clicking on the **Add source message** and **Add target message** buttons, as depicted in Figure 3.8.

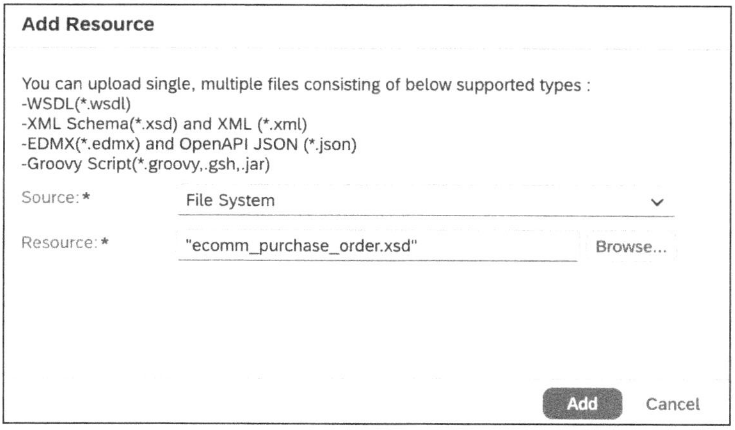

**Add Resource**

You can upload single, multiple files consisting of below supported types :
-WSDL(*.wsdl)
-XML Schema(*.xsd) and XML (*.xml)
-EDMX(*.edmx) and OpenAPI JSON (*.json)
-Groovy Script(*.groovy,.gsh,.jar)

Source: *     File System                                                    ⌄

Resource: *     "ecomm_purchase_order.xsd"                    Browse...

[ Add ]   Cancel

**Figure 3.6**  Design Journey: Uploading XSD

| Name | Type | Size | Actions |
|------|------|------|---------|
| ⌄   Schemas (2) | | | |
|     ecomm_purchase_order | XSD | 2 KB | ↓ |
|     legacy_erp_purchase_order | XSD | 1 KB | ↓ |

**Figure 3.7**  Design Journey: Uploaded XSD Resources

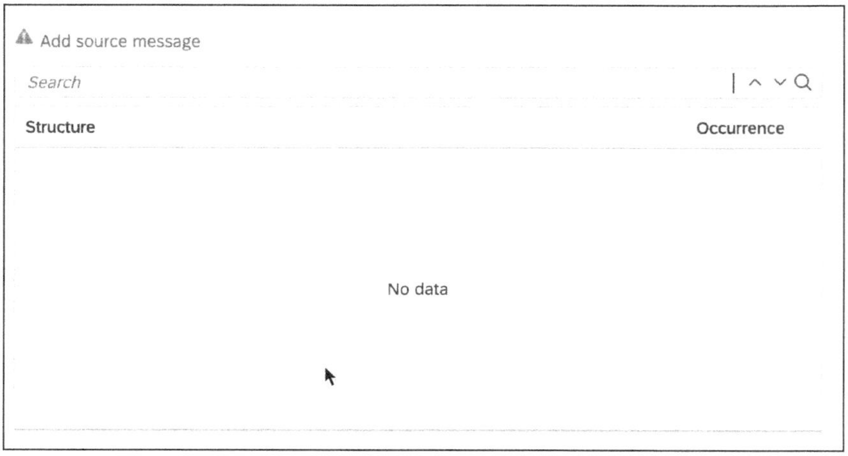

⚠ Add source message

Search                                                          |  ∧  ∨  Q

**Structure**                                                    Occurrence

No data

**Figure 3.8**  Design Journey: Empty Source Message Panel

Next, you'll be prompted with a selection dialog that shows your previously uploaded XSD files (see Figure 3.9). For the source side, you want to select the **ecomm_purchase_order.xsd** (which represents the format of the e-commerce platform where the message originates from) and confirm by clicking **OK**. You also need to do this for the target side, where you'll select **legacy_erp_purchase_order.xsd**.

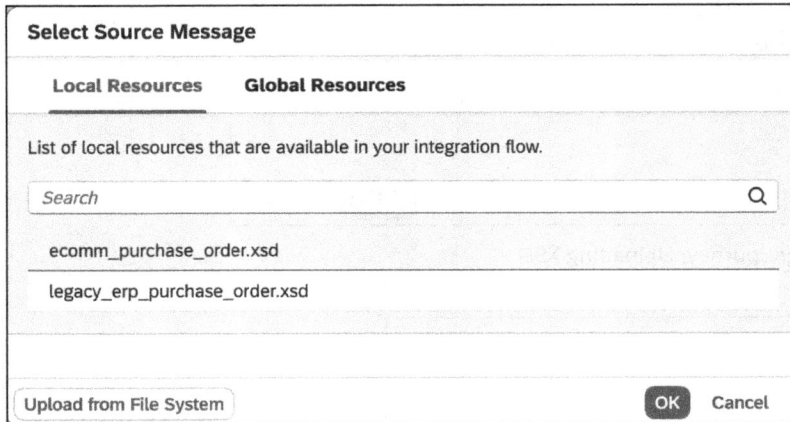

**Select Source Message**

**Local Resources**      **Global Resources**

List of local resources that are available in your integration flow.

Search                                                          Q

ecomm_purchase_order.xsd

legacy_erp_purchase_order.xsd

Upload from File System                    OK   Cancel

**Figure 3.9**  Design Journey: Selecting Source and Target Definitions

Once you've finished assigning the corresponding message compositions to each side of the mapping editor, you'll be presented with each message structure (see Figure 3.10). Notice that no fields are connected to each other yet. This is where the real magic happens. Since it's an interactive and graphic editor, you can leverage the drag-and-drop feature of our mouse to connect the fields to each other. It's also worth mentioning that the bright-red glowing elements on the target side indicate a missing value. You can also see this in the warning icon next to the target side's **PurchaseOrder** text at the top of the list.

You can start with the first item on your purchase order—the **OrderID**—by clicking on the left-hand side of the source panel, dragging it to the right-hand side (the target panel), and releasing it. This creates a connection line (see Figure 3.11) that indicates a field mapping. It also opens the field function panel at the bottom of the same screen, where you can introduce field operations, transformations, and other functions that will potentially modify the outcoming content.

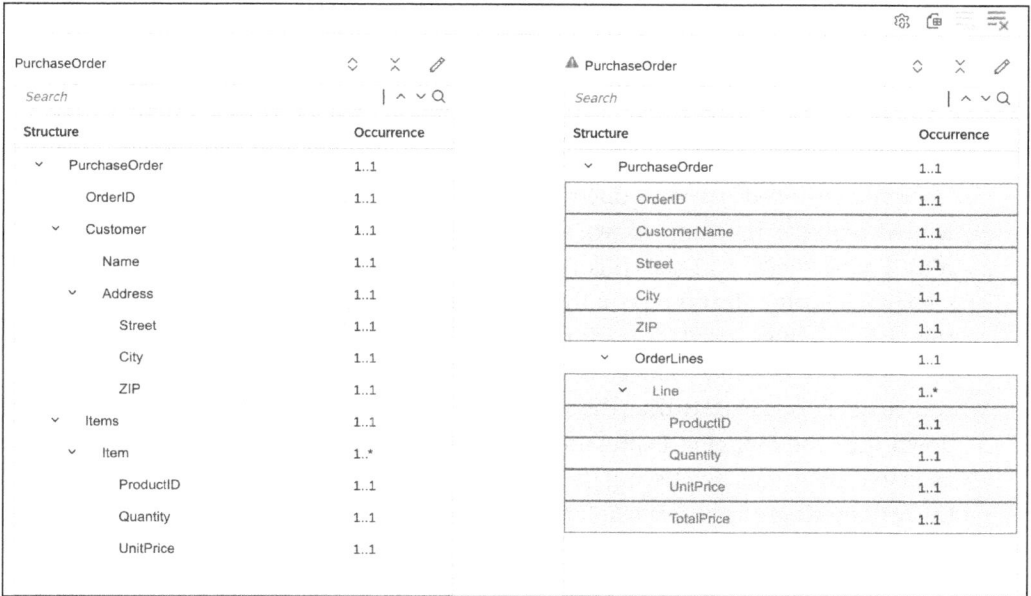

**Figure 3.10** Design Journey: Unmapped Field Definitions from Source to Target

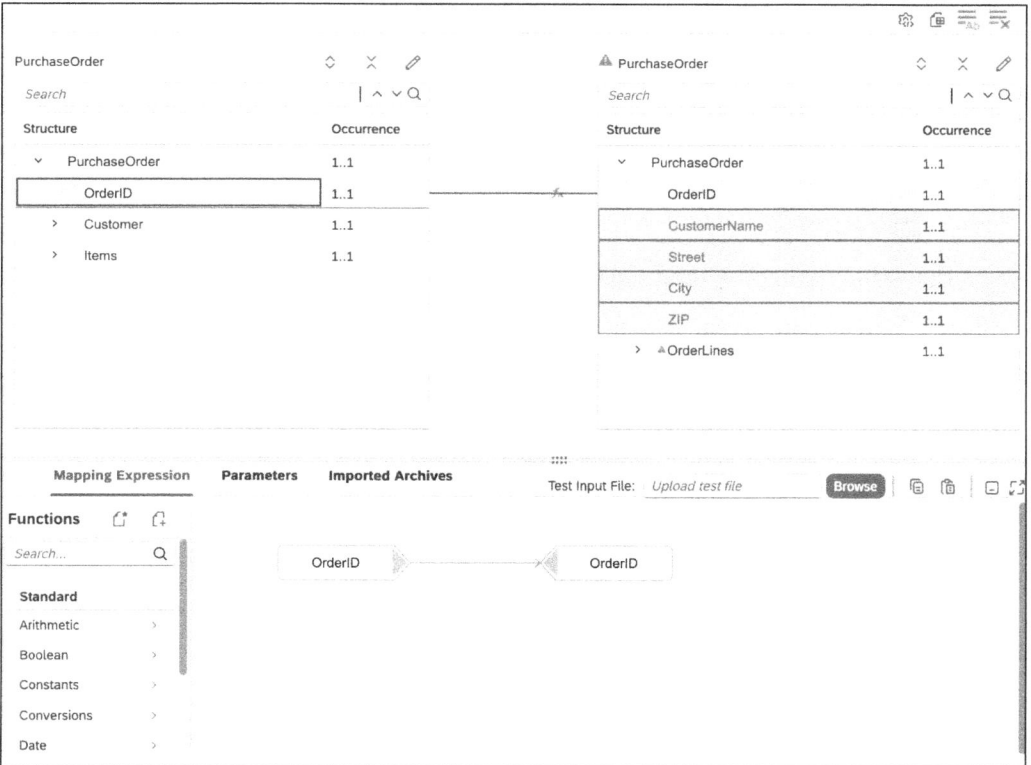

**Figure 3.11** Design Journey: First Field Mapping of OrderID

After your first successful field mapping, you'll see how the **OrderID** of the target side changed color to white color to indicate a value connection. Then, you can continue with the other fields, connecting each source item to its correlated target equivalent. You can skip the **Name/CustomerName** connection for now, since it requires a custom function that you'll tackle soon.

Once you've linked all simple one-to-one fields, you can go to the **TotalPrice** target field, which combines the value of two source values—the **Quantity** and the **UnitPrice**—by multiplying them. To make this function happen, you need to drag one of the two source fields onto the target—in this case, you take the **Quantity**—and click the **fx** button on the connection. This loads the **Mapping Expression** editor at the bottom of the screen and opens the connected function flow. Since it only shows one of the two source fields, you need to drag the second field—the **UnitPrice**—from the left-hand side down into the **Mapping Expression** whitespace. Then, you'll have both input fields available, but you'll be missing the multiplication. You can find this procedure under the **Arithmetic** group and drag it into the canvas. After you have all the necessary nodes for your workflow, you just need to connect them to each other as shown in Figure 3.12—and you'll have a working mathematical transformation mapping.

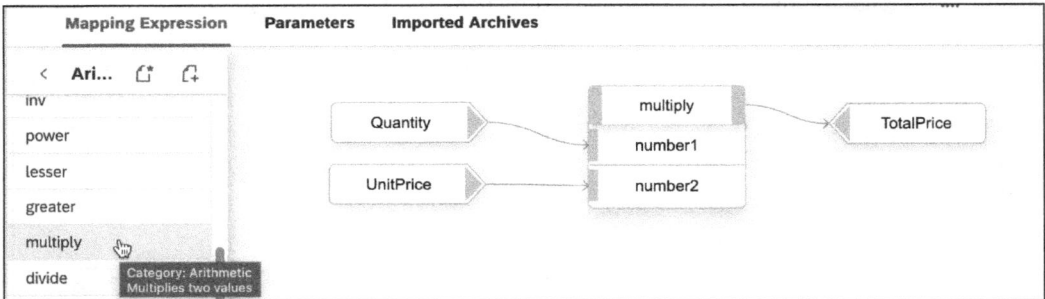

**Figure 3.12** Design Journey: Total Price Mapping Function

The last piece of your message mapping puzzle is to create and connect the UDF for normalizing the customer's name. To introduce this function, start with a simple one-to-one connection to the **CustomerName** and open the function editor by clicking on the **fx** button again. After focusing on the **Mapping Expression** section, you create a UDF by clicking on the **Create** button, as depicted in Figure 3.13.

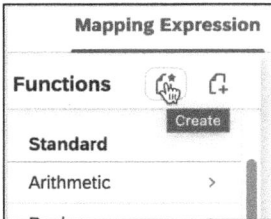

**Figure 3.13** Design Journey: Creating UDF

This will open a dialog where you can name the connected Groovy script that contains the function implementation (see Figure 3.14). You continue the generation of the script by filling in the **Name** field with a meaningful name for the script—in this case, "trimAndFormatCustomerName"—and you continue by clicking **OK**.

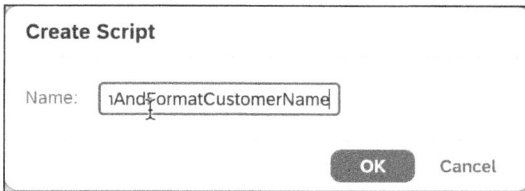

**Figure 3.14** Design Journey: Inputting UDF Script Name

The UDF script you create will then be opened in the Groovy editor and prefilled with some demo function code. You can replace the entire file's content with the Groovy snippet that's mentioned in Listing 1.73. The procedure defines three parameters: name, which is a list of input strings; output, which is the UDFs output content; and the mapping context, which is not used in this case.

```
import com.sap.it.api.mapping.*;

public void formatName(String[] name, Output output, MappingContext context) {
    if (name == null || name.length == 0 || name[0] == null) {
        output.addValue("");
        return;
    }
    String rawName = name[0].trim().toLowerCase();
    StringBuilder formatted = new StringBuilder();
    for (String part : rawName.split(" ")) {
        if (!part.isEmpty()) {
            formatted.append(Character.toUpperCase(part.charAt(0)))
                    .append(part.substring(1))
                    .append(" ");
        }
    }
    output.addValue(formatted.toString().trim());
}
```

**Listing 3.73** Design Journey: UDF for Name Trimming and Reformatting

You can save the script by clicking the **OK** button (see Figure 3.15), and then, you can close the script editor and navigate back to the mapping view.

PurchaseOrder_eCommerce_ERP-legacy / trimAndFormatCustomerName.groovy /       OK     Cancel     ⑦

# trimAndFormatCustomerName.groovy

```
1   import com.sap.it.api.mapping.*;
2
3 ▾ public void formatName(String[] name, Output output, MappingContext context) {
4 ▾     if (name == null || name.length == 0 || name[0] == null) {
5           output.addValue("");
6           return;
7       }
8       String rawName = name[0].trim().toLowerCase();
9       StringBuilder formatted = new StringBuilder();
10 ▾    for (String part : rawName.split(" ")) {
11 ▾        if (!part.isEmpty()) {
12              formatted.append(Character.toUpperCase(part.charAt(0)))
13                  .append(part.substring(1))
14                  .append(" ");
15          }
16      }
17      output.addValue(formatted.toString().trim());
18  }
19
```

**Figure 3.15** Design Journey: Creating UDF Function Code

Then, back on the mapping panel, you can shift your focus to the **Mapping Expression** view of the **Name/CustomerName** connection. To incorporate your UDF into the field transformation, you follow the same principle as with the **TotalPrice** workflow, with the only differences being that you use only one input field and you find the mapping function at the bottom of the section list, below the **Custom** header in the group that's named like your Groovy script. Next, you add the **formatName** item from your custom group to the canvas and connect the nodes as presented in Figure 3.16.

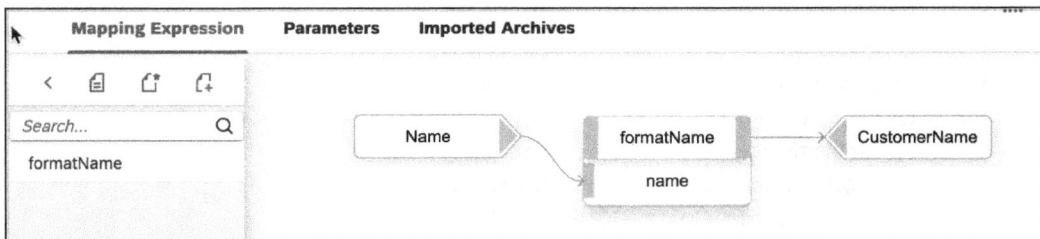

| **Mapping Expression** | **Parameters** | **Imported Archives** | | |
|---|---|---|---|---|

```
<   ⊟   ⟳*   ⟳+
Search...            Q          Name  ▷          formatName  ──×  CustomerName
formatName                                          name
```

**Figure 3.16** Design Journey: Introducing Naming UDF in Field Mapping

With all fields mapped on the target side—and in our case, the source side as well—our mapping is finished, and it shows all connections (see Figure 3.17) from left to right, without any errors or warnings.

In this practical design journey, we tackled a common enterprise integration scenario: transforming incoming, nested XML purchase orders into a flattened XML format that's suitable for legacy ERP ingestion. The input XML includes hierarchical customer and item structures, while the output requires that all relevant data—order, customer, address, and line items—be presented in a straightforward, flat document.

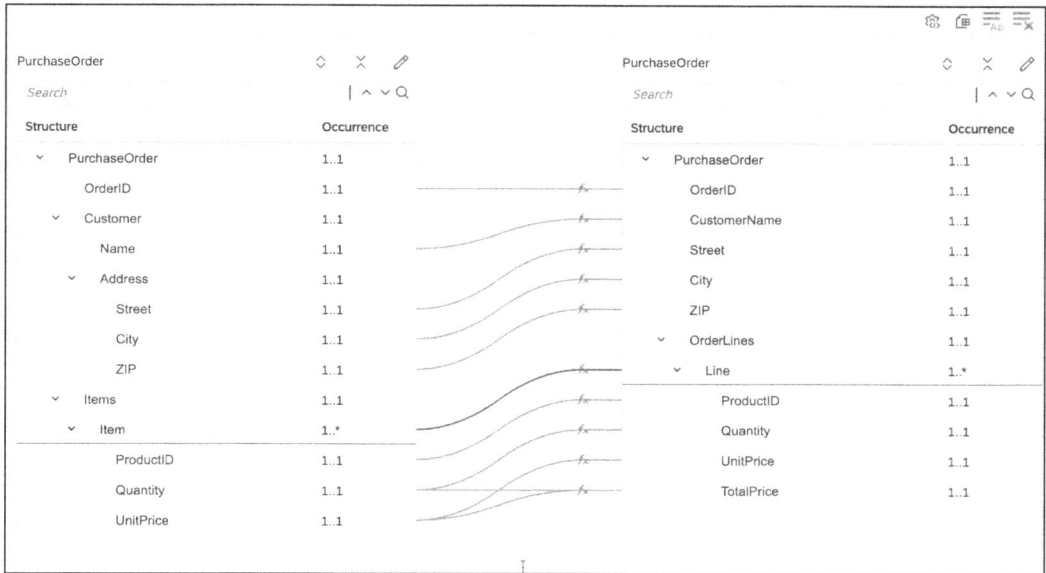

**Figure 3.17** Design Journey: Completed Mapping

This example showcases how message mapping, combined with UDFs, empowers integration designers to deliver clean, normalized, and value-added data and thus bridges the gap between technical requirements and real-world business outcomes.

## 3.6  Summary

This chapter presented the key concepts, tools, and practical considerations involved in Groovy scripting, graphical message mapping, and XSLT mapping. It also showed you how to effectively combine them. Groovy scripting is the procedural, free-form logic layer of Cloud Integration, and it lets developers implement transformation and control logic that they can't efficiently express by using purely declarative tools such as graphical mapping and XSLT. Groovy operates within the JVM to provide access to the full Java ecosystem (within SAP's runtime restrictions) and deep integration with the Cloud Integration message API.

Graphical message mapping in Cloud Integration offers a model-driven approach to transforming XML payloads, so it's highly accessible to developers and consultants who don't have deep programming expertise. The mapping editor visualizes source and target structures and thereby enables field-to-field connections with drag-and-drop actions.

XSLT is a W3C standard that's designed for XML document transformation. In Cloud Integration, XSLT mapping is a declarative, rule-based alternative to graphical mapping, and it's particularly suited to template-driven or deeply recursive XML transformations.

The design journey you went on in this chapter makes one point abundantly clear: no single mapping method—be it Groovy scripting, graphical message mapping, or XSLT mapping—is universally superior in all scenarios. Each method offers unique strengths and, inevitably, certain limitations. You must therefore deliberately choose a tool, based on the specific technical and organizational requirements of your integration project.

# Chapter 4
# Adapters

*Adapters are the backbone of connectivity in SAP Integration Suite.*

Nowadays, in the digital age, companies must seamlessly integrate numerous systems, applications, and data sources. To help them achieve this, SAP Integration Suite connects heterogeneous landscapes, designs end-to-end processes, and implements future-proof integration scenarios—and adapters are a central element of this integration architecture. They form the bridge between a wide variety of technologies, protocols, and formats, and they're essential for modern integration scenarios. They handle communication among systems, transform data, guarantee security standards, and enable reliable connections across system boundaries.

An *adapter* is a special software component within SAP Integration Suite (Cloud Integration capability) that serves as an interface for connecting different systems and applications. Adapters enable communication between SAP and non-SAP systems, both in the cloud and on-premise, by supporting different protocols, formats, and technologies. They're a critical connection between SAP Integration Suite and external applications, data sources, and services.

Some key features and functions of adapters are as follows:

- **Communication interface**
  Adapters facilitate the transfer of data and processes among disparate systems, irrespective of those systems' local or cloud-based operation.

- **Predefined connectors**
  There are a variety of standardized adapters for widely used systems and services, such as IDoc, SAP SuccessFactors, SAP Ariba, X (formerly Twitter), and Facebook.

- **Flexibility**
  Adapters abstract technical details, such as protocols, authentication mechanisms, and data formats, to make it easier to integrate complex system landscapes.

- **Extensibility**
  If you have special requirements, you can develop your own adapters by using SAP's Adapter SDK.

The adapter landscape of SAP Integration Suite is the key building block for system integration in hybrid landscapes. Through adapters, it seamlessly connects cloud offerings like SAP S/4HANA Cloud, on-premise systems like SAP S/4HANA, and non-SAP applications like ServiceNow.

Nowadays, standardized communication is essential, and many adapters support industry standards such as HTTP, OData, SOAP, and SFTP. This simplifies integration and increases reusability. Security and compliance are also vital, and adapters provide essential options for authentication, encryption, and access control in security-critical integration projects. By using adapters, you can build the foundation of a future-proof integration, and SAP's continuous development of adapters ensures seamless integration of new technologies and protocols.

This chapter provides a comprehensive review of the most critical adapter groups in SAP Integration Suite. It offers practical use cases, in-depth technical background information, and best practices derived from real customer projects. The structure follows a pragmatic approach based on typical requirements in SAP projects.

In Section 4.1, we provide a systematic introduction to the adapter concept. What types of adapters are there? You need to know how they work technically. What are the differences between sender and receiver adapters? You need to know which adapters support synchronous communication and which work purely asynchronously. We also explain how adapters are configured in the SAP Integration Suite cockpit and which advanced options are available, such as timeout handling, header manipulation, and certificate verification.

In Section 4.2, we go into file-based processing, which remains a prevalent practice in numerous industries, including mechanical engineering, logistics, and retail. We'll show you how to efficiently implement file imports and exports with SFTP, FTP, and file adapters.

Section 4.3 introduces the adapters that were developed for SAP's protocols and applications, which are at the heart many customers' IT landscape. These include the IDoc adapter, RFC adapter, and SuccessFactors adapter. We demonstrate typical integration scenarios and address challenges such as mapping complexity and performance.

Section 4.4 is dedicated to adapter types that specialize in API communication. We devoted a whole section to them because APIs are the backbone of modern cloud architectures.

Section 4.5 focuses on adapters that enable database access and messaging-based integrations. These adapters are crucial in modern microservice architectures, especially for connecting legacy systems or implementing event-driven scenarios. And with the use of ProcessDirect adapters, you can modularize iFlows.

This chapter will teach you everything you need to know about adapters and how to use them for your integration requirements. You must use the right adapter in the right place to ensure robust, maintainable, and future-proof integrations, regardless of whether they're file-based, API-driven, SAP specific, or event driven.

## 4.1   Overview of Adapters

The Cloud Integration capability of the SAP Integration Suite seamlessly integrates a wide variety of systems in hybrid IT landscapes. Adapters are essential building blocks that enable communication between SAP and non-SAP systems. From a technical perspective, adapters in cloud integration correspond to Apache Camel components. *Apache Camel* is the go-to open-source framework for implementing EIPs. Cloud integration is based on Apache Camel, and in this context, an adapter is a Camel component that acts as a sender or receiver endpoint in an iFlow. The adapter architecture is open, expandable, and extremely stable and high-performing, thanks to its Camel base — which is a decisive advantage in modern integration scenarios.

Adapters play a critical role in the integration process. They establish connections to external systems, implement protocols (e.g., REST, SOAP, IDoc, OData, SFTP), optionally convert data formats (e.g., XML, JSON, flat file, CSV, binary), and integrate authentication mechanisms such as OAuth2 and certificate authentication and security features like TLS. They seamlessly integrate into Camel-based flows using enterprise integration patterns. There's also a technical classification into Transmission Control Protocol (TCP)–based, HTTP-based, and non-TCP/non-HTTP based adapters.

Sender adapters fall into two categories: polling based and push based. *Polling-based adapters* (e.g., SFTP and mail adapters) check defined sources for new messages at regular intervals, which typically range from seconds to minutes or hours. New data is transferred to the iFlow as soon as it's detected, and this method is perfect for systems that don't support active push functionalities. On the other hand, *push-based adapters* are initiated by external calls, such as HTTP and OData senders. Here, an external system actively calls a defined endpoint to transfer a message. This method performs better because it doesn't generate unnecessary queries, but it requires corresponding trigger logic in the sending system.

A key difference between this and classic SAP PO is the handling of the message format. In PO, every incoming message is automatically transformed into the internal XI protocol. But Cloud Integration doesn't have a native standard format, binary formats are passed on unchanged, and functions such as XPath are only available if a conversion (e.g., to XML) is explicitly performed. Converters are used here to prepare messages for downstream processing.

The *OData receiver adapter* is a prime example of a highly functional adapter. It offers a wizard for configuring OData APIs, supports paging and batch processing for large amounts of data, automatically removes namespaces from XML responses, and handles token renewal for OAuth2-based authentication. These features make the adapter efficient and user-friendly.

Security is a key aspect of adapter configuration. Adapters support TLS/SSL, OAuth2, certificate authentication, basic authentication, and integration with SAP's own keystore. Payload encryption and the masking of sensitive data in monitoring are key security features.

If no suitable standard adapter is available, you can develop your own adapter. SAP's *Adapter SDK* is the solution. You do development in Java by using Apache Camel and Open Services Gateway Initiative (OSGi), and you integrate your adapters into SAP Integration Suite and operate them there. You can use them to connect legacy systems, special protocols, and user-defined security mechanisms.

To ensure successful and high-performance use of adapters in Cloud Integration, you need to adhere to key best practices.

- Use standard adapters whenever possible.
- Activate tracing and error handling at an early stage.
- Explicitly model conversion rules.
- Use performance parameters (e.g., for paging or message splitting) in a targeted manner.
- Use converters for XML-based routing, filtering, and mapping.

## 4.2   File-Based Communication

*File-based communication* plays a central role in many integration projects and remains highly relevant, despite the widespread use of APIs and event-based architectures. One reason for this is the widespread use of legacy systems that don't support modern interfaces and can only be integrated via file imports or exports. When you're collaborating with external partners, such as logistics, finance, or production service providers, the exchange of structured files (CSV, XML, EDIFACT, etc.) is often the lowest common denominator and thus a de facto standard.

File-based interfaces offer a high degree of robustness and traceability because files can be easily versioned, archived, and reprocessed in the event of errors. Also, batch-oriented processes that process large amounts of data outside business hours often rely on file-based integration because they're independent of real-time availability. Finally, files are the connecting element between the cloud and on-premise in hybrid scenarios. Modern cloud services provide APIs, but many internal systems are only connected via files. Cloud Integration offers various adapters for this purpose, and you can use them to reliably read, process, and write files securely, performantly, and flexibly.

Cloud Integration provides the following adapters that support different file transfer protocols. Each of these adapters addresses specific technical and organizational requirements in a modern integration landscape.

- **SFTP adapter**
  This uses the *Secure File Transfer Protocol* (SFTP) to securely transfer files. It's preferred in cloud and production scenarios because the transfer is fully encrypted, and

it supports both password and public key authentication to enable access to external systems with high security and stability. SFTP is the de facto standard for secure file transfers, thanks to its broad support in cloud environments.

- **FTP adapter**
  This is a classic *File Transfer Protocol* (FTP) adapter. Data transfer is unencrypted, so you should only use this adapter in protected networks or for noncritical data. FTP is the go-to solution for legacy systems that don't support more modern technology.

- **FTPS adapter**
  In the File Transfer Protocol Secure (FTPS) adapter, the FTP protocol is extended with transport encryption that uses SSL/TLS. While FTPS is more secure than FTP, it poses challenges in terms of firewall configuration and port management. Cloud Integration natively supports FTPS.

The differences among FTP, FTPS, and SFTP are shown in Table 4.1.

| Protocols | Encryption | Authentication | Ports |
|---|---|---|---|
| FTP | No | User name/password | 21 |
| FTPS | Yes (TLS/SSL) | ■ User name/password<br>■ Certificate | 990/21 |
| SFTP | Yes (SSH) | ■ User name/password<br>■ Public key | 22 |

**Table 4.1** FTP Versus FTPS Versus SFTP

In the following sections, we'll walk through adapter roles, testing your connection to the FTP/SFTP server, the SFTP sender adapter, and the SFTP receiver adapter.

### 4.2.1    Adapter Roles

It's clear that a sender (pulling) is different from a receiver (pushing). File-based adapters are used in two ways in iFlows: as senders that initiate messages and as receivers that process and store messages. This distinction is crucial for the design of integration scenarios, and it has a significant impact on the architecture, error handling, and performance of a flow.

*Sender adapters* detect files through cyclic polling and then start an iFlow. On the other hand, *receiver adapters* act exclusively reactively, and they're activated by the execution of an already started flow (e.g., to write a processing result to a file). Cloud Integration and both adapter types guarantee the reliable, flexible, and secure connection of systems that lack a modern API interface. File-based interfaces are the most practical solution, especially in migration projects where legacy systems must continue to be used. The following sections describe both roles. We'll also describe the Poll Enrich component, which you can use to read (poll) messages from an external component and add the content to the original message in the middle of the message-processing sequence.

## Sender Adapters

In integration scenarios, you need to use sender adapters to continuously monitor external data sources and trigger data processing when changes are detected. They're perfect for use cases where systems don't offer event-based push mechanisms and therefore need to be queried regularly. The central task of a sender adapter is to bridge the gap between the external world (e.g., an SFTP server) and the internal integration process. The adapter checks for new files via a configurable time interval, and if it detects a new file, it reads the file and transfers the contents to the iFlow for further processing.

A sender adapter has the following features:

- It operates in *polling mode*, which means it regularly monitors a target directory on a server (e.g., via SFTP) without requiring an external trigger. As soon as it detects a file, it starts the read process.
- Polling is time-controlled, which means the system checks for new files at fixed intervals. You can freely configure these intervals (e.g., every 60 seconds).
- The sender adapter only processes files that match a predefined name pattern. This enables the processing of specific file types or content.
- After the adapter successfully reads the file, it can delete the file, move it, or leave it unchanged, depending on the postprocessing configuration.
- The adapter serves as the starting point for the iFlow: as soon as it detects a file, it reads it and starts processing automatically in the iFlow.
- In combination with the Poll Enrich component, you can use a sender adapter in an enriched context, where the entire flow isn't triggered but the file provides additional information to an existing message.
- The adapter detects new files based on a file name pattern (e.g., *.csv)

## Receiver Adapters

*Receiver adapters* are the counterparts to sender adapters. Sender adapters initiate iFlows, while receiver adapters are responsible for the final storage of data, usually in the form of a file. You'll always use them when a flow serves to prepare content and transfer it to a target system that works on a file basis.

Receiver adapters have the following properties:

- They're called on demand by an iFlow. This means they only become active when an incoming message in the flow issues a command to write a file.
- They're perfect for transferring data as a file to a target system (e.g., SFTP or FTP server), especially as the result of a transformation or aggregation.

- An adapter creates a file with a defined or dynamically generated name, and the content of the file can consist of the message body or a combination of the body and properties.

- You can flexibly configure target directories, character encodings, file name patterns, suffixes, prefixes, and replacement options.

- You can define error handling for situations like missing write permissions and connection interruptions.

- Receiver adapters are essential for integrating with legacy systems that rely on file access and for transferring documents or reports in automated processes.

- You can combine a receiver adapter with routing mechanisms in the flow, it can dynamically control different file destinations based on context data, and an iFlow will call it on demand.

- It reliably writes a file to a target directory on an SFTP/FTP server or on-premise file sharing service.

**Poll Enrich Component**

The *Poll Enrich component* is a special integration pattern in Cloud Integration. It allows you to enrich a message from a flow with additional data from an external source, which is typically a file. There's a clear difference between Poll Enrich and classic sender adapters: those actively start a flow, while Poll Enrich acts within a running flow and is used exclusively for targeted data enrichment.

Currently, you can only use the Poll Enrich functionality with an SFTP server, and you can only use the SFTP sender adapter with it. Compared to the regular start adapter configuration, the Poll Enrich component offers less functionality, including reduced post-processing options and limited authentication options. Nevertheless, the Poll Enrich component is a powerful tool for dynamically extending flows with external information, whether through configuration files, lookup values, or supplementary master data.

The component works as follows:

1. An existing iFlow is triggered by a main message (e.g., via HTTP, a timer, or a message queue).

2. The Poll Enrich component is used within the flow to read a file via the SFTP adapter.

3. The read content is embedded directly into the existing message as enrichment (e.g., in the message body, in the header, or as a property).

The Poll Enrich component's key advantages are its ability to make iFlows more modular and easier to maintain. The external file doesn't trigger the start of the flow; it merely provides additional data. This creates a decoupled architecture that increases reusability and design flexibility. It enables the querying of multiple external sources in parallel or sequentially to dynamically enrich a main message, and that's perfect for complex scenarios with client logic, individual customer settings, and flexible configuration data that's maintained outside the core backend.

## 4.2.2   Testing the Connection to the FTP/SFTP Server

Some of the most common errors in file-based interfaces occur when you're trying to establish a connection to an FTP, FTPS or SFTP server and complete the authentication process. One common issue is the inability to connect to the server, which is often caused by an unresolved hostname. Another potential problem is missing server certificates within Cloud Integration, and incorrect credentials can also lead to failed connections. Fortunately, you can test the connection directly in SAP Integration Suite before deploying an iFlow.

To do this, navigate to the **Monitor • Integrations and APIs** section of the side menu and then click the **Connectivity Tests** tile in the **Manage Security** section (see Figure 4.1).

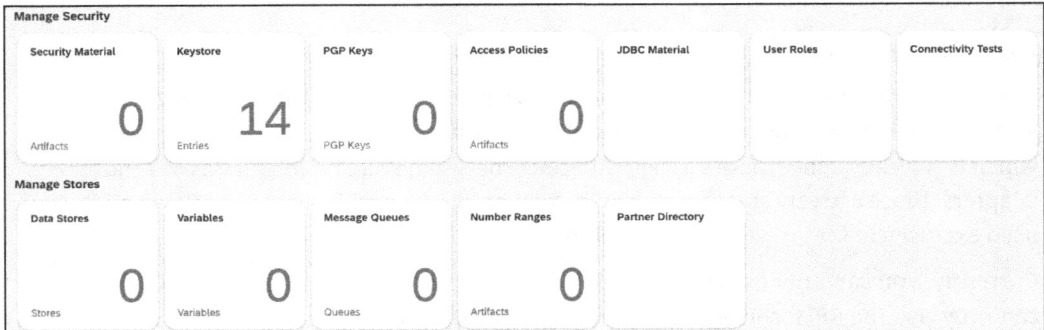

**Figure 4.1**  Opening Connectivity Test

To connect to an SFTP server, you must maintain and provide the known hosts file, and you can obtain the entry for this file via a connection test.

**Known Hosts File**

The *known hosts file* is central to secure communication via SFTP. It contains the public keys of trusted servers with which a connection will be established. When you connect to a server for the first time, the server's public key will be stored so that future connections can be verified. This prevents man-in-the-middle attacks and ensures that the connection is made to an authentic server. In cloud integration scenarios, you need to store all relevant server certificates in the known hosts file; otherwise, the connection may fail.

To perform a connection test to an SFTP server, click on the **SSH** tab and enter the SFTP server's host name in the **Host** field, as shown in Figure 4.2. You can also adjust the **Port** and select a **Proxy Type** from the dropdown list. For publicly accessible SFTP servers, select the **Internet** proxy type, or if you're accessing the SFTP server via the cloud connector, select the **OnPremise** proxy type. Then, in the **Authentication** area, select the **None** option and click the **Send** button. The response will appear on the right side of the screen, where you can copy the host key to the clipboard by clicking **Copy Host Key**.

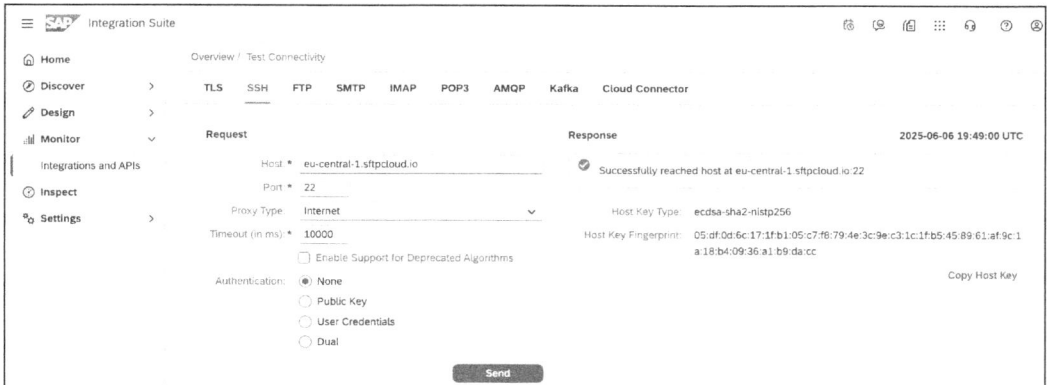

**Figure 4.2** SFTP Connection Test

Then, you can add the entry to the known hosts file. To do so, navigate to the **Monitor •  Integration and APIs** section of the side menu and open the **Security Material** tile under **Manage Security**. If a known hosts file already exists, download it to your local computer and add the entry. If there are no known hosts, as shown in Figure 4.3, you must create a file locally and insert the host key. Then, click the **Upload** button, select **Known Hosts (SSH)** from the context menu, and upload the file.

**Figure 4.3** Uploading Known Hosts File

After you upload the file successfully, you should find it in the **Security Material** section, as shown in Figure 4.4. The file type is **SSH Known Hosts**, and as you'll see, the user credentials for the SFTP server will also have been uploaded to this section.

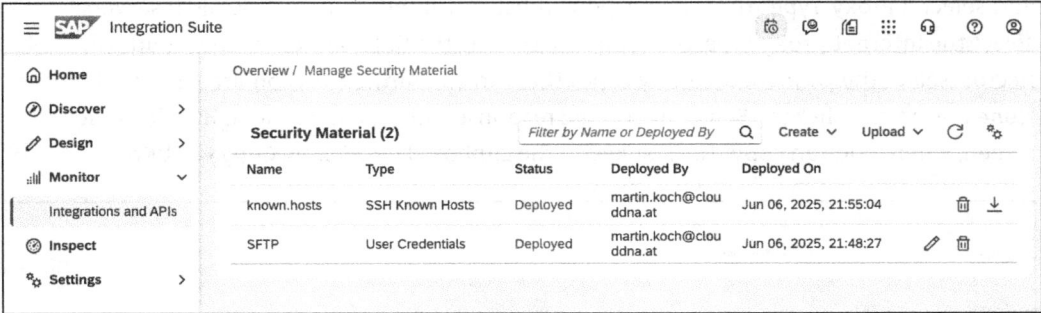

**Figure 4.4** Security Material with Known Hosts

Next, you must verify that the identification works with the SFTP server, and you'll also have the option to test access to the desired directories. To do so, navigate back to the **SSH** tab where you performed the connection test, as shown in Figure 4.5. Enter the SFTP server data as before, and in the **Authentication** area, select the authentication option you want, which in this case is **User Credentials**. Then, select a deployed user credential in the **Credential Name** field, and for the **Host Key Verification** field, select **Against Tenant**. Optionally, you can check the **Check Directory Access** box to verify your access to the directory you want. You can enter the directory you want in the **Directory** field, or you can leave the field blank, which will mean the user's root directory will be checked. Then, click **Send**, and the contents of the directory you've specified will appear on the right-hand area of the screen.

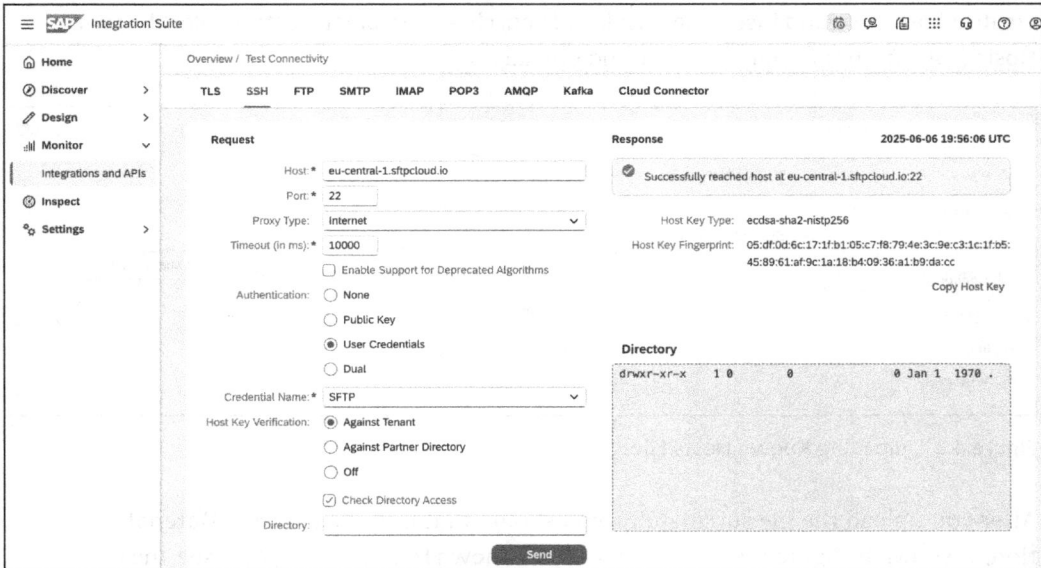

**Figure 4.5** Connection Test with Authentication

### 4.2.3    Sender Adapters

SFTP sender adapters connect Cloud Integration tenants to remote systems by using the SSH File Transfer Protocol to read files from that system. The SSH File Transfer Protocol is also known as the Secure File Transfer Protocol (SFTP). When an SFTP sender adapter is configured, message processing occurs as follows during runtime: the tenant sends a request to an SFTP server (which can be considered the sender system), but the data flow goes in the opposite direction—from the SFTP server to the tenant. In other words, the tenant reads files from the SFTP server (in a process called *polling*).

In the **Source** section of the SFTP sender adapter configuration (see Figure 4.6), you specify the **Directory**, which defines the relative path from which to read files (e.g., "parentdirectory/childdirectory"). The **File Name** field allows you to specify which files to read, and if you leave it blank, all files in the directory you specify will be processed. You can use simple expressions, such as *file*.txt* or *file?.txt*, or you can enable regex filtering starting from adapter version 1.17 to interpret the file name as a true regular expression. Simple expressions support wildcards, such as * (any number of characters) and ? (one character). When working with subdirectories, consider the relative path in your pattern. Note that complex regex patterns may impact performance because the evaluation timeout is set to five seconds by default.

---

**Regular Expressions**

*Regular expressions* (regexes) are tools you can use to describe character patterns in text. They allow you to search for, filter, and validate character strings in a targeted manner. For example, you can use them to select specific file names and check inputs, such as email addresses. Simple placeholders—such as . (any character), * (any number of repetitions), + (at least one repetition), and \d (a digit)—help define flexible patterns. For instance, the expression file\d+\.txt finds files such as *file1.txt* and *file123.txt*. However, regexes require precise syntax; even small errors can lead to unexpected results.

---

Note that filenames longer than 100 characters, including the relative path, are subject to restrictions. Only one such file is processed at a time, even if multiple workers are available. If processing fails repeatedly, other files with the same initial 100 characters won't be processed without manual intervention. Additionally, the **Keep File and Mark as Processed in Idempotent Repository** option doesn't apply to these long filenames.

In the connection settings, the **Address** defines the hostname or IP address of the SFTP server, optionally including a port (e.g., eu-central-1.sftpcloud.io:22). The **Proxy Type** setting determines how the adapter connects to the target system, as follows:

- Internet
  This is used for direct SFTP server connections.
- Manual
  This allows manual entry of proxy host, proxy port, and proxy protocol (HTTP and SOCKS v4 or v5), including an optional proxy credential name for authentication.

- **On-Premise**

  You can enable this connectivity through the cloud connector by specifying a location ID to match the destination configuration in the SAP BTP cockpit.

The adapter includes parameters to control connection stability. The **Timeout** parameter (in milliseconds) defines how long the system waits when establishing a connection or reading a file. The default is 10,000 ms, and the maximum is 299,999 ms. **Maximum Reconnect Attempts** (the default value of which is 3) specifies how many retries the adapter will perform when establishing the initial connection before giving up. The **Reconnect Delay (in ms)** parameter determines the length of the pause between reconnect attempts. (The default is 1,000 ms.) These retries only apply to the initial connection setup, not to connection loss during message processing. Optionally, you can configure the adapter to automatically disconnect after each message.

For legacy environments, you can check the box to **Enable Support for Deprecated Algorithms**, such as `diffie-hellman-group1-sha1` and `ssh-rsa`, to allow compatibility with older SFTP server configurations.

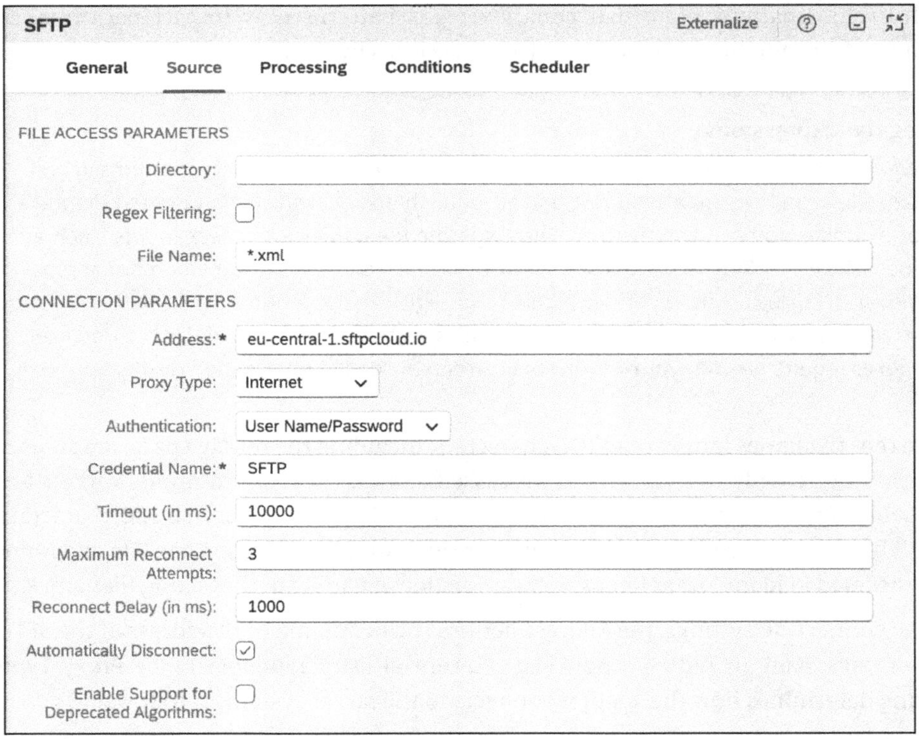

| **SFTP** | | | | Externalize |
| --- | --- | --- | --- | --- |
| General | Source | Processing | Conditions | Scheduler |

FILE ACCESS PARAMETERS

| Directory: | |
| --- | --- |
| Regex Filtering: | ☐ |
| File Name: | *.xml |

CONNECTION PARAMETERS

| Address: * | eu-central-1.sftpcloud.io |
| --- | --- |
| Proxy Type: | Internet ⌄ |
| Authentication: | User Name/Password ⌄ |
| Credential Name: * | SFTP |
| Timeout (in ms): | 10000 |
| Maximum Reconnect Attempts: | 3 |
| Reconnect Delay (in ms): | 1000 |
| Automatically Disconnect: | ☑ |
| Enable Support for Deprecated Algorithms: | ☐ |

**Figure 4.6** SFTP Source Settings

In the **Processing** tab (see Figure 4.7), you can specify how files should be handled when polling the SFTP server. The **Read Lock Strategy** prevents the reading of files that are still being written. The options are **None** (the default setting, which reads immediately), **Content Change** (which waits for the file to stop changing), **Done File Expected** (which waits for a *.done* file), and **Rename** (which renames the file before reading). When you're using *.done* files, use the *${filename}.done* pattern and match it carefully to avoid errors.

The **Empty File Handling** option determines how to handle zero-byte files. You can choose to process, skip, or log and postprocess them without running the main logic. When the iFlow runs on multiple nodes, you can check the **Poll on One Worker Only** box to ensure that only one node polls files at a time. This is useful for preserving order or limiting parallel connections. If you check that box, you can also check the **Stop on Exception** box to halt polling for the current batch on errors. **Sorting** allows you to order by file name, size, or timestamp, and **Sorting Order** lets you define ascending or descending sorting.

The **Max. Messages per Poll** parameter defines how many files are picked up per polling cycle. The default is 20, and the maximum is 500. In multinode setups, this number is multiplied by the number of nodes. To avoid reprocessing in case of errors or to ensure strict ordering, you can reduce this value to 1; however, this impacts performance. The **Lock Timeout (in mins)** parameter (the default value of which is 15 minutes) determines how long the system will wait before attempting to reprocess a file, particularly after a system outage.

Other options include checking the **Change Directories Stepwise** box, which switches folders one level at a time, and checking the **Include Subdirectories** box, which allows polling from all nested folders. If you check that box, you can also check **Flatten File Names** to ignore the directory structure and treat all files as flat. The **Use Fast Exists Check** option enables faster file existence checks directly on the server, and it's checked by default.

**Post-Processing** controls what happens after a file is successfully read. These are the options you can choose from:

- **Delete File**
  This is the default.

- **Keep File and Mark as Processed in Idempotent Repository**
  This prevents reprocessing without deletion.

- **Keep File and Process Again**
  This processes the file with every poll and is useful for testing.

- **Move File**
  This moves files to a target folder after processing.

When you're using the **Idempotent Repository**, you can store processed file references in either a **Database** (which is recommended for multiple nodes) or **In-Memory** (which is not recommended for multinode environments). When you're using the **Move File** option, you must define the archive directory as a relative path, and it can include dynamic placeholders, such as timestamps.

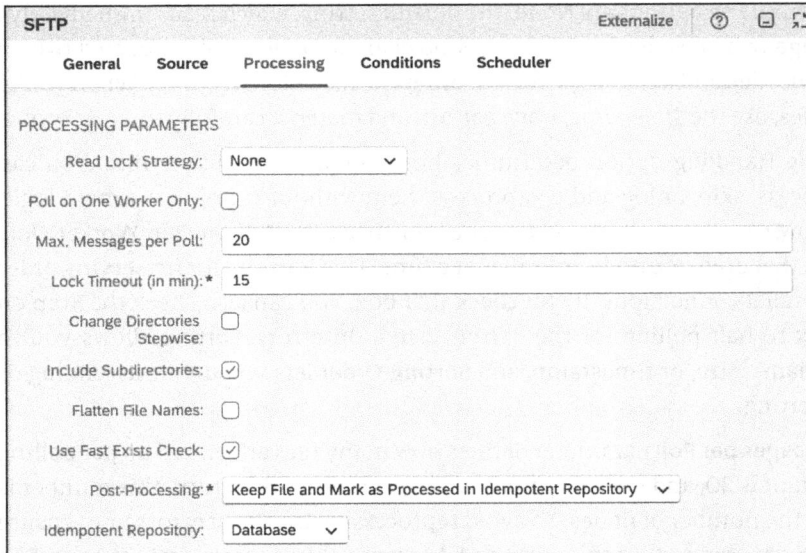

**Figure 4.7**  SFTP Processing Settings

In the **Scheduler** tab (see Figure 4.8), you can configure the frequency with which the iFlow polls the SFTP server. You can choose between one-time execution and recurring schedules. For one-time execution, select **Schedule on Day** and set the **On Date** and **At Time** fields to define the exact execution date and time. To define intervals for repeated execution within a time window, such as every hour or every 15 minutes, use the **Every** field along with a defined time zone to ensure correct timing across regions.

For recurring schedules, select **Schedule to Recur**, where you can define the following:

- Daily execution at a specific time or repeated interval
- Weekly execution, by selecting specific weekdays and setting the time or interval
- Monthly execution on a particular day of the month, by specifying the time or interval

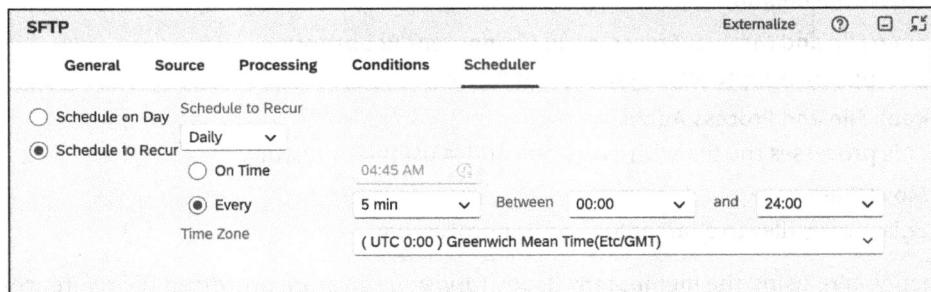

**Figure 4.8**  SFTP Scheduler Settings

You should thoughtfully configure the scheduler: polling the SFTP server too frequently, especially with intervals shorter than one minute, can overload the server and degrade performance. Therefore, you should only use shorter polling intervals if you need to in your business scenario.

The FTP sender adapter in Cloud Integration is functionally like the SFTP adapter. The main difference lies in how transport **Encryption** is handled. The SFTP adapter uses the SSH protocol for secure file transfer, while the FTP adapter offers the following encryption options based on the FTPS standard (see Figure 4.9):

- **Explicit FTPS**
  This starts with an unencrypted connection and upgrades to TLS via the AUTH TLS command. The default port is 21, unless otherwise specified.

- **Implicit FTPS**
  TLS encryption is initiated immediately when the connection is established. The default port is 990.

- **Plain FTP - no encryption**
  No encryption is applied, so this option isn't recommended for productive use. The default port is 21.

Unlike the SFTP adapter, the FTP adapter doesn't support manual proxy configuration. This means that options such as **Proxy Type: Manual** and specifying the proxy host and port are unavailable. However, all other configuration options—such as file name filtering, scheduling, processing, and post-processing—remain largely the same as in the SFTP adapter.

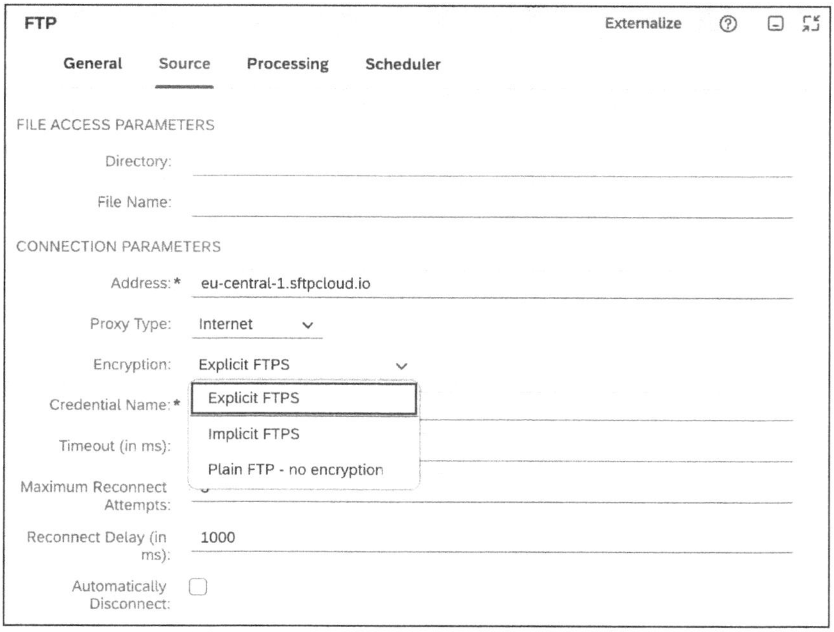

Figure 4.9  FTP Sender Adapter Source Settings

### 4.2.4   Receiver Adapters

SFTP receiver adapters enable a Cloud Integration tenant to transfer files to a remote system by using SFTP. At runtime, when a receiver SFTP adapter is configured, the iFlow sends data from the Cloud Integration tenant to the external SFTP server—meaning the tenant writes files to the target system. This allows the communication partner to retrieve the files directly from the SFTP server.

In the **Target** tab of the SFTP receiver adapter (see Figure 4.10), you can specify where and how the file should be written to the target SFTP server. In the **Directory** field, you can specify the relative path on the server, and in the **File Name** field, you can define the name of the output file. You can set both statically or dynamically by using property or header expressions (e.g., "${property.myProp}"). If you don't provide a file name, the system will use the CamelFileName header, and if that header is absent, the system will use the exchange ID instead. Wildcard patterns (e.g., *.txt) are not supported.

You can check the **Append Timestamp** box to add a GMT-based timestamp to the file name, but you must not use this option with dynamic file names via expressions because it leads to invalid configurations and errors. Note that files created in quick succession may end up with identical names.

You can fill in the **Address** field to define the SFTP server (the host and optional port) and the **Proxy Type** field to determine how the connection will be routed. You can select from the following proxy types:

- **Internet**
  This is for direct access.
- **On-Premise**
  This is for cloud connector–based access (and it requires a location ID).
- **Manual (Edge runtime only)**
  This allows you to specify the proxy host, port, protocol (HTTP or SOCKS4/5), and optional proxy credentials.
- **Dynamic**
  This uses the SAP_FtpProxyType property (values: internet or onPremise) to decide at runtime.

**Authentication** requires a user name (with a restricted character set) and, depending on the mode, a private key alias that references a key in the keystore. Timeout settings include the following:

- **Timeout**
  The default value is 10,000 ms.
- **Maximum Reconnect Attempts**
  The default value is 3.
- **Reconnect Delay**
  The default value is 1,000 ms.

These values can also be controlled via dynamic iFlow properties, such as SAP_FtpTime-out, SAP_FtpMaxReconnect, and SAP_FtpMaxReconDelay. By checking the **Automatically Disconnect** box, you can choose to close the SFTP connection after each transfer. Finally, you can check the **Enable Support for Deprecated Algorithms** box to allow compatibility with older SFTP server configurations (e.g., SSH-RSA, Diffie-Hellman-Group1-SHA1).

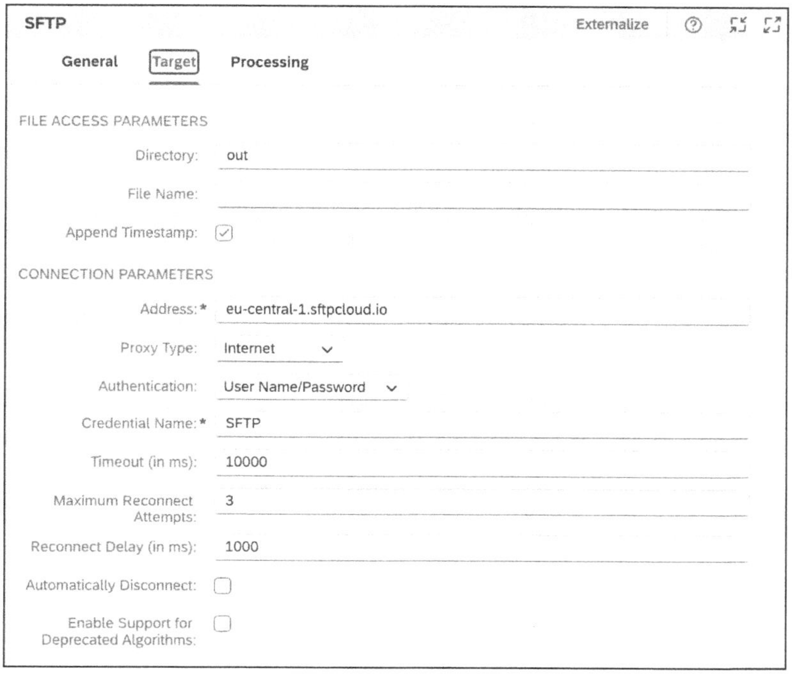

**Figure 4.10**  SFTP Receiver Target Settings

The **Processing** tab (see Figure 4.11) contains several options that control how files are written to the SFTP server. The **Change Directories Stepwise** box is checked by default, and it makes adapter navigate each directory level individually. The **Create Directories** box is also checked by default, and it ensures that any missing directory levels in the file path are created automatically. You can check the **Flatten File Names** box to remove directory structures from the file path, write all files into a single directory, and use only file names.

To prevent unauthorized path traversal, you can check the **Prevent Directory Traversal** box to halt processing if the file path includes relative backtracking (e.g., ../). To help you avoid such security issues, we strongly recommend that you define both the directory and the file name explicitly.

The **Use Fast Exists Check** option (which is checked by default) performs file existence checks directly on the SFTP server, but you should uncheck this box if it isn't supported by the server. The **Handling for Existing Files** setting controls what happens if a file with the same name already exists, and you can choose from the following options:

- **Append**
  This adds to the existing file (which is not recommended with parallel processing).
- **Fail**
  This raises an error.
- **Ignore**
  This skips writing.
- **Override**
  This replaces the file.

When you're using **Fail**, **Ignore**, or **Override**, you can check the **Use Temporary File** box to allows the system to first write to a temporary file (e.g., *target_${exchangeId}.temp*) before renaming it to the final target name. This reduces the risk of overwriting in scenarios with high parallel processing. The temporary file name should include dynamic elements to ensure uniqueness.

All settings can also be controlled dynamically via iFlow properties, such as SAP_Ftp-Stepwise, SAP_FtpCreateDir, SAP_FtpFlattenFileName, and SAP_FtpAfterProc.

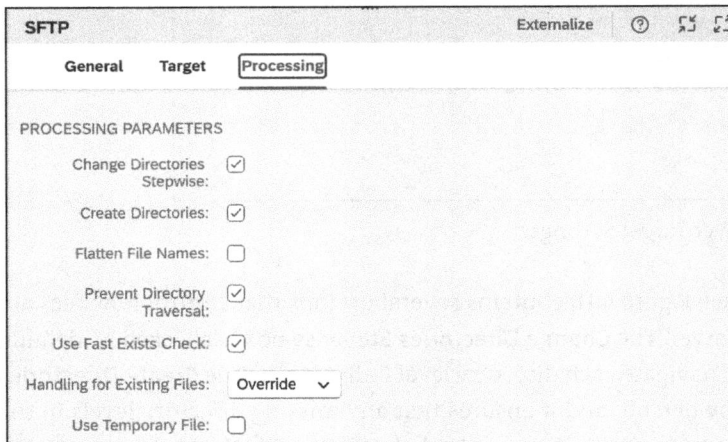

**Figure 4.11** SFTP Receiver Processing Settings

The FTP receiver adapter in Cloud Integration differs from the SFTP receiver primarily in two areas: encryption options and proxy configuration (see Figure 4.12). While the SFTP receiver exclusively uses the SSH-based SFTP, the FTP receiver supports multiple encryption types based on FTPS. You can choose from the following options in the **Encryption** field:

- **FTP**
  This is plain FTP without encryption, so it's not recommended for productive use). The default port is 21.

- **FTPES**
  This is explicit FTPS, and the connection starts unencrypted and then switches to TLS by using the AUTH TLS command. The default port is 21.
- **FTPS Implicit FTPS**
  TLS encryption is initiated immediately upon connection. The default port is 990.
  **Dynamic**
  The encryption method is selected at runtime by using the SAP_FTP_ENCryption property (FTP, FTPS, or FTPSE).

If the property isn't defined, a runtime error will occur.

Another key difference lies in the proxy configuration. Both the FTP and SFTP receivers support the following options in the **Proxy Type** field:

- **Internet**
  This is a direct connection to the FTP/SFTP server.
- **On-Premise**
  This is a connection via the cloud connector (which requires a location ID).
- **Dynamic**
  The proxy type is determined at runtime via the SAP_FTP_PROXY_TYPE property (via the internet or on-premise).

However, the FTP receiver doesn't support manual proxy configuration (i.e., specifying the proxy host, port, and protocol). This simplifies setup but reduces flexibility in more complex network environments.

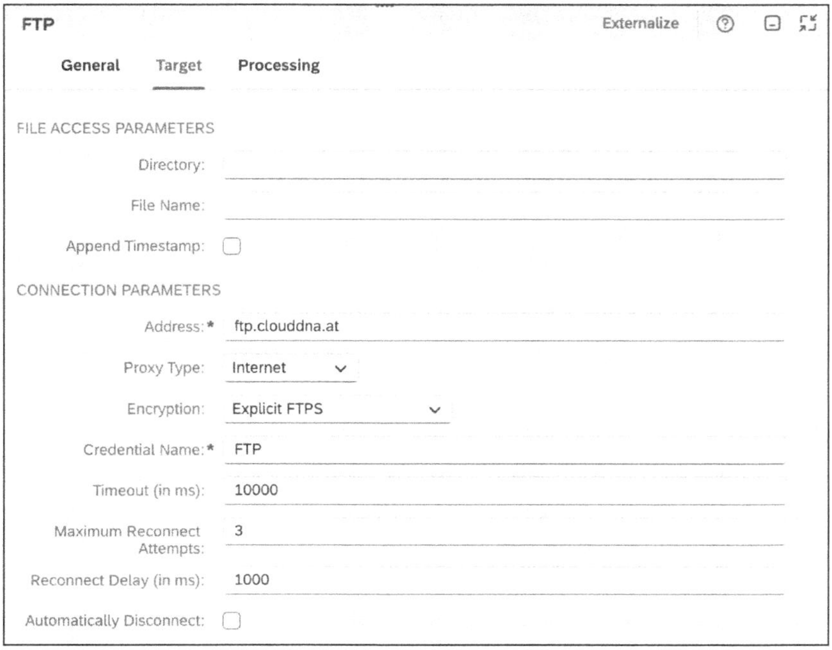

| FTP | | | Externalize | ⑦ | ▭ | ⤢ |
|---|---|---|---|---|---|---|
| **General** | Target | **Processing** | | | | |

FILE ACCESS PARAMETERS

| | |
|---|---|
| Directory: | |
| File Name: | |
| Append Timestamp: | ☐ |

CONNECTION PARAMETERS

| | |
|---|---|
| Address: * | ftp.clouddna.at |
| Proxy Type: | Internet ⌄ |
| Encryption: | Explicit FTPS ⌄ |
| Credential Name: * | FTP |
| Timeout (in ms): | 10000 |
| Maximum Reconnect Attempts: | 3 |
| Reconnect Delay (in ms): | 1000 |
| Automatically Disconnect: | ☐ |

**Figure 4.12** FTP Receiver Settings

## 4.3   SAP-Specific Adapters

SAP Integration Suite provides SAP-specific adapters that enable seamless, standards-compliant integration with SAP systems. These adapters are tailored to the specific requirements of the SAP world, they play a central role in hybrid integration scenarios, and they support proprietary protocols, formats, and authentication mechanisms. The IDoc adapter, for example, is used to exchange structured business data in asynchronous mode and is based on SOAP communication with the SAP backend. The IDoc adapter is particularly suitable for processing large volumes of data, such as distributing master data, purchase orders, and delivery notifications. The adapter processes IDoc messages in a flat structure and automatically converts them to XML format, which enables easy further processing within iFlows. Typically, you carry out configuration in conjunction with partner profiles and port definitions in the SAP ERP or S/4HANA system.

The RFC adapter enables synchronous access to RFC-Enabled Function Modules (RFMs) and Business Application Programming Interfaces (BAPIs) in ABAP-based on-premise systems, and, depending on the scenario, in SAP S/4HANA Cloud as well. You can establish the connection via the cloud connector for internal systems or directly through the internet for cloud-based instances. The adapter allows functions to be called directly with import, export, and table parameter transfers. The data is displayed in SAP Integration Suite as XML, and you can flexibly transform and process it further. This makes the RFC adapter ideal for use cases such as validations, individual queries, and transactional processes with direct feedback.

There's a special SAP SuccessFactors adapter that supports modern cloud APIs. It provides access to OData V2, OData V4, SOAP, and REST-based interfaces, and it covers both older and newer SAP SuccessFactors modules. The adapter handles central tasks such as CSRF token handling, pagination, and metadata processing, and it's particularly suitable for integrating employee lifecycle processes, master data management, and organizational changes in HR-related end-to-end processes.

These adapters form the backbone of many mission-critical integration scenarios in hybrid SAP landscapes. Leveraging these adapters' built-in intelligence and protocol support allows organizations to accelerate their integration projects, reduce maintenance overhead, and ensure consistency with SAP's architectural standards and best practices.

### 4.3.1   IDoc Adapter

The IDoc adapter allows Cloud Integration to exchange Intermediate Document (IDoc) messages with systems that communicate via SOAP web services. To this end, the adapter comes in both sender and receiver variants, which we'll discuss in the following sections.

The sender adapter allows Cloud Integration to receive IDocs from an SAP backend system, and the receiver adapter enables Cloud Integration to send IDocs to the SAP system. This bidirectional support makes the IDoc adapter suitable for a wide range of integration scenarios, including asynchronous master data replication, transactional document exchange, and B2B message processing.

---

### IDoc 101

An *Intermediate Document* (IDoc) is a standard SAP format for exchanging structured business data between systems in an asynchronous manner. IDocs play a central role in enabling communication between SAP and non-SAP systems and between different SAP systems without requiring immediate responses. IDocs are commonly used in business processes such as sending sales orders, distributing material master data, and receiving shipping notifications from external partners.

At its core, an IDoc is a text-based data structure that consists of three main parts: a control record that contains metadata such as the sender and receiver and the IDoc type, multiple data records that contain the actual business content, and optional status records that log processing information. The structure of each IDoc is defined by its IDoc type. For example, the ORDERS05 type is used for sales orders, and the MATMAS05 type is used for material masters.

In traditional scenarios, communication via IDocs is typically handled through Transactional Remote Function Call (tRFC), which ensures reliable data transfer. In cloud scenarios that use SAP Integration Suite, the communication is based on SOAP. In the SAP backend, IDoc processing is closely linked with partner profiles, port definitions, and the Application Link Enabling (ALE) framework.

Although IDocs are not the newest technology, they remain highly relevant due to their robustness, standardized format, and native integration into SAP's business processes. IDocs are particularly well-suited for mass data exchange and system decoupling. When you're working with IDocs, you need to identify the appropriate message type (e.g., ORDERS, INVOIC), determine the correct IDoc type (e.g., ORDERS05), and ensure the backend is configured accordingly. With this foundation, the IDoc adapter in SAP Integration Suite will enable seamless, reliable integration.

---

**4**

### IDoc Senders

In Cloud Integration, the IDoc sender adapter serves as an entry point for structured business data that's sent from an SAP ERP or SAP S/4HANA system in the form of IDocs. Unlike classic SAP PI/SAP PO scenarios, which use TRFC, communication here takes place via the IDoc SOAP protocol—that is, via HTTPS with SOAP-encapsulated IDoc payloads. The adapter listens on a specific URL that's defined during adapter configuration (e.g., /orders05), and this address is the central endpoint to which the ERP system sends IDocs. The IDoc message is transmitted as XML and can be forwarded, cached, or analyzed directly without modification. It's also important to note that, like many other Cloud Integration adapters, the adapter is available as both a sender and a receiver. In this scenario, the adapter acts as a sender because it's the starting point of the iFlow. The following steps will guide you through the process of configuring the system in SAP.

Configuring the IDoc sender adapter is easy. Simply assign a unique address in the **Address** field, as shown in Figure 4.13.

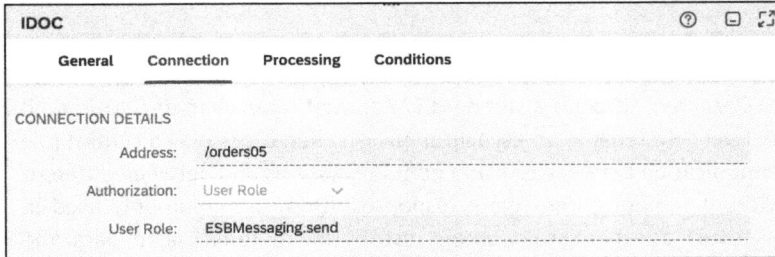

**Figure 4.13** IDoc Sender Configuration

After deploying the iFlow, you can view the endpoint for the deployed artifacts (see Figure 4.14). You'll need this information later, when creating the SM59 destination in the SAP S/4HANA system.

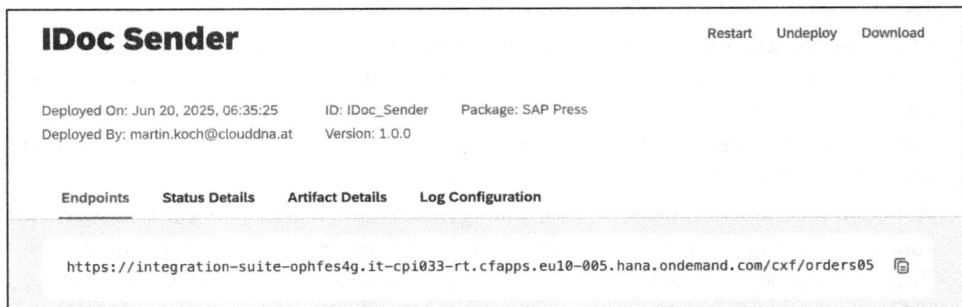

**Figure 4.14** IDoc Endpoints

Next, you create a logical system that the ERP system will use to identify communication with the external target (i.e., Cloud Integration). Use Transaction BD54 for configuring the logical system (see Figure 4.15), which is referenced in the partner agreement. Choose a name that's unique and descriptive.

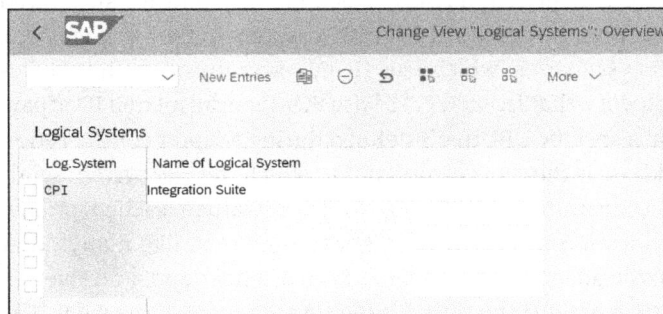

**Figure 4.15** Creating Logical System

**Logical Systems 101**

A *logical system* is a unique technical designation for an SAP or external system within a distributed landscape of systems. You use it to identify communication partners during integration via IDoc or RFC, for example. You use Transaction BD54 to perform maintenance and Transaction SCC4 to perform assignment to the client. Typical uses of this are as follows:

- Definition of sender and receiver systems for IDoc communication
- Basis for partner agreements (WE20)
- Prerequisite for ALE distributions and data replication

An example is using ERPCLNT100 for the SAP ERP system and CPI for Cloud Integration.

To establish an SSL-secured connection between SAP S/4HANA and Cloud Integration, you must import the public SSL certificate of your Cloud Integration tenant into SAP S/4HANA. This step is mandatory; otherwise, SSL-secured communication with the cloud system won't be possible. You do this by using Transaction SM59. Perform the following steps:

1. Open Transaction STRUST and select **SSL Client (Anonymous)**.
2. Click **Import Certificate**.
3. Select the locally saved certificate file from your Cloud Integration tenant.
4. Add the certificate to the list of certificates.
5. Save your changes.

You can use Transaction SM59 to define destinations that use RFC, HTTP, and other protocols. You can use these technical connection profiles to access SAP systems or external systems, such as Cloud Integration, web services, and databases. Each destination contains all the necessary connection data, including the host name, port, path, authentication information, and security settings. Since communication via SOAP takes place over HTTP, you must create an HTTP destination of type G (see Figure 4.16). Define the following properties:

- **Connection Type**
  Enter "G" and "HTTP Connection to External Server."
- **Host**
  Enter the target host part of the IFlow endpoint URL (e.g., *abc123.hana.ondemand.com*).
- **Port**
  Enter 443 (the default HTTPS port).
- **Path Prefix**
  For example, you can enter */cxf/orders05*, but it must match the IDoc sender adapter address exactly.

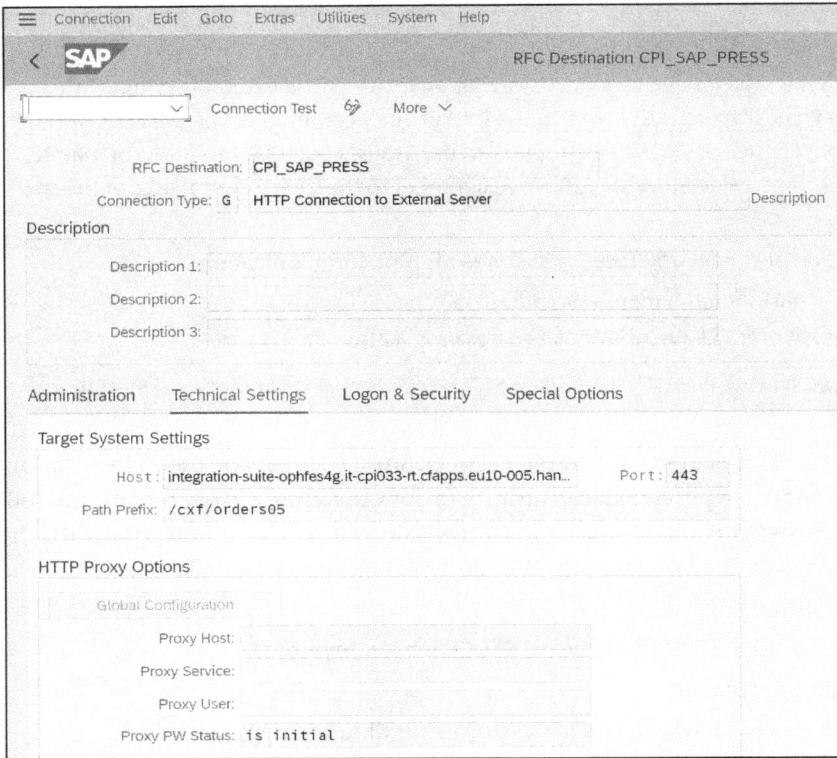

**Figure 4.16** Transaction SM59 HTTP Destination

You must also configure the authentication. To do this, open the **Logon & Security** tab, choose **Basic Authentication** in the **Logon with User** section, and input the SAP Integration Suite user's user name and password (see Figure 4.17). Alternatively, you can use the client credentials from a service key.

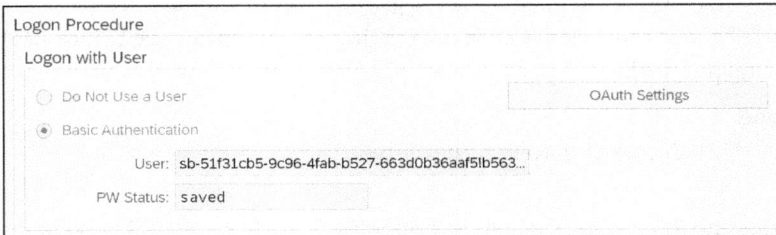

**Figure 4.17** Configuring Credentials

The connection to SAP Integration Suite uses the HTTPS protocol, and you need to configure it for the destination. Navigate to the **Security Options** section, select the **Active** option for **SSL**, and select the **DEFAULT SSL Client (Standard)** option for the **SSL Client PSE ID** (see Figure 4.18).

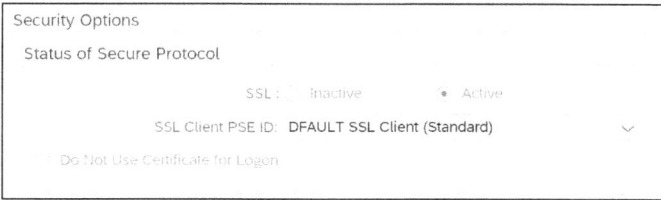

**Figure 4.18**  Configuring SSL

Finally, perform a connection test. A successful message will return **HTTP 500**, incorrect credentials will return **HTTP 401 Unauthorized**, and missing SSL certificates will return **ICM_HTTP_SSL_ERROR**.

Since Cloud Integration expects SOAP messages via HTTP, you must set up an XML HTTP port. A sample configuration is shown in Figure 4.19. Use the following settings:

- **Port**
  For example, you can enter "CIPORT."
- **RFC destination**
  Enter the HTTP destination that you created previously in Transaction SM59.
- **Content Type**
  Select **Text/XML**.
- **SOAP Protocol**
  Check the box. If you don't, the IDoc won't be transferred in the correct format, and Cloud Integration won't be able to process it.

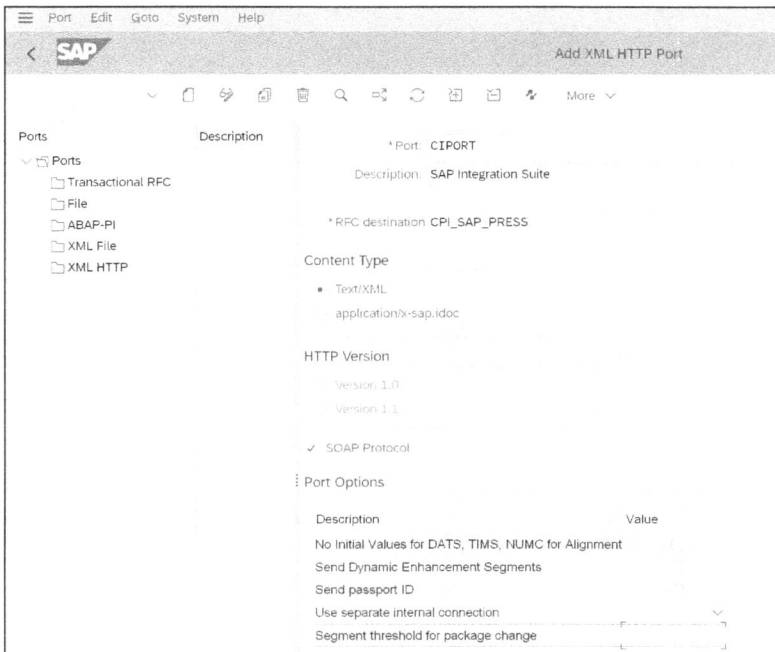

**Figure 4.19**  Configuring the XML HTTP Port

In the SAP system, the partner agreement in Transaction WE20 is a central control element for exchanging IDocs. It specifies the type of message to be exchanged with each external or internal partner, including technical details such as the port, IDoc type, and processing settings. Partners may include another SAP system, a cloud system such as Cloud Integration, or an EDI service provider.

Figure 4.20 shows a sample WE20 configuration. Use the following properties:

- **Partner No**
  Enter "CPI" since it corresponds to the name from Transaction BD54.
- **Partner Type**
  Enter "LS" for logical system.
- **Message Type**
  For example, you can enter "ORDERS," which corresponds to IDoc type ORDERS05.
- **Receiver port**
  Enter "CIPORT" since it's the XML HTTP port from WE2.1.

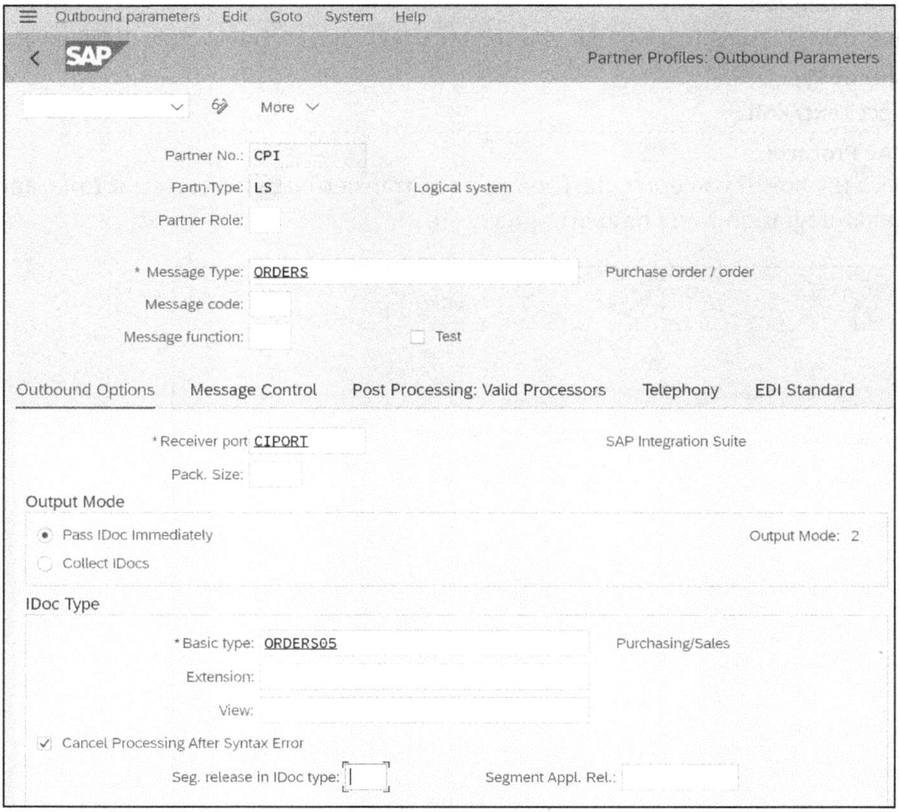

**Figure 4.20** Partner Profile

**IDoc Receivers**

To activate the IDoc reception service for SOAP communication, you must activate the corresponding ICF service in SAP S/4HANA. Go to Transaction SICF (see Figure 4.21), then enter the **Service Path** as "/sap/bc/srt/idoc" and activate the service by using the context menu.

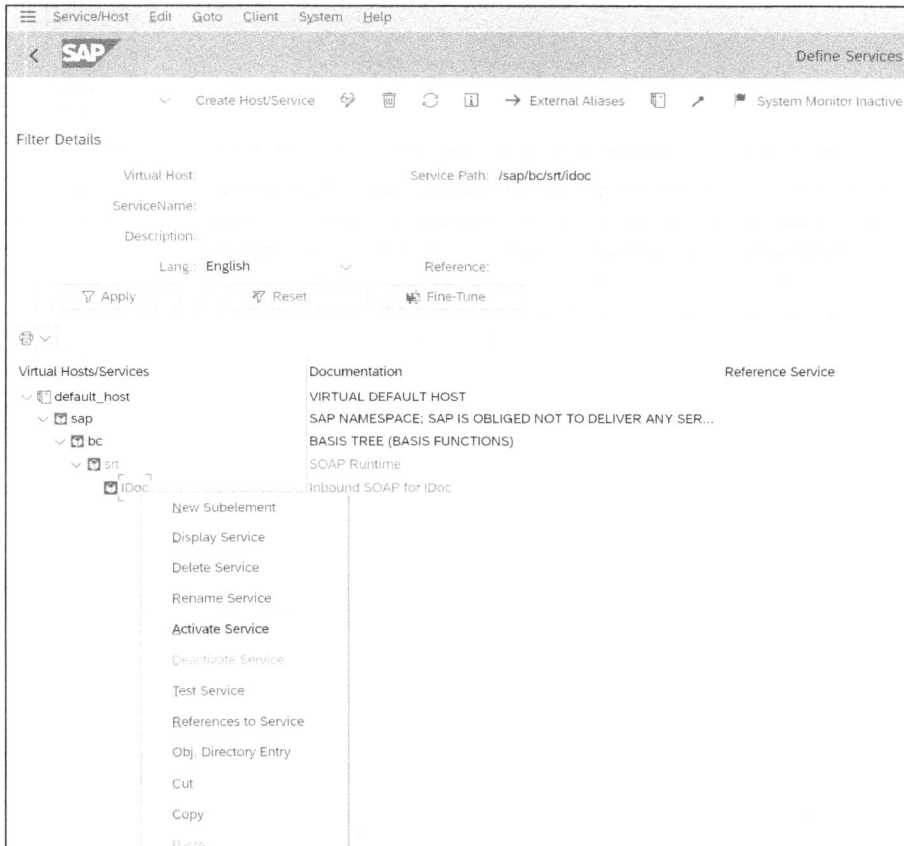

**Figure 4.21** Activating ICF Node

After you activate the /sap/bc/srt/idoc ICF service, you must register the IDoc web service in the SAP system to ensure that the system will correctly assign incoming SOAP messages to IDoc processing. Go to Transaction SRTIDOC to manage and register IDoc web services (see Figure 4.22), select the **Register Service** option in the top area, and execute the report. To make the SAP system act correctly as a communication partner, you must define a logical system and assign it to the current client. You must do this in all ALE and IDoc-based scenarios. Perform the following steps:

1. Call Transaction SALE.
2. Create the logical system.
3. Assign the logical system to the clients.

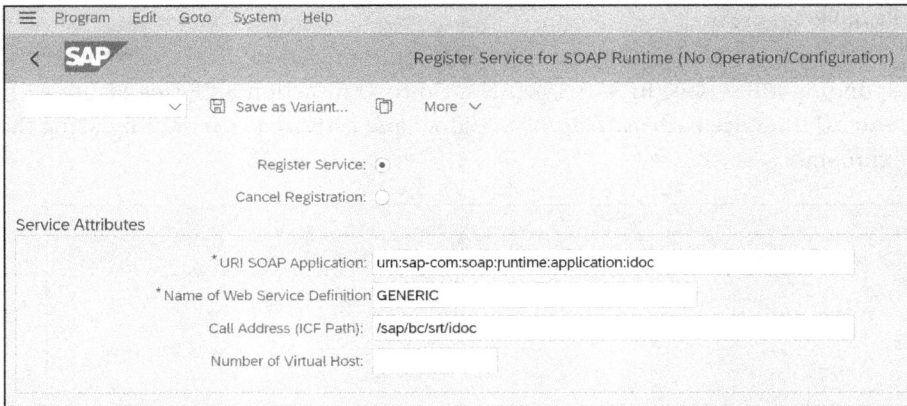

**Figure 4.22** Service Registration

The next step is to set up the partner profile, which defines how the system processes incoming or outgoing IDocs for a specific partner. In your case, the partner is your own logical system because you want to receive messages from the Cloud Integration iFlow. Perform the following steps (see Figure 4.23):

1. Call Transaction WE20.

2. Check the box for **Partner Type** LS (which stands for Logical System).

3. Add inbound parameters.

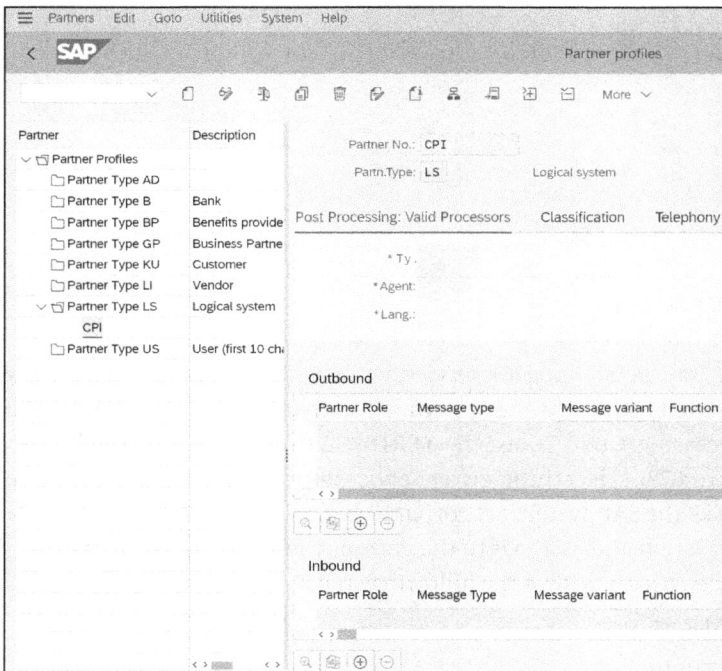

**Figure 4.23** Partner Profile

IDocs are typically sent to an SAP system via the cloud connector, but you can also send them to a publicly accessible SAP system. Enter the endpoint you want in the **Address** field; for a connection via the cloud connector, this is the virtual host name. You must also specify the path of the ICF node, which is *sap/bc/srt/idoc*. Then, supplement the address with the *sap-client* URL parameter. Figure 4.24 shows an example of a complete URL.

For a connection via the cloud connector, select **On-Premise** in the **Proxy Type** field. If the system is publicly accessible, select **Internet**. You should select **Text/XML** as the **IDoc content type**, and you must also specify how **Authentication** should take place.

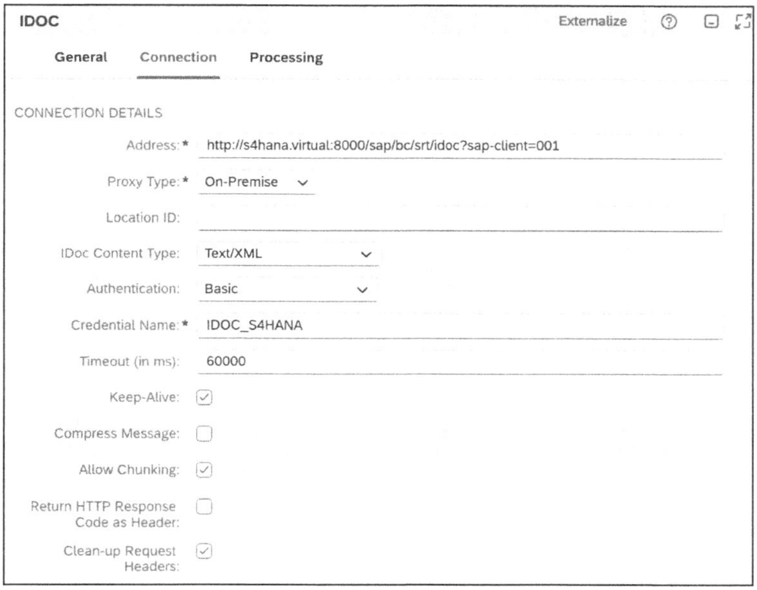

| IDOC | | Externalize  ⑦  ☐  ⌗ |
|---|---|---|
| **General**   **Connection**   **Processing** | | |

CONNECTION DETAILS

| | |
|---|---|
| Address: * | http://s4hana.virtual:8000/sap/bc/srt/idoc?sap-client=001 |
| Proxy Type: * | On-Premise   ⌄ |
| Location ID: | |
| IDoc Content Type: | Text/XML   ⌄ |
| Authentication: | Basic   ⌄ |
| Credential Name: * | IDOC_S4HANA |
| Timeout (in ms): | 60000 |
| Keep-Alive: | ☑ |
| Compress Message: | ☐ |
| Allow Chunking: | ☑ |
| Return HTTP Response Code as Header: | ☐ |
| Clean-up Request Headers: | ☑ |

**Figure 4.24** IDoc Receiver Adapter Configuration

## 4.3.2   RFC Receiver

The RFC adapter in Cloud Integration facilitates communication with on-premise ABAP systems via the RFC protocol. This specialized adapter is used to call remote-enabled function modules directly from the iFlow.

---

**Remote Function Calls 101**

RFC is a core SAP technology that enables a system to remotely call and execute a function module in another SAP or non-SAP system as if it were a local procedure. RFC enables synchronous communication, meaning the sender waits for a response from the called system. This makes RFC ideal for real-time integration scenarios, such as reading data, validating input, and triggering business processes.

SAP uses RFCs extensively to connect distributed systems, interface with external applications, and modularize complex logic across system boundaries. The called function module must be marked as **Remote-Enabled** in the ABAP backend and follow specific rules, such as avoiding the use of global GUI elements.

SAP Integration Suite provides an RFC adapter that allows you to call remote-enabled function modules directly from your iFlows. The adapter establishes a connection to the on-premises SAP backend, usually via the cloud connector, and allows you to define input and output parameters for the RFC call. Communication is done using CPIC or TRFC, depending on the scenario.

RFCs are useful for tightly coupled, transactional, and real-time communication. Common use cases include calling BAPIs, retrieving real-time master data, and triggering business logic in SAP ERP or SAP S/4HANA.

An RFC call structurally resembles a method call in that it has importing parameters (inputs), exporting parameters (outputs), tables, and optional exceptions. When they're integrated via SAP Integration Suite, the input and output structures are represented as XML and can be mapped and processed within the iFlow like any other message.

Despite the increasing shift toward APIs and OData services, RFC remains a foundational integration method in SAP landscapes, especially when legacy systems, performance-sensitive processes, or complex business logic are involved. The RFC adapter in SAP Integration Suite allows developers to reliably standardize the connection between modern, cloud-based iFlows and the proven backend logic of traditional SAP systems.

You can use RFC to integrate on-premise ABAP systems with the cloud connector. You can also use it with the SAP S/4HANA Cloud system, which is available on the public internet.

As a prerequisite, you must define the required RFC destination in the SAP BTP cockpit for your application. These settings are used by SAP Java Connector (SAP JCo) to establish and manage connections with on-premises and the SAP S/4HANA Cloud systems that are available on the public internet. When you're configuring a destination, select **Internet** as the **Proxy Type** to establish a connection with an application over the Internet.

Configuring the RFC receiver adapter is very easy. All you need to do is enter the name of the **Destination** (see Figure 4.25).

To successfully implement an RFC-based integration, you need the corresponding XSD of the function module. This definition is provided in the form of a WSDL file.

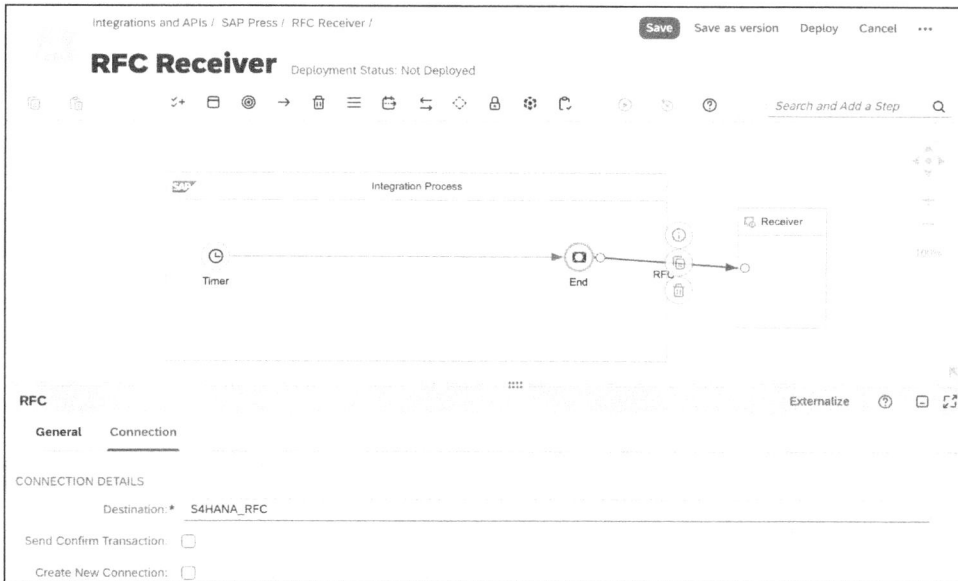

**Figure 4.25**  Configuring iFlow RFC Receiver Adapter

You can generate the WSDL yourself, provided that you have access to the target back-end system (e.g., an SAP ERP or SAP S/4HANA system) where the function module is available. There is an ICF service that provides the WSDL for a function module, and you can retrieve the WSDL file for the function module you want via the following URL:

*https://<backend_host>:<backend_https_port>/sap/bc/soap/wsdl11?services=<function_module_name>*

Replace the *<backend_host>*, *<backend_https_port>*, and *<function_module_name>* placeholders with the corresponding values for your system.

To execute an RFC call against an on-premises system, you must create an RFC protocol mapping in cloud connector, as shown in Figure 4.26. The virtual host is needed in the destination, and you must also list the desired function modules as resources.

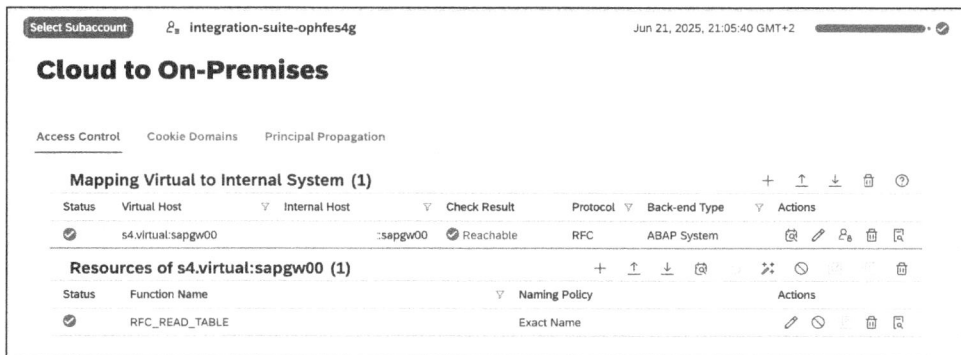

**Figure 4.26**  Cloud Connector Mapping to Internal System

Next, create an RFC destination in the subaccount in SAP BTP (see Figure 4.27). Select **OnPremise** for the **Proxy Type** and choose the authentication mechanism you want for the **Authorization Type** field.

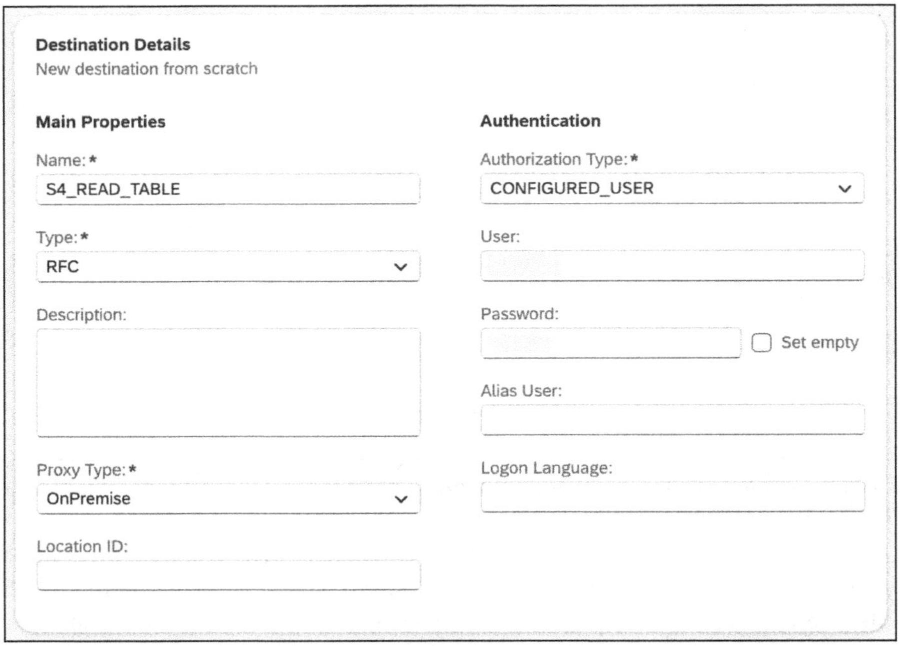

**Figure 4.27** SAP BTP Destination

In the **Target System Configuration** area, enter the virtual host name in the **Application Server Host** field and the system number in the **System Number** field. Also, enter the client of the target system in the **Client** field (see Figure 4.28).

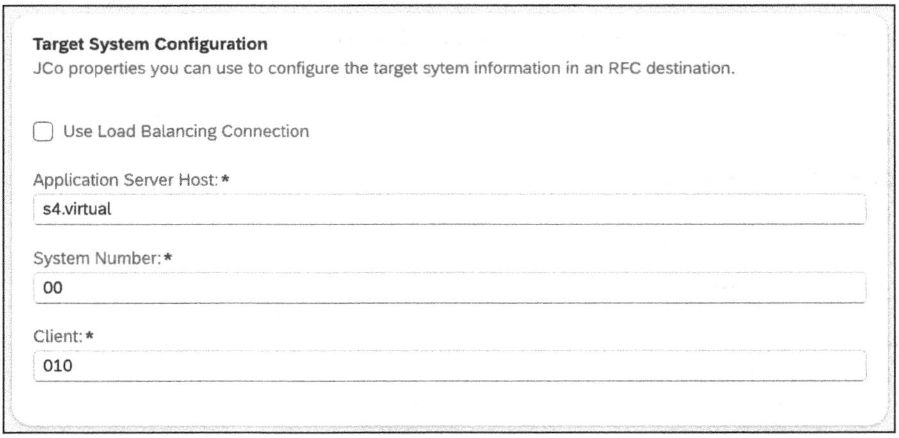

**Figure 4.28** Destination Target System Configuration

### 4.3.3   SAP SuccessFactors Adapter

SAP SuccessFactors is the leading cloud-based human capital management (HCM) solution in the SAP portfolio. It integrates with other SAP systems, such as SAP S/4HANA and SAP ERP, as well as third-party applications, via standardized APIs that you can connect by using various adapters in the SAP Integration Suite. Four central adapter types are available for SAP SuccessFactors: OData V2, OData V4, REST, and SOAP. However, depending on the module and API technology, the area of application is subject to various restrictions.

For OData and REST APIs, you perform authentication via either HTTP basic authentication or OAuth 2.0. We recommend using OAuth 2.0 because HTTP basic authentication is deprecated. As you'll see in the following sections, the following adapters are available:

- OData v2 receiver
- OData v4 receiver
- REST sender and receiver
- SOAP sender and receiver

First, however, we'll discuss OAuth 2.0 configuration.

#### OAuth 2.0 Configuration

To use OAuth2 for authentication for SAP SuccessFactors, you need to configure it correctly. To do this, you must first create a key pair in Cloud Integration by going to the **Manage Keystore** app and clicking **Create · Key Pair**, as shown in Figure 4.29.

**Figure 4.29**  Creating Key Pair

Then, maintain the necessary parameters as shown in Figure 4.30. The **Common Name (CN)** is particularly important because it's used to identify the user. Finally, click the **Create** button.

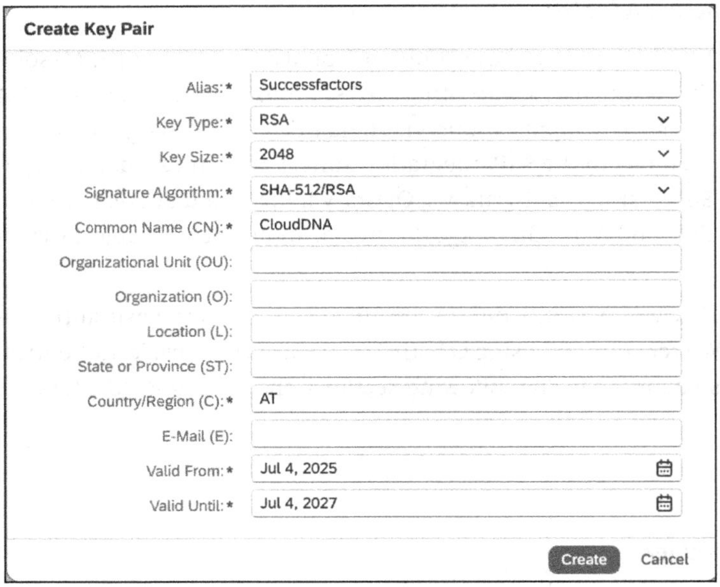

**Figure 4.30** Key Pair Configuration

Then, download the certificate by clicking on **Download Certificate** in the entry you created earlier (see Figure 4.31).

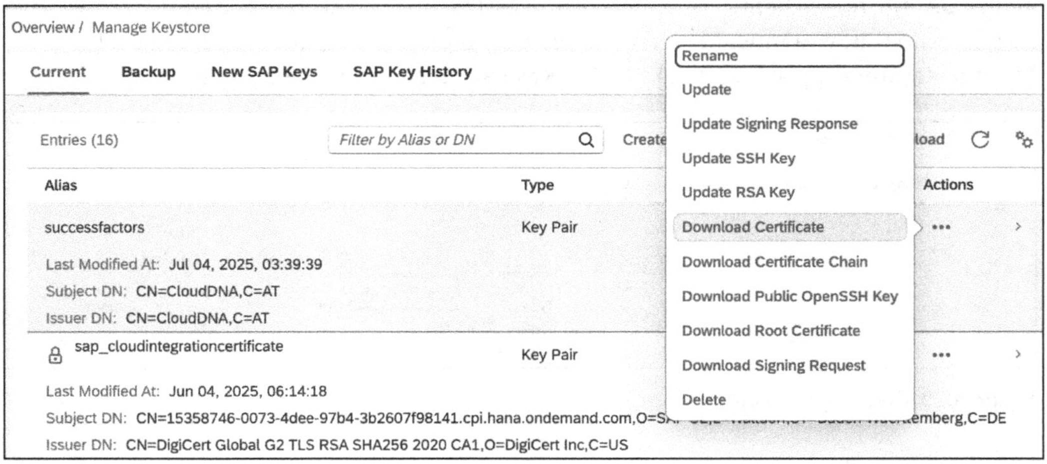

**Figure 4.31** Downloading Certificate

Next, you must complete the configuration in SAP SuccessFactors. Open the Manage OAuth2 Client Applications app (see Figure 4.32), assign an **Application Name**, and enter the URL of the tenant management node in the **Application URL** field. Then, copy the contents of the X.509 certificate into the **X.509 Certificate** field, remove the lines that say -----**BEGIN CERTIFICATE**----- and -----**END CERTIFICATE**-----, and click **Register**.

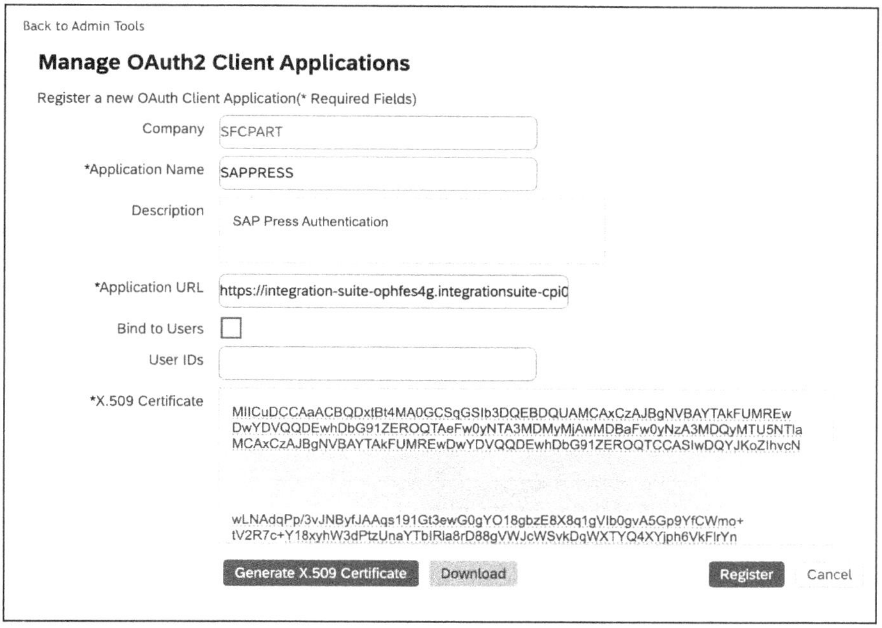

**Figure 4.32**  Registering New OAuth Client Application

Then, you'll see an additional field called **API Key** (see Figure 4.33). Copy the API key to the clipboard for further configuration.

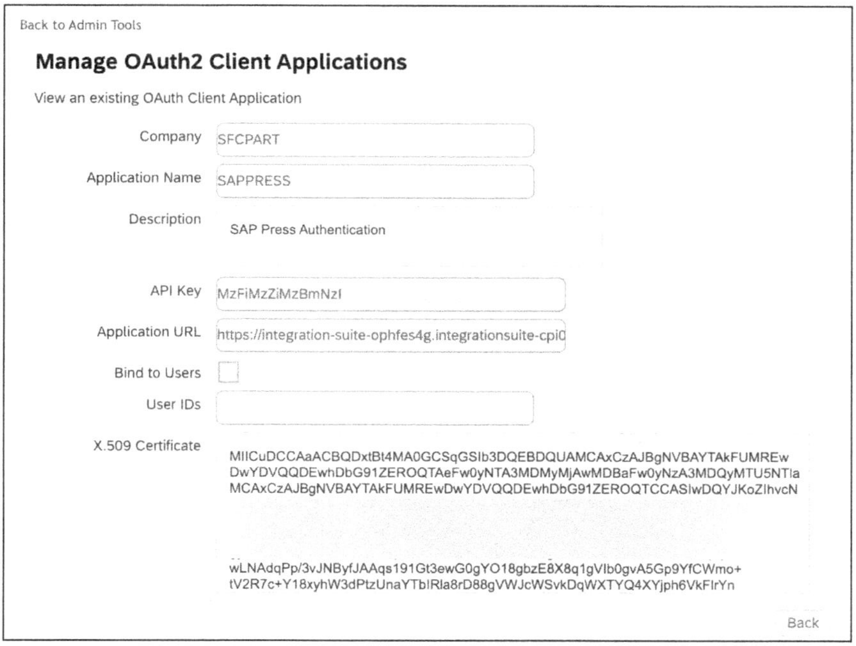

**Figure 4.33**  Copying API Key

AT that point, you can complete the configuration by triggering the creation of a new entry in Cloud Integration in the Manage Security Material app (see Figure 4.34). Do this by clicking the **Create** arrow and selecting **OAuth2 SAML Bearer Assertion** from the dropdown list.

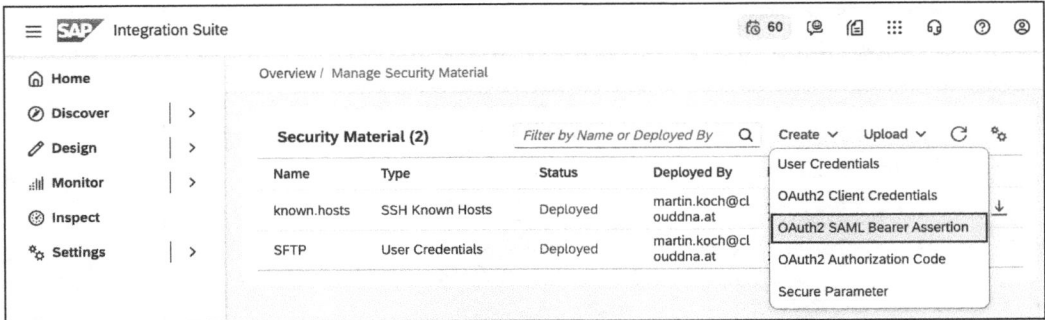

**Figure 4.34** Creating OAuth2 SAML Bearer Assertion

Then, enter "www.successfactors.com" in the **Audience** field and copy the API key from the clipboard into the **Client Key** field. Maintain the **Token Service URL** for the API server you're using, select **SuccessFactors** as the **Target System Type**, enter the SAP SuccessFactors company ID in the **Company ID** field, and select **Key Pair Common Name (CN)** for the **User ID** attribute. Then, in the **Key Pair Alias** field, enter the name of the key pair you previously created (making sure that the value is case sensitive), and finally, click **Deploy** (see Figure 4.35).

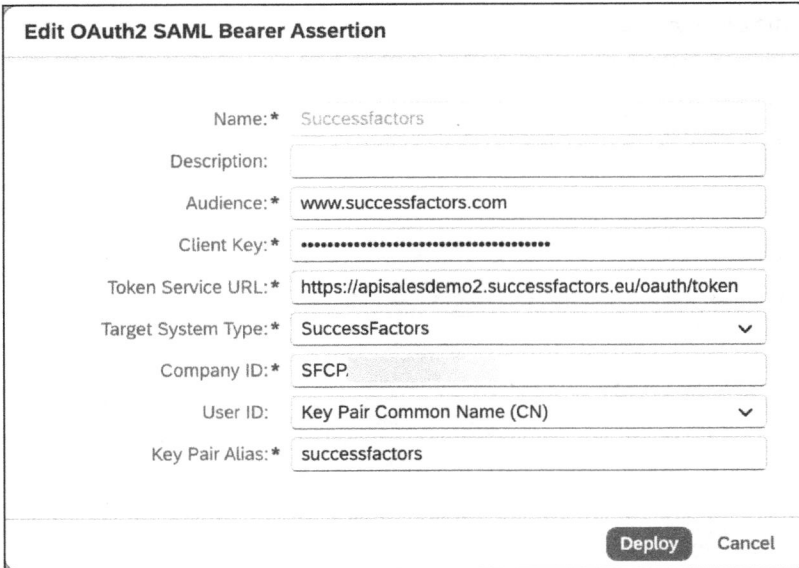

**Figure 4.35** Configuring OAuth2 SAML Bearer Assertion

## OData Version 2 Receiver

The OData V2 adapter is the standard for integrating with SAP SuccessFactors. Many SAP SuccessFactors APIs, particularly those that are related to SAP SuccessFactors Employee Central, use the OData V2 protocol. It's intended to be used for accessing master data such as users, organizational units, positions, and time data, and it's ideal for bidirectional scenarios with SAP ERP or SAP S/4HANA. Typical use cases include the following:

- Integration of employee master data from SAP S/4HANA into SAP SuccessFactors Employee Central
- Synchronization of recruiting data
- Integration with the onboarding module
- Custom use cases with SAP SuccessFactors extensions

On the **Connection** tab, you need to define the following parameters (see Figure 4.36):

- **Address**
  Specify the base URL of the SAP SuccessFactors OData API endpoint (e.g., "https://api4.successfactors.com/").
- **Address Suffix**
  Specify the address suffix for the OData v2 endpoint. The default value is **/odata/v2**.
- **Credential Name**
  Reference a preconfigured credential (**Basic** auth or **OAuth2**) from the SAP BTP credential store.

**Figure 4.36**  SAP SuccessFactors Receiver Adapter Connection Settings

In the **Processing** tab, you need to define the following parameters (see Figure 4.37):

- **Operation Details**
  Select the type of operation to perform on the entity, such as **Query (GET)**, **Create (POST)**, **Update (PUT)**, **Read (GET)**, **Delete**, or **Upsert (POST)**.

- **Resource Path**
  Specify the entity or resource on which the operation is performed. You can use the **Model Operation** wizard to help you make the right selection.

- **Fields (for PUT and POST operations only)**
  Define which entity fields will be included in the operation.

- **Query Options (for GET operations only)**
  Add parameters like $top and orderby, which must follow SAP SuccessFactors query structure guidelines.

- **Enable Batch Processing (for POST – Upsert only)**
  Check this box to allow multiple operations to be sent together in a $batch request.

- **Custom Query Options**
  Select this to supports additional query parameters that are specific to SAP SuccessFactors, such as purge.

- **Content Type (for PUT and POST only)**
  Choose **Atom** or **JSON** as the payload format.

- **Content Type Encoding (for PUT and POST only)**
  Choose **None** or **UTF-8 encoding** for the payload.

- **Pagination (for GET – Query only)**
  Define how query results are paginated. Options include **Server Snapshot-Based** (default), **Client**, or **Server Cursor-Based**.

- **Page Size (for GET only)**
  Set the maximum number of records per page. The default is 1,000 if you leave this field empty.

- **Retry on Failure**
  Check this box to enable automatic retries to handle intermittent network issues. Retries are based on specific HTTP status codes and vary by operation type.

- **Process in Pages (for GET; query only)**
  Check this box to enable batch processing of large data sets in a looped local process that's defined by **Page Size** and loop conditions.

- **Timeout**
  Set the maximum waiting time for a response before terminating the connection.

- **Request Headers**
  Enter a list of headers to send to SAP SuccessFactors (pipe-separated or "*" for all).

- **Response Headers**
  Enter a list of headers to read from the response (pipe-separated or "*" for all).

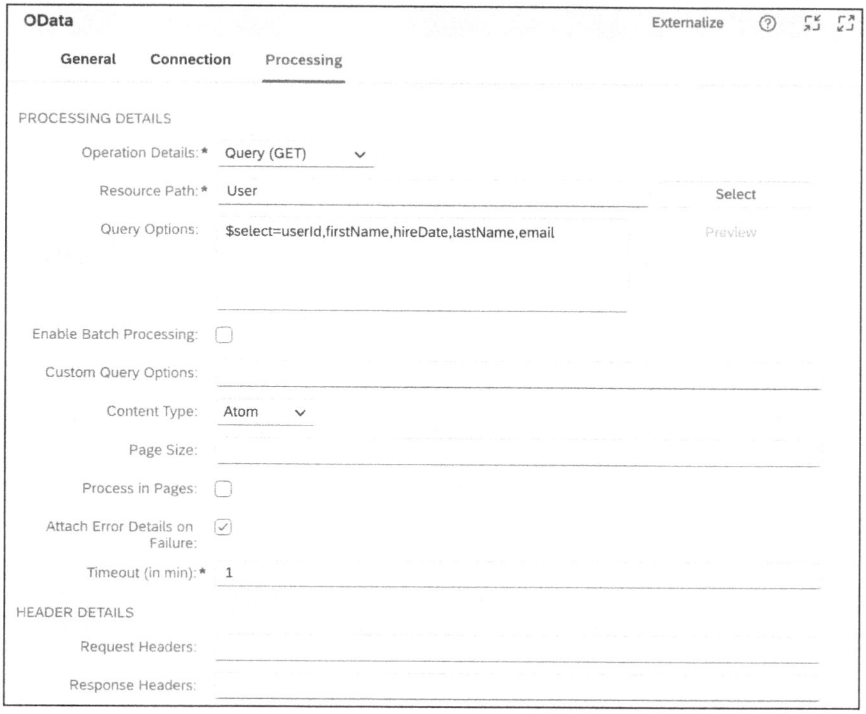

**Figure 4.37**  SAP SuccessFactors Receiver Adapter Processing Settings

### OData Version 4 Receiver Adapter

The OData V4 adapter supports the modern OData protocol, version 4, but it's currently used very little in the SAP SuccessFactors world. It only supports SAP SuccessFactors Learning OData V4 entities, and it can't be used with other SAP SuccessFactors modules or APIs.

When you're configuring the adapter, you must set up details in two main tabs: **Connection** and **Processing**. In the **Connection** tab, you define the following parameters:

- **Address**
  Enter the base URL of the SAP SuccessFactors OData V4 service endpoint. For SAP SuccessFactors Learning, this points to the SAP SuccessFactors Learning OData API endpoint.

- **Authentication**
  Define the authentication method. The adapter supports OAuth 2.0 client credentials and OAuth2 SAML bearer assertion.

- **Credential Name**
  Enter the name of the deployed security artifact.

- **Proxy Type**
  Specify whether to connect directly or via the cloud connector (on-premise). The default is **Internet**.

- **Reuse Connection**
  Check this box to reuse connection objects from the internal connection pool. This improves the network turnaround time for multiple communications to the same endpoint.

In the **Processing** tab, you define the following parameters:

- **Operation Details**
  Select the adapter for **Create (HTTP POST Request)**, **Update (HTTP PUT Request)**, or **Query (HTTP GET Request)**.

- **Resource Path**
  Enter the resource path of the entity that you want to access.

- **Query Options**
  Specify the query options and operation details that you want to send to the OData V4 service. This is only relevant for the query operation.

### REST Sender and Receiver

SAP SuccessFactors provides a variety of REST-based APIs, particularly for modern and mobile applications within SAP SuccessFactors Learning. You can use the REST sender adapter inbound and the receiver adapter outbound. However, the REST adapter only supports SAP SuccessFactors Learning REST APIs. It isn't suitable for any other modules.

The SAP SuccessFactors REST sender adapter allows Cloud Integration to periodically poll SAP SuccessFactors by using the provided REST APIs. Unlike with a push-based approach, this adapter uses a scheduler to call SAP SuccessFactors endpoints at defined intervals and retrieve data. It's especially useful for scenarios where event-based triggers are not available or not required.

In the **Connection** tab, you define the following parameters:

- **Address**
  Enter the base URL of the SAP SuccessFactors Learning REST API endpoint that you want to poll.

- **Address Suffix**
  Enter the URL suffix for SAP SuccessFactors Learning.

- **Credential Name**
  Enter the name of the deployed security artifact (OAuth credentials) that will be used to authenticate with SuccessFactors. The adapter supports OAuth 2.0 client credentials only.

- **Proxy Type**
  Specify whether to connect directly or via the cloud connector (on-premise). The default is Internet.

- **Operation**
  The adapter supports GET requests only.

- **Entity**
  Specify the SAP SuccessFactors Learning entity you want to access.
- **Parameters**
  Specify the OData query parameter you want to send to SAP SuccessFactors.
- **Page Size**
  Enter the maximum number of records on each page of a response.

In the **Scheduler** tab, you define when and how often the SAP SuccessFactors REST sender adapter should execute its requests to the SAP SuccessFactors API. You can configure it by using these options:

- **Run Once**
  Select this option to execute the operation immediately after deploying the iFlow.
- **Schedule on Day**
  - **On Date**
    Specify the exact date when the operation should run.
  - **On Time**
    Specify the exact time on that date for execution.
  - **Every**
    Define a fixed interval (for example, every 15 minutes) at which the operation should run repeatedly.
  - **Time Zone**
    Choose the time zone to use for the preceding settings.
- **Schedule to Recur**
  - **Daily**
    Specify a daily recurrence, including the time or interval and time zone.
  - **Weekly**
    Select specific weekdays for execution and define the time or interval.
  - **Monthly**
    Choose a specific day of the month when you want to execute the operation, along with the time or interval.

### SOAP Sender and Receiver

Some older SAP SuccessFactors interfaces, particularly those in legacy modules such as SAP SuccessFactors Onboarding 1.0 and SAP SuccessFactors Learning reports, are based on SOAP. The SAP SuccessFactors SOAP API is also known as SFAPI, and you can accessed these APIs by using the generic SOAP adapter of SAP Integration Suite. The purpose of the APIs is to provide access to older SOAP-based services, mainly reporting, onboarding, and parts of learning. However, SOAP isn't recommended for new integrations and is only used to support existing legacy processes. Please also be aware that the SFAPI was deprecated as of August 1, 2018.

The SAP SuccessFactors SOAP sender adapter in Cloud Integration polls data from SAP SuccessFactors by using SOAP-based APIs. Like the REST sender adapter, it operates on a scheduler basis, triggering requests at configured intervals instead of receiving data that's pushed from SuccessFactors.

In the **Processing** tab, you define the technical connection and data retrieval details:

- **Address**
  Enter the URL of the SuccessFactors data center you want to connect to. You can select it using a built-in browsing option.

- **Address Suffix**
  This field is automatically filled in as */sfapi/v1/soap* when you choose the SOAP message protocol.

- **Authentication**
  Choose one of the following authentication methods:
  - Basic Authentication
  - OAuth2 SAML bearer assertion, which supports either principal propagation or technical user propagation.

- **Credential Name**
  Enter the name of the deployed credential artifact on your tenant used for authentication.

- **Proxy Type**
  Select the type of proxy you want to use to connect to SAP SuccessFactors. If you select **Manual**, you must also enter one of the following:
  - **Proxy Host**
    This is the proxy server hostname.
  - **Proxy Port**
    This is the port number of the proxy server.

- **Call Type**
  Define the type of request:
  - The default is **Synchronous Query**.
  - Alternatively, you can choose **Asynchronous/Adhoc Query** for ad hoc operations.

- **Operation**
  Only GET is supported, and it's used to read information from SAP SuccessFactors.

- **Entity**
  Select or modify the entity in SAP SuccessFactors that you want to access.

- **Query**
  Specify your query criteria to filter data from the desired table or field.

- **Parameters**
  Enter additional key-value pairs that are required by SuccessFactors to retrieve the relevant information.

- **Page Size**
  Define the maximum number of records to fetch per page in the response.
- **Timeout**
  Set the maximum time (in minutes) to wait for a response from SuccessFactors before timing out.

In the **Scheduler** tab, define when and how often the adapter should execute the SOAP request:

- **Run Once**
  Select this to execute the request once, immediately after deployment.
- **Schedule on Day**
  - **On Date**
    Enter the date on which to execute the request.
  - **On Time**
    Enter the time of day when to execute.
  - **Every**
    Enter the interval at which to repeat the request.
  - **Time Zone**
    Enter the time zone to use for scheduling.
- **Schedule to Recur**
  - **Daily**
    Specify the time or interval and time zone for daily execution.
  - **Weekly**
    Select the days of the week and specify the time or interval and time zone.
  - **Monthly**
    Choose a day of the month plus the time or interval and time zone.

You use the SAP SuccessFactors SOAP receiver adapter in Cloud Integration to send data from your iFlow to SAP SuccessFactors by using SOAP APIs. It enables you to perform operations such as creating, updating, or querying data in SAP SuccessFactors via SOAP-based services.

On the **Connection** tab, you set up how to technically connect to SAP SuccessFactors:

- **Address**
  Enter the URL of the SAP SuccessFactors data center endpoint you want to connect to. You can browse and select the correct data center URL.
- **Address Suffix**
  This is automatically populated as *sfapi/v1/soap* when you're using the SOAP protocol.
- **Authentication**
  You can choose between the following:
  - **Basic Authentication**
  - **OAuth2 SAML Bearer Assertion** (with principal or technical user propagation).

- **Credential Name**
  Enter the name of the deployed security artifact (credential) on your tenant.
- **Proxy Type**
  Specify how to connect to SAP SuccessFactors, either via the internet or on premise (via the cloud connector). If you select **Manual**, you must also enter the following:
  - **Proxy Host**
    This is the hostname of the proxy server.
  - **Proxy Port**
    This is the port number of the proxy server.

In the **Processing** tab, you define how data should be sent and processed:

- **Call Type**
  - The default is **Synchronous Query**.
  - Select **Asynchronous/Adhoc Query** if you want to execute an ad hoc operation.
- **Operation**
  Select the operation to perform on the entity in SAP SuccessFactors. Supported operations are **Query, Delete, Insert, Update, and Upsert**.
- **Entity**
  Select the SuccessFactors entity you want to access, such as **CompoundEmployee, JobPosting, or PerEmergencyContacts**.
- **Query**
  Define your query to filter or specify the data you want. For example, you can fetch only job postings with **postingStatus = open**.
- **Parameters**
  Enter additional key-value pairs as needed for your query or operation.
- **Page Size**
  Enter the maximum number of records to fetch in one page of the response.
- **Timeout**
  Enter the maximum time (in minutes) the system should wait for a response.
- **Retry on Connection Failure**
  Select this if you want the adapter to retry if there's a connection failure.
- **Process in Pages**
  Select this to enables batch (paged) processing of messages, where page size is defined by the **Page Size** value. Note the following:
  - You can't use this with query operation if the process call step is in a multicast branch.
  - You must use this inside a local integration process that's invoked by a looping process call step.

## 4.4   API and Web Service Integration

Cloud Integration's adapters connect a wide range of systems and protocols. The HTTP, SOAP and OData adapters are especially crucial in this context, so we'll discuss each of them in this section. Each adapter is designed to meet a specific technical use case, and it ensures flexible, secure, and standardized communication with external systems.

The *HTTP adapter* is the most flexible of the three, and you can use it for everything from simple REST calls to complex, custom-designed interfaces. It's the ideal solution for communicating with third-party systems and user-defined APIs for which no special adapter is available. You can define headers, URL parameters, and payloads via HTTP.

The *SOAP adapter* integrates using the widely used SOAP protocol (which is primarily used in classic enterprise applications) and older web services. It's ideal for scenarios that require strongly typed XML messages, fixed WSDL definitions, and WS-Security. SOAP is a staple in the SAP world, particularly for integrating SuccessFactors, Ariba, and older on-premise systems.

The *OData adapter* is the best choice for modern, REST-based integrations with SAP systems. Open Data Protocol (OData) is a standardized protocol that's used primarily for Fiori apps, S/4HANA APIs, and SuccessFactors OData Services. The adapter supports both OData V2 and OData V4 to give developers maximum flexibility to efficiently implement create, read, update, delete (CRUD) operations.

---

**Cross-Site Request Forgery**

*Cross-site request forgery* (CSRF) is a security measure that protects against unwanted or malicious actions being performed on a server on behalf of a logged-in user. In Cloud Integration (e.g., in the HTTPS sender adapter), CSRF protection ensures that only requests with a valid CSRF token are accepted, especially for changing operations (e.g., POST, PUT, DELETE). This protection is essential because without it, attackers could easily trigger actions by misusing the browser of a logged-in user. You should activate CSRF protection to ensure that every request is deliberately authorized and has not been injected by third parties. This is a central component of modern web security.

---

### 4.4.1   HTTP Adapter

In this section, we'll start by exploring the HTTP sender adapter, which support HTTPS only. Then, we'll dive into the specifics of the HTTP receiver adapter, and finally, we'll discuss typical HTTP error codes.

**HTTP Sender Adapter**

The *HTTPS sender adapter* in Cloud Integration is used to expose iFlows so that external systems can send data into your integration scenario via HTTPS. It ensures secure inbound communication using standard *HTTP methods* (GET, POST, PUT, DELETE, etc.).

Cloud Integration supports Hypertext Transfer Protocol Secure (HTTPS) exclusively for inbound communication by using the HTTPS sender adapter. Plain (i.e. unencrypted) HTTP isn't allowed, and this restriction is for security reasons: HTTPS ensures that all data exchanged between the sender system and your iFlow is encrypted and protected from interception or manipulation. If there's no encryption (as in plain HTTP), sensitive business data—such as personal information, credentials, and financial details—is susceptible to exposure during transmission.

In the **Connection** section (see Figure 4.38), you configure how external systems access your iFlow:

- **Address**
  Specify a relative path to define the endpoint for your iFlow (e.g., "/myIntegrationFlow"). When you deploy this path, it will be appended to the service instance URL with */http* to form the complete endpoint (e.g., *https://<service-instance>/http/myIntegrationFlow*). The address supports characters such as ~, -, ., $, *, and alphanumeric values. There are certain restrictions on where you can place these characters, but you can also use wildcards like */\** for dynamic endpoint support.

- **CSRF Protected**
  Check this box to enable CSRF protection so that the adapter will validate modifying requests by requiring an X-CSRF-Token header. If a GET or HEAD request is received to fetch a token, it will return the token and stop further processing.

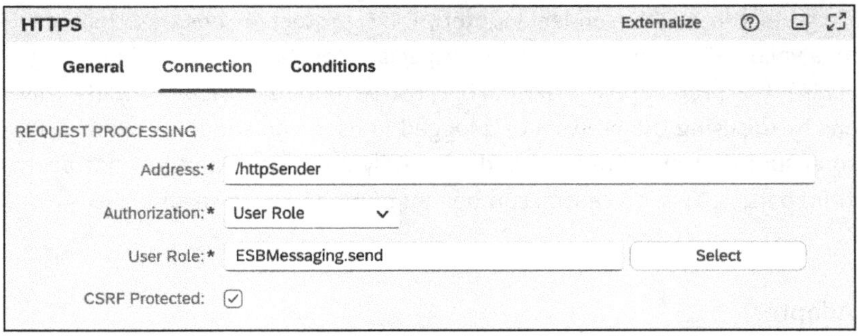

**Figure 4.38** HTTP Sender Adapter Connection Settings

**HTTP Receiver Adapter**

You use the HTTP receiver adapter in Cloud Integration to send HTTP requests to external target systems. It supports a wide range of use cases, from simple REST calls to complex API integrations. It offers extensive configuration options for HTTP methods, authentication, and headers.

In the **Connection** tab (see Figure 4.39), you need to configure the technical connection to the target system:

- **Address**

  Enter the URL of the target system (e.g., *"https://mysystem.com"*). Please keep in mind that HTTPS is strongly recommended, especially when you're using basic or client certificate authentication. You shouldn't include URL parameters here—they belong in the **Query** field.

- **Query**

  Enter a string of parameters to be appended to the HTTP request. It must be URL-encoded if it contains special characters.

- **Proxy Type**

  - **Internet**

    Select this for cloud systems.

  - **On-Premise**

    Select this for systems accessed via the cloud connector.

  - **Manual**

    Select this to allow manual configuration of the proxy host and port. (This is only available with SAP Process Orchestration.)

- **Method**

  The supported HTTP methods you can select include **POST**, **GET**, **HEAD**, **PATCH**, **TRACE**, and **Dynamic** (meaning dynamically defined at runtime, using expressions).

- **Send Body**

  Check this box to enable sending a message body for GET, DELETE, HEAD, or dynamic methods. By default, these don't include a body.

- **Expression**

  You can use this with the dynamic method to specify the HTTP method via a header or property (e.g., "${header.methodType}").

- **Authentication**

  Select the authentication method against the target system. The supported options are as follows:

  - **None**

    You'll select this when setting the authorization header manually in a **Content Modifier** or **Script** step.

  - **Basic**

    This is a user name and password.

  - **OAuth2 Client Credentials**

  - **OAuth2 SAML Bearer Assertion**

  - **Client Certificate**

  - **Principal Propagation**

    Only use this with on-premise systems.

- **Credential Name**
  This references a deployed credential artifact (user credentials or OAuth credentials), and you use it when **Authentication** is set to **Basic**, **OAuth2**, or **SAML**.

- **Private Key Alias**
  This is required when you're using client certificate authentication; you specify the private key from the keystore.

- **Timeout**
  Specify the maximum time in milliseconds to wait for a response. The default is 60,000 ms.

- **Throw Exception on Failure**
  This box is checked by default, and it throws an exception if the HTTP response indicates failure (with a non-2xx status code).

- **Attach Error Details on Failure**
  When you check this box, it attaches request/response headers and a body to the message log in case of failure.

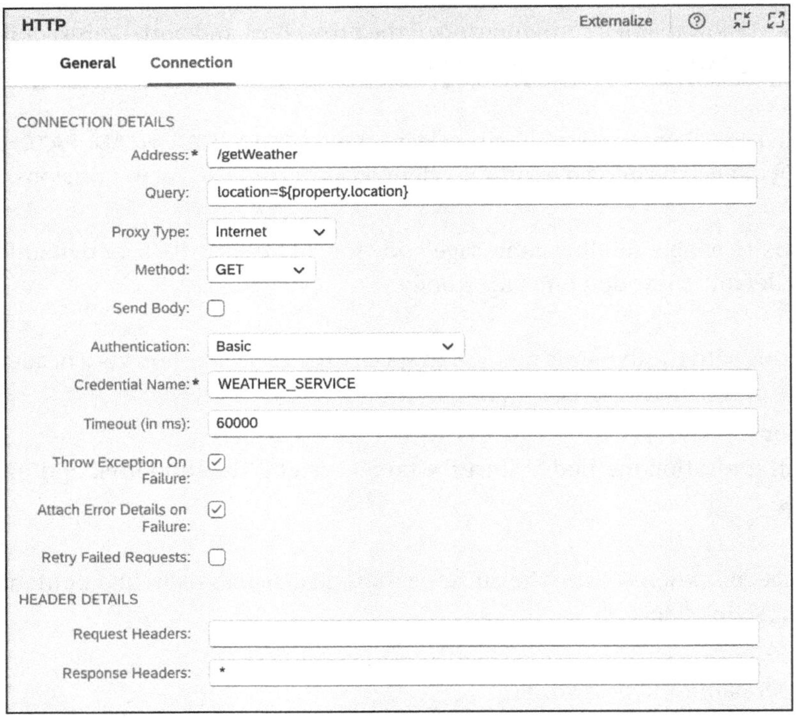

**Figure 4.39**  HTTP Receiver Connection Settings

- **Retry Failed Requests**
  Check this box to allow retrying of failed requests. Additional related options include the following:

- **HTTP Error Response Codes**
  You can enter a comma-separated list of HTTP codes that should trigger retries.

- **Retry Interval**
  You can enter the time (in seconds) to wait between retries (up to 60 seconds).

- **Retry Iterations**
  You can enter the number of retry attempts (up to 3).

**Typical HTTP Errors in Cloud Integration**

Communication via HTTP(s) is a central element in cloud-based integration scenarios, and that makes it all the more important for you to understand the typical HTTP error codes and their causes. Only then can you design iFlows that are stable and error-resistant and analyze problems quickly. Here is an overview of the most important HTTP error codes that you may encounter in cloud integration:

- **400 Bad Request**
  This error usually indicates that the request is syntactically incorrect. Common causes are incorrectly formatted JSON or XML payloads, invalid headers, and missing mandatory parameters in the URL. Tip: check the payload and header carefully in tools such as Postman or directly in message monitoring.

- **401 Unauthorized**
  This error indicates authentication has failed. In most cases, the stored user information (e.g., basic authorization credentials or OAuth tokens) is incorrect or the target API expects additional headers (e.g., API keys).

- **403 Forbidden**
  This error indicates that authentication is OK but authorization is missing. This is typical for APIs with granular roles or restricted access to certain resources. Check whether the user or token used has the necessary permissions.

- **404 Not Found**
  This error indicates that the requested endpoint doesn't exist or is misspelled. This often happens with dynamically composed URLs or when API versions are specified incorrectly.

- **405 Method Not Allowed**
  This error indicates that the HTTP method (e.g., **GET**, **POST**, **PUT**, **DELETE**) isn't allowed for the requested endpoint. For example, it will happen when an endpoint only accepts GET requests and the iFlow tries to access it with POST.

- **500 Internal Server Error**
  This is the classic error, and it indicates that a problem on the server side exists and isn't specified in detail. Causes can include faulty business logic, technical errors in the backend system, and overload. Logs and monitoring details are essential here.

- **503 Service Unavailable**
  This error indicates that the server is temporarily unavailable. It can be caused by maintenance windows, peak loads, and network problems. In practice, we recommend that you use a retry mechanism to deal with this error.

In Cloud Integration, you should not only monitor HTTP errors but also handle them appropriately, through retry policies, alternative routes, or dedicated error subprocesses in iFlow. Using detailed log outputs (at the trace level) and the iFlow trace tool can also help to quickly narrow down the source of errors.

### 4.4.2   SOAP Adapter

*Simple Object Access Protocol* (SOAP) is the widely used protocol for exchanging structured information in the implementation of web services. SOAP is message oriented and built on strict standards, and that makes it especially popular in enterprise environments. REST is resource oriented and often simpler. SOAP messages are XML-based and are typically sent over HTTP or HTTPS, and SOAP also supports other transport protocols, such as SMTP. SOAP boasts major strengths, including strong typing and a formal contract defined via WSDL. This makes integrations predictable and robust. SOAP also supports advanced features such as WS-Security, which enables message-level security (signing and encryption) and transaction handling. This makes it a preferred choice for scenarios that require strict security, reliability, and compliance—such as financial services, HR systems (like SAP SuccessFactors), and other mission-critical enterprise applications.

Thanks to its formal specifications and rich capabilities, SOAP remains a key integration technology in many SAP landscapes and beyond, despite the rise of more lightweight REST APIs. In SAP landscapes, two variants are often discussed: SOAP 1.x and SOAP SAP RM. Both are supported in Cloud Integration with dedicated adapters. SOAP 1.x is the standard for XML-based web service communication. SOAP SAP RM extends SOAP 1.x with SAP's reliable messaging capabilities, and that guarantees delivery and maintains strict message order in mission-critical integrations.

SOAP 1.x refers to the standard versions of SOAP (1.1 and 1.2) as defined by the W3C. It focuses on point-to-point communication, uses an XML-based message format, and usually runs over HTTP or HTTPS. SOAP 1.x supports extensions like WS-Security for message-level security and WS-Addressing for routing, but it doesn't inherently guarantee message delivery or ordering. Reliability features are optional and depend on implementation.

SOAP with SAP Reliable Messaging (SOAP SAP RM), on the other hand, builds on standard SOAP 1.x but adds a specific SAP implementation of reliable messaging. SOAP SAP RM's primary objective is to guarantee that messages are delivered exactly once (EO) and, when necessary, exactly once in order (EOIO), which means in a strict sequence. This is especially critical for business processes where data consistency and guaranteed delivery are essential, such as financial transactions and order processing. Reliable messaging is firmly integrated into SAP PI and SAP PO environments, and it provides built-in support for message persistence, automatic retries, and recovery mechanisms in case of temporary failures. That ensures that no data is lost and no duplicates occur.

### SOAP SAP RM Sender Adapter

When configuring a SOAP SAP RM sender adapter in Cloud Integration, you define key settings in the **Connection** and **Conditions** tabs. These settings determine how the iFlow exposes its SOAP endpoint and handles incoming messages. When configuring the **Connection** tab (see Figure 4.40), you provide the following properties:

- **Address**
  Enter the relative endpoint address at which the iFlow listens (e.g., "/HCM/GetEmployeeDetails").

- **URL to WSDL**
  **Enter t**he URL that points to the WSDL that defines the web service interface of the receiver. You can select a WSDL from an on-premise ES Repository or upload it from your local workspace. There are more features and restrictions on the WSDL:
  - If you don't provide a WSDL, a generic one will automatically be generated.
  - Each WSDL file you upload must contain exactly one service and be one-way (input-only).
  - You can also upload an archive (*.zip*) that contains WSDLs and XSDs if there are dependencies.
  - Download options for WSDL are available under **Manage Integration Content** in the **Monitor** view, but you can't parse WSDLs with external references.

- **Processing Settings**
  These are only visible in older adapter versions or certain product profiles. Select one of the following:
  - **Standard**
    If you select this option, messages will be processed with standard WS mechanisms and errors will not be returned to the consumer.
  - **Robust**
    If you select this option, service will be invoked synchronously and processing errors will be returned to the consumer.

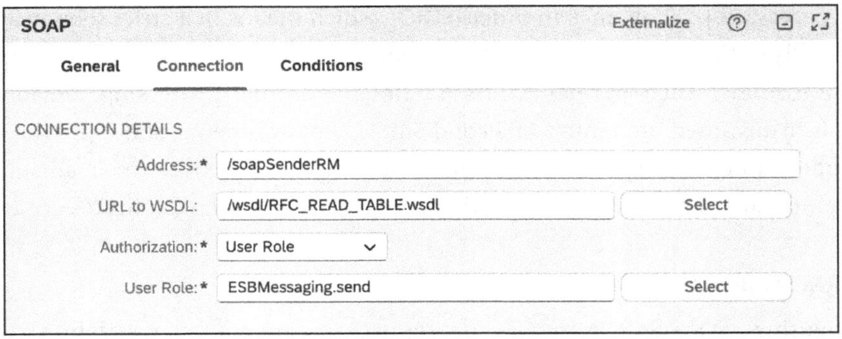

**Figure 4.40** SOAP Sender SAP RM

### SOAP 1.x Sender Adapter

When configuring the Connection tab (see Figure 4.41) for the SOAP 1.x sender adapter, you define several more key parameters than you do in the SOAP SAP RM sender:

- **Use WS-Addressing**
  This is an additional option that's unique to SOAP 1.x: you can check this box to enable WS-Addressing to accept addressing information directly from message headers provided by the sender at runtime.

- **Service and Endpoint**
  You provide the name of the service and the corresponding endpoint as defined in the referenced WSDL. Filling out these fields is required because it ensures precise interface binding.

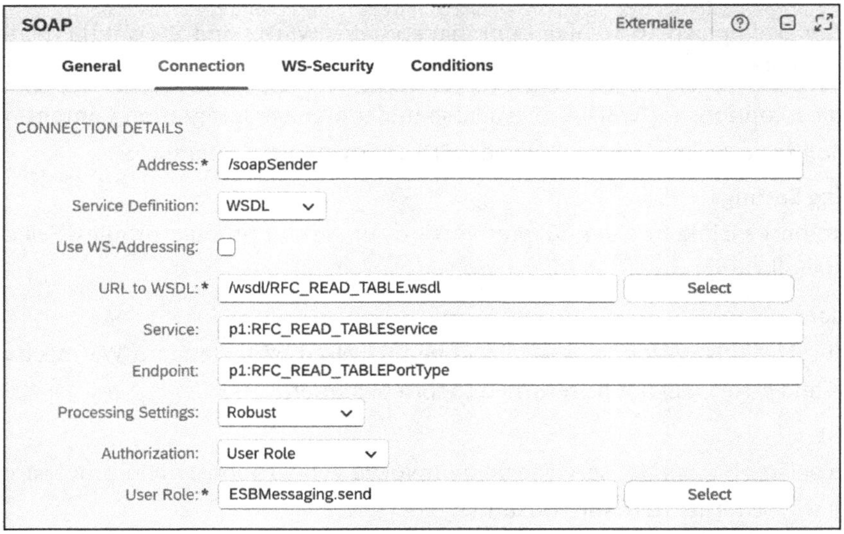

**Figure 4.41** SOAP Sender Connection Settings

- **Processing Settings**

  This is similar to the SOAP SAP RM adapter, where robust processing is implicitly activated in recent versions. The following options are available:

  – **WS Standard**

    If you select this option, messages will be processed with standard WS mechanisms and errors will not be returned to the consumer.

  – **Robust**

    If you select this option, the service will be invoked synchronously and any processing errors will be returned to the sender.

The **WS-Security** tab (see Figure 4.42) provides extensive options for securing messages, which is a major difference from the SOAP SAP RM adapter. Enter or select the following selections:

- **WS-Security Configuration**

  Choose between no security and manual channel configuration.

- **WS-Security Type**

  Select a security combination, like one of the following:

  – **Verify Message and Sign Response** (the default)

  – **Verify and Decrypt Message and Sign and Encrypt Response**

- **Private Key Alias for Response Signing**

  Enter an alias for the private key that will be used to sign response messages.

- **Public Key Alias for Response Encryption**

  You use this to encrypt the response if encryption is enabled.

- **Signing Order**

  Specify whether to sign before encrypting or vice versa.

- **Algorithm Suite Assertion**

  Define which cryptographic algorithms are required by the consumer.

- **Save Incoming Signed Message**

  Check this box if you want to store an incoming signed (and encrypted) message.

- **Check Time Stamp**

  Check this box only if the sender (the WS consumer) includes a time stamp with the message.

- **Sender is Basic Security Profile Compliant**

  Check this box if the sender system supports a dedicated security level that's described by the basic security profile (as most systems do). Only uncheck this box if the sender system doesn't support this security profile.

- **INITIATOR TOKEN**

  Make selections from the following dropdown lists:

- **Include Strategy**
  **Always to Recipient** is the default, and it guarantees that the WS consumer will send the certificate along with the message.
- **X509 Token Assertion**
  Specify the format of the certificate sent by the WS consumer, along with the message.

■ **RECIPIENT TOKEN**
  You can select from the following options:

- **Never**

- **Always to Initiator**
  Select this to add the certificate to the response message.

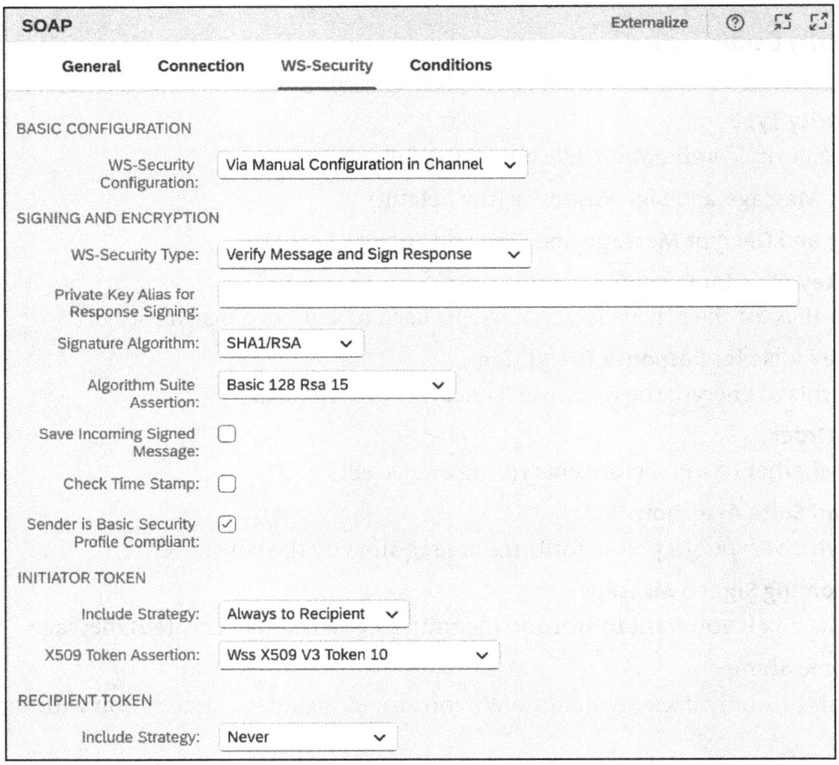

**Figure 4.42**  SOAP Sender WS-Security Settings

**SOAP SAP RM Receiver Adapter**

For the SOAP SAP RM receiver adapter, you configure where and how to deliver the outgoing SOAP message by making following entries and selections in the **Connection** tab:

■ **Address**
  Enter the endpoint URL of the target system, for example, "http://<host>:<port>/

payment." You can dynamically set this address at runtime by using headers or properties (e.g., "${header.a}"). The actual resolved endpoint URL is shown in the message processing log (MPL) under **RealDestinationUrl**.

- **URL to WSDL**
Enter the URL to the WSDL that defines the receiver's service interface. You can upload it from your local workspace or from an on-premise ESR. Please note that the adapter supports only one-way WSDLs (input only, with no synchronous responses).

- **Service and Endpoint**
Enter the names of the service and port in the referenced WSDL. In older adapter versions, using the same port name across multiple receivers isn't supported—you'll need a separate copy of the WSDL in those cases.

- **Operation Name**
Enter the specific operation to be called in the referenced service.

- **Proxy Type**
Define the connection path by selecting one of the following:

  - **Internet**
  Select this for cloud targets.

  - **On-Premise**
  Select this for on-premise systems via the cloud connector. If you select **On-Premise**, you can specify a **Location ID** to connect to a particular cloud connector instance.

- **Authentication**
Select one of the supported options:

  - **None**

  - **Basic**
  This requires a **User Credentials** artifact to be deployed on the tenant.

  - **Client Certificate**
  This requires a keystore and a private key alias.

- **Private Key Alias**
If you select **Client Certificate** under **Authentication**, enter the alias of the private key that's used for authentication. You can also set this dynamically by using a header or property.

- **Credential Name**
If you select **Basic** under **Authentication**, enter the name of the **User Credentials** artifact.

- **Timeout**
Enter the maximum time (in milliseconds) to wait for a response before the connection times out. The default is 60,000 ms.

- **Compress Message**
  Check this box to enable sending compressed requests and indicate that compressed responses are supported.

- **Allow Chunking**
  Check this box to enables HTTP chunked transfer encoding. (It's checked by default.)

- **Return HTTP Response Code as Header**
  Check this box to write the HTTP response code into the **CamelHttpResponseCode** header, which is useful for trace logs and postprocessing logic.

- **Clean Up Request Headers**
  Check this box to remove adapter-specific headers after the receiver call and preventing them from propagating further.

In the **Processing** tab, you define how to determine the message ID for reliable messaging:

- **SAP RM Message ID Determination**
  Here, you control how the target message ID is set to ensure EO semantics. Choose one of the following options:

  - **Generate**
    A new message ID is always generated.

  - **Reuse**
    This is the default option, and it makes the system reuse the message ID from the SapMessageIdEx header. If that message ID is not available, the system will generate a new ID.

  - **Map**
    This option maps a source message ID (from a header or property) to a new target message ID. It also creates a globally unique identifier (GUID) that links back to the original ID.

- **Source for SAP RM Message ID**
  If you select **Map** under **SAP RM Message ID Determination**, enter the header or property to dynamically provide the source message ID. If no header or property is found at runtime, a new message ID will be generated automatically. Mapping entries are stored for 90 days, and only the first 120 characters of the source ID are used.

### SOAP 1.x Receiver Adapter

In the **Connection** tab, you configure where and how to deliver the outgoing SOAP message (see Figure 4.43) for the SOAP 1.x receiver adapter. The main differences from configuring the SOAP SAP RM adapter are as follows:

- **Service**
  Specify the name of the service that's contained in the referenced WSDL.

- **Endpoint** (port)
  Specify the name of the endpoint of the service you selected in the **Service** field. This name is referenced in the specified WSDL.

- **Operation Name**
  Enter the name of the operation of a selected service, which you provide in the Service field. It's contained in the referenced WSDL.

- **Authentication**
  **OAuth 2.0 SAML Bearer Assertion Grant, Basic, and Client Certificate** are all supported. This allows dynamic configuration of **Credential Name** and **Private Key Alias** via headers or properties

- **Proxy Type**
  This allows you to specify the **Proxy Host** and **Proxy Port** manually. When you select **On-Premise**, you must use HTTP instead of HTTPS for virtual addresses and you can't use client certificate authentication.

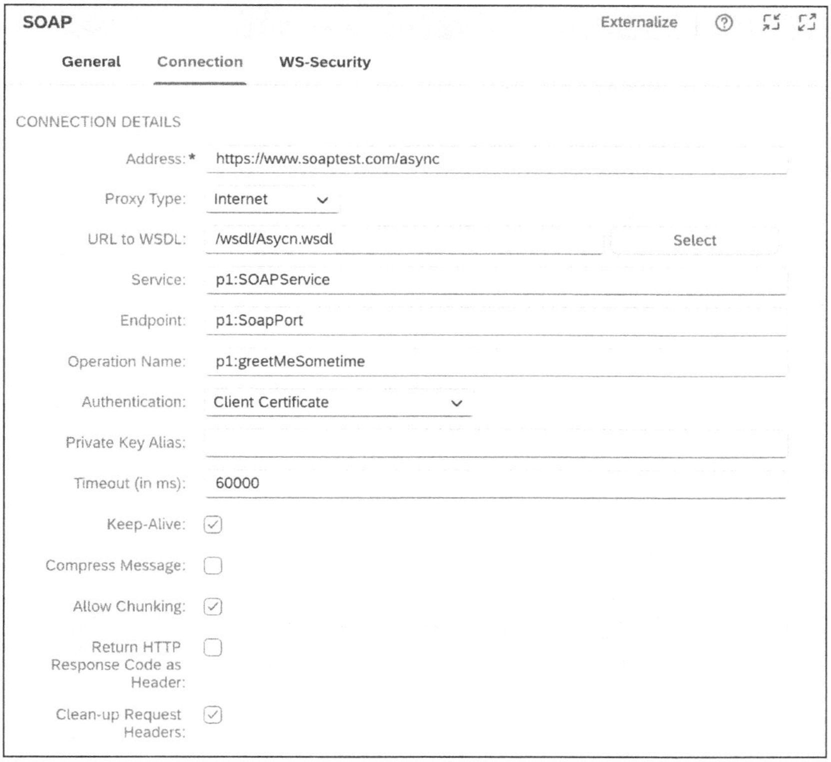

| SOAP | | Externalize | ⑦ | ⌜⌝ | ⌜⌝ |
|---|---|---|---|---|---|
| **General** | **Connection** | **WS-Security** | | | |

CONNECTION DETAILS

| | |
|---|---|
| Address: * | https://www.soaptest.com/async |
| Proxy Type: | Internet ⌄ |
| URL to WSDL: | /wsdl/Asycn.wsdl     Select |
| Service: | p1:SOAPService |
| Endpoint: | p1:SoapPort |
| Operation Name: | p1:greetMeSometime |
| Authentication: | Client Certificate ⌄ |
| Private Key Alias: | |
| Timeout (in ms): | 60000 |
| Keep-Alive: | ☑ |
| Compress Message: | ☐ |
| Allow Chunking: | ☑ |
| Return HTTP Response Code as Header: | ☐ |
| Clean-up Request Headers: | ☑ |

**Figure 4.43** SOAP Receiver SOAP 1.x Connection Settings

The **WS-Security** tab configuration in the SOAP 1.x receiver adapter offers the following options for securing messages at the message level and authenticating the sender (see Figure 4.44). Make the following entries and selections on this tab:

- **WS-Security Configuration**
  Choose how to configure security by selecting one of the following options:
  - **Via Manual Configuration in Channel**
    Security settings are defined manually in the adapter.
  - **Based on Policies in WSDL**
    Security settings are taken from WS-Policy definitions in the receiver's WSDL.
  - **None**
    No WS-Security will be applied.
- **USERNAME TOKEN**
  This allows authentication with a username and password (inside the SOAP header). The options are to use a plain text password, a hashed password (when using manual configuration), or none.
- **Credential Name**
  You must provide this if you use a token is used.
- **WS-Security Type**
  If you use manual configuration, you must define how to protect the message by choosing one of the following options:
  - **Sign Message**
    The message will be signed.
  - **Sign and Encrypt Message**
    The message will be signed and encrypted.
  - **None**
    No security will be applied.
- **Private Key Alias for Signing**
  Enter the alias of the tenant's private key (from the keystore) that will be used to sign the message.
- **Public Key Alias for Encryption**
  If you select **Sign and Encrypt Message WS-Security Type,** enter the alias of the receiver's public key that will be used for encryption.
- **Set Time Stamp**
  Check this box to add a timestamp to the message; if you're using a request-response pattern, a timestamp will be expected in the response as well.
- **Receiver is Basic Security Profile Compliant**
  This box is checked by default, and you should uncheck it if the receiver doesn't support the basic security profile.
- **Layout**
  Here, you control the order of elements in the security header. Select one of the following options:

– **Strict**
Elements will be added in a strict "declare before use" order.

– **Lax**
This option allows more flexible ordering by following general WSS standards.

■ **Algorithm Suite Assertion**
Define which algorithms the receiver is allowed to use.

■ **Initiator Token**
If you're using manual configuration, use this to control certificate-related policies for the WS consumer, including the following:

– The include strategy (e.g., "Always to Recipient")

– X.509 token formats

– Additional references like the issuer serial, key identifier, and thumbprint

■ **Recipient Token**
If you're using manual configuration), use this to define policies for returning the certificate to the receiver, including the following:

– The include strategy (e.g., "Always to Initiator")

– X.509 token formats and references

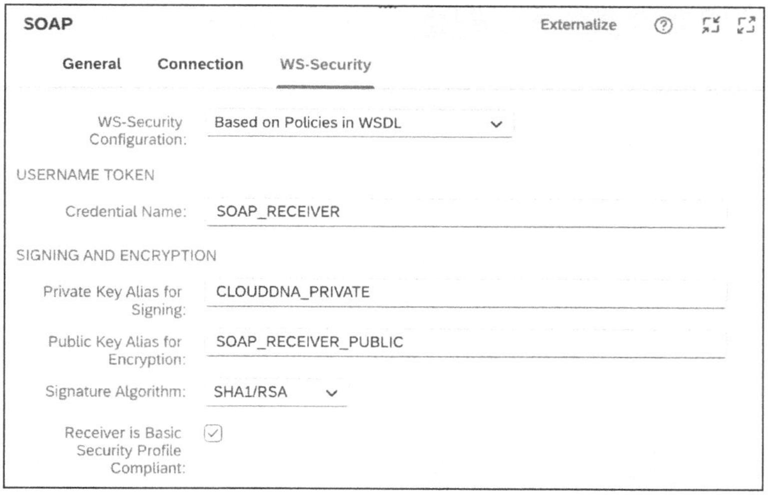

**Figure 4.44** SOAP Receiver SOAP 1.x WS-Security Settings

### 4.4.3   OData Adapter

In this section, we'll start by having a look at the OData sender adapter, which allows communication partners to send messages via the OData protocol to Cloud Integration. Thereafter, we'll explore the OData receiver adapter, which allows you to send data to a receiver system by using the OData protocol.

**OData Sender Adapter**

The OData sender adapter in Cloud Integration allows you to expose an iFlow as an OData service and thus enable external systems to call it using OData requests. You'll perform the configuration mainly in the **Adapter Specific** tab (see Figure 4.45) by entering or selecting the following values:

- **Authorization**
  Define how to authorize clients calling the OData service by selecting one of the following options:

  - **User Role**
    This option authorizes clients based on a predefined user role (ESBMessaging. send). You select the required role from the tenant's available user roles.

  - **Client Certificate**
    This authorizes clients by using a client certificate.

- **EDMX**
  Upload and select the EDMX file, which defines the OData service metadata (the data model and service operations).

- **Operation**
  Choose the OData operation to be performed (for example, **Create**, **Read**, **Update**, or **Delete**) on the target entity set.

- **Entity Set**
  Select the entity set from the EDMX definition on which the operation you choose will be executed.

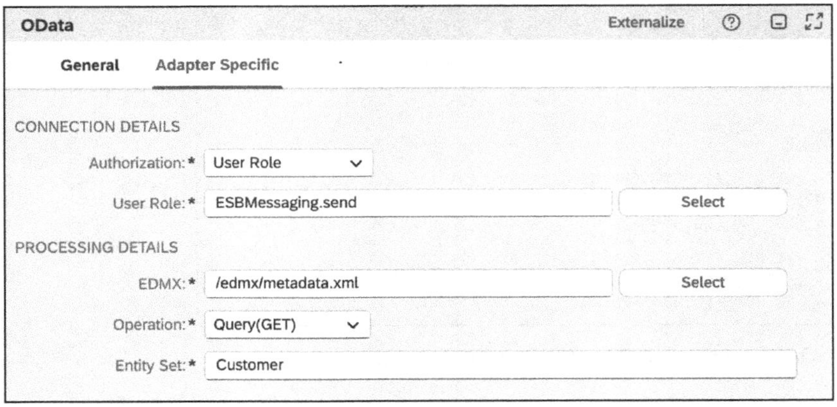

**Figure 4.45** OData Sender Adapter

**OData Receiver Adapter**

Because the OData v2 and the OData v4 adapter configuration is nearly the same, we'll discuss the OData v2 receiver adapter in this section. The **Connection** tab hosts the technical connection details (see Figure 4.46), and you must make the following entries and selections there:

- **Address**
  Enter the URL of the target OData V2 service.

- **Proxy Type**
  Choose between **Internet** (the default) and **On-Premise**.

- **Authentication**
  Choose from among these supported methods:
  - **None**
  - **Basic**
  - **Principal Propagation**
  - **Client Certificate**
  - **OAuth2 Client Credentials**
  - **OAuth2 SAML Bearer Assertion**

- **Credential Name**
  If you chose **Basic**, **OAuth2 Client Credentials**, or **OAuth2 SAML Bearer Assertion** for **Authentication**, specify the credential artifact you want to deploy.

- **Private Key Alias**
  If you choose **Client Certificate** for **Authentication**, specify the private key alias.

- **CSRF Protected**
  This box is checked by default to protect against cross-site request forgery.

- **Reuse Connection**
  This box is checked by default to improve performance by reusing connections.

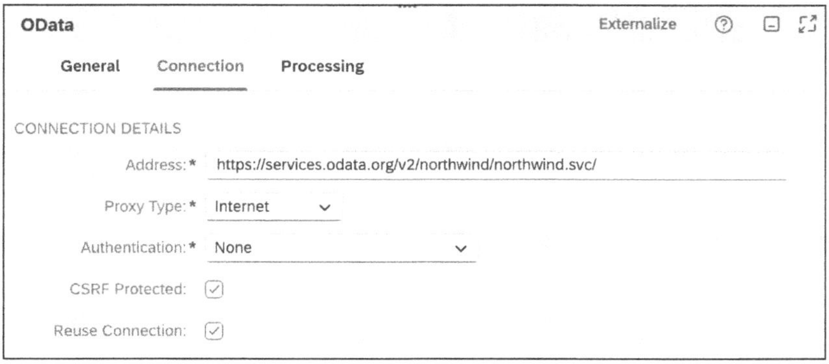

**Figure 4.46** OData Receiver Connection Settings

Next, you configure the OData operations in the **Processing** tab (see Figure 4.47). Select or enter the following:

- **Operation Details**
  Select the operation to perform.

- **Resource Path**
  Enter the path to the OData entity; you can also set it dynamically.

- **Query Options**
  These are available for query and read operations to pass additional query parameters.

- **Fields**
  You use this for **Create, Merge, Patch, and Update** to specify entity fields.

- **Enable Batch Processing**
  Check this box to enable $batch processing to perform multiple operations in one request.

- **Custom Query Options**
  Enter additional query options that aren't covered by standard settings.

- **Content Type**
  Select either **Atom** (the default) or **JSON** for payload formatting.

- **Content Type Encoding**
  Define encoding for **Create, Merge, Patch, Update, and Function Import**.

- **Page Size**
  Set the maximum records per page. This field is empty by default.

- **Process in Pages**
  Check this box when you're using **Query** to process large result sets in batches.

- **Timeout**
  Enter the maximum wait time in minutes for a response from the OData service.

- **Attach Error Details on Failure**
  This box is checked by default to attach request and response details to failure reports.

- **HEADER DETAILS**
  Enter the **Request Header** and **Response Header** that should be passed from / to Cloud Integration. Use a pipe-separated list of HTTP headers or a wildcard (*) to pass all headers.

- **METADATA DETAILS**
  Provide the headers to include in the internal $metadata call and additional parameters for the $metadata call.

Use the wizard to model the **Resource Path** and **Query** options, then click the **Select** button next to the **Resource Path** input field to open the wizard.

Define the connection to the target system in the first step (see Figure 4.48).

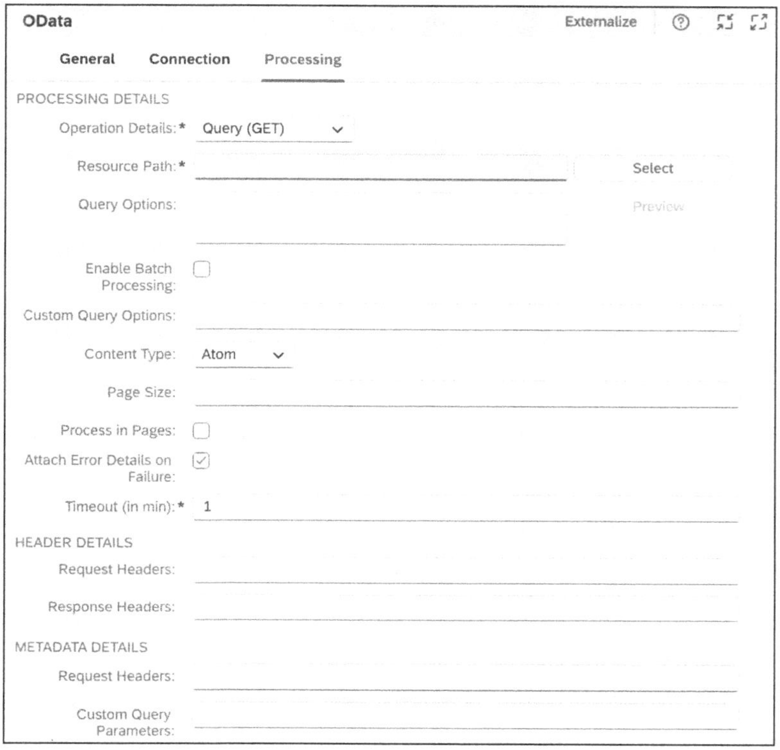

**Figure 4.47**  OData Receiver Processing Settings

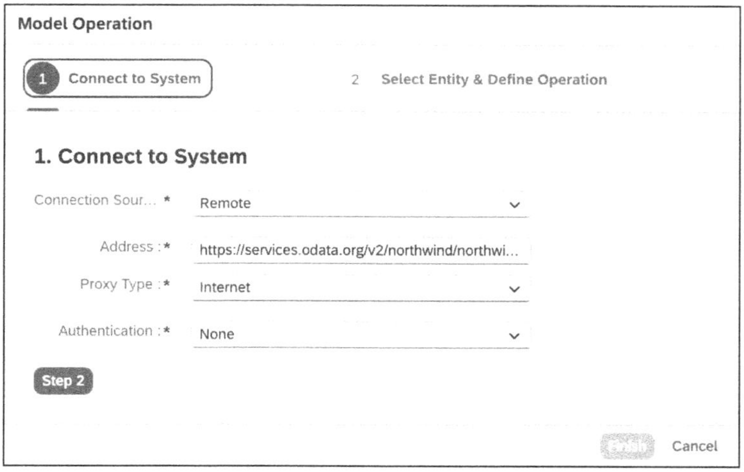

**Figure 4.48**  OData Wizard Connection Settings

In the second step, you can model the operation as shown in Figure 4.49. By checking the **Generate XML Schema Definition** box, you can create an XSD for the modeled operation.

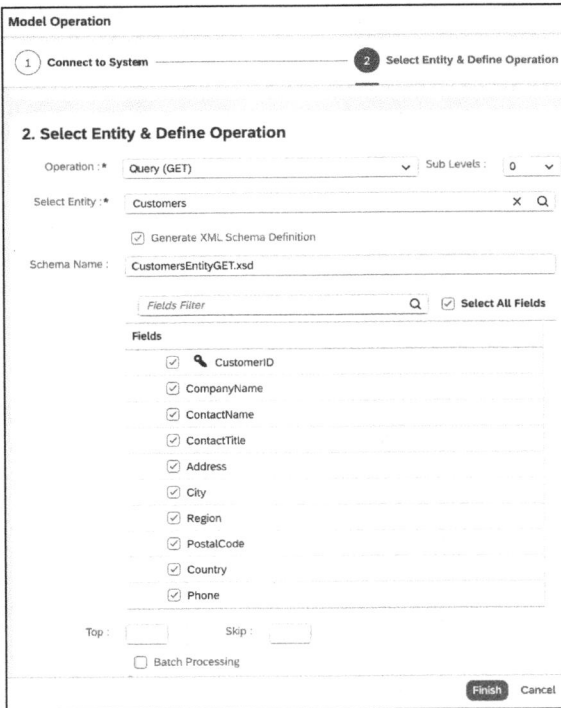

**Figure 4.49** OData Wizard Operation

In the third step of the wizard (see Figure 4.50), you can define the filter and sorting options.

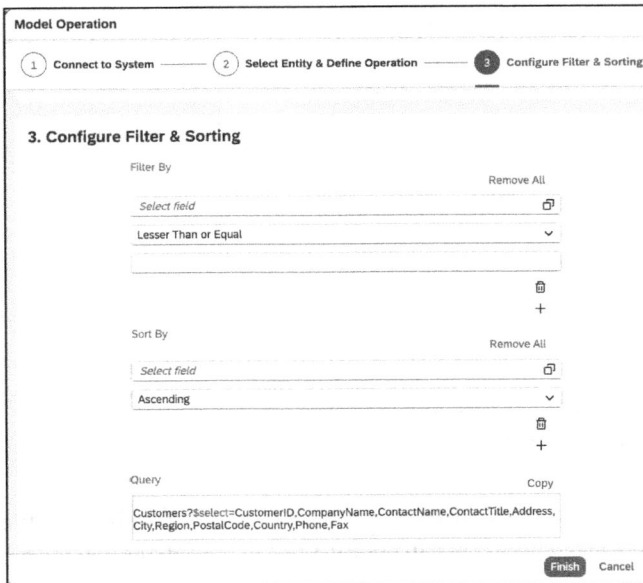

**Figure 4.50** OData Wizard Filter and Sorting

## 4.5 Database and Messaging Adapters

Modern integration scenarios demand more than just the exchange of data between applications; they require seamless interaction with databases and reliable messaging systems. We recommend that you reuse iFlows for multiple scenarios and common tasks. Cloud Integration provides specialized adapters to address these needs: the JDBC, JMS, and Process Direct adapters.

The Java Database Connectivity (JDBC) adapter (see Section 4.5.1) lets iFlows interact directly with relational databases to enable operations like reading, writing, and updating data. This is key for situations where you're pulling messages from a database or storing transactional data.

The Java Message Service (JMS) adapter (see Section 4.5.2) is designed for robust, asynchronous messaging to allow iFlows to exchange messages reliably, using queue-based or topic-based communication patterns. It's important for decoupling systems and supporting high-throughput, event-driven architectures.

The ProcessDirect adapter (see Section 4.5.3) lets you communicate directly in-memory between iFlows in the same tenant. It makes modular design easier, improves reusability, and reduces overhead by allowing flows to call each other without additional network communication.

### 4.5.1 JDBC Receiver Adapter

*Java Database Connectivity* (JDBC) is a standard Java API that enables applications to connect to and interact with relational databases. It provides a set of interfaces and classes for sending SQL queries, updating data, and retrieving results in a database-agnostic way. In the context of Cloud Integration, the JDBC adapter employs this technology to execute SQL statements (such as SELECT, INSERT, UPDATE, and DELETE) directly from iFlows. This feature enables seamless integration of business processes with database systems, whether they're cloud-based or on-premise, without requiring additional middleware or database-specific coding. JDBC supports transactional processing, parameter mapping, and secure connectivity, making it a powerful tool for scenarios that require direct database access, data enrichment, or data persistence as part of an integration.

The JDBC receiver adapter in Cloud Integration allows iFlows to directly execute SQL operations against a database. This product is designed for modern, cloud-native connectivity to a wide range of relational databases, both in the cloud and on premises. It's the perfect solution for a multicloud environment.

The JDBC receiver adapter in Cloud Foundry supports the following wide range of databases, both in the cloud and on premise:

- SAP HANA (in the cloud)
- SAP HANA Platform (on premise)

- SAP ASE service (in the cloud)
- SAP ASE, platform edition (on premise)
- DB2
- Oracle (in the cloud)
- Oracle (on premise)
- Microsoft SQL Server (in the cloud)
- Microsoft SQL Server (on premise)
- PostgreSQL (in the cloud)
- PostgreSQL (on premise)
- MariaDB (in the cloud)

When connecting to on-premise databases, the JDBC receiver adapter uses the cloud connector, which securely connects Cloud Integration running in Cloud Foundry with your on-premise network. This allows the adapter to access databases behind your corporate firewall without exposing them directly to the internet. This ensures secure, controlled connectivity and enables organizations to integrate their cloud-based processes seamlessly with existing on-premise data landscapes.

Before using the JDBC receiver adapter in Cloud Foundry, you must complete several setup steps. First, upload and deploy the appropriate JDBC driver in Cloud Integration. Drivers for these databases are already included, so you don't have to perform this step when you're working with SAP HANA, SAP ASE, or PostgreSQL. To upload the JDBC driver, navigate to the **Monitoring · Integration and APIs** menu and click the **JDBC Material** tile (see Figure 4.51).

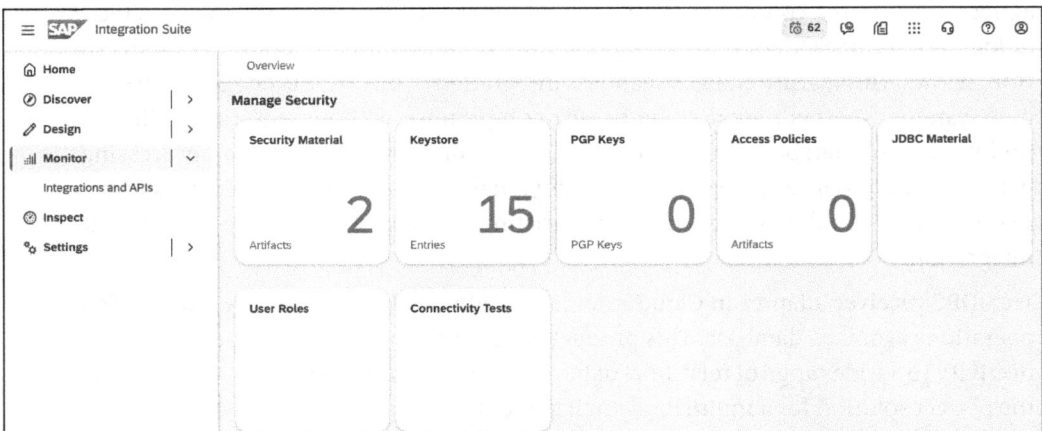

**Figure 4.51** JDBC Material Management

Then, in the **Manage JDBC Materials** section, click the **Add** button to open the dialog for uploading a JDBC driver (see Figure 4.52). Next, select the appropriate **Database Type**, upload the corresponding **JDBC driver file**, and click **Deploy** to make the driver available for use in your integration scenarios.

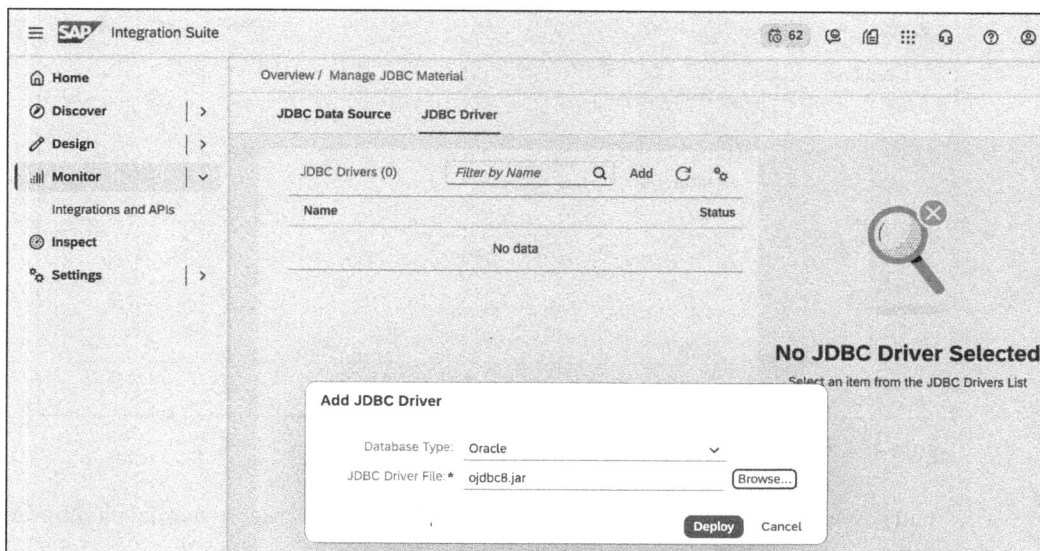

**Figure 4.52** Uploading JDBC Driver

Next, create the JDBC data source, which defines the connection details to your target database. If you plan to connect to an on-premise database, you must also configure the cloud connector to establish a secure TCP connection between the cloud tenant and your internal network. A sample configuration of a JDBC data source is shown in Figure 4.53. If you're connecting to an on-premise database, you must use the virtual hostname in the **JDBC URL** field.

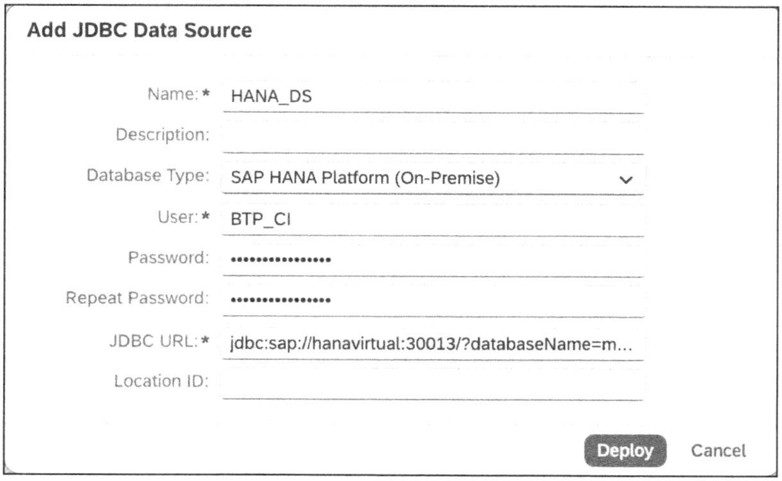

**Figure 4.53** Configuring Data Source

Then, design and configure your iFlow. Configuring the adapter's is straightforward, and you provide the **JDBC Data Source Alias** that you created previously (see Figure 4.54).

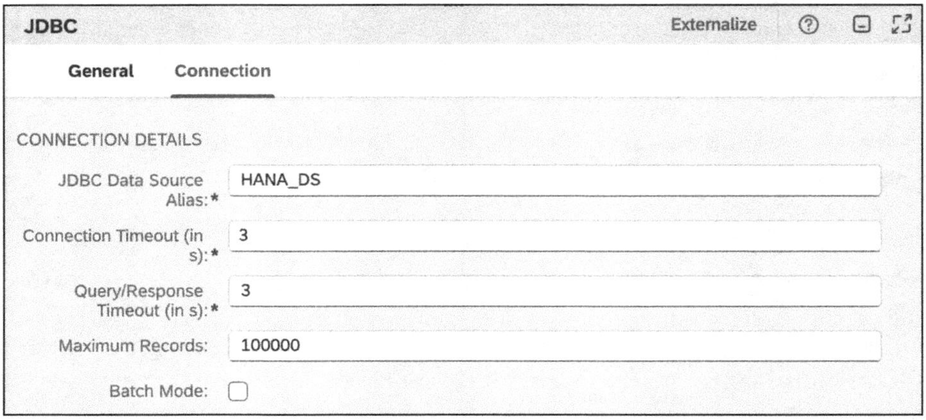

**Figure 4.54** Configuring JDBC Receiver Adapter

Finally, you must define the SQL statements and map any dynamic parameters required for your scenario. The payload is the data you're sending through the JDBC receiver adapter, and you can perform various operations on this XML payload, including INSERT, UPDATE, and DELETE. Please keep in mind that SAP provides the XML structure. A sample XML payload for an insert operation is shown in Listing 4.1.

```
<root>
<Insert_Statement1>
<dbTableName action="INSERT">
<table>bupa</table>
<access>
<bupa_id>121</bupa_id>
<bupa_name>test121</bupa_name>
</access>
<access>
<bupa_id>122</bupa_id>
<bupa_name>test122</bupa_name>
</access>
<access>
<bupa_id>123</bupa_id>
<bupa_name>test123</bupa_name>
</access>
</dbTableName>
</Insert_Statement1>
</root>
```

**Listing 4.1** JDBC Receiver Insert XML Payload

### 4.5.2   JMS Adapter

*Java Message Service* (JMS) is a standard messaging API that enables reliable, asynchronous communication between distributed systems and applications. With JMS, messages are stored temporarily in queues or topics, which allows for decoupled integration. This means that the sender and receiver don't need to be online at the same time, which makes JMS especially useful for handling surges in message load, supporting failover scenarios, and integrating systems that operate at different speeds.

To understand *asynchronous decoupling*, imagine a sender transmitting a message to Cloud Integration for inbound processing. If an error occurs during outbound processing—for instance, if the target system is temporarily unavailable—then the middleware can automatically retry processing the message, which eliminates the need for the sender to resend it. In this setup, the sender relies entirely on the middleware to handle retries and ensure eventual delivery once the error is resolved. To enable this process, the incoming message is stored safely in a queue using the JMS receiver adapter. Outbound processing is modeled in a separate iFlow that begins by consuming the message from the queue using the JMS sender adapter. This outbound flow continuously retries fetching and processing the message from the queue as long as the error persists, so it provides a robust, decoupled error-handling mechanism.

You can configure the JMS adapter as both a sender and a receiver to allow iFlows to receive messages from and send messages to message queues. You don't need to operate or maintain a separate JMS server because SAP Integration Suite already includes a fully managed, built-in JMS service that is ready to use out of the box.

---

**Supported Message Brokers**

Please note that the adapter only supports a fully managed and provided SAP message broker based on the Event Mesh capability. The adapter is designed to work seamlessly within the SAP Integration Suite environment, and it doesn't support connections to external or customer-managed message brokers. This ensures a fully integrated, secure, maintenance-free messaging backbone that doesn't require additional third-party broker infrastructure.

---

When you're using JMS, you must understand the resource boundaries so you can design reliable and scalable integrations. By default, each tenant is allocated the following JMS resources:

- Number of JMS queues: 30
- Queue capacity: 9.3 GB
- Number of transactions: 150
- Number of consumers: 150
- Number of providers: 159

Customers who need more capacity can increase these resources to a defined maximum limit through self-service adjustments. When you're looking at individual queue allocations, each JMS queue includes a defined set of resources to ensure predictable behavior:

- Queue capacity: 300 MB
- Transactions: 5
- Consumers: 5
- Providers: 5

You must also consider additional technical constraints when designing message flows:

- A single queue can use up to 95% of the total tenant queue capacity.
- Each transaction can process up to 256 messages.
- Messages larger than 5 MB are automatically split internally by Cloud Integration before they are stored in the queue.
- Consequently, the maximum possible message size (including attachments) that can be stored in a queue is 1,280 MB (256 × 5 MB).
- Furthermore, headers and exchange properties set before storing the message must not exceed 4 MB in total.

In the following sections, we'll discuss both the JMS sender adapter and the JMS receiver adapter.

**JMS Sender Adapter**

The sender adapter consumes messages from a JMS queue. In the **Connection** tab of the JMS sender adapter, you configure how messages are sent and processed from the queue (see Figure 4.55). You define the following parameters:

- **Queue Name**
  Define the name of the message queue (letters, numbers, underscores, max. 80 characters, no spaces).

- **Access Type**
  You must choose between the following options:

  - **Non-Exclusive**
    This allows parallel message processing by multiple consumers, and it's suitable when order isn't critical.

  - **Exclusive**
    This ensures strictly ordered processing by allowing only one consumer at a time.

  Please be aware that the access type must match in both sender and receiver adapters and can't be changed after creation.

- **Number of Concurrent Processes**

  If you choose **Non-Exclusive** as the **Access Type**, specify how many parallel processes per worker node can process messages. The default is 1, and we recommend that you to keep it low (from 1 to 5) to avoid resource or memory issues.

- **Retry Interval**

  Define how long to wait before retrying a failed message delivery.

- **Exponential Backoff**

  Select this option if you want to double the retry interval after each failed attempt.

- **Maximum Retry Interval**

  If you select **Exponential Backoff,** set a limit for the maximum retry interval. The minimum is 10 minutes, and the default is 60 minutes.

- **Dead-Letter Queue**

  If you select **Non-Exclusive** as the **Access Type**, check this box (or leave it checked as the default) to mark messages that cause repeated out-of-memory errors as **Blocked** and place them in a dead-letter state after two failed retries. Administrators can manually release these blocked messages in the Monitoring app to trigger retries again. In high-load scenarios with only small messages, you can disable this option to improve performance, but at the risk of potential system outages if large messages cause failures.

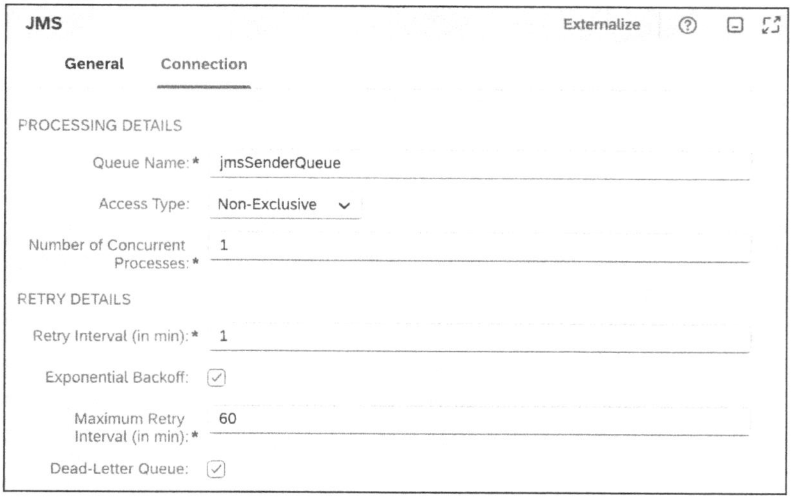

| JMS | | Externalize | ⑦ | ☐ | ⌗ |
|---|---|---|---|---|---|
| **General**  Connection | | | | | |

PROCESSING DETAILS

| Queue Name: * | jmsSenderQueue |
|---|---|
| Access Type: | Non-Exclusive ⌄ |
| Number of Concurrent Processes: * | 1 |

RETRY DETAILS

| Retry Interval (in min): * | 1 |
|---|---|
| Exponential Backoff: | ☑ |
| Maximum Retry Interval (in min): * | 60 |
| Dead-Letter Queue: | ☑ |

**Figure 4.55** JMS Sender Adapter Configuration

---

**Dead-Letter Queue**

A *dead-letter queue* (DLQ) is a special type of message queue that's used to hold messages that the main queue consumers can't process successfully, even after multiple retry attempts. A message ends up in a DLQ if it continuously causes errors, which may be due to invalid data, configuration issues, or repeated out-of-memory problems. Rather than blocking the main processing or retrying endlessly, these problematic messages are moved to the DLQ for manual inspection and corrective action. This mechanism improves system stability and reliability by preventing a single "bad" message from stopping the entire flow. Administrators can review, fix, or reprocess these messages later through monitoring tools.

### JMS Receiver Adapter

In the **Processing** tab, you define how messages are handled when they're stored in a JMS queue (see Figure 4.56). Define the following parameters:

- **Queue Name**
  Define the name of the message queue. You can use letters, numbers, and underscores, the maximum length is 80 characters, and spaces are not allowed. You can also set the name dynamically by using headers or properties, but be aware that queues that are created dynamically are not precreated at deployment.

- **Access Type**
  You can choose between the following options:

  - **Non-Exclusive**
    This allows parallel consumption by multiple workers, and there's no guaranteed order.

  - **Exclusive**
    This guarantees strict message order by allowing only one consumer.

  Please note that the access type must match between sender and receiver adapters and can't be changed later.

- **Retention Threshold for Alerting**
  Define when to trigger an alert if messages remain unprocessed for too many days. The default value is two days.

- **Expiration Period**
  Set how many days messages stay in the queue before expiring, which should be at least twice the retention threshold. The default is 30 days, and the maximum is 180 days.

- **Compress Stored Message**
  Check this box to reduce disk space and network usage by compressing messages in the queue.

- **Encrypt Stored Message**
  Check this box to add security by encrypting stored messages.

- **Transfer Exchange Properties**
  Check this box to transfer exchange properties into the JMS queue. This is not generally recommended, due to potential size limitations that can cause errors.

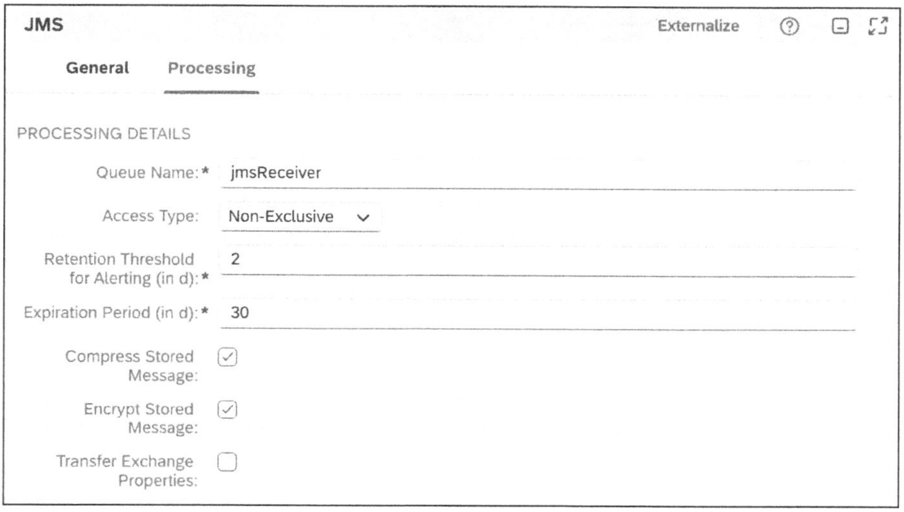

**Figure 4.56** JMS Receiver Adapter Configuration

### 4.5.3   ProcessDirect Adapter

Use the ProcessDirect adapter (sender and receiver) to establish fast and direct communication between iFlows. This will reduce latency and network overhead. The Process-Direct adapter supports the following use cases that improve maintainability, modularity, and collaboration:

- **Decomposing large iFlows**
  Don't build a single, complex iFlow with hundreds of steps—break it down into smaller, manageable subflows that can interact seamlessly through ProcessDirect. This maintains in-process message exchange without introducing additional network latency.

- **Customizing standard content flexibly**
  Cloud Integration empowers developers to enhance or customize standard integration content by adding custom logic at specific touchpoints. HTTP and SOAP adapters are often used, but the ProcessDirect adapter is the more efficient, low-latency alternative for connecting these plug-in touchpoints within the same tenant.

- **Enabling parallel development and teamwork**
  Large iFlows are bound to become bottlenecks if they're maintained by a single developer, so the ProcessDirect adapter allows multiple developers to work in parallel on different sections of the integration by splitting the flow into independent subflows. This modular approach ensures better collaboration, faster development, and easier ownership.

- **Promoting reuse across projects and scenarios**
  You'll often need to implement common logic—such as error handling, logging, mapping extensions, and retry mechanisms—repeatedly across multiple integration

scenarios. The ProcessDirect adapter allows you to define these shared flows once and call them from various integration processes, even across different projects. This will reduce redundancy, simplify maintenance, and improve consistency throughout your integration landscape.

- **Dynamic endpoints**
  The ProcessDirect adapter supports dynamic routing, thus enabling a producer iFlow to direct messages to different consumer iFlow endpoints at runtime. You can look up the target endpoint address dynamically from the message headers, body, or exchange properties. The message is routed to the appropriate consumer flow based on this resolved value, thus enabling highly flexible, content-based routing scenarios without the need for static configurations.

You should also note some hints and limitations that are related to the following topics:

- **Multiple MPLs**
  When you're using the ProcessDirect adapter, each producer and consumer iFlow generates its own MPL. These logs are linked together with correlation IDs, which makes it possible to trace related executions across flows.

- **Transaction processing**
  You can support transactions within each iFlow by using the ProcessDirect adapter. However, the transaction scope is restricted to a single iFlow and doesn't extend across multiple flows connected via ProcessDirect.

- **Header propagation**
  By default, headers are not automatically propagated from the producer to the consumer iFlow. If header forwarding is needed, you must explicitly list the specific headers in the <Allowed Header(s)> configuration of the runtime settings.

- **Property propagation**
  Properties are not propagated across producer and consumer iFlows. The only exception is the SAP_SAPPASSPORT property, which SAP uses internally and which users must therefore not modify.

You use the ProcessDirect receiver adapter to send data to other iFlows. In this case, the iFlow is considered a producer iFlow and the iFlow has a receiver ProcessDirect adapter. The ProcessDirect sender adapter consumes data from other iFlows, which makes the iFlow a consumer iFlow. In this case, the iFlow has a sender ProcessDirect adapter. The configuration of both the sender adapter and the receiver adapter is simplified to an **Address** property only (see Figure 4.57).

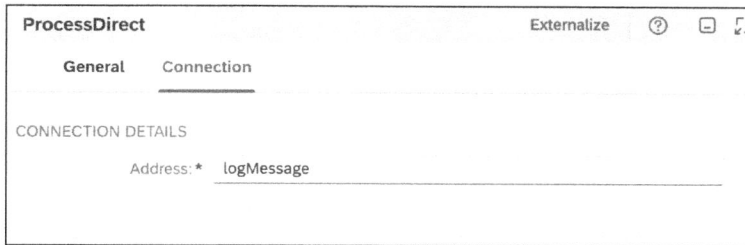

**Figure 4.57** Process Direct Adapter Configuration

## 4.6 Summary

In today's digital world, it's very important for businesses' different systems, applications, and data sources to work well together. SAP Integration Suite is the main tool in the SAP BTP portfolio, and it's used to connect different types of IT systems, design processes that follow a logical sequence, and implement integration scenarios that can be used in the future.

Adapters are at the core of this architecture. They act as bridges between different technologies, protocols, and formats. They allow communication between SAP and non-SAP systems, take out the technical details, ensure security standards, and provide a reliable connection across system boundaries.

This chapter gives a thorough overview of the most important adapter groups in SAP Integration Suite. It has a practical focus, and it includes in-depth technical background information, best practices from customer projects, and specific use cases.

Section 1.1 started with an introduction to the adapter concept, technical functionality, differences between sender and receiver adapters, and advanced configuration options. Section 1.2 explained file-based scenarios, which help you easily import and export files using SFTP, FTP, and file adapters. Section 1.3 discussed SAP-specific adapters, which are essential in the SAP environment, and it focused on IDoc, RFC, and SAP SuccessFactors adapters. Section 1.4 discussed adapters for modern, API-driven architectures, which are essential for cloud integration. Finally, Section 1.5 covered database and messaging adapters, which are adapters for database access, messaging integration, and event-driven scenarios. They include modular iFlow design with ProcessDirect.

Choosing the right adapter is crucial for implementing strong, easy-to-maintain, and flexible integrations—whether they are based on files, APIs, SAP solutions, or events.

# Chapter 5
# Loosely Coupled Interfaces

*In the rapidly evolving landscape of enterprise IT, agility, scalability, and resilience have become pillars of integration architecture. Traditional integration models, which often rely on tightly coupled systems and synchronous communication, struggle to meet the demands of modern digital ecosystems. When systems are directly dependent on one another's availability, performance, and interface consistency, the risk of bottlenecks, single points of failure, and cascading outages increases significantly. In response to these limitations, organizations are increasingly turning to asynchronous messaging patterns and loosely coupled designs to build robust, future-proof integration solutions.*

This chapter explores the principles and practices of implementing loosely coupled interfaces by using *Java Message Service (JMS)* within SAP Business Technology Platform (SAP BTP)—specifically through SAP Integration Suite. Messaging via JMS is a cornerstone of decoupled integration design because it offers an architecture where producers and consumers of data operate independently and are temporally separated. This architectural paradigm not only boosts system reliability and scalability but also aligns closely with event-driven architectures and microservices principles.

SAP Integration Suite, which is a core component of SAP BTP, provides native support for JMS-based messaging. It enables the use of queues and topics to manage communication among systems that may not always be available at the same time or operate at the same throughput levels. By buffering messages in durable queues and handling them asynchronously, integration developers can design flows that are resilient to downstream failures, network fluctuations, and variable workloads.

The intention of this chapter is to equip you with a practical and conceptual understanding of how to leverage JMS for loose coupling within Cloud Integration scenarios. Whether you are designing interfaces between SAP S/4HANA and SAP SuccessFactors, integrating with third-party cloud services, or creating custom APIs that require fault tolerance and queue-based buffering, the patterns we discuss here are applicable across a wide range of use cases.

Section 5.1 introduces the fundamental concepts of messaging and loose coupling. By decoupling the sender and receiver, messaging systems allow applications to evolve independently, communicate asynchronously, and maintain functionality even under partial failures. This chapter highlights the benefits and trade-offs of this integration style.

Section 5.2 presents the structure of a messaging architecture by describing the essential elements such as message producers, consumers, queues, and topics. It also discusses the roles of brokers and message persistence to clarify how complex systems exchange information reliably without direct method calls.

Section 5.3 turns to the configuration of the JMS adapter, which serves as the bridge between applications and the messaging infrastructure. Proper configuration enables seamless delivery of messages and sets the foundation for secure, performant communication across different technical environments.

Finally, Section 5.4 examines error handling and retry strategies, which are crucial aspects of building fault-tolerant systems. From dead letter queues (DLQs) to backoff policies, this section demonstrates how to design messaging solutions that not only exchange data but also withstand operational challenges gracefully.

## 5.1   Introduction to Messaging and Loose Coupling

In modern enterprise architectures, integration scenarios often involve multiple systems with different technologies, lifecycles, and availability constraints. Traditional point-to-point or synchronous communication patterns, such as direct HTTP and SOAP calls, tightly couple the sender and receiver systems, and this approach can lead to system bottlenecks, cascading failures, and scalability issues.

*Loosely coupled interfaces* address these challenges by decoupling the communication between systems. Instead of via direct communication, messages are sent through an intermediary—typically, a message broker or queue—that allows systems to operate independently. In this model, the sender system is only responsible for placing a message in the queue while the receiver processes messages from the queue at its own pace.

SAP BTP offers robust capabilities to support loose coupling through SAP Integration Suite—particularly via the use of JMS, which enables *asynchronous messaging*, in which systems exchange information without direct dependencies. This not only increases fault tolerance and scalability but also improves maintainability and flexibility of integration landscapes.

To help you better understand these concepts, this section explores several of their foundational aspects. We begin with asynchronous processing and explain how decoupled communication patterns help systems achieve resilience and scalability. Next, we introduce the JMS as the standard API that enables such messaging within SAP BTP. Building on this, we outline the key benefits of applying JMS in integration scenarios, highlighting its role in reliability and flexibility. At the same time, it's important to recognize JMS' limitations, so we also discuss when not to use JMS to help you pick the right tools for the right scenario. The section concludes with best practices for designing loosely coupled interfaces to provide you with practical guidance for building iFlows that are robust, efficient, and maintainable.

### 5.1.1    Asynchronous Processing

At the heart of loosely coupled interfaces lies the principle of *asynchronous processing*— a communication model that breaks the rigid dependencies between systems by decoupling the sender and receiver in both time and state. Unlike in synchronous interfaces, where the sender waits for a response before proceeding, asynchronous messaging allows each system to operate independently. This approach introduces greater flexibility, fault tolerance, and scalability into enterprise integration landscapes, and it's a foundational concept for modern, resilient architectures.

In a traditional synchronous scenario—such as an HTTP request or a remote function call—the sender and receiver are tightly bound. If the receiving system is slow, unavailable, or experiencing errors, the sender is immediately impacted. This not only affects performance but also propagates instability across the system. Conversely, asynchronous processing introduces a buffer, often in the form of a message queue or event stream, that stores the message temporarily until the receiving system is ready to consume it. This architecture allows for much looser coupling between services, thus improving system uptime and enabling nonblocking communication flows.

JMS plays a crucial role in enabling asynchronous communication within Cloud Integration. JMS queues act as durable message brokers that allow iFlows to send and receive data independently of each other. A producer flow can generate and enqueue a message without needing to know whether the consumer is currently online or available. The consumer flow, on the other hand, can process the message later—when resources become available or at a controlled rate that matches the backend system's capacity. This decoupling of message producers and consumers is one of the key enablers of scalability, as it allows system components to scale independently and handle traffic bursts more efficiently.

Another major benefit of asynchronous processing is improved fault tolerance. Temporary outages in downstream systems no longer result in immediate interface failures. Instead, messages are retained in the queue, retried according to configurable strategies, and optionally redirected to a DLQ if they fail beyond defined thresholds. This enables operational teams to recover gracefully from failures without losing data, while also providing visibility into unresolved errors.

Asynchronous processing also allows for parallelism and concurrency, especially on the consumer side. Multiple consumer instances or threads can process queued messages simultaneously to increase throughput without overloading the source system.

This makes asynchronous messaging ideal for high-volume integrations such as order ingestion, data replication, and system-to-system handoffs.

It's important to note that asynchronous processing introduces new design considerations. For example, since the response isn't immediate, developers must use additional correlation mechanisms (e.g., correlation IDs, reference tokens) track end-to-end message flow. They must also consider message ordering (using EOIO, for example), idempotency, and state management, especially when building interfaces with business-critical requirements.

This approach also serves as a stepping-stone toward more advanced patterns like event-driven architecture (EDA), which we cover in detail in a later chapter of this book. While asynchronous messaging relies on queues or channels for point-to-point or broadcast delivery, event-driven systems elevate this concept by using events to signal changes in system state. In an EDA model, systems react to business events—such as a sales order being created or a delivery status changing—rather than being directly told what to do. This enables even looser coupling, greater extensibility, and real-time responsiveness across distributed systems.

In many enterprise landscapes, asynchronous processing via JMS acts as a bridge between traditional, tightly coupled interfaces and more modern, event-based models. It offers a pragmatic and powerful way to incrementally move toward decoupled architectures without overhauling legacy systems all at once. By adopting asynchronous processing patterns within Cloud Integration, organizations can gain immediate benefits in reliability, performance, and maintainability—while also laying the groundwork for more dynamic and intelligent architectures powered by events.

### 5.1.2   Java Messaging Service

As organizations move toward building more decoupled and resilient integration landscapes, JMS emerges as a key enabler of asynchronous, loosely coupled communication within Cloud Integration. JMS provides a standardized messaging protocol that allows distributed applications to communicate through message queues in a reliable and scalable way—regardless of whether the sender and receiver are available at the same time.

In the context of SAP Integration Suite, JMS acts as a *message broker* that facilitates asynchronous communication between iFlows. The core concept is simple but powerful: a producer flow sends a message to a JMS queue, where it's stored persistently until a consumer flow retrieves and processes it. This mechanism decouples the lifecycle and availability of the producing and consuming systems to ensure that neither side is directly dependent on the other being online or immediately responsive. This is particularly valuable for handling system outages, batch processing, and integrations with systems that operate at different speeds or volumes.

JMS offers the following two delivery modes that cater to different business requirements:

- *Exactly once* (EO) ensures that a message is delivered one time and only one time, which is essential for ensuring data consistency across systems.
- *Exactly once in order* (EOIO) provides both single delivery and strict order guarantees for messages that belong to the same sequence. This is particularly useful for business transactions in which the order of events matters, such as financial postings and shipment updates.

One of the key features of JMS in Cloud Integration is its persistence. Messages stored in a JMS queue are saved in a durable store, which means they are not lost even in the event of system restarts or crashes. This level of reliability makes JMS suitable for mission-critical enterprise processes where data integrity and recovery are non-negotiable.

Beyond their basic message handling, JMS queues offer a range of advanced features that support fault-tolerance and operational flexibility. For example, failed messages can be retried automatically based on configurable retry intervals and attempt counts. If a message still can't be processed after the defined number of retries, it can be redirected to a DLQ for later analysis and manual reprocessing. This built-in error-handling framework ensures that temporary failures don't disrupt the overall message flow and that irrecoverable errors can be isolated without data loss.

Another important advantage of JMS is its ability to handle high volumes and enable parallelism. You can configure consumer flows with multiple concurrent threads to allow the processing of several messages in parallel. This is especially useful in high-throughput scenarios like order processing, master data replication, and IoT data ingestion. Combined with its ability to monitor queue metrics, retry patterns, and message traces, JMS offers both scalability and deep operational insight.

From an architectural perspective, JMS isn't only a messaging engine but also a design abstraction that supports the principles of loose coupling, modularity, and fault isolation. By placing a JMS queue between systems, integration architects introduce a layer that can absorb fluctuations in load, break dependency chains, and improve the overall resiliency of the solution. It becomes possible to evolve systems independently, schedule updates without downtime, and respond more gracefully to failure scenarios.

Moreover, JMS can act as a bridge toward event-based design, especially when used with message enrichment and header-based routing. Although JMS is fundamentally point-to-point, its ability to decouple message producers from consumers makes it conceptually aligned with event-driven patterns, where systems respond to occurrences rather than relying on tightly coupled request/response cycles.

### 5.1.3    Key Benefits

The adoption of loosely coupled interfaces marks a significant shift in how modern enterprise systems communicate, integrate, and scale. Unlike tightly coupled architectures—in which system components are interdependent and sensitive to changes—loosely coupled interfaces introduce flexibility, independence, and resilience, all of which are vital in the dynamic environments businesses operate in today.

One of the most immediate benefits of loose coupling is *asynchronous communication*, which removes the need for systems to be available at the same time. This temporal decoupling ensures that if a target system is offline, under maintenance, or experiencing load issues, the sending system can still transmit data via queues without interruption. This improves the uptime and stability of the overall landscape, thus allowing each system to operate according to its own availability and performance profile. Systems become more autonomous, and the likelihood of cascading failures caused by a single point of disruption is significantly reduced.

Another core advantage is *scalability*. By decoupling producer and consumer flows with technologies such as JMS, each part of the system can be scaled independently. For instance, during periods of high message volume, additional consumer threads can be added without impacting the sending system. Likewise, message producers can operate at high speeds without overwhelming the receivers, thanks to buffering through queues. This enables organizations to handle seasonal spikes, high-load periods, and bursts of activity without costly overprovisioning or architectural rework.

*Resilience and fault tolerance* are also greatly improved. Loosely coupled interfaces can gracefully handle transient failures through built-in retry mechanisms, and messages that repeatedly fail can be redirected to DLQs for later inspection. This prevents data loss and ensures that errors don't go unnoticed or silently degrade business processes. With additional support for error categorization and manual reprocessing, teams are better equipped to identify, respond to, and resolve issues quickly.

A key benefit that's often overlooked is *change agility*. In tightly coupled systems, changes to one interface often require coordinated updates across multiple systems—which is a slow, risky, and error-prone process. Loosely coupled interfaces mitigate this risk by isolating changes. A new version of a consumer flow can be introduced independently, or new consumers can be added without impacting the producer at all. This modularity allows teams to innovate faster, deploy incrementally, and respond more quickly to new requirements or regulations.

Moreover, observability and monitoring are enhanced in a decoupled setup. With clear separation of message producers, queues, and consumers, administrators can track message flow through each layer and identify bottlenecks and errors with greater precision. Metrics such as queue size, message age, retry frequency, and throughput provide insights that are useful not only for operational monitoring but also for long-term forecasting and capacity planning.

Another strategic benefit is support for hybrid and multicloud architectures. Loosely coupled messaging systems can span across on-premise and cloud environments to bridge gaps between legacy systems and modern platforms. JMS-based messaging within SAP BTP, for example, allows for reliable and asynchronous communication among SAP S/4HANA, third-party APIs, and cloud-native services, regardless of where they are hosted.

Finally, loosely coupled interfaces lay the groundwork for EDAs. While this book includes a dedicated chapter on EDAs, it's important to note here that the principles of loose coupling, asynchronous flow, and queue-based design naturally extend into event-driven patterns. As businesses move toward real-time, reactive systems that respond to business events rather than procedural triggers, loosely coupled messaging becomes the architectural stepping-stone toward future readiness.

### 5.1.4   When Not to Use Java Messaging Service

While JMS-based messaging and loosely coupled interfaces offer significant advantages in terms of scalability, resilience, and decoupling, they are not universally suitable for every integration scenario. Like all architectural patterns, their effectiveness depends on the business context, functional requirements, and performance expectations. You must understanding when not to use JMS or loose coupling to avoid overengineering, unnecessary complexity, and unintended side effects.

One of the most common reasons to avoid JMS is because you need real-time, synchronous feedback. If a client system or end user expects an immediate response—such as in login validation, pricing checks, or payment authorizations—a decoupled, asynchronous model introduces unacceptable latency. Synchronous protocols like HTTP or SOAP are better suited to such cases because they support request–response patterns with strict timing guarantees. Implementing JMS for these use cases would complicate the architecture without providing meaningful benefit and could even lead to a degraded user experience.

Additionally, simple, one-off integrations that don't involve complex workflows, long processing chains, or retry mechanisms often don't warrant the use of a JMS queue. For example, you may be able to better implement a basic file transfer between two stable systems or a low-volume API call that rarely fails by using direct point-to-point integration. Adding JMS to these flows could result in increased development effort, monitoring overhead, and operational complexity, without significantly improving reliability or scalability.

Data consistency requirements can also be a constraint. In tightly coupled transactional processes—especially those that require atomicity, consistency, isolation, and durability (ACID)–compliant operations across multiple systems—asynchronous processing introduces a risk of data desynchronization. For instance, in multistep order processing—where each step must succeed or roll back as one atomic unit—synchronous processing may be necessary to preserve transactional integrity. While JMS provides delivery guarantees like EO and EOIO, it doesn't natively support distributed transactions across systems, which could lead to complex compensation logic or inconsistent states if not carefully managed.

Another consideration is monitoring and troubleshooting complexity. Asynchronous flows with loose coupling make it harder to trace a single transaction end-to-end, especially in environments without mature observability tools or disciplined correlation practices. Teams that are unfamiliar with asynchronous design may struggle to track down errors, which can lead to longer resolution times. In such environments, simpler synchronous flows may be easier to manage and support.

Organizations should also consider skill set and operational maturity. Those that are just beginning their journey with SAP Integration Suite or that don't have experience in managing queues, retries, and dead letter processing might initially benefit from building synchronous flows. Jumping into loosely coupled designs without a solid foundation in message handling and exception strategies can lead to poorly configured retry loops, overflowing DLQs, or message duplication—which can ultimately reduce reliability rather than enhancing it.

Finally, regulatory or security constraints can influence the decision of whether or not to use JMS. In some industries, such as finance and health care, strict audit trails and sequencing requirements may be easier to enforce in synchronous transactions that provide immediate feedback and tightly bound state changes. While JMS can support message logging and traceability, meeting specific compliance mandates may be more straightforward if you use tightly controlled synchronous processes.

### 5.1.5   Best Practices for Designing Loosely Coupled Interfaces

Based on both conceptual and technical guidance, several best practices have emerged as vital to building effective JMS-based integration solutions:

- Clearly separate producer and consumer flows. Designing producer and consumer flows as independent artifacts allows for cleaner deployments, better maintenance, and flexible scaling. This modularity also simplifies testing and versioning.

- Use EOIO for sequence-sensitive data. Where business transactions—such as financial postings or status updates— depend on strict message order, EOIO ensures that messages are processed in the sequence in which they were sent. This avoids race conditions and state inconsistencies.

- Implement robust error handling. You should incorporate exception subprocesses to manage failures at each critical point in the iFlow, use DLQs as a fallback to capture unresolved errors without halting message processing, and ensure that all failure scenarios are logged and monitored appropriately.

- Optimize resource usage. Be mindful of JMS resource quotas in your SAP BTP tenant. Use shared queues with routing logic when possible to reduce queue count, compress payloads to minimize memory usage, and offload large files to external storage like SAP Document Management.

- Configure smart retries. You should tune retry intervals and counts based on the expected recovery time of the target system, and you should use exponential back-off to avoid overwhelming systems that are struggling or recovering.

- Implement monitoring and alerting. Proactively track message health, queue utilization, and retry behavior. Set up alerts for thresholds such as queue size, DLQ growth, or retry saturation. Integrate with existing ITSM platforms to ensure that operational teams can act quickly.

- Understand when not to use JMS. JMS is ideal for asynchronous, decoupled use cases. However, for real-time interactions, synchronous request–response needs, and transactionally bound operations, a different integration model—such as direct HTTP or OData—may be more appropriate. Use JMS where it adds the most value.

## 5.2   Structure of Messaging Architecture

*Messaging architecture* provides the backbone for reliable, asynchronous communication among distributed systems. By decoupling producers and consumers, it allows applications to scale independently, ensures message durability, and supports fault-tolerant data exchange across services. A well-designed messaging architecture enables flexibility, resilience, and observability in complex enterprise environments.

This section explores the structure of such an architecture in four parts. Section 5.2.1 covers the producer iFlow, which describes how messages are created and published into the system. Section 5.2.2 introduces the JMS queue, which is a central component for storing and routing messages. Section 5.2.3 examines the Consumer iFlow by focusing on how applications receive, process, and acknowledge messages. Finally, Section 5.2.4 covers monitoring and metrics, emphasizing the importance of visibility and performance tracking to ensure reliable message delivery and system health.

As shown in Figure 5.1, the JMS channel is used to connect the source system with the target system. The queue breaks the two sides into separate threads that don't directly connect to each other. This means that the source system can publish its updates uninterruptedly into an inbound heap of events, which are then processed simultaneously while also introducing the possibility of multiplexing the event for multiple target systems at the same time, without duplicating the event.

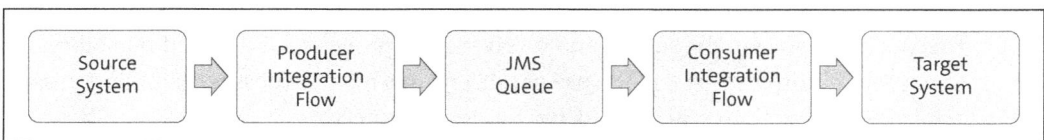

**Figure 5.1**  End-to-End Messaging Architecture Using JMS Queue between Source and Target System

### 5.2.1   Producer iFlow

The producer iFlow serves as the starting point of any JMS-based message process in Cloud Integration. Its primary responsibility is to receive data from upstream systems, standardize and enrich the message content, and deliver it into a JMS queue for asynchronous processing. This decouples the sending system from the consuming process to enable better scalability, error resilience, and system independence.

Typically, the producer flow begins by ingesting input from a source system, which could be exposed through a RESTful API, an IDoc interface, a SOAP web service, or an SFTP trigger. This flexibility allows SAP Integration Suite to connect to a wide variety of both SAP and non-SAP applications and services. Once the message is received, it usually undergoes transformation into a standardized internal format so downstream consumers receive data in a consistent and expected structure. You can achieve this transformation by using graphical mapping tools, Groovy scripts, or XSLT mappings, depending on complexity and format requirements.

Next, the message is enriched with essential metadata, often using header properties. These might include a unique message or correlation ID, priority level, business context information, or routing hints—which are attributes that are crucial for traceability, error handling, and message filtering in later stages. Enrichment ensures that each message carries the contextual information that's needed for intelligent routing and robust monitoring.

Finally, the message is handed off to the JMS adapter, which acts as the bridge between the iFlow and the JMS queue. The adapter pushes the message into the defined queue, where it will remain in a durable, persistent state until a consumer flow retrieves it for further processing. By offloading the message into the queue, the producer flow completes its role, freeing up system resources and enabling true asynchronous processing—which is an essential architectural pattern in modern, decoupled enterprise integrations.

This iFlow, as depicted in Figure 5.2, represents an inbound process where a message is received via HTTPS and routed to a JMS queue. The flow begins with a sender system initiating communication through HTTPS. The first step is **Define request**, in which the inbound message is structured. The flow then moves to the **Send to JMS queue** step, which enqueues the message for asynchronous processing. A JMS receiver component subscribes to this queue, and meanwhile, the flow continues by defining a response in the **Define response** step.

It concludes with the **End** event by sending a reply to the sender. This design allows decoupling between the sender and receiver to improve scalability and reliability. The clear separation of steps and the use of JMS ensures reliable message handling in asynchronous integration scenarios using the cloud connector.

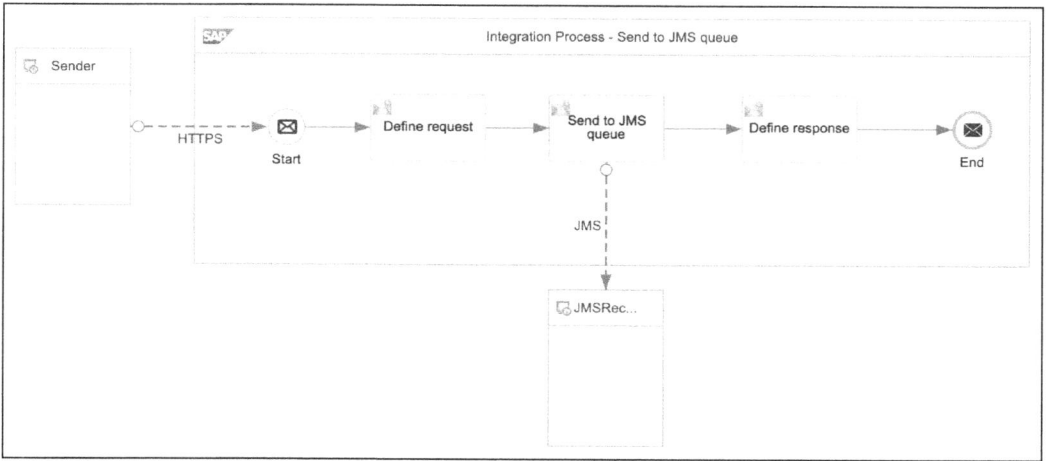

**Figure 5.2** Inbound Message Process

Next, you can configure the JMS sender step (see Figure 5.3) by clicking the JMS arrow that points to the JMS Adapter at the bottom. This takes you to the **Processing** tab of the JMS configuration within an iFlow, where you can enter "DecoupleFlows" as the **Queue Name** of the step. Enter "2" in the **Retention Threshold for Alerting (in d)** field to trigger alerts if messages remain for over 2 days. You can set messages to expire after 90 days by entering "90" in the **Expiration Period (in d)** field. Check the **Encrypt Stored Message** box to enable message encryption and ensure secure storage, but leave **Transfer Exchange Properties** unchecked.

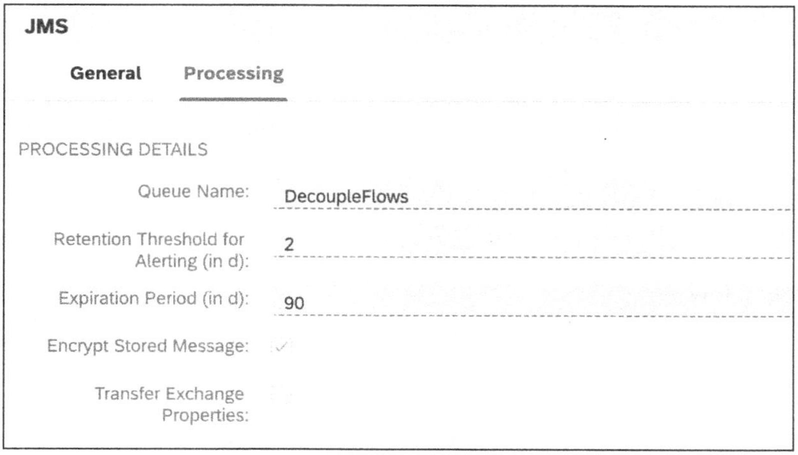

**Figure 5.3** Inbound Process: JMS Settings

### 5.2.2   Java Messaging Service Queue

The JMS queue plays a central role in Cloud Integration's asynchronous messaging architecture. It serves as a reliable buffer between the producer and consumer flows that enables decoupled processing and ensures message durability, even in cases of system downtime or processing delays. Once a message is handed off by the producer iFlow, the JMS queue holds it until a consumer is ready to retrieve and process it. This introduces a critical layer of resilience and load management.

One of the defining characteristics of the JMS queue is its persistent message storage. Messages placed into the queue are stored durably, which means they survive system restarts and remain intact until they are successfully processed or manually removed. This persistence guarantees delivery reliability—which is especially important in enterprise environments where data loss is unacceptable.

JMS queues also support two critical delivery modes: EO and EOIO. EO ensures that each message is delivered a single time, regardless of system failures or retries, while EOIO maintains the sequence of message delivery within a specified group (such as a business transaction or customer ID). These delivery guarantees allow architects to meet strict business and regulatory requirements for consistency and order.

In Cloud Integration, administrators must explicitly configure JMS queues under the JMS resources section of the tenant. There, they define queue names, retention settings, and quotas. It's also important to note that queues are subject to quota and memory limitations, which vary depending on the selected service plan—standard plans allow fewer queues and lower throughput than advanced or enterprise plans. Therefore, you must do careful capacity planning to prevent queue saturation and ensure sustainable operation.

The SAP monitoring dashboard provides real-time visibility into each queue's status, including message count, age, and health indicators. With this tool, integration teams can identify backlogs, monitor throughput, and proactively manage queue behavior, thus reinforcing the critical role JMS queues play in supporting scalable, resilient, and asynchronous integration processes.

### 5.2.3   Consumer iFlow

The consumer iFlow in a JMS-based architecture is responsible for retrieving messages from the queue and performing the business logic required to deliver them to their final destination. It acts as the second half of the decoupled integration pattern by processing messages independently of the producer flow and offering the flexibility to scale, enrich, and route messages, based on dynamic business needs.

The flow begins with the JMS adapter configured as a sender that continuously listens to a specific queue and pulls incoming messages. As messages are retrieved, the flow typically applies filtering or routing logic, often using JMS header properties or specific fields within the payload to direct messages along different processing paths. This step is essential when multiple message types or business domains share a queue because it ensures that each message reaches the correct downstream system.

Once routed, messages are passed through transformation components where the data is enriched, validated, or mapped into a format that's required by the target system. This transformation can involve lookups, value conversions, or even complex scripting, depending on the business scenario. Finally, the message is forwarded by the appropriate outbound adapter—such as HTTP, SOAP, or SAP SuccessFactors—to complete its journey to the target application or service.

A defining strength of the consumer flow is its support for parallel processing. By configuring multiple concurrent threads, the system can handle higher message volumes and improve throughput without impacting message order when EOIO isn't required. To ensure robust reliability, the flow typically includes exception subprocesses that handle errors gracefully, thus enabling retries or triggering alert workflows. If a message can't be processed even after multiple attempts, it's automatically routed to a DLQ for post-failure analysis and manual recovery.

In this way, the consumer flow acts as the intelligent processing layer of JMS-based integration. It delivers messages efficiently while providing the resilience, scalability, and fault tolerance that are needed for enterprise-grade scenarios.

The "Poll from JMS queue" iFlow in Figure 5.4 illustrates a JMS-driven processing pipeline within SAP Integration Suite that's optimized for decoupling and asynchronous message handling.

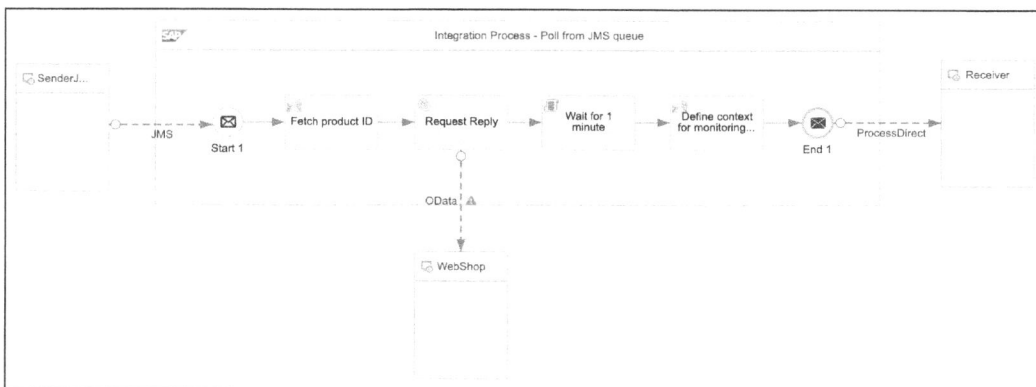

**Figure 5.4** Outbound JMS Processing Flow

It begins with a JMS sender adapter that polls messages from a JMS queue, triggering the process at **Start 1**. From there, it proceeds as follows:

1. The **Fetch Product ID** integration step involves extracting or constructing a product ID from the incoming message. This is likely a content modifier or script step that's used to prepare the payload for the downstream service call.

2. Using the product ID derived earlier, the **Request Reply** step sends a request to the WebShop component to retrieve the most recent product data. It follows a request-reply pattern over an OData connection to ensure that the iFlow always processes current and accurate product information.

3. The deliberate **Wait for 1 Minute** simulates a high-compute or time-consuming backend process in real-world scenarios. In practical terms, it represents steps like intensive data transformations, aggregation, and third-party API calls that take substantial time. The actual use might involve a complex computation that can't be shown in a simplified integration design.

4. The **Define context for monitoring** step enriches the message with context properties, which are critical for traceability, alerting, and custom monitoring in runtime tools.

5. The process ends at **End 1**, where the enriched message is handed off via the **Process-Direct** adapter to another iFlow or processing block and is finally sent to the **Receiver** system.

This modular design ensures reliability and traceability in asynchronous, resource-intensive scenarios.

### 5.2.4    Monitoring and Metrics

Effective monitoring is essential for maintaining the health, performance, and reliability of JMS-based iFlows in Cloud Integration. The **Monitoring** section within SAP Integration Suite provides administrators and support teams with real-time visibility into message flow execution, queue usage, and error conditions. This observability layer plays a critical role in proactively identifying issues, enforcing service-level agreements (SLAs), and ensuring smooth end-to-end operations.

One of the key capabilities is the ability to track messages throughout their lifecycle—from the moment they enter the producer flow, through the JMS queue, and finally to the consumer flow. Administrators can drill into individual message traces to inspect payloads, headers, adapter logs, and retry history. This end-to-end traceability is invaluable when troubleshooting delayed or failed messages because it provides full transparency into where and why processing may have broken down.

The monitoring dashboard also makes it easy to identify stuck or failed messages. Integration support teams can filter messages by status (e.g., **Failed, Retried, Successful**) and isolate bottlenecks, such as messages that are stuck in retry loops and flows that repeatedly send content to a DLQ. By analyzing patterns in failure rates and exception types, teams can take targeted actions to strengthen specific integration components.

In addition, Cloud Integration enables users to monitor message retries, processing durations, and throughput at both the flow and tenant level. This data is crucial for detecting abnormal load spikes, degraded performance, and underutilized resources. Advanced users can further correlate these insights with business events (e.g., promotions, financial closings) to forecast future capacity needs.

Finally, the system provides visibility into JMS quota usage and threshold breaches. Alerts can notify operations teams when a queue is nearing capacity, retry limits are being hit frequently, and DLQs are growing unexpectedly. These proactive alerts enable faster response times and help prevent queue exhaustion and message loss.

## 5.3    Configuration of the Java Messaging Service Adapter

The JMS adapter in Cloud Integration supports both sending and receiving messages to and from queues. It's configured within iFlows to either publish messages to a queue or consume messages from it. Correct configuration is a critical factor in ensuring reliable message delivery, efficient resource usage, and alignment with overall integration requirements.

In the following sections, we'll examine the most important aspects of JMS configuration in the SAP BTP context. Section 5.3.1 looks at the configuration properties for the sender and receiver, which define how messages are published and consumed. Section 5.3.2 highlights the limitations of the JMS queue in SAP BTP to help you make design decisions that account for platform-specific constraints. Section 5.3.3 helps you understand transactional behavior and concurrency so you can build reliable flows that can handle parallel processing without data inconsistencies. Section 5.3.4 discusses queue design strategy, including how to structure queues for scalability and maintainability, and Section 5.3.5 covers the optimization techniques that improve throughput and performance. Finally, Section 5.3.6 addresses common pitfalls that developers should avoid to prevent issues in production scenarios. Together, these sections provide a comprehensive guide to configuring JMS effectively for robust and scalable integration solutions.

### 5.3.1    Configuration Elements

When using the JMS adapter, choose one of the following access type options to determine how messages are processed, based on your integration needs.

- **Non-Exclusive**
  Choose this access type if you want messages to be processed in parallel across multiple worker nodes, which is suitable for scenarios where message order isn't essential. The JMS adapter won't serialize messages, which means that there will be no guarantee of the order in which the messages will be consumed by Cloud Integration. For additional information, also check out Table 5.1.

- **Exclusive**
  Choose this access type for cases where preserving the order of message processing is crucial. This guarantees that messages will be processed in the exact order they are received by allowing only one consumer to have access to the queue at any time.

**Sender Adapter**

When configuring how the sending side produces messages into the JMS queue, you can configure some parameters to fine-tune the behavior of the queue. The following list covers how to configure these parameters, which you do on the screen shown in Figure 5.5 (as you'll see when you open an iFlow).

- **Queue Name**
  Enter the name of the message queue.
- **Access Type**
  Select one of the following options:
  - **Non-Exclusive**
    This allows multiple consumers or workers to process messages from the queue in parallel. This is suitable for scenarios where message order isn't critical and parallel processing is beneficial.
  - **Exclusive**
    This ensures that only one consumer or worker has access to the queue at a time. This is useful in scenarios that require messages to be processed in the order they were received.
- **Number of Concurrent Processes**
  If you select **Non-Exclusive** for the **Access Type**, enter the number of concurrent processes for each worker node. The recommended value depends on the number of worker nodes, the number of queues on the tenant, and the incoming load.
- **Retry Interval**
  Enter a value for the time to wait before retrying message delivery.
- **Exponential Backoff**
  Check the box to double the retry interval after each unsuccessful retry
- **Maximum Retry Interval (in min)**
  If you select **Exponential Backoff**, enter a value for the maximum number of minutes to wait before retrying message delivery.
- **Dead-Letter Queue**
  If you select **Non-Exclusive** for **Access Type**, check this box to take the message out of processing and mark it as Blocked in the queue if message processing has been stopped due to an out-of-memory error in the worker node, the message has later been retried twice by the JMS sender adapter, and each retry has again led to an out-of-memory error in the worker node.

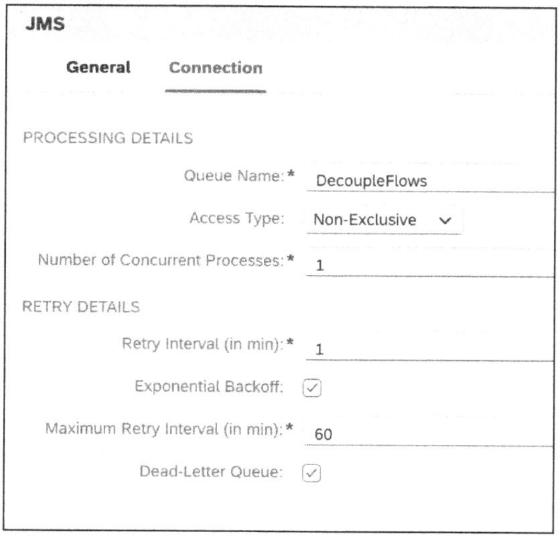

**Figure 5.5** JMS Sender Adapter Configuration Properties

**Receiver Adapter**

In addition to producing messages into a queue, you'll need to consume and process messages from a queue. This is where the receiver adapter comes into play. The following covers how to configure these parameters to alter the processing power of the receiver adapter, which you do on the screen shown in Figure 5.6.

- **Queue Name**
  Enter the name of the message queue.

- **Access Type**
  Select one of the two types of access to the JMS queue:

  - **Non-Exclusive**
    This allows multiple consumers or workers to process messages from the queue in parallel. It's suitable in scenarios where message order isn't critical and parallel processing is beneficial.

  - **Exclusive**
    This ensures that only one consumer or worker has access to the queue at a time. This is useful in scenarios that require messages to be processed in the order they were received.

- **Retention Threshold for Alerting (in d)**
  Enter the time period (in days) by which the messages have to be fetched. The default value is 2.

- **Expiration Period (in d)**
  Enter the time period (in days) by which the messages have to be fetched, which must be no less than the number of days you entered for the **Retention Threshold for**

285

**Alerting**. We recommend that you enter a number of days that's at least twice the retention threshold. The default is set to 30 days, and the maximum possible value is 180.

- **Compress Stored Message**
  Check this box to compress the message in the JMS queue, which reduces disk space usage and network traffic.

- **Encrypt Stored Message**
  Check this box to encrypt the message in the JMS queue.

- **Transfer Exchange Properties**
  Check this box to transfer the exchange properties to the JMS queue. We don't recommend using this option because headers and exchange properties are subject to size restrictions, which can result in problems or errors.

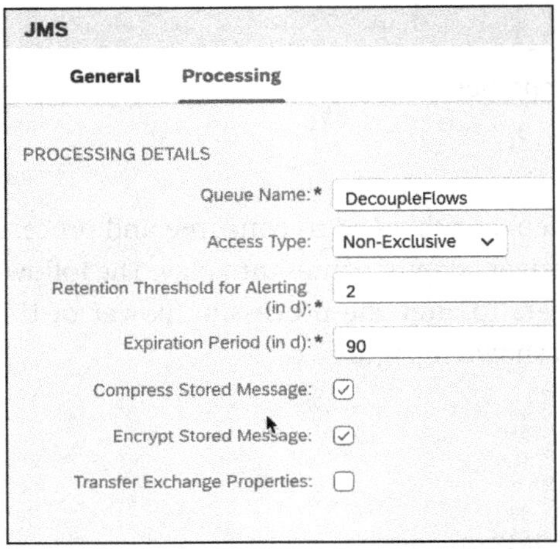

**Figure 5.6** JMS Receiver Adapter Configuration Properties

### 5.3.2   Java Messaging Service Adapter Limitations

Modern integration landscapes demand scalable, fault-tolerant architectures that are capable of handling varying message volumes without downtime or data loss. Asynchronous communication via JMS is one of the cornerstones of such architectures, and in SAP Integration Suite, JMS queues are used to decouple producers and consumers, buffer peak traffic, and recover gracefully from failures.

However, JMS resources are not unlimited. SAP imposes certain quotas and thresholds to maintain system stability, especially in shared cloud environments. You must understanding and optimize these resources to avoid failures, ensure throughput, and build sustainable interfaces.

This section focuses on JMS resource limitations in SAP BTP and strategies for optimizing their usage in production-grade integration scenarios.

**Understanding Java Messaging Service Resource Quotas in SAP Business Technology Platform**

SAP Integration Suite provides predefined resource quotas for JMS usage, which vary depending on the service plan you select (e.g., standard or advanced). These quotas govern the maximum number of queues, message throughput, concurrent consumers, and overall storage capacity that are available to a tenant. You must understanding these limits so you can plan scalable, high-volume integrations and avoid unexpected failures due to resource exhaustion.

Fortunately, SAP BTP offers self-service entitlement configuration, which allows administrators to request additional capacity through the **Entitlements and Quotas** section of the SAP BTP cockpit. This enables organizations to scale their JMS usage in alignment with growing business needs without requiring custom provisioning or support tickets. Table 5.1 summarizes the typical default values and how they can be expanded. We recommend that you to review these quotas regularly—especially during peak load periods or architectural changes—to ensure sufficient buffer and avoid disruptions in asynchronous message processing.

| Resource Types | Standard Plan Limits | Self-Service Maximums |
|---|---|---|
| Number of JMS queues | 30 | 100 |
| Queue capacity | 9.3 GB | 30 GB |
| Number of transactions | 150 | 500 |
| Number of consumers | 150 | 500 |
| Number of providers | 159 | 500 |

**Table 5.1** JMS Resource Limitations

**Message Size Considerations**

One of the most critical yet often overlooked limitations in JMS usage within SAP Integration Suite is the size of individual messages. Although there's no fixed, hard limit that's imposed per message, each JMS queue operates under a shared quota that governs the total volume of stored data. This means that a small number of overly large payloads can quickly consume the available capacity, which will leave less room for other messages and potentially cause queues to reach their maximum thresholds. When queues become full, message delivery will fail and potentially disrupt business-critical processes.

To avoid this, you need to implement the following best practices for managing JMS message size:

- Compress payloads if feasible (e.g., JSON → GZIP). Applying compression significantly reduces the byte size of structured data formats like JSON and XML, and this conserves queue space and can improve throughput by reducing transmission time.
- Use external storage for large attachments and include only metadata in the JMS message.
  You should upload documents, images, and binary files to external systems such as SAP Document Management or cloud object storage. The JMS message should then carry a reference link or document ID to minimize in-queue payload weight.
- Keep JMS headers concise since they consume space too. Custom headers are useful for routing and traceability, but you should use them sparingly. Overuse of headers adds overhead and inflates message size.
- Avoid unnecessary XML/JSON verbosity. Trim redundant fields, whitespace, and deeply nested structures to streamline the payload and make better use of JMS quotas.

### 5.3.3   Transactional Behavior and Concurrency

Each consumer iFlow in Cloud Integration that retrieves messages from a JMS queue operates in transactional mode to ensure reliability and message integrity. In this mode, a message is only considered successfully processed once the entire flow completes without error. If there's a failure of any part of the flow—such as a transformation step, external system call, or routing logic—the transaction is rolled back and the message is returned to the queue for retry. This transactional behavior ensures that no message is lost or partially processed, and that's crucial for maintaining consistency in enterprise data flows.

You can use the following key transactional controls to fine-tune performance and stability:

- **Batch size**
  This defines how many messages are fetched and processed in a single transactional unit. The default batch size is up to 256 messages, and larger batches may lead to processing delays, memory strain, and timeouts. In high-load environments, tuning this parameter helps balance throughput and resource efficiency.

- **Concurrency**
  You can configure the number of parallel threads that process messages simultaneously. Higher concurrency increases processing speed and is ideal for large message volumes, but it also demands more CPU and memory resources. The optimal concurrency setting depends on system capacity and the complexity of the processing logic.

- **Visibility timeout**
  This parameter controls how long a message remains invisible to other consumers once it's picked up. If the flow crashes or hangs during processing, the message

becomes visible again after the timeout so it to be retried. This mechanism prevents message loss in unforeseen failure scenarios.

Together, these controls provide a flexible, high-performance foundation for building resilient, transactional message processing flows.

### 5.3.4   Queue Design Strategy and Optimization

When you're designing JMS-based messaging architectures in Cloud Integration, you may be tempted to create a dedicated queue for each integration scenario. While this provides clear separation, it can quickly exhaust the limited number of queues that are available per tenant, especially in standard service plans. A more scalable and resource-efficient approach is to use shared queues with internal routing logic inside the iFlows. In this model, messages of different types or use cases are published to the same queue, and the consumer flow applies filters, content-based routing, or header-based selectors to direct each message down the appropriate processing path.

There are several benefits to using shared queue models:

- It reduces total queue count and associated resource consumption. Consolidating use cases onto fewer queues helps them stay within JMS quota limits, leaves room for future growth, and reduces administrative overhead.

- It makes monitoring and management easier. Having fewer queues makes it simpler to track message health, throughput, and failures, so it improves visibility and operational control.

- It facilitates dynamic scaling of consumer flows. With one shared queue, it becomes easier to manage concurrency, tune performance, and respond to spikes in load without adjusting multiple queues individually.

However, this approach requires thoughtful design, and you must still consider message isolation and error handling. For high-risk or business-critical processes where data privacy, compliance, or strict SLA enforcement is required, it may be better to segregate those flows into dedicated queues. This prevents noisy neighbors from interfering with performance, and it simplifies incident resolution for isolated domains. Shared queues, when designed carefully, offer a powerful pattern for efficient and scalable integration design.

### 5.3.5   Optimization Techniques

Efficient use of JMS queues is a cornerstone of building scalable, resilient, and cost-effective integrations within SAP Integration Suite. As messaging demands grow across systems and business processes, careful management of JMS resources becomes increasingly important. By strategically optimizing queue design, limiting payload size, and fine-tuning processing patterns such as concurrency and batching, integration teams can ensure high throughput, avoid system bottlenecks, and maintain message integrity. Just as important, these optimizations help organizations stay within defined platform resource limits to prevent unexpected disruptions and enable sustainable growth across the integration landscape. Let's look at some key techniques for accomplishing this.

### Limiting Queue Creation

In complex integration landscapes, it's common to see a proliferation of JMS queues—often one per use case or interface. While this offers clean separation, it quickly leads to resource exhaustion and operational overhead. An effective alternative is JMS queue consolidation, in which multiple integration scenarios share a common queue. You can make this possible by unifying the logic of multiple flows into a single, modular iFlow that routes and processes messages dynamically, based on metadata such as JMS headers, message type, or business context.

Using constructs like message routers, content-based filters, and external configuration tables, developers can maintain functional separation within a unified flow. This strategy reduces queue count, optimizes resource usage, and simplifies monitoring, while still maintaining flexibility and extensibility. However, it requires careful design to ensure routing logic is transparent, testable, and maintainable.

### Minimizing Message Volume

JMS queues in SAP Integration Suite are subject to overall quota limits, which means excessive message sizes can quickly exhaust available capacity, trigger retries, and degrade performance. To avoid this, you need to keep payloads lean and purpose-fit for message-based communication. One effective strategy is to offload large attachments or bulk data—such as PDFs, images, and data exports—to external storage systems like SAP Document Management service or cloud object stores (e.g., Azure Blob Storage, Amazon S3). The JMS message then only needs to carry metadata or a reference (such as a URL or document ID), and that drastically reduces size while preserving access to full content.

Additionally, content compression techniques, such as applying GZIP to JSON or XML payloads, can further reduce message footprint without sacrificing structure. Implementing size validation early in the iFlow helps reject oversized messages before they consume queue resources. By combining external storage with compression and validation, you ensure that JMS remains a high-performance, cost-effective messaging layer.

### Using DLQs Effectively

DLQs are further described in Section 5.4.3, and they're essential components of robust JMS-based integration, particularly for critical business flows where message loss is unacceptable. If you configure a DLQ for such flows, failed messages that exceed retry limits or encounter unrecoverable errors will be automatically redirected for later inspection. This ensures that vital data isn't silently discarded and can be recovered or corrected. To use DLQs effectively, you need to implement a regular monitoring routine—through either SAP's monitoring tools or external alerting systems—to track DLQ growth and identify recurring failure patterns. Periodic manual or automated cleanup and reprocessing should be part of your operational best practices. You can handle reprocessing through a dedicated error-handling iFlow that's capable of enriching or transforming the message before resending it to the original queue or target system. This proactive approach safeguards data integrity, improves system resilience, and provides valuable insights into the long-term quality and stability of the integration landscape.

### Scaling Consumers Dynamically

To help you handle fluctuating workloads efficiently, SAP Integration Suite allows for dynamic scaling of JMS consumers, which enables more processing threads during high-traffic periods such as month-end closings, promotional events, and batch imports. By increasing the number of concurrent consumers on a JMS-based iFlow, the system can process more messages in parallel, thus reducing backlog and improving responsiveness. However, indiscriminate scaling can overwhelm downstream systems or saturate memory and CPU resources.

To manage this risk, you should consider implementing *condition-based throttling*, in which the message consumption rate adapts to external triggers such as system availability, queue length, and business hours. You can achieve throttling by using process control parameters, delay patterns, or dynamic configurations fetched from external sources. When combined, consumer scaling and throttling enable iFlows to remain both responsive and controlled, thus ensuring stable performance under pressure without compromising system integrity.

### Monitoring Routinely

Consistent monitoring of JMS queues is crucial for maintaining system health and anticipating capacity issues before they impact operations. A well-structured monitoring routine should include weekly utilization checks to track queue depth, processing throughput, and message age, because they help to detect anomalies such as processing delays and stuck messages. Additionally, monthly reporting and forecasting based on historical usage patterns enables teams to predict growth trends, plan for upcoming peak loads, and adjust configurations or upgrade service plans proactively. Combining short-term operational oversight with long-term capacity planning ensures sustainable performance and avoids unexpected service interruptions

### 5.3.6   Common Pitfalls and Their Mitigation

In this section, we'll walk you through some common pitfalls that organizations may encounter when using a JMS adapter, and we'll show you how to counter them:

- **Too many queues created per use case**
  - **Impact**
    Creating a dedicated JMS queue for each iFlow may initially seem a well-organized process, but it quickly leads to exhaustion of the allowed queue quota, especially in standard SAP BTP service plans. With only about 30 queues available by default, this approach limits scalability and increases administrative overhead.
  - **Mitigation strategy**
    Instead of one queue per interface, you can design shared queues with intelligent message routing. Use JMS header properties (e.g., type, businessArea) and message selectors in the consumer flows to filter and process only relevant messages. This approach reduces the total number of queues required while maintaining logical separation. It also simplifies monitoring and resource allocation to help maintain healthy queue capacity over time.

- **Unbounded payload growth**
  - **Impact**
    Large message payloads consume more space in the queue, reducing available capacity. Since queue retention is shared across all messages, a few large messages can exhaust the queue and prevent new messages from being published. Additionally, large payloads may cause timeouts, retries, and downstream processing delays.
  - **Mitigation strategy**
    To prevent payload-related bottlenecks, adopt a disciplined payload design. Compress data by using algorithms like GZIP, strip unnecessary metadata and whitespace, and consider storing attachments or documents in external repositories (like SAP Document Management or Azure Blob Storage) and referencing them via a link or ID in the message. Also, implement data contracts that enforce message size limits and validation checks early in the producer flow to filter oversized messages before they enter the queue.

- **Overlapping retries due to high concurrency**
  - **Impact**
    Setting high concurrency on a consumer flow can lead to multiple instances trying to process the same message simultaneously, particularly when messages fail and reenter the queue rapidly. This can result in message duplication, increased load, and inconsistent system behavior.
  - **Mitigation strategy**
    Use EOIO configuration if the processing requires strict sequencing or transactional integrity. Limit concurrency where messages must be processed serially to

preserve state consistency. Introduce idempotent logic in your iFlows so that repeated processing doesn't lead to duplicated records or actions. For critical flows, you can use a retry pattern that includes delay and exponential backoff to minimize collision and processing contention.

- **No cleanup of failed DLQ messages**
  - **Impact**
    DLQs accumulate messages that can't be processed due to repeated failures. If these are not monitored and cleaned up, they consume valuable queue space, clutter monitoring dashboards, and may mask real-time issues by obscuring alert visibility.
  - **Mitigation strategy**
    Schedule regular DLQ reviews as part of your operational routine and implement alerting rules when DLQ size exceeds a threshold (e.g., 100 messages). Develop a dedicated error-handling flow to read from DLQs, enrich error messages, and either retry them under modified conditions or escalate them to manual resolution. Document error causes and recurring patterns to improve the robustness of your main processing flows over time.

- **Missing alerts for near-exhausted queues**
  - **Impact**
    Without timely alerts, you may not realize that a queue is nearing its maximum capacity until new messages are rejected. This can lead to business-critical message loss or backpressure on the producing systems, which in turn can cause upstream delays or failures.
  - **Mitigation strategy**
    Proactively configure JMS monitoring alerts in the Integration Suite. Set thresholds for alerting at around 80% usage so you can take action before hard limits are reached. Use the SAP Alert Notification service to forward alerts to operations teams or integrate with external monitoring systems like Splunk or Microsoft Teams. Pair these alerts with automated dashboards that provide visibility into all active queues and their status, so operators can react swiftly to anomalies.

Table 5.2 serves as a quick summary of these common pitfalls, their impacts, and mitigation strategies.

| Pitfalls | Impacts | Mitigation Strategies |
|---|---|---|
| Too many queues created per use case | They exhaust the queue quota. | Use shared queues with selectors. |
| Unbounded payload growth | It hits storage limits and increases retries. | Compress and split large payloads. |

**Table 5.2** JMS Pitfalls and Mitigations

| Pitfalls | Impacts | Mitigation Strategies |
|---|---|---|
| Overlapping retries, due to high concurrency | Message processing gets duplicated. | Use EOIO and limit concurrency per flow. |
| No cleanup of failed DLQ messages | It occupies space and increases noise. | Schedule regular DLQ reviews and cleanup routines. |
| Missing alerts for near-exhausted queues | They cause sudden interface failure. | Set up alert rules in the monitoring dashboard. |

**Table 5.2** JMS Pitfalls and Mitigations (Cont.)

## 5.4   Error Handling and Retry Strategies

In asynchronous messaging architectures, error handling and retry strategies are crucial for ensuring reliability and integrity. This is required to keep the entire system and all its components in a consistent state, and it also renders transactional operations completed. Database and other persistent stores need to stay synchronized and may not contain loose, dangling or other inappropriate content. Cloud Integration provides a comprehensive toolkit for managing various error scenarios that arise during message processing.

To address these challenges effectively, you need to understand the different building blocks and strategies that are available in the SAP BTP context. To help you with this, in Section 5.4.1, we explore the exception subprocesses, which provide a structured way to catch and manage errors within iFlows. In Section 5.4.2, we'll look at retry mechanisms, which help ensure that temporary issues don't lead to message loss. In Section 5.4.3, we look at cases where messages can't be processed even after retries and the DLQ offers a safe fallback. In Section 5.4.4, we cover manual message reprocessing, which allows administrators to intervene and resolve issues directly. In Section 5.4.5, we cover error categorization, which makes troubleshooting more effective by helping to distinguish between recoverable and nonrecoverable problems. Finally, Section 5.4.6 covers alerts and monitoring Integration, which ensure that critical issues are surfaced in real time and can be acted upon proactively. Together, these approaches form a holistic framework for building resilient, fault-tolerant integrations with JMS.

### 5.4.1   Exception Subprocesses

At the core of any resilient integration design lies the principle that failures should be expected and handled deliberately. In Cloud Integration, *exception subprocesses* provide a structured and maintainable way to capture and react to runtime errors during message processing. Instead of allowing a failure to terminate an iFlow abruptly, an exception subprocess offers a controlled path where the error can be logged, categorized, retried, or rerouted as needed. Developers can define these subprocesses for the entire iFlow or scope them more narrowly around specific steps, such as a content modifier, a message mapping, or an external service call. This flexibility enables developers to differentiate between technical issues—like HTTP timeouts or transformation failures—and business rule violations that may require different handling strategies. Exception subprocesses not only improve system robustness but also provide vital observability, and that makes them foundational building blocks of any effective retry and recovery mechanism in JMS-based integration scenarios.

To implement exception handling effectively, you should focus on granularity, reusability, and traceability. One proven design pattern is to create localized exception subprocesses near the point of failure—for example, wrapping external API calls—so that error handling is contextual and focused. These subprocesses should log essential diagnostic information—such as error type, message ID, timestamp, and custom JMS headers—to support root cause analysis. For recurring logic like error logging or alert generation, you can extract the subprocess into a reusable integration artifact, such as a shared logging flow or centralized monitoring endpoint. Categorization is also key: you should differentiate between transient errors (which are suitable for retry), data-related issues (which are suitable for escalation), and unexpected faults (which are suitable for alerts or manual intervention). You should also avoid implementing generic catch-all subprocesses that simply swallow errors without meaningful action, because that can obscure critical failures and delay response times. By designing exception subprocesses thoughtfully, you as an integration developer can create resilient, maintainable, and transparent error-handling frameworks that strengthen the overall stability and observability of asynchronous messaging flows.

## 5.4.2   Retry Mechanisms

An essential component of fault-tolerant integration is the ability to recover gracefully from temporary failures without manual intervention. In Cloud Integration, the JMS adapter's built-in *retry mechanisms* are specifically designed to address transient issues such as network instability, temporary backend unavailability, and intermittent timeouts. Rather than immediately marking a message as failed, the adapter pauses and attempts to resend the message based on a preconfigured retry policy. This ensures that message delivery remains reliable without overburdening the source system or requiring real-time availability from downstream systems. The retry logic operates at the queue level, which means the message remains safely in the JMS queue and is only removed upon successful processing. This asynchronous behavior allows for natural buffering during spikes in system load or short-term outages, and that makes the retry mechanism a critical safeguard in loosely coupled architectures.

To configure retries effectively, developers and architects must balance recovery speed with system stability. The most fundamental settings include the **Retry Interval**, which is the pause duration between attempts, and the **Retry Count**, which limits how many times the adapter will attempt delivery before escalating the issue or forwarding the message to a DLQ. For more nuanced control, the adapter also supports *exponential backoff*, which is a strategy that increases the wait time between retries after each failure. This helps keep fragile downstream systems from being overwhelmed while still allowing time for recovery. A well-tuned retry mechanism should align with the availability characteristics of the target system; for example, retrying every five minutes for up to an hour might suit a CRM system that experiences nightly downtime. Additionally, it's important to monitor retry patterns in production: if certain queues frequently hit retry limits, it could signal deeper issues that require architectural changes. Ultimately, when paired with exception subprocesses and proper alerting, JMS retries form a resilient first line of defense against temporary faults that enables seamless and automated recovery within asynchronous integration scenarios.

### 5.4.3   Dead Letter Queue

In a robust integration landscape, not all errors can or should be resolved through automated retries. In cases where all retry attempts are exhausted, DLQs provide a critical safety net by isolating problematic messages from the main processing path. Instead of letting failed messages block or clog the queue, the system routs them to a designated DLQ for post-mortem analysis and controlled recovery. Each failed message in the DLQ retains important metadata—such as headers, timestamps, error codes, and the original payload—that offers valuable context for debugging. This mechanism not only prevents queue saturation but also ensures that messages requiring manual correction or escalation are retained for later handling. By separating retriable errors from terminal failures, DLQs enable teams to maintain throughput in production environments without sacrificing data visibility or reliability.

From an operational standpoint, effective DLQ management is key to sustainable integration operations. Routine monitoring of DLQ content should be part of your weekly or even daily health checks, particularly for critical business processes. You can manually reprocess messages in the DLQ via the Cloud Integration **Monitoring** UI, either as-is or after applying corrections based on error diagnostics. To streamline this process, integration teams should define escalation thresholds—for example, by triggering an alert when more than 100 messages accumulate or when a specific message type repeatedly fails. These thresholds can trigger alerts via SAP Alert Notification service, or they can be integrated with external ITSM platforms like ServiceNow and Jira. Furthermore, by analyzing recurring error types in DLQs, teams can proactively improve error handling in upstream flows, which can lead to more resilient and self-healing integrations. Also, businesses should never treat DLQs as passive storage—they are powerful operational tools when paired with strong visibility, analytics, and governance strategies. With well-maintained DLQs in place, businesses can ensure graceful degradation, preserve message integrity, and uphold SLAs, even during failure conditions.

### 5.4.4   Manual Message Reprocessing

A key advantage of using DLQs in Cloud Integration is the ability to manually reprocess failed messages after root cause analysis or corrective action. This feature ensures that even unrecoverable errors don't result in data loss, provided that appropriate follow-up is performed. You handle reprocessing through the **Monitoring** section of SAP Integration Suite, which offers a user-friendly interface that you can use to review and act on failed transactions. To begin, you navigate to **Monitor • Message Queues** (see Figure 5.7) and open the message queues table. This will bring up a list of configured JMS message queues that you can select, and you can enumerate all messages in the given queue (see Figure 5.8). When a message hasn't been processed successfully, it will show a status of **Failed**, and you can select that message and click **Retry** on the top bar to retry the processing, or you can click **Move** to move the message into a different (e.g., DLQ) queue.

This capability empowers support teams to resolve integration issues efficiently without requiring redeployment or flow modifications. That makes it a powerful tool for operational continuity and SLA compliance.

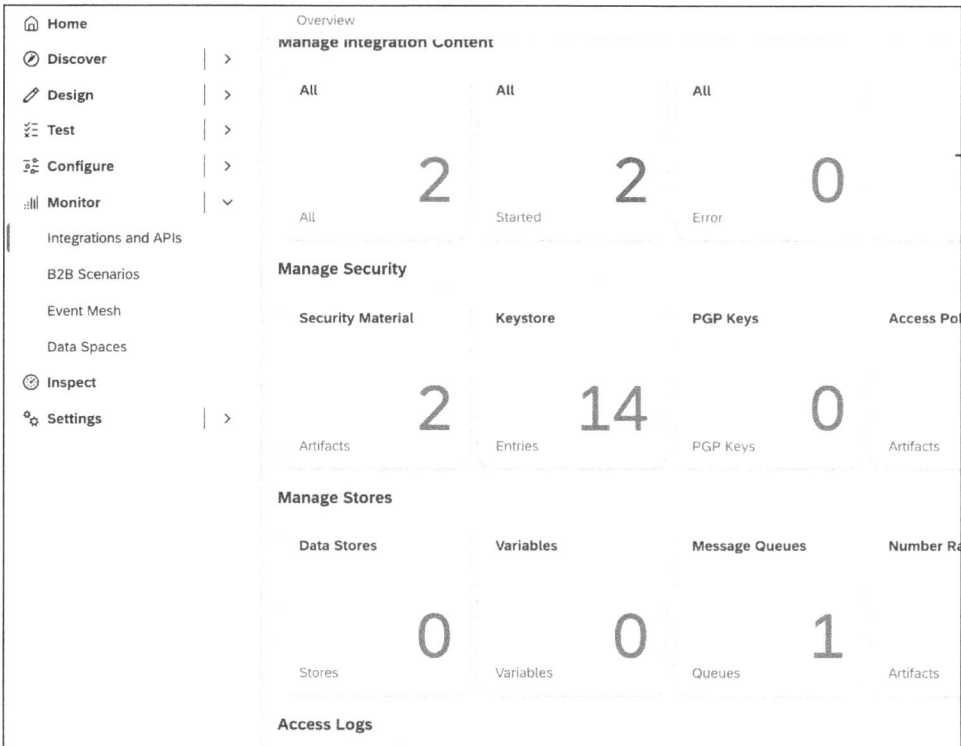

**Figure 5.7** Cloud Integration Monitoring Overview

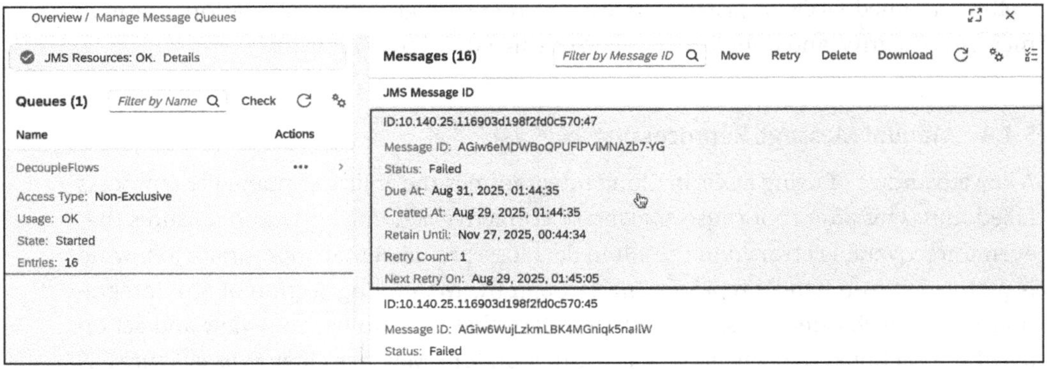

**Figure 5.8** Cloud Integration Message Queue Table for Message Management

### 5.4.5   Error Categorization

Effective error handling in JMS-based iFlows begins with the ability to categorize errors accurately, as not all failures are created equal. Broadly, errors encountered during JMS processing can be classified into three categories: transient, persistent, and system level.

*Transient errors* are typically temporary issues such as network delays, backend service timeouts, and temporary unavailability of external systems. These are retriable, and the JMS adapter's built-in retry mechanism is well-suited to handle them automatically because it allows the system to recover once conditions normalize.

*Persistent errors*, on the other hand, stem from data-related problems such as invalid formats, missing fields, and failed validations. These errors are nonretriable and will persist regardless of the number of attempts. In such cases, forwarding the message to a DLQ is the most appropriate response because it enables manual inspection and correction.

*System errors* are the third and often most critical error category. They include internal integration issues, such as memory exhaustion, configuration errors, code bugs, and unavailable resources like JMS quota overflows. These errors require immediate operational intervention and are typically accompanied by system logs or alerts.

You need to understand which category an error belongs to so you can determine the correct remediation strategy—whether to let the retry mechanism handle it, forward it to a DLQ, or escalate it to the support or infrastructure team. For practical implementation, you should designate flows to tag errors with category-specific identifiers and include detailed exception handling logic that's tailored to each error type. Additionally, having analytics and monitoring tools that report on error trends by category can greatly enhance decision-making and issue triage. Categorization transforms error handling from a reactive to a strategic capability, and that helps organizations build more intelligent, self-aware integrations that maintain operational continuity, even under failure conditions.

### 5.4.6 Alerts and Monitoring Integration

Proactive monitoring is a cornerstone of operational excellence in enterprise integration, and Cloud Integration provides robust options to support real-time alerting and observability. To detect and react to issues in JMS-based flows, integration developers can leverage native integrations with SAP Alert Notification service, SAP BTP cockpit logs, and third-party external monitoring tools like Splunk, Dynatrace, and Azure Monitor. These integrations allow for seamless tracking of key system events and thresholds, which helps ensure that critical failures or bottlenecks are not only recorded but also immediately actionable. Whether they alert operations teams to an unexpected spike in queue depth or a surge in retry activity, centralized alerts notify teams promptly so they can take corrective action before user-facing impacts occur.

Administrators can define alert triggers based on a range of metrics and thresholds that are tied to JMS queue activity. For example, administrators can configure alerts when message retries exceed a set threshold, which indicates that a downstream system may be unresponsive or overloaded. Similarly, the population of DLQs can trigger a warning, especially if recurring errors begin to accumulate. Another vital trigger is queue size utilization—when a queue approaches 80% or more of its capacity, an alert can prevent service degradation by prompting early investigation. These alerts can be consumed through multiple channels, including email, Slack, Microsoft Teams, and webhook-based integrations with ITSM systems like ServiceNow.

By embedding alerting into your monitoring landscape, you can enable continuous integration health checks, streamline root cause analysis, and enhance the resilience of asynchronous messaging architectures. Ultimately, well-configured alerts not only support faster incident response but also create the transparency needed to uphold SLAs and reduce downtime across business-critical integrations.

## 5.5 Summary

Loosely coupled interfaces built on JMS queues within SAP BTP offer a proven and powerful architectural approach to designing scalable, fault-tolerant, and resilient enterprise integrations. In increasingly complex and distributed IT landscapes, where system uptime, elasticity, and adaptability are crucial, the ability to decouple producers and consumers of data is no longer a luxury—it's a necessity.

This chapter has demonstrated how asynchronous messaging using JMS forms the foundation for such loosely coupled systems. By removing the direct dependency between sending and receiving systems, asynchronous processing ensures that integrations can continue operating smoothly, even when downstream systems are unavailable, undergoing maintenance, or experiencing performance degradation. The message queue serves as a persistent buffer that enables the system to absorb fluctuations in load, retry transient failures, and continue functioning under unpredictable conditions.

With proper planning and implementation, JMS messaging in SAP BTP empowers organizations to build enterprise-grade integration solutions that are not only reliable and scalable but also flexible, maintainable, and future-proof. As businesses embrace cloud-native design, microservices, and API-first thinking, the role of loosely coupled interfaces becomes increasingly important.

JMS is more than a transport protocol—it's a building block for resilient architecture. By embracing the principles and practices covered in this chapter, integration architects and developers can confidently design systems that respond gracefully to change, recover from failure, and scale to meet the demands of modern enterprise operations.

# Chapter 6
# Artifact Reusability

*Artifact reusability is the capacity to design integration components—such as value mappings, message mappings, integration flows (iFlows), scripts, and other configuration elements—in a modular and shareable manner. Instead of reinventing the wheel for each integration scenario, reusability promotes the creation of standardized, configurable artifacts that can be leveraged across multiple projects or business units. This approach aligns closely with the principles of agile development and DevOps, and it enables faster time to value and improved governance.*

SAP Business Technology Platform (SAP BTP) enhances this vision by offering tools and services that support reusable content. SAP Integration Suite's capabilities—such as the Integration Advisor, Open Connectors, SAP Process Integration runtime, and content modifier templates—make it easier to create and manage reusable artifacts. Furthermore, the package management feature allows developers to bundle and transport reusable content across tenants and environments with minimal effort.

This chapter delves into the practical strategies, tools, and architectural patterns for building reusable artifacts within Cloud Integration. You will learn how to abstract business logic, design parameterized iFlows, manage versioning, and promote content through the lifecycle in a controlled and scalable way. Real-world examples and use cases will illustrate how organizations can establish reusability as a cornerstone of their integration strategy and thus drive not only technical excellence but also tangible business outcomes. Ultimately, mastering artifact reusability empowers integration developers and architects to work smarter—not harder. It transforms integration from a tactical necessity into a strategic enabler, thus fostering innovation, agility, and resilience in the digital enterprise.

## 6.1 Introduction to Reusability

Integration has become the backbone of digital transformation. Every modern enterprise operates in a complex ecosystem of applications, cloud services, APIs, and on-premise systems that exchange vast amounts of data daily. Whether it's synchronizing orders from e-commerce platforms with ERP systems, consolidating financial transactions across regions, or connecting IoT devices to analytics platforms, integration is no longer an isolated technical task—it's a strategic capability.

Yet as organizations expand their digital landscapes, the complexity of integration grows exponentially. In the early days, it was common practice to build monolithic iFlows—self-contained processes that handled every aspect of a use case, from message validation and transformation to error handling and logging. This approach worked when the number of flows was small, but in today's environment, it introduces severe limitations.

Imagine an enterprise running hundreds of iFlows, each of which contains hardcoded scripts for logging, embedded mappings for data transformation, and its own logic for token management. What happens when a security policy changes and every flow needs a new logging format? What if a tax calculation rule changes and impacts multiple flows across regions? Without a reusable architecture, the cost and risk of implementing these changes skyrocket. This scenario illustrates a fundamental truth: reusability is no longer optional—it's essential.

At its core, *reusability* in SAP Integration Suite is the capacity for integration components to be maintained and applied across multiple scenarios without duplication, due to the way they're designed. Creating reusability means creating *shared building blocks*—such as scripts, mappings, and modular flows—that encapsulate specific logic and expose it for reuse in other integration processes.

Instead of copying and pasting a Groovy script into twenty different flows, developers store it in a script collection and reference it externally. Instead of recreating message mappings for each integration, they build a shared mapping artifact and apply it wherever the transformation is needed. Instead of embedding logging or validation logic in every flow, they modularize these functions into subflows that are invoked via ProcessDirect.

This concept aligns closely with long-established principles in software engineering, such as don't repeat yourself (DRY) and separation of concerns (modularization). In integration design, *DRY* means that no business rule, mapping logic, or technical routine should exist in more than one place unless absolutely necessary. Modularization, on the other hand, ensures that flows remain manageable by isolating distinct responsibilities into reusable components.

It's tempting to view reusability as merely a way to reduce development effort. While that's certainly true, the implications run much deeper. Reusability influences cost efficiency, consistency, agility, and innovation—all of which are critical for business success in the digital era:

- **Cost efficiency**
  By eliminating duplication, organizations can significantly reduce the time and resources required for development and maintenance. A single update to a shared artifact can propagate across dozens of flows.

- **Consistency**
  Reusable artifacts ensure uniform application of business rules and technical standards. This consistency is vital for compliance with regulatory requirements, such as GDPR and financial reporting standards.

- **Agility**
  In dynamic business environments, requirements change frequently. Reusability allows organizations to adapt quickly and implement updates in centralized components without disrupting every dependent flow.
- **Innovation**
  When teams are not bogged down by repetitive tasks, they can focus on innovation—exploring new services, APIs, and integration patterns that drive competitive advantage.

To fully appreciate the value of reusability, you need to understand how integration design has evolved. In traditional middleware systems, integrations were often implemented as large, monolithic processes that combined validation, mapping, enrichment, error handling, and logging into a single artifact. While convenient for initial development, this design introduced several drawbacks:

- Any change required touching multiple unrelated parts of the flow, which increased the risk of unintended side effects.
- Common logic, such as logging or authentication, was duplicated across flows, which in turn made updates labor-intensive.
- Testing and debugging were difficult, due to the size and complexity of the flows.

Modern integration architectures embrace modularity and reusability as antidotes to these problems. By decomposing integration logic into smaller, purpose-driven components, organizations achieve greater maintainability and flexibility. SAP Integration Suite supports this paradigm through features such as script collections, reusable message mappings, and ProcessDirect-based modularization, which we'll explore in detail throughout this chapter.

As illustrated in Figure 6.1, this chapter is anchored on three foundational pillars that drive artifact reusability within Cloud Integration: script collections, reusable message mappings, and modularization with ProcessDirect.

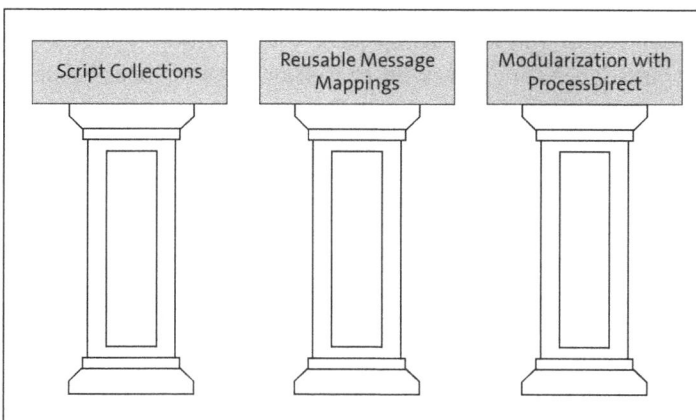

**Figure 6.1** Pillars of Artifact Reusability

These pillars represent the core techniques that enable integration developers and architects to build scalable, maintainable, and efficient solutions across diverse landscapes.

Let's look at each pillar in a little more detail:

- **Script collections**
  These allow centralized storage and reuse of custom Groovy scripts, which reduces duplication and enhances governance. Custom logic is often essential for handling advanced scenarios, but embedding scripts in multiple flows leads to duplication. Script collections allow developers to store Groovy scripts centrally and reference them across flows, which ensures consistent logic, simplifies updates, and strengthens governance.

- **Reusable message mappings**
  These promote consistency in data transformation by externalizing mapping logic, which facilitates managing and sharing across multiple iFlows. Data transformation is at the heart of integration, and instead of re-creating mapping logic for each scenario, SAP Integration Suite enables developers to externalize mappings as standalone artifacts. You can share these mappings across flows, combine them with value mappings for discrete value conversions, and enhance them through user-defined functions for complex logic.

- **Modularization with ProcessDirect**
  This introduces a service-like architecture that lets you break down iFlows into smaller, callable subflows that streamline complexity and improve maintainability. ProcessDirect is perhaps the most transformative capability because it allows flows to call other flows internally within the tenant. This enables architects to design modular subflows for logging, validation, and token management.

Together, these pillars form a comprehensive strategy for building integration landscapes that are scalable, robust, and future ready.

While the benefits of reusability are compelling, achieving them requires more than just technical features. Governance plays a crucial role, because without version control, naming conventions, and dependency tracking, shared artifacts can become liabilities rather than assets. Similarly, you must not overlook performance considerations, and since modularization introduces orchestration overhead, you must balance it against the benefits of separation.

Integration with CI/CD pipelines further strengthens the value proposition. Automated testing of shared artifacts, combined with controlled transport mechanisms like SAP Cloud Transport Management, ensures that changes propagate safely and predictably across environments. In this way, reusability becomes a natural extension of DevOps practices that supports agile delivery without compromising stability.

**Building with LEGO versus Pouring Concrete**

To illustrate the power of reusability, consider two construction approaches. The first involves pouring a massive slab of concrete and carving every feature into it—a process that's rigid, labor-intensive, and unforgiving of change. The second approach uses LEGO bricks—standardized, reusable components that you can assemble in countless configurations. When a design change occurs, you can swap out individual bricks without rebuilding the entire structure.

In integration, monolithic flows are like concrete slabs: hard to change and prone to cracking under pressure. On the other hand, reusable artifacts are the LEGO bricks of integration—flexible, composable, and resilient to change. SAP Integration Suite provides the toolbox for building with LEGO rather than concrete slabs, and this chapter will show you how to use it effectively.

## 6.2   Using Script Collections

Reusable Groovy scripts are some of the most effective mechanisms for improving maintainability and standardization in SAP Integration Suite. Rather than embedding logic directly into each iFlow, you can externalize scripts and organize them into script collections—which are modular artifacts that support centralized management and cross-scenario reuse.

This section explains what script collections are, how they're structured, where and how they're referenced, and what best practices you should follow to ensure governance, consistency, and long-term sustainability.

### 6.2.1   Overview of Groovy Scripting in SAP Integration Suite

As detailed in Chapter 3, Groovy is a dynamic scripting language that's built on the Java platform. Within SAP Integration Suite, it serves as the primary language for expressing advanced logic and runtime behavior that can't be easily modeled with standard graphical tools. By embedding Groovy scripts in iFlows, developers gain low-level control over message processing. Groovy scripts enable the following:

- Reading and manipulating payloads (e.g., JSON, XML, text)
- Accessing and modifying headers and exchange properties
- Performing conditional checks and exception handling
- Calling external APIs using Java libraries

Unlike static adapters and configuration values, Groovy scripts can respond to dynamic runtime conditions. This makes them ideal for complex scenarios such as content-based routing, token handling, data validation, and message transformation.

Scripts can be embedded directly into a flow or referenced from a shared script collection. While inline scripts offer quick prototyping, they're not maintainable at scale. Shared script collections provide a structured alternative that fosters consistency and reuse.

Here's a useful analogy: script collections are to iFlows what shared libraries are to application development. They encapsulate reusable logic into centrally maintained, testable components—thus reducing redundancy and enforcing standards.

As we'll explore next, script collections allow you to manage logic centrally and apply it consistently across many integration projects. You can add these collections to integration packages as artifacts, deploy them separately, and reference them as you need to.

### 6.2.2   Creating and Managing Script Collections

Now, let's now look at how you create a script collection. Follow these steps:

1. Navigate to an integration package in your design workspace.
2. Click **Add • Script Collection**, as shown in Figure 6.2.

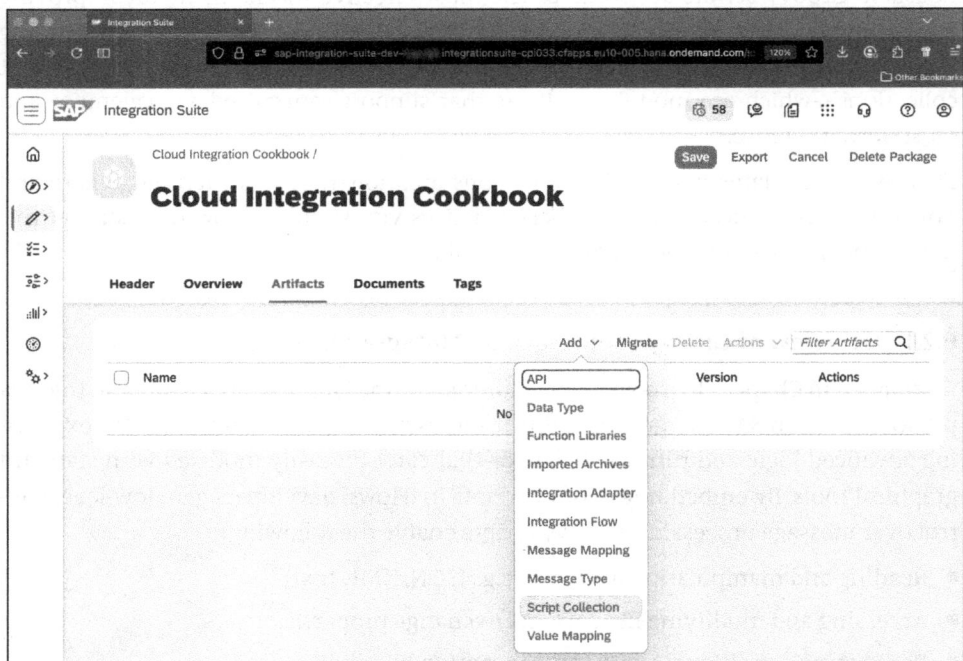

**Figure 6.2**  Creating Script Collection in Integration Package

3. Enter a meaningful **Name** and **Description** (e.g., "core-utils.groovy.scripts") in the boxes of the screen shown in Figure 6.3 and then click **OK**. You can keep the default **ID** and **Runtime Profile** values.

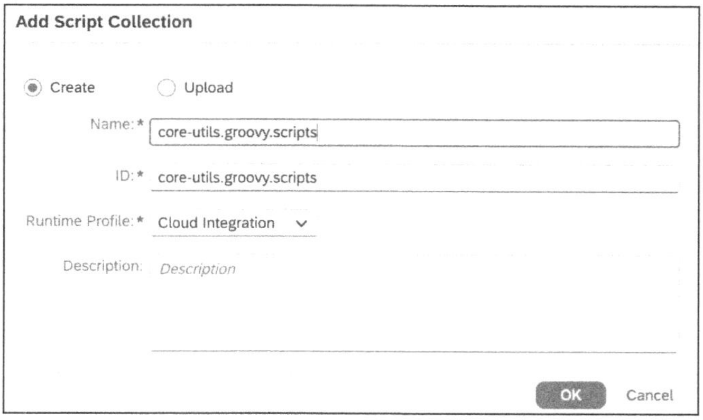

**Figure 6.3**  Naming Script Collection

4. Create or upload *.groovy* files into the collection. You can directly create scripts in SAP BTP by clicking the **Create** button depicted in Figure 6.4.

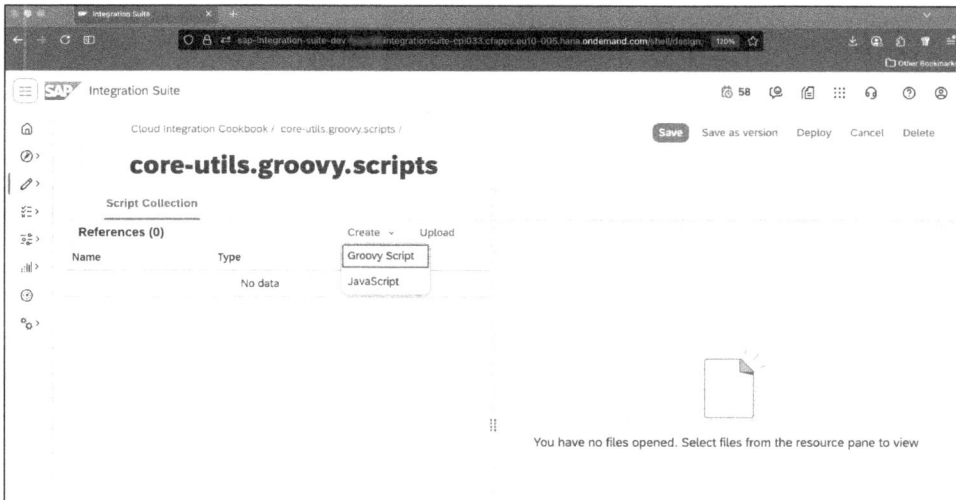

**Figure 6.4**  Creating Script in Script Collection

5. If you click on **Create • Groovy Script**, you'll need to give the script a descriptive file **Name** (as shown in Figure 6.5) and then click **Create**.

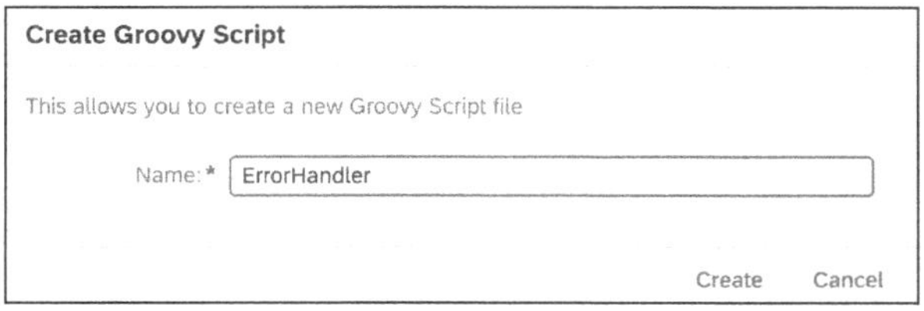

**Figure 6.5**  Giving Created Script Descriptive Name

6. Deploy the collection to make it available to other flows in the workspace, by clicking the **Deploy** button depicted in Figure 6.6.

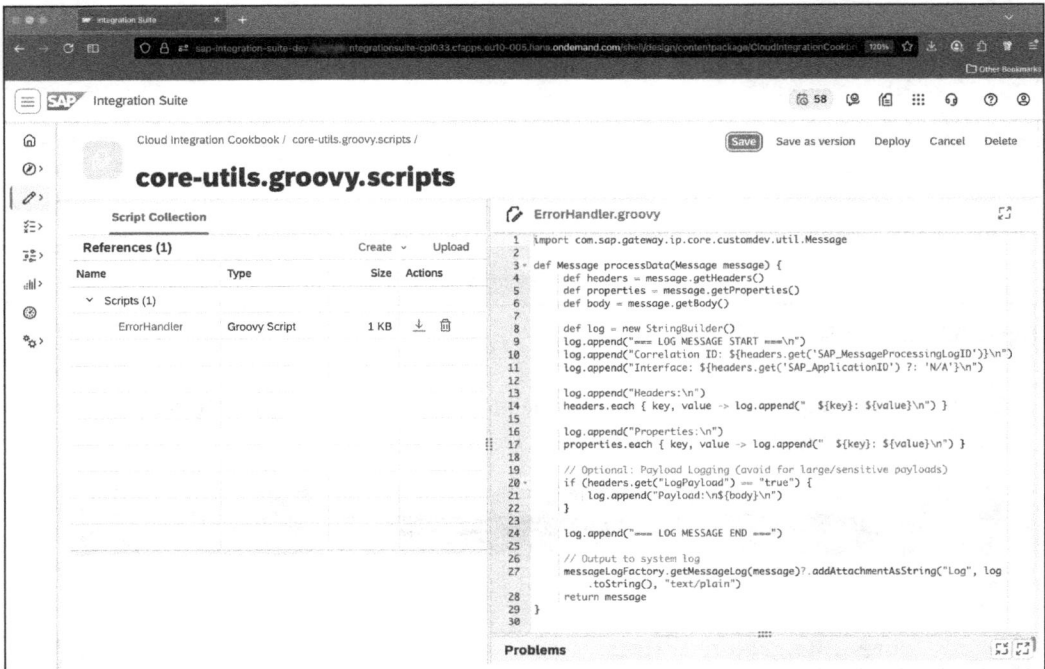

**Figure 6.6** Preview Created Script in Collection Overview

Each script file inside a script collection should contain the following:

- A descriptive filename (e.g., *ErrorHandler.groovy*)
- A Groovy class or method with clearly defined responsibilities
- Documentation in the form of Javadoc-style or inline comments
- No hard-coded values or context-specific logic

As the number of reusable scripts grows, managing collections becomes crucial. A well-structured script library might separate files into the following domains:

- *auth/TokenHelper.groovy*
- *logging/LogUtil.groovy*
- *date/DateFormatter.groovy*

You can organize collections functionally (by business domain) or technically (by operation type). Avoid overly large collections and modularize based on context and scope. Table 6.1 shows an example naming strategy.

| Purposes | Collection Names |
|---|---|
| Authentication | *auth-scripts* |
| Logging and audit | *core-logging-utils* |
| General tools | *integration-helpers* |

**Table 6.1** Example Naming Strategy

Finally, some benefits of central management are as follows:

- Consistent updates across flows
- Avoidance of logic duplication
- Easier onboarding and knowledge sharing
- Better test coverage and CI/CD integration

### 6.2.3   Referencing Scripts in iFlows

Once you deploy a script collection, its contents become available to all iFlows in the same tenant and package. You can insert these scripts into flows without copying any code. Here's how to reference a shared script:

1. Open your iFlow.
2. Insert a **Script** step in the appropriate section of the flow (e.g., between content modifier and message mapping).
3. Choose **External Script** as the source.
4. Select the deployed script collection and pick a script file.

The runtime behavior of a reused artifact or script collection is as follows:

- The integration runtime loads the script from the shared artifact.
- Any updates to the Script Collection affect all flows that reference it (after redeployment).
- The script can access headers, properties, and payloads via standard message and exchange APIs.

Some tips for parameterization are as follows:

- Pass values via message headers or exchange properties.
- Avoid hard-coded paths, credentials, and business logic.

Let's look at an example: a script called *LogMessage.groovy* is referenced in twelve different flows. When the logging format needs to change (e.g., it needs to include a correlation ID), you modify it in one place and all twelve flows benefit after redeployment.

The Groovy script in Listing 6.1 is designed for use within Cloud Integration to generate detailed runtime logs during message processing. It enhances visibility into the contents and context of a message as it passes through an iFlow, which is especially valuable for debugging and operational monitoring.

```groovy
import com.sap.gateway.ip.core.customdev.util.Message
import java.util.HashMap

def Message processData(Message message) {
    def headers = message.getHeaders()
    def properties = message.getProperties()
    def body = message.getBody()

    def log = new StringBuilder()
    log.append("=== LOG MESSAGE START ===\n")
    log.append("Correlation ID: ${headers.get('SAP_MessageProcessingLogID')}\n")
    log.append("Interface: ${headers.get('SAP_ApplicationID') ?: 'N/A'}\n")

    log.append("Headers:\n")
    headers.each { key, value -> log.append("  ${key}: ${value}\n") }

    log.append("Properties:\n")
    properties.each { key, value -> log.append("  ${key}: ${value}\n") }

    // Optional: Payload Logging (avoid for large/sensitive payloads)
    if (headers.get("LogPayload") == "true") {
        log.append("Payload:\n${body}\n")
    }

    log.append("=== LOG MESSAGE END ===")

    // Output to system log
    messageLogFactory.getMessageLog(message)?.addAttachmentAsString("Log",
log.toString(), "text/plain")
    return message
}
```

**Listing 6.1** Reusable Logging Script

It starts by importing necessary classes and defining the processData function, which takes a Message object as input. Within the function, it retrieves key components of the message—headers, properties, and the body—by using standard methods that are provided by the Cloud Integration runtime.

A `StringBuilder` object named `log` is then initialized to build a structured, human-readable log entry. The script begins the log with a clearly marked start section and follows it up with the correlation ID (`SAP_MessageProcessingLogID`) and the interface ID (`SAP_ApplicationID`). If the application ID isn't available, it defaults to `N/A`.

The script then iterates through all headers and properties, appending each key-value pair to the log. This gives developers or support engineers a full snapshot of the message's context at that point in time, including metadata such as authentication tokens, content types, and custom-defined parameters.

An optional block handles payload logging, which is conditionally executed based on the presence of a header named `LogPayload` with a value of `true`. This safeguard helps prevent unnecessary logging of large or sensitive message bodies, which can affect performance or expose confidential data.

Finally, the script concludes the log with a closing line and uses the `messageLogFactory` to attach the compiled log text to the message in the *message processing log*. This attachment can be viewed in the Cloud Integration monitoring tools, so it's easy for developers and operations teams to trace message flow and diagnose issues.

Overall, this script is a versatile and reusable logging utility that you can plug into various integration scenarios to improve observability, support troubleshooting, and enforce logging standards across your integration landscape.

### 6.2.4   Common Reusable Script Scenarios

Reusable scripts solve recurring challenges and abstract common integration patterns. This section discusses key categories of scripts that frequently deliver value when shared: logging utilities, error handling and wrapping, security token management, header normalization, and time and date formatting.

**Logging Utilities**

You can create scripts that write structured logs containing timestamps, payload IDs, and custom markers. You can use them to do the following:

- Debug message flow issues.
- Track business events.
- Integrate with external observability platforms.

A logging Groovy script in Cloud Integration plays a crucial role in enhancing observability and traceability during message processing. Such a script is typically inserted into iFlows to capture runtime information such as headers, properties, payloads, and metadata. The goal is to provide developers and support teams with a clear view of the data as it moves through each stage of the integration, without impacting the business logic.

Listing 6.2 begins by extracting key elements from the Message object, including headers, properties, and the message body. It then builds a structured log entry, often using a StringBuilder for efficiency. Typical log entries include details such as correlation IDs (e.g., SAP_MessageProcessingLogID), interface names, and timestamps. These elements help link logs to specific integration runs for easier troubleshooting.

```
import com.sap.gateway.ip.core.customdev.util.Message
import java.util.HashMap

def Message processData(Message message) {
    def headers = message.getHeaders()
    def properties = message.getProperties()
    def payload = message.getBody(String)

    def logBuffer = new StringBuilder()
    logBuffer.append("=== INBOUND MESSAGE LOG START ===\n")
    logBuffer.append("Timestamp: ${new Date()}\n")
    logBuffer.append("Correlation ID: ${headers.get('SAP_MessageProcessin-
gLogID')}\n")
    logBuffer.append("Sender: ${headers.get('SAP_Sender')} | Receiver: ${head-
ers.get('SAP_Receiver')}\n")

    // Log key headers
    logBuffer.append("\n-- Headers --\n")
    headers.each { key, value ->
        logBuffer.append("${key}: ${value}\n")
    }

    // Log key properties
    logBuffer.append("\n-- Properties --\n")
    properties.each { key, value ->
        logBuffer.append("${key}: ${value}\n")
    }

    // Conditional payload logging
    if (headers.get("LogPayload")?.equalsIgnoreCase("true")) {
        logBuffer.append("\n-- Payload --\n")
        logBuffer.append(payload)
    }

    logBuffer.append("\n=== INBOUND MESSAGE LOG END ===")

    // Add log as attachment (visible in MPL)
    def messageLog = messageLogFactory.getMessageLog(message)
```

```
    if (messageLog != null) {
        messageLog.addAttachmentAsString("InboundLog", logBuffer.toString(),
"text/plain")
    }

    return message
}
```

**Listing 6.2** Common Scripts: Custom Detailed Logging with Unified Formatting

One best practice is conditional payload logging, which is often controlled by a custom header like LogPayload. This approach avoids logging large or sensitive payloads unless it's explicitly required, so as to improve performance and protect sensitive data.

The script may also use the messageLogFactory to attach the log as a plain text file to the message log. This attachment can be viewed in the Cloud Integration monitoring tools, so it provides a nonintrusive yet comprehensive audit trail.

By centralizing and standardizing logging logic within a reusable script, teams can ensure consistent logging behavior across flows, simplify troubleshooting, and improve operational transparency—all without cluttering individual iFlows with repetitive logic.

**Error Handling and Wrapping**

A shared script can enrich exceptions with additional details, create consistent error messages, and forward fault information to monitoring systems.

In Cloud Integration, shared Groovy scripts are powerful tools for handling errors in a structured and consistent manner. Rather than allowing exceptions to propagate unhandled—or relying on inconsistent error formats across iFlows—a centralized error-handling script can enrich, standardize, and communicate fault information effectively. Listing 6.3 shows an example of custom enriched error handling.

```
import com.sap.gateway.ip.core.customdev.util.Message

def Message processData(Message message) {
    try {
        // Simulate downstream logic
        def payload = message.getBody(String)
        if (payload.contains("error")) {
            throw new Exception("Simulated processing error")
        }
        return message
    } catch (Exception e) {
        def errorDetails = [
            timestamp : new Date().toString(),
```

```
        errorCode : "INT-500",
        message   : e.getMessage(),
        traceId   : message.getHeader("SAP_MessageProcessingLogID")
    ]

    def errorJson = new groovy.json.JsonBuilder(errorDetails).toPrettyString()
    message.setBody(errorJson)
    message.setHeader("Content-Type", "application/json")
    message.setProperty("error_occurred", true)
    return message
  }
}
```

**Listing 6.3** Common Scripts: Centralized Error Format Creation

Such a script typically intercepts exceptions during message processing and augments them with contextual details such as correlation IDs, interface names, timestamps, and root cause summaries. This added context is invaluable for support teams who need to trace the origin of an error quickly and understand its business impact.

In addition to enrichment, these scripts can format error messages in a standardized structure. This helps maintain consistency across all interfaces and simplifies error parsing for downstream systems and monitoring tools. For example, the script might output a JSON error payload that contains a unique error code, a human-readable message, and relevant technical details—all in a format that's agreed upon by consuming applications.

Moreover, advanced error-handling scripts can forward fault details to external systems, such as SAP Alert Notification service, SAP Solution Manager, and third-party monitoring platforms. This enables proactive alerting and accelerates resolution times.

By encapsulating this logic in a shared, reusable script, integration architects ensure that all flows adhere to the same error-handling standards. This promotes better governance, improves observability, and significantly reduces time spent on diagnosing and resolving issues. Overall, centralized error handling transforms a reactive process into a proactive, structured part of integration design.

### Security Token Management

OAuth token acquisition and caching can be abstracted into a reusable script. In Cloud Integration, secure communication with external systems often requires the use of OAuth 2.0 for authentication. While the platform provides built-in support for OAuth through its HTTP and Open Connectors adapters, there are scenarios in which these mechanisms may fall short—particularly when it comes to token renewal and handling token expiration errors gracefully. In such cases, creating a custom reusable script for OAuth token acquisition and caching becomes a practical and robust alternative, as shown in Listing 6.4.

```groovy
import com.sap.gateway.ip.core.customdev.util.Message
import java.util.HashMap
import groovy.json.*

def Message processData(Message message) {
    def serviceURL = "https://example.identityprovider.com/oauth/token"
    def clientId = message.getProperty("client_id")
    def clientSecret = message.getProperty("client_secret")

    def connection = new URL(serviceURL).openConnection()
    connection.setRequestMethod("POST")
    connection.setRequestProperty("Content-Type", "application/x-www-form-urlencoded")
    connection.doOutput = true

    def payload = "grant_type=client_credentials&client_id=${URLEncoder.encode(clientId)}&client_secret=${URLEncoder.encode(clientSecret)}"
    connection.outputStream.write(payload.getBytes("UTF-8"))

    def response = connection.inputStream.text
    def json = new JsonSlurper().parseText(response)

    message.setProperty("access_token", json.access_token)
    return message
}
```

**Listing 6.4** Common Scripts: Access Token Management

This script-based approach allows developers to programmatically retrieve access tokens from an authorization server by using client credentials or other grant types. Once they acquire it, they can cache the token within an iFlow or in external storage (like a data store or tenant-level property) to avoid unnecessary requests for each call. By checking token validity and expiration proactively, the script can refresh it only when needed, thus reducing latency and improving overall efficiency.

One of the key reasons for implementing a custom OAuth script is to improve reliability. The native Cloud Integration OAuth adapter, while convenient, can sometimes fail to renew tokens correctly—especially under high load, unexpected responses, and timeout conditions. These failures can lead to **401 Unauthorized** errors and disrupt downstream processes. A custom script provides full control over the logic, error handling, and retry behavior, thus ensuring more predictable and maintainable integration behavior.

Additionally, encapsulating this logic in a shared script promotes reuse across multiple flows and services that connect to the same OAuth-protected endpoint. This not only standardizes authentication but also simplifies maintenance, improves traceability, and reduces duplication. As a result, organizations can ensure more secure, stable, and scalable integration with OAuth-based APIs.

### Header Normalization

This process ensures that all outbound calls include standardized headers (e.g., authorization, language, correlation ID). When building iFlows in Cloud Integration, you need to ensure that all outbound HTTP calls consistently include a standardized set of headers. These headers play a critical role in ensuring secure communication, maintaining language consistency, and enabling end-to-end traceability across distributed systems. This can be seen in Listing 6.5 in a small summarizing script to denylist classified headers and include others.

```
import com.sap.gateway.ip.core.customdev.util.Message
import java.util.HashMap

def Message processData(Message message) {
    def headers = message.getHeaders()
    def newHeaders = new HashMap()

    // Clean sensitive headers
    def blacklist = ["Authorization", "Cookie"]
    headers.each { key, value ->
        if (!blacklist.contains(key)) {
            newHeaders.put(key, value)
        }
    }

    // Set default headers if not present
    newHeaders.putIfAbsent("Accept-Language", "en-US")
    newHeaders.put("X-Correlation-ID", headers.get("SAP_MessageProcessin-
gLogID") ?: UUID.randomUUID().toString())

    newHeaders.each { key, value -> message.setHeader(key, value) }

    return message
}
```

**Listing 6.5** Common Scripts: Header Normalization

One of the most important headers is the Authorization header, which typically carries access tokens such as OAuth 2.0 bearer tokens. Including this header correctly ensures that target systems can authenticate the request and grant appropriate access to protected resources.

Another recommended header is the `Accept-Language` header or a custom `Language` header, which communicates the preferred language for responses. This is particularly useful in global scenarios where localized content or error messages are returned based on language preferences.

Equally important is the inclusion of a *correlation ID*—a unique identifier that's used to trace a request across multiple systems and integration layers. This header helps operational and support teams diagnose issues by linking logs, traces, and monitoring events to a specific transaction or message instance.

Standardizing these headers across all outbound flows improves governance, simplifies troubleshooting, and supports consistent behavior across integrations. Using a reusable script or policy to enforce this standard can ensure reliability and reduce the risk of missing critical headers.

### Time and Date Formatting

Time and date formatting is frequently used for legacy system integration or localizations. In many integration landscapes, especially those involving legacy systems or localized business processes, certain logic patterns are used repeatedly. These may include date and number formatting, character encoding conversions, and transformations that are tailored to regional standards and older protocols. Rather than embedding such logic directly into each flow, you should encapsulate it in shared Groovy scripts as shown in Listing 6.6.

```
import com.sap.gateway.ip.core.customdev.util.Message
import java.text.SimpleDateFormat
import java.util.TimeZone

def Message processData(Message message) {
    def inputFormat = new SimpleDateFormat("yyyy-MM-dd'T'HH:mm:ss'Z'")
    inputFormat.setTimeZone(TimeZone.getTimeZone("UTC"))

    def outputFormat = new SimpleDateFormat("dd.MM.yyyy HH:mm:ss")
    outputFormat.setTimeZone(TimeZone.getTimeZone("Europe/Berlin"))

    def inputDate = inputFormat.parse(message.getBody(String))
    def formattedDate = outputFormat.format(inputDate)

    message.setBody(formattedDate)
    return message
}
```

**Listing 6.6** Common Scripts: Date and Time Formatting

These reusable scripts act as adapters or converters, and they bridge the gap between modern cloud-native integrations and legacy- or region-specific system requirements. For instance, a script might format timestamps into a legacy ERP-compatible format, sanitize special characters, or apply localized tax or currency rules.

By centralizing this logic, developers can ensure consistency across flows, reduce redundancy, and simplify maintenance. They only need to make any update or fix in one place to instantly benefit all dependent integrations. This approach enhances flexibility, supports compliance with local standards, and extends the reach of modern integration platforms to older or specialized systems.

### 6.2.5   Best Practices for Script Maintenance

Reusable scripts are vital assets in any integration landscape. They promote consistency, reduce duplication, and improve maintainability across multiple iFlows. However, the value of shared scripts quickly diminishes when maintenance practices are neglected. Without clear governance, scripts can become outdated and undocumented or even introduce security risks. The following best practices help ensure that script collections remain robust, secure, and efficient throughout their lifecycle:

- **Versioning**
  Every script collection should follow a consistent and meaningful versioning scheme, and semantic versioning (MAJOR.MINOR.PATCH) is highly recommended. For example, you can increment the MAJOR version when breaking changes are introduced, MINOR for backward-compatible improvements, and PATCH for bug fixes. Each script file should include a changelog comment block at the top that outlines recent modifications, authorship, and rationale. This documentation ensures traceability and helps avoid regressions during updates. Developers should tag releases with explicit version identifiers (e.g., v1.2.0) so that other teams can reference stable versions confidently. As already mentioned, different types of changes do affect the artifacts at different levels of severity. These can be categorized in different groups as listed in Table 6.2.

| Types of Change | Descriptions | Affected Segments |
|---|---|---|
| Breaking changes | These are changes that are incompatible with previous versions (e.g., removing functions or altering parameters). | **MAJOR** (e.g., 2.0.0 → 3.0.0) |
| New backward-compatible feature | It adds new functionality without impacting existing users (e.g., a new utility function). | **MINOR** (e.g., 2.1.0 → 2.2.0) |

**Table 6.2** Version Numbering: Types of Changes

| Types of Change | Descriptions | Affected Segments |
|---|---|---|
| Bug fix or patch | It fixes a defect or issue without changing existing functionality or interfaces. | **PATCH** (e.g., 2.1.1 → 2.1.2) |
| Refactoring with no external impact | These are internal code changes that don't affect the script's public interface. | **PATCH** (e.g., 2.1.3 → 2.1.4) |
| Documentation-only updates | These are updates to comments, changelogs, or READMEs with no code changes. | **PATCH** (e.g., 2.1.2 → 2.1.3) |
| Deprecated feature notice (without removal) | It marks a feature as deprecated but still functional. | **MINOR** (e.g., 2.2.0 → 2.3.0) |
| Performance optimizations (non-breaking) | These are enhancements that improve speed or resource usage without interface changes. | **MINOR** or **PATCH** (context-dependent) |
| Complete rewrite with the same functionality | This is a code overhaul that maintains the same inputs and outputs. | **MINOR** (if interface unchanged) or **MAJOR** (if changed) |

**Table 6.2**  Version Numbering: Types of Changes (Cont.)

- **Testing**

  Testing is crucial to script reliability, particularly when scripts are shared across multiple flows. For every reusable script, you should create dedicated test iFlows that simulate realistic scenarios and edge cases. These test flows should validate expected input/output behavior and capture how the script handles invalid or unexpected data. Using tools like Postman or built-in SAP simulation environments can help automate and repeat tests during deployment cycles, and consistent testing reduces the risk of production incidents and builds trust in the shared code base.

- **Documentation**

  Well-documented scripts are easier to use, modify, and troubleshoot, so you should include inline comments throughout your code to explain logic, especially for complex sections. At the top of each script file, use a Javadoc-style comment block to describe its purpose, input parameters, dependencies, and return values. Additionally, maintain a central *README.md* or equivalent documentation file within the script collection directory. This file should outline the overall purpose of the collection, usage guidelines, supported parameters, and known limitations. Clear documentation empowers other developers to confidently reuse scripts without constant support.

- **Change control**

  Effective change control processes help manage the risk associated with modifying shared scripts, and you should use version control systems like Git and SAP Cloud Transport Management to track changes, compare versions, and roll back when necessary. You should define a structured approval workflow that includes peer reviews and testing before a script is promoted to production, and when breaking changes are introduced, you should notify all flow owners who rely on the script so they can validate compatibility. Proactive communication and structured change management prevent downstream disruptions.

- **Security**

  You must build security into every stage of script development. Never hardcode credentials or tokens in your scripts. Instead, use secure parameterization techniques such as environment variables or secure parameter storage. You should always sanitize and escape inputs to scripts to prevent injection attacks and malformed data errors. Avoid logging sensitive data and be cautious about exposing implementation details in error messages. These practices will protect your integration environment from vulnerabilities and unauthorized access.

- **Collaboration**

  Reusable scripts are often developed and consumed by multiple teams, so to manage collaboration effectively, you should assign clear ownership for each script collection. Owners are responsible for quality, documentation, and approval of changes, so you should encourage peer reviews to help owners catch issues early and share knowledge. It's also beneficial to include usage examples within the script comments or documentation, to make it easier for new users to understand and adopt the script. Collaboration fosters a shared sense of responsibility and encourages continuous improvement.

- **Governance**

  To ensure scripts meet the required standards, you should establish a governance checklist that you can used during reviews or deployments to verify script readiness. There are several key questions to include on the checklist: Is there a version tag and changelog? Which flows use this script? Is the script properly documented, with input parameters clearly described? Is a test flow available and up-to-date? By systematically validating these criteria, teams can maintain high script quality and prevent issues from slipping into production.

## 6.3  Reusing Message Mappings

It's impossible to overstate the challenge of transforming data across heterogeneous systems in modern integration landscapes. Businesses today operate in complex digital ecosystems where multiple applications, platforms, and services exchange data continuously. These systems often use different structures, naming conventions, and even semantic meanings for similar data elements. For example, while one application may represent a customer address as a single concatenated string, another system may store it in separate fields for street, city, and postal code. This lack of uniformity creates the need for robust transformation mechanisms that can reliably convert data from one format to another without introducing errors or inconsistencies.

SAP Integration Suite provides several options you can use to achieve these transformations, with message mapping emerging as the most widely adopted technique due to its intuitive design and powerful features. At its core, *message mapping* allows developers to define relationships between source and target data elements by using a graphical interface, which eliminates the need for manual coding in many scenarios. However, while creating a new mapping for each iFlow might seem like the fastest way forward, this approach quickly becomes unsustainable as the number of flows grows. Duplication of mapping logic across multiple flows introduces maintenance overhead, increases the likelihood of inconsistencies, and complicates governance. Therefore, the concept of reusability becomes essential.

*Reusable message mappings* allow multiple iFlows to share a single transformation logic. This principle significantly reduces redundancy, accelerates development, and improves the overall reliability of integrations. By managing mappings as shared artifacts rather than isolated components, organizations can establish a scalable integration architecture that's aligned with modern best practices. In this section, we'll examine the key aspects of reusing message mappings in SAP Integration Suite. This includes understanding the available mapping types, creating and structuring reusable mappings, leveraging value mappings for discrete value conversion, implementing user-defined functions for advanced scenarios, and applying governance principles to maintain control throughout the lifecycle.

### 6.3.1 Understanding Mapping Types in SAP Integration Suite

Before we explore the concept of reuse, you need to understand the mapping options that are available in SAP Integration Suite. The platform supports several mapping techniques, each of which is designed for different scenarios. The most common type is message mapping, which provides a graphical editor that allows users to map fields between source and target structures. This approach is particularly suitable for XML-to-XML transformations and is widely adopted due to its simplicity and effectiveness. Developers can use drag-and-drop functionality to establish relationships between elements and apply built-in functions for complex logic, such as conditional mappings and mathematical calculations.

Another mapping type that's available in the platform is *XSLT mapping*, which relies on XSL transformations for data conversion. XSLT is a declarative language for transforming XML documents, and it offers powerful capabilities for advanced scenarios. However, it requires specialized knowledge of XML, XPath, and XSLT syntax, so it's less accessible to business-oriented integration developers. XSLT mappings are best suited for use cases where precise control over transformation logic is required or when organizations already have a strong investment in XSLT-based tooling.

Finally, *operation mapping* is a concept that was inherited from SAP Process Integration (SAP PI), and it combines multiple mapping steps into a single operation. While it's supported in Cloud Integration, it's primarily relevant for organizations that are migrating from legacy SAP PI/SAP PO environments, and isn't commonly used in greenfield projects.

Among these options, message mapping stands out as the preferred choice for most SAP Integration Suite projects. It combines the ease of use of a visual interface with the ability to implement advanced logic through functions and user-defined code. Unlike XSLT, which demands technical depth, message mapping lowers the bar for entry and accelerates development cycles without sacrificing flexibility. For this reason, the remainder of this section will focus on strategies and practices for making message mapping reusable, efficient, and easy to maintain.

### 6.3.2   Creating and Structuring Message Mappings for Reuse

Designing reusable mappings begins with thoughtful planning and structuring. A mapping artifact isn't just a technical component; it's an asset that will be referenced by multiple iFlows over time. Therefore, it must be easy to identify, understand, and maintain. The first step in designing reusable mappings is to follow a clear naming convention that reflects both the source and target message formats, as well as the functional domain. For instance, you should name a mapping that's designed to transform a sales order from an XML structure into an SAP IDoc something like *MM_SalesOrder_XML_to_IDoc*, rather than using a generic name like *Mapping1* or *OrderMapping*.

When creating the mapping, you should start by importing the appropriate schemas for the source and target messages. You can provide these as XSD files, WSDL definitions, or EDMX models. Once you load the schemas, the graphical editor will allow users to link fields by simply dragging and dropping from the source tree to the target tree. For complex mappings, you can apply standard functions such as string concatenation, substring extraction, and conditional logic. When the built-in functions are insufficient, users can create user-defined functions by using Java syntax, which is a capability we'll explore later in this chapter.

Documentation is another critical aspect of creating reusable mappings. Each mapping artifact should include a description of its purpose, the systems it connects, and any special transformation rules that apply to it. Inline comments within the mapping logic can further enhance clarity, especially for functions and conditional expressions. Well-documented mappings not only flatten the learning curve for new developers but also make audits and governance easier to manage.

### 6.3.3   Cross-Flow Reuse of Message Mappings

Once you've created and deployed a mapping, multiple iFlows can reference it within the same tenant. This is where the concept of reuse truly delivers value. Instead of duplicating logic across flows, you can configure their mapping steps to point to an existing mapping artifact. This ensures that all flows that apply similar transformations remain consistent over time.

The process of referencing a reusable mapping is straightforward. In the iFlow designer, you insert a mapping step at the appropriate stage of the flow and select the option to use an externalized mapping. The system then provides a list of available mappings, from which you can choose the artifact you want. Once linked, the mapping operates as if it were defined within the flow itself, but without duplicating any configuration.

The benefits of this approach are significant. First, it ensures consistency across all iFlows using the same transformation logic. Second, it improves maintenance efficiency, as updates to the shared mapping propagate to all dependent flows automatically, provided they're redeployed. Finally, it enhances governance because fewer artifacts means fewer opportunities for errors and discrepancies.

However, this convenience introduces a new responsibility: impact management. When a shared mapping is modified, every flow that references it could potentially be affected. Therefore, organizations must implement rigorous change control procedures, including dependency analysis and regression testing, before deploying updates to production. Maintaining a list of consumer flows for each shared mapping is a recommended best practice you should follow to facilitate this process.

### 6.3.4   Leveraging Value Mappings for Discrete Conversions

Structural transformations often need to be complemented by value-level conversions. For instance, while mapping an order message from an e-commerce platform to SAP, developers might need to convert country codes from two-letter ISO format to three-letter format or map currency codes to numeric values. Hardcoding these conversions inside message mappings can quickly become unmanageable and introduces a high risk of inconsistencies.

SAP Integration Suite addresses this challenge through *value mappings*, which act as centralized lookup tables for key-value pairs. A value mapping artifact can contain multiple groups, each of which represents a specific context such as a sender-receiver pair. This allows the same mapping table to serve different scenarios while maintaining clarity and governance.

Consider the example of country codes. A value mapping could define that for the ECOM → SAP context, the US value would map to USA, DE would map to DEU, and so on. Similarly, a currency mapping could translate USD to 840 and EUR to 978. By referencing these mappings within message mappings by using the ValueMapping() function, developers achieve a clean separation of structural and value transformations. This not only simplifies the mapping logic but also makes updates easier, as changes to value mappings don't require redeploying the message mapping itself.

Value mappings also enhance governance by providing a single point of truth for commonly used conversions. They support versioning, can be transported across environments, and can be managed by business users without technical intervention. That makes them powerful tools for agile integration.

### 6.3.5   User-Defined Functions for Advanced Logic

While standard mapping functions cover most scenarios, complex transformations sometimes require custom logic. *User-defined functions* (UDFs) enable developers to implement such logic directly within the mapping artifact by using Java syntax. UDFs are ideal for tasks like advanced date formatting, string manipulation, and implementing business rules that can't be expressed through graphical functions.

For example, consider a scenario where a date in the source message must be reformatted from yyyy-MM-dd to dd.MM.yyyy. You can do this by using the UDF in Listing 6.7.

```
public String formatDate(String input) {
  try {
    java.text.SimpleDateFormat sourceFmt = new java.text.SimpleDateFormat("yyyy-
MM-dd");
    java.text.SimpleDateFormat targetFmt = new java.text.SimpleDateFor-
mat("dd.MM.yyyy");
    return targetFmt.format(sourceFmt.parse(input));
  } catch (Exception e) {
    return "";
  }
}
```

**Listing 6.7** User-Defined Mapping Function to Format Date

When using UDFs, you should adhere to best practices such as keeping functions small and focused, handling exceptions gracefully, and documenting their purpose and parameters. Reusable UDF libraries can further enhance consistency by providing a common set of functions for multiple mappings, such as the one provided in Listing 6.7.

### 6.3.6   Performance Optimization and Error Handling

As mappings become more complex, performance considerations come into play. Large payloads, deep hierarchical structures, and multiple nested loops can impact processing time. To optimize performance, you should minimize unnecessary loops, preprocess data where possible, and offload heavy logic to Groovy scripts or external services when appropriate. You should also test with realistic payload sizes to avoid surprises in production.

Error handling is another critical aspect of robust mapping design. Mappings should validate mandatory fields, apply default values for optional fields, and log errors in a structured manner. In scenarios where data integrity is critical, mappings can implement controlled error propagation and thus allow iFlows to handle exceptions gracefully, rather than failing abruptly.

**6**

### 6.3.7   Governance, Lifecycle Management, and Continuous Integration/ Continuous Delivery

Reusable artifacts introduce governance challenges that organizations must address proactively. Version control is essential to prevent accidental overwrites and ensure traceability. We recommend that you use semantic versioning to increment major, minor, and patch numbers based on the nature of changes. Impact analysis tools and dependency documentation help identify which flows consume a given mapping and thus help you make informed decisions during updates.

Integration with CI/CD pipelines further strengthens governance. Automated testing with mock payloads can validate mapping logic before deployment, thus reducing the risk of regressions. You can also use transport management service or CTS+ to move mappings across environments in a controlled manner and thus ensure alignment with organizational change management processes.

---

**Two Practical Scenarios That Illustrate Reuse**

To help you understand the concepts we've discussed, consider two practical examples. In the first scenario, a company integrates multiple e-commerce platforms with SAP S/4HANA, and rather than creating separate mappings for each platform, it develops a single reusable mapping for order-to-IDoc transformation. By combining this with value mappings for currency and region codes, the company achieves faster rollout and consistent logic across all channels.

In the second scenario, a global enterprise must harmonize tax codes across dozens of jurisdictions. A centralized value mapping, combined with UDFs for region-specific rules, ensures compliance while reducing maintenance complexity. Updates to tax logic can be deployed once and instantly applied to all dependent flows, which saves significant time and effort.

---

## 6.4   Modularization with ProcessDirect

In modern integration landscapes, modular design has evolved from a best practice into a necessity. As enterprises scale their digital ecosystems, iFlows grow in number and complexity, and they often incorporate multiple systems, services, and business processes. Building these flows as monolithic entities leads to several challenges: maintainability becomes cumbersome, reusability is minimal, and adapting to changes requires significant effort.

The key to overcoming these issues lies in *modularization*—the art of decomposing large integration logic into smaller, reusable components. In SAP Integration Suite, the ProcessDirect adapter plays a pivotal role in enabling this modular approach. ProcessDirect provides a mechanism for direct, high-performance communication between iFlows within the same tenant, without exposing the communication channel to external systems. Unlike HTTP or SOAP-based inter-flow calls, which introduce network overhead and require authentication handling, ProcessDirect offers an internal, lightweight, and secure pathway that's optimized for intra-tenant communication. This capability allows developers to structure integration logic into reusable building blocks—subflows that can be invoked by multiple parent flows for common tasks such as logging, validation, enrichment, and error handling.

To fully leverage the benefits of modularization with ProcessDirect, you need to understand not only what it is but also how to apply it effectively in real-world integration scenarios. Section 6.4.1 will therefore provide a closer look at the fundamental concept of ProcessDirect and its role in promoting modular design within SAP Integration Suite. Section 6.4.2 will examine how to implement ProcessDirect in practice, including typical usage patterns and scenarios where it adds the most value. Finally, Section 6.4.3 will discuss both recommended practices and common pitfalls to avoid when designing modular iFlows. By covering these aspects, this chapter aims to provide a comprehensive overview of how ProcessDirect can transform integration projects from rigid, monolithic designs into flexible, reusable, and maintainable architectures that align with the needs of scalable enterprise landscapes.

### 6.4.1   Modularization and ProcessDirect

The concept of modularization is deeply rooted in software engineering principles such as separation of concerns, single responsibility, and DRY. Applying these principles in integration design means that each flow or subflow should handle a specific concern, rather than combining unrelated logic in a single artifact. For instance, a monolithic iFlow that processes an order, validates input, enriches data, performs error handling, and logs execution details isn't only complex to maintain but also difficult to scale. Also, if a change in logging policy is required, developers must modify every iFlow where logging is implemented. This approach introduces duplication, increases the risk of inconsistencies, and slows down innovation.

Modular flows address these limitations by centralizing common functionality into dedicated components that can be reused across multiple integration scenarios. A logging subflow, for example, can be designed once and invoked by any iFlow that requires audit tracking. Similarly, a validation subflow can enforce consistent data quality rules across processes. ProcessDirect makes this possible by enabling synchronous and asynchronous invocation of such reusable components within the same tenant, with minimal latency and without external dependencies.

*ProcessDirect* is an internal communication mechanism that's provided by SAP Integration Suite to link multiple iFlows securely and efficiently. It acts as a virtual connector that binds a sender channel in one flow to a receiver channel in another, using an endpoint URI as the addressing mechanism. Unlike HTTP and SOAP, where network calls traverse the internet and incur authentication overhead, ProcessDirect operates entirely within the tenant. This design ensures lower latency, reduced resource consumption, and simplified security management.

A ProcessDirect endpoint is identified by a unique address, which is typically structured as `/ProcessDirect/<endpointName>`. An iFlow that wants to expose functionality for reuse includes a ProcessDirect sender adapter that's configured with this endpoint. Any other flow in the tenant can invoke this subflow by adding a ProcessDirect receiver adapter and specifying the same endpoint uniform resource identifier (URI). At runtime, the message is routed internally to the target flow and executed, and the response (if any) is returned to the caller.

ProcessDirect supports both synchronous and asynchronous communication patterns to give developers flexibility in designing modular interactions. For example, synchronous calls are ideal for validation and enrichment tasks where the parent flow depends on the subflow's output. Asynchronous calls, on the other hand, are better suited for scenarios like logging and monitoring, where the subflow operates independently and the caller doesn't need to wait for completion.

One of the core principles behind modularization is loose coupling, which is the process of minimizing dependencies among components. In the context of integration, loose coupling ensures that changes in one flow have minimal or no impact on other flows that consume its functionality. ProcessDirect supports this principle by allowing subflows to expose well-defined interfaces—so they essentially act as internal APIs within the integration platform. Each subflow is responsible for a single concern, such as performing a security check, retrieving an authentication token, or formatting a message for downstream systems.

When designing modular flows, developers should define clear input and output contracts for each subflow. These contracts specify which headers, properties, and payload structures the subflow expects and what it returns to the caller. Adhering to these contracts prevents unintended side effects and makes the architecture easier to maintain.

### 6.4.2   Implementing ProcessDirect in iFlows

Implementing ProcessDirect involves configuring sender and receiver channels in the relevant flows. The flow that provides a reusable service—which we refer to as the *provider flow*—contains a ProcessDirect sender adapter that's configured with a unique endpoint address. The flow that consumes this service—which is called the *caller flow*—uses a ProcessDirect receiver adapter that points to the same address.

For example, imagine creating a subflow that validates incoming order messages. This flow would start with a ProcessDirect sender channel that's configured as follows:

```
Endpoint Address: /validate/order
Quality of Service: Exactly Once (EO)
Exchange Pattern: Synchronous
```

The main iFlow that processes orders would include a receiver channel with these settings:

```
Address: /validate/order
Adapter Type: ProcessDirect
Exchange Pattern: Synchronous
```

When the parent flow reaches the ProcessDirect receiver step, it sends the message to the validation subflow, waits for the response, and then continues its processing logic based on the validation result.

### 6.4.3   Using ProcessDirect

One of the advantages of ProcessDirect is its ability to propagate message context between flows. This includes headers, properties, and payload. Developers can leverage this feature to pass dynamic values—such as correlation IDs, authentication tokens, and error codes—between parent and child flows. For example, a caller flow might set a property named `transactionId` before invoking the ProcessDirect subflow, and the subflow can read this property, include it in its logs, and return a status message to the caller. This mechanism allows for rich interactions between flows without hardcoding values, so it enhances flexibility and maintainability.

**Synchronous Versus Asynchronous Communication**

Choosing between synchronous and asynchronous invocation is a key design decision. In synchronous mode, the caller flow waits for the subflow to complete before proceeding. This approach is suitable for tasks that influence subsequent processing steps, such as validation, data enrichment, and token retrieval. However, synchronous calls can introduce latency if the subflow performs resource-intensive operations.

Asynchronous mode, by contrast, decouples the caller from the subflow's execution. The message is handed off to the subflow, and the caller continues immediately, without waiting for a response. This pattern is ideal for logging, auditing, and notification scenarios where the outcome of the subflow doesn't affect the caller's logic.

Developers should also consider error-handling implications. In synchronous mode, errors in the subflow can propagate back to the caller, thus allowing immediate handling. On the other hand, in asynchronous mode, error handling must be implemented within the subflow itself or through monitoring tools.

The versatility of ProcessDirect makes it suitable for a wide range of scenarios. Common use cases include centralized logging, data validation, enrichment, security token management, and error handling. For example, a dedicated logging subflow can collect metadata, headers, and payload snippets for audit purposes. Instead of implementing logging logic in every iFlow, developers simply invoke the logging subflow via ProcessDirect. Similarly, developers can centralize authentication flows that retrieve and refresh tokens to reduce redundancy and ensure compliance with security policies.

Error handling in modularized architectures requires special attention. When a ProcessDirect subflow fails, the way the error propagates depends on the exchange pattern. In synchronous mode, the error can be returned to the caller, which then decides how to proceed. This allows for centralized error management logic in the parent flow. In asynchronous mode, the subflow must handle errors independently, by either retrying or by sending alerts to monitoring systems.

SAP Integration Suite provides robust monitoring tools to track ProcessDirect interactions. Developers can use the message processing logs to trace the entire call chain and thus leverage correlation IDs to link related transactions across flows. This capability simplifies troubleshooting and audit activities, especially in complex architectures involving multiple subflows.

Additionally, while ProcessDirect eliminates the overhead of external calls, performance optimization remains critical in high-volume environments. Each ProcessDirect invocation incurs context-switching costs, so modularization should strike a balance between granularity and efficiency. Breaking flows into too many small subflows can lead to excessive orchestration overhead, and conversely, overly large flows undermine the benefits of modularization. A pragmatic approach is to modularize by business function—such as validation, logging, or enrichment—rather than splitting at an excessively fine-grained level.

To further optimize performance, you should do the following:

- Use asynchronous mode for noncritical subflows.
- Minimize payload size when passing data between flows.
- Avoid deep nesting of ProcessDirect calls.

You also need to testing with realistic load scenarios to identify bottlenecks and validate performance assumptions before going live.

Modularization also introduces governance challenges that you must address systematically. Each ProcessDirect endpoint should have a clear naming convention and documentation describing its purpose, input/output contracts, and dependencies. You must also implement version control to prevent disruptions when interfaces change. We recommended that you include version identifiers in endpoint names, such as /validate/order/v2, while maintaining backward compatibility for legacy consumers.

Security is inherently simpler with ProcessDirect than with external protocols, as communication remains within the tenant. However, you should still validate inputs and enforce authorization checks where necessary to prevent unintended misuse of shared subflows.

Finally, organizations should integrate reusable subflows into their organization's CI/CD pipeline to ensure consistency and quality across environments. Automated testing can validate ProcessDirect interactions, while transport mechanisms such as transport management services and CTS+ provide controlled deployment across development, test, and production landscapes. Implementing these practices reduces the risk of runtime failures and aligns modularization with enterprise governance frameworks.

---

**Practical Scenarios**

Consider a company that operates multiple iFlows for order processing, invoicing, and shipment tracking. Each flow must log critical events for compliance and monitoring, and rather than duplicating logging logic, the organization creates a reusable logging subflow that's exposed via ProcessDirect at /log/event. Every main flow invokes this endpoint asynchronously and passes relevant metadata in headers and properties. The logging subflow writes these details to a centralized monitoring system to ensure consistency and reduce maintenance effort. When the logging format changes, developers update the subflow once and all flows benefit immediately.

In another scenario, several iFlows interact with APIs that require nonstandardized OAuth tokens. Instead of embedding token management logic in every flow, the organization creates a dedicated authentication subflow that handles token retrieval, caching, and refresh operations. Main flows invoke the subflow synchronously via ProcessDirect to obtain a valid token before making API calls. This design centralizes security logic, simplifies compliance audits, and reduces the likelihood of errors due to inconsistent token handling.

---

The benefits of modularization extend beyond technical efficiency. By reducing duplication and simplifying maintenance, organizations can lower total cost of ownership, accelerate project timelines, and improve agility in responding to changing business requirements. Modular architectures also facilitate team collaboration by enabling parallel development of independent components.

Looking ahead, modularization through ProcessDirect aligns with the trend toward event-driven architectures and API-based ecosystems. As SAP Integration Suite evolves, ProcessDirect will likely integrate more closely with event meshes and API management capabilities to enable even greater flexibility and scalability.

## 6.5 Summary

The journey through artifact reusability in SAP Integration Suite reveals a fundamental truth about modern integration: efficiency, scalability, and maintainability are no longer optional—they're essential. In increasingly interconnected business landscapes, iFlows must evolve beyond isolated, monolithic constructs toward modular, reusable components that support rapid change and long-term sustainability.

This chapter explored the guiding principles, techniques, and best practices for implementing reusability, focusing on three key pillars: script collections, message mappings, and modular integration using ProcessDirect. Across all these areas, certain themes consistently emerge. Governance is paramount, and reuse without governance risks introducing fragility instead of resilience. Therefore, practices such as version control, dependency tracking, and automated regression testing must be integral to any reuse strategy. Similarly, developers must not overlook performance optimization. Modularization and reusability, while beneficial, introduce orchestration layers that require careful tuning to maintain throughput in high-volume environments. Finally, the integration of reusable artifacts into CI/CD pipelines ensures that these components can evolve safely and predictably across the software delivery lifecycle.

Looking ahead, the principles we discussed in this chapter align closely with broader industry trends. As enterprises transition to API-first and event-driven architectures, the demand for modular, reusable integration components will only increase. Technologies such as event meshes, serverless functions, and low-code automation will further expand the landscape of possibilities, but the underlying philosophy remains the same: build once, use everywhere, govern consistently.

In conclusion, artifact reusability isn't merely a technical convenience; it's a strategic enabler of digital agility. By mastering the techniques outlined in this chapter—leveraging script collections, reusing message mappings, and implementing modular flows with ProcessDirect—organizations can achieve a level of scalability, maintainability, and operational excellence that positions them for success in an ever-changing digital economy.

## Summary

# Chapter 7
# Data Management and Security

*This chapter is dedicated to the central aspects of data security and data management in Cloud Integration. It shows how data is protected during transmission and storage, how identities are verified, and how technical user access is controlled. The focus isn't only on theoretical principles but also on concrete implementation options within the cloud environment.*

Section 7.1 explains how various authentication methods—such as basic authentication, OAuth 2.0, and client certificates—are used in practice. Section 7.2 covers the role the secure store plays as a protected storage location for sensitive information. Finally, Section 7.3 and Section 7.4 demonstrate how you can use data variables, number ranges, and persistent data storage with the data store to realize stable and traceable iFlows.

Security and data management are not secondary issues—they form the backbone of a reliable and auditable integration architecture. This chapter provides you with the knowledge you need to make your interfaces not only functional but also robust, secure, and compliant with regulations.

## 7.1 Authentication and Authorization

The terms *authentication* and *authorization* are at the heart of every modern IT architecture, especially in the area of system integration. In the context of Cloud Integration, they're of central importance for secure, controlled, and rule-based access to interfaces, services, and data sources—whether by users, applications, or other systems.

Especially with cloud-based integration platforms that exchange data across system, company, and even country boundaries, the clean separation and implementation of authentication and authorization is essential. Without clearly defined access concepts and identity mechanisms, integrations not only are prone to errors but also pose a significant security risk.

In the following sections, you'll learn how the various authentication methods—such as basic authentication, OAuth 2.0, and client certificates—are used in practice. But first, we'll provide you with an introduction to authentication and authorizations.

### 7.1.1   What Is Authentication and Authorization?

The first step in accessing a resource—such as an API, an application, or a database—is to identify the requesting subject. *Authentication* answers the question, "Who is requesting access?" In Cloud Integration, the requesting subject can be a user (a human user) or a system (a technical client). Authentication mechanisms are used to uniquely identify these entities, for example, via a user name and password, tokens, or certificates. The platform offers a wide range of authentication options that you can flexibly combined and tailor to your company's security strategy. It makes a distinction between the following:

- **User-controlled authentication**
  For example, users can employ this when logging in to a web interface.

- **Systemic authentication**
  For example, this can be for machine communication between two endpoints.

Cloud Integration provides different mechanisms for both scenarios, from simple basic authentication and token-based procedures (such as OAuth 2.0) to certificate-based authentication.

While authentication determines who performs an action, *authorization* defines what this person or system is allowed to do. Within Cloud Integration, this is done using a fine-grained role concept that's based on the mechanisms of SAP BTP.

You'll usually assign authorizations via role collections, which in turn bundle groups of business roles. By assigning a user to such a collection, you give them access to exactly those functions that are necessary for them to do their tasks—no more and no less. This role-based access control is essential for several reasons:

- It ensures that only authorized persons can read, change, and delete data.
- It supports the separation of responsibilities (e.g. developers, administrators, operations).
- It makes it possible to log security-relevant actions in a traceable manner.
- It forms the basis for audits, certifications, and regulatory requirements.

In practice, this means that even if a user has been correctly authenticated, they're not automatically allowed to access every API or function. Only with a valid role assignment does the user receive the necessary authorization—for example, to model integration flows (iFlows), access the secure store, or operate monitoring functions.

In Cloud Integration, authentication and authorization are closely linked. Every access—whether by a developer in the user interface, an external service via HTTP, or a business application via an API—is first authenticated, and then, the authorization is checked. We'll illustrate the interaction with the following example:

1. An external service wants to call an HTTP endpoint in Cloud Integration.
2. The service authenticates itself by using a previously issued OAuth access token.

3. When the request arrives, Cloud Integration checks the token and validates the signature.

4. At the same time, it checks whether the token contains the necessary scopes to call up the requested resource.

5. If both are given, the request is accepted and processed; otherwise, it's rejected.

This pattern also applies in a similar way to user interactions, for example when a user logs in via Identity Authentication or accesses the web interface. Here, too, the user is first uniquely identified and then checked to see whether the assigned roles allow the desired access.

Identity management is an elementary component of a secure authentication and authorization strategy. In SAP BTP, this is supplemented by the Identity Authentication service and the Identity Provisioning service.

*Identity Authentication* as a central identity provider that can bring together various authentication sources, including SAP's own users, external directories such as Microsoft Entra ID, and even SAML identities. Identity Authentication is linked to the appropriate Cloud Integration subaccount via trust configurations. This allows central login pages to be implemented and single sign-on functionalities to be activated.

*Identity Provisioning*, in turn, makes it possible to automatically transfer users, groups, and role assignments from external sources to SAP BTP. This reduces manual effort, prevents errors, and ensures governance-compliant user structures.

Cloud Integration supports a variety of common authentication methods, depending on the use case and security requirements:

- **Basic authentication**
  This is the use of a user name and password, which is suitable for simple scenarios.

- **OAuth 2.0**
  This is token-based authentication, which is preferred for APIs and cloud scenarios.

- **Client certificates**
  This is certificate-based authentication, which is ideal for highly secure system communications.

- **SAML 2.0**
  This is assertion-based authentication, especially in conjunction with corporate directories.

- **JSON Web Token (JWTs)**
  These are alternative token mechanics that support certain scenarios in API management.

It's crucial to use these methods in the right context. For example, you should never secure a publicly accessible API endpoint with basic authentication; instead, you should use a token-based method such as OAuth. Conversely, basic authentication in combination with TLS can be completely sufficient for simple system communication on the intranet.

On the other hand, authorization within Cloud Integration is based on the SAP BTP role model. You can assign authorization at the following three levels:

- **Individual roles (business roles)**
  For example, you can enter "PI_Integration_Developer" as a role name and assign authorization to it.

- **Role collections**
  You can combine several roles into one functional unit

- **Users and user groups**
  You can do this manually or automatically via Identity Authentication/Identity Provisioning.

This structure allows granular control of access. For example, a developer can have access to the modeling of iFlows but no access to the administration of security materials. An administrator, on the other hand, can have full access to subaccount settings but no access to development functions. In addition, role assignments can also be limited in time, activated depending on context (e.g. via a policy engine), or controlled via group memberships. This creates a highly flexible, auditable authorization model.

In many industries—especially in finance, industry, and the public sector—access to systems and data is subject to strict regulatory requirements. Cloud Integration supports these requirements with the following functions:

- Audit logs and access logs
- Logging of failed authentications
- Centralized role definition and version control
- Support for identity federation and single sign-on
- Integration with external security information and event management (SIEM) systems

This enables companies to document and analyze every access transparently and, if necessary, process it in an audit-proof manner. At the same time, it enables a rapid response to security-relevant incidents—for example, by blocking compromised users or revoking tokens.

As powerful as the mechanisms for authentication and authorization are, you shouldn't underestimate their complexity and the related challenges. In practice, the following challenges often arise:

- Lack of consistency between subaccounts (e.g., different role assignments)
- Hard-coded access data in iFlows or scripts
- Insufficient separation between productive and test environments

- Missing or outdated certificates in the trust store
- Insufficiently documented authorization processes

The following best practices have proven effective in mitigating these risks:

- Centralization of user administration via Identity Authentication and Identity Provisioning
- Use of secure store and security materials, instead of plain text data
- Regular role reviews and recertification processes
- Use of policy templates for new subaccounts
- Documentation and automation of trust and role configuration

With these measures, you can set up a consistent, secure, and maintainable authentication and authorization concept that will not only meet today's requirements but also withstand future scaling and security requirements.

### 7.1.2 Basic Authentication

*Basic authentication* (also known as HTTP Basic Auth) is one of the simplest and most widely used methods of authentication in the context of HTTP-based communication protocols. It's a direct method of securing access to web resources by transmitting the user name and password in the header of each request.

With basic authentication, the user name and password are combined in the username:password format, encoded in Base64, and then transmitted in the HTTP header of the request. The client sends this header to the server with every request:

```
Authorization: Basic dXNlcjpwYXNzd29yZA==
```

The character string after Basic is the Base64-encoded form of the access data. It's also important to emphasize that Base64 isn't encryption—the data is merely encoded, not protected. The server, such as an iFlow endpoint of Cloud Integration, checks the access data and then grants or denies access.

In Cloud Integration, you can use basic authentication for both inbound and outbound connections, as follows:

- **Inbound**
  External systems or applications call an HTTP endpoint of an iFlow and transmit the user name and password in the header. SAP Integration Suite validates these via the user that's stored in the subaccount.
- **Outbound**
  Cloud Integration calls up a REST service from a third-party provider, for example.

The required access data is configured and used via a destination or as security material (a user name and password) in the secure store.

For secure use of basic authentication, we recommend that you implement central administration of access data in the secure store. The user data is stored as a security artifact and only referenced in the iFlow. This prevents direct access to sensitive data within the iFlow source code. A typical application scenario is, for example, the use of the HTTP receiver adapter in an iFlow, where the authentication procedure is set to **Basic** and reference is made to a configured credentials object.

The advantages of basic authentication are as follows:

- **Simplicity**
  It can be implemented quickly and requires no additional authentication services.
- **Compatibility**
  Almost every system and every programming language supports this procedure.
- **Fast testing**
  It's ideal for development or test environments, where fast connections are required.

However, despite its widespread use, basic authentication is now considered technically outdated and security-critical, especially in productive cloud environments, for the following reasons:

- **No encryption**
  Without HTTPS, credentials can be transmitted in plain text—and that's a significant security risk.
- **Static login data**
  There are no mechanisms for token lifetime, automatic blocking, or expiration.
- **Lack of delegation**
  In contrast to modern methods such as OAuth, Basic Auth doesn't allow any separation between the user and the application context.

For these reasons, you should only use basic authentication in secure, controlled networks (e.g. VPN, intranet) or temporary tests. In productive scenarios, especially in the public cloud environment, we strongly recommend that you use token-based procedures such as OAuth 2.0 and client certificates.

Now, let's look at a practical example of basic authentication. In this example, you'll use an OData V2 adapter, but basic authentication works in the same way with any other adapter. To start, create a new OData V2 connection by using a **Request Reply** step. Then, connect it to a receiver, select the **OData** adapter, and select the **V2** version (see Figure 7.1)

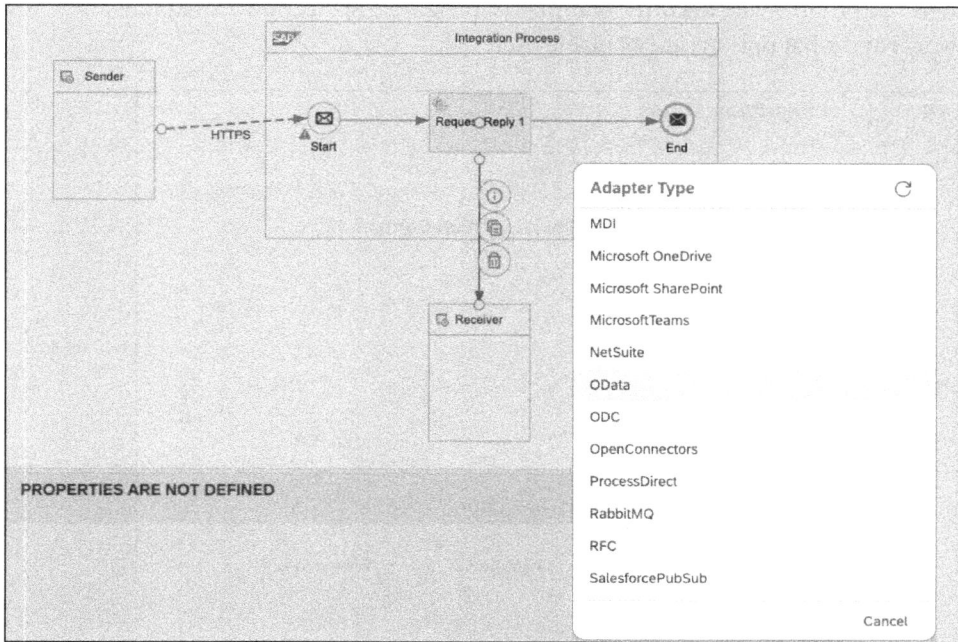

**Figure 7.1**  Creating Connection for Basic Auth

Then, switch to the **Connection** tab and select **Basic** for **Authentication** (see Figure 7.2).

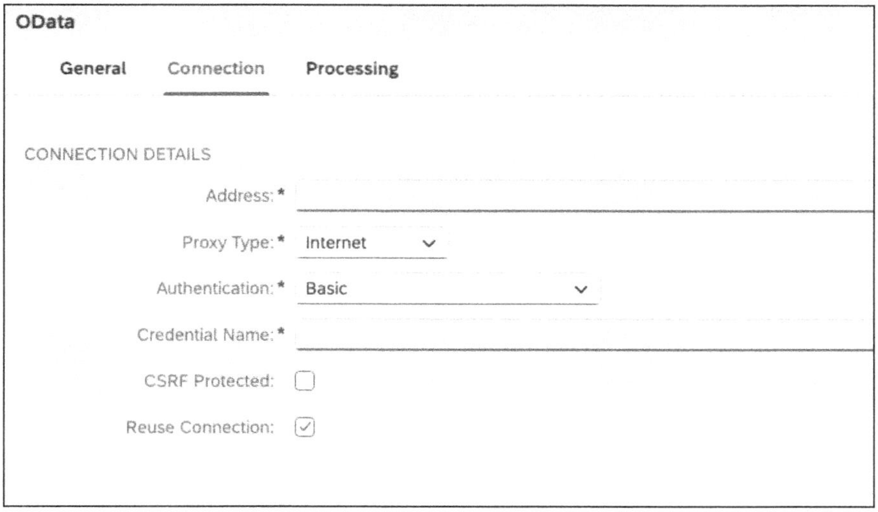

**Figure 7.2**  Selecting Basic as Authentication

Then, you need to create a **Credential Name.** To do this, switching to the security material in Cloud Integration, navigate on the left-hand side to **Monitor · Integration and**

**APIs,** and then select **Security Material** (see Figure 7.3). Section 7.2 explains how this works and what options are available there.

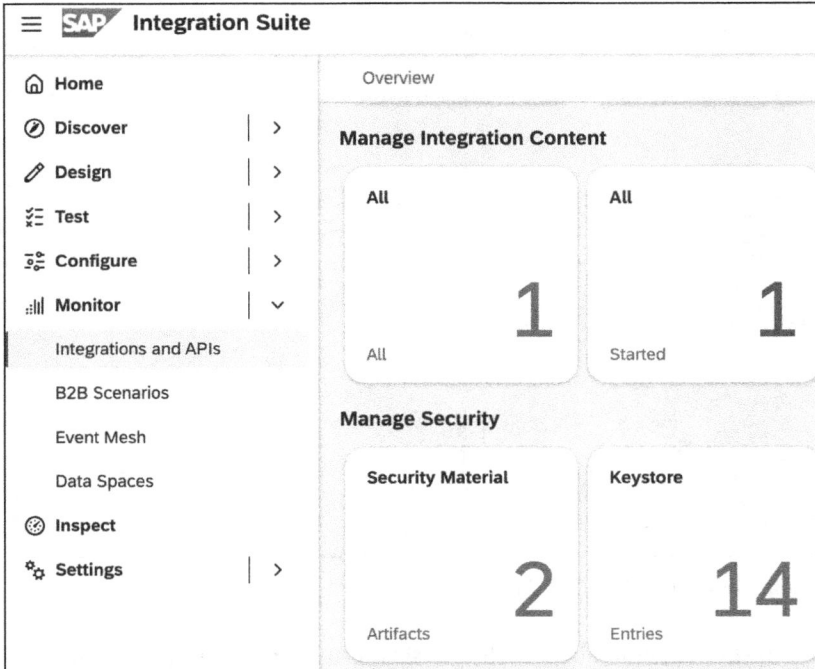

**Figure 7.3**  Switching to Security Material

From the **Security Material** menu, select **Create · User Credentials** at the upper right, as shown in Figure 7.4.

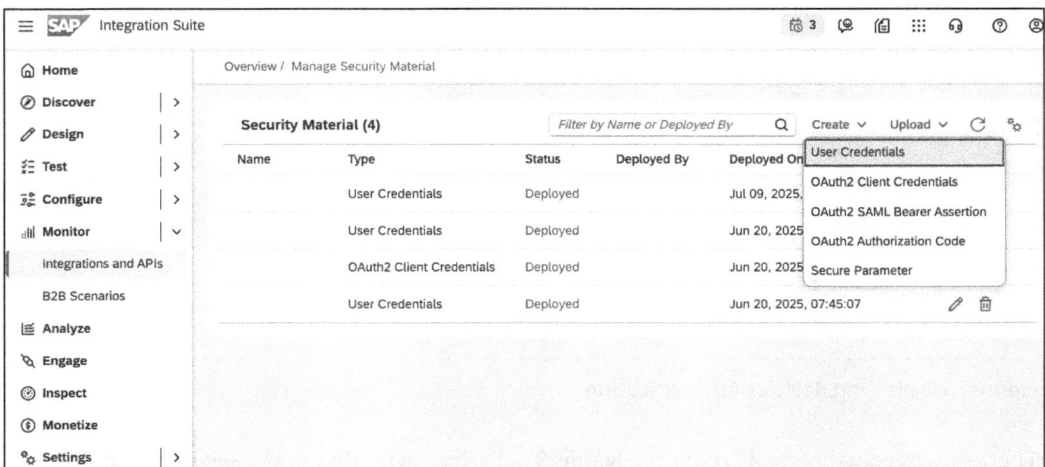

**Figure 7.4**  Selecting Create User Credentials

Then, you assign a **Name**, enter the **Username** and **Password**, and click **Deploy** (see Figure 7.5). Cloud Integration will save these credentials, and in your iFlows, you'll only use the **Name** of this credential. If someone subsequently tries to find out the user's password in Cloud Integration, they'll be out of luck because the password will no longer be visible but Cloud Integration will have saved it.

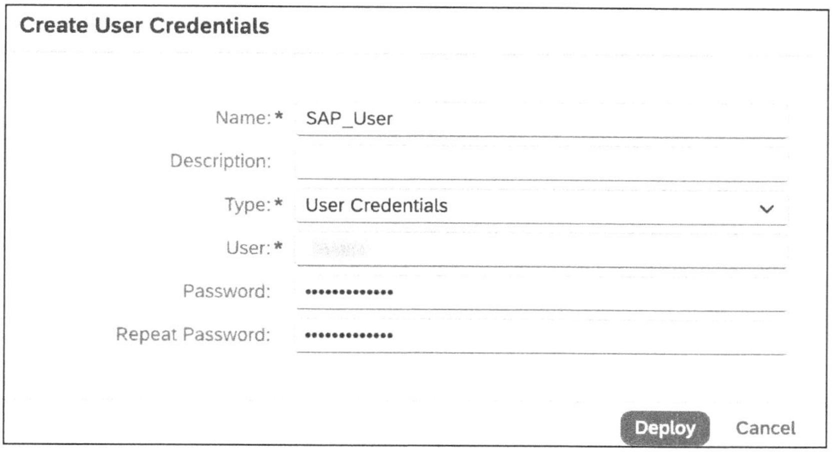

**Create User Credentials**

| | |
|---|---|
| Name: * | SAP_User |
| Description: | |
| Type: * | User Credentials ⌄ |
| User: * | |
| Password: | •••••••••••• |
| Repeat Password: | •••••••••••• |

[ Deploy ]  Cancel

**Figure 7.5** Creating User Credentials

Then, switch back to your iFlow, enter the name of the user credential that you've just created in the **Credential Name** field (see Figure 7.6), and click on **Deploy**. With that, you'll have successfully configured basic authentication.

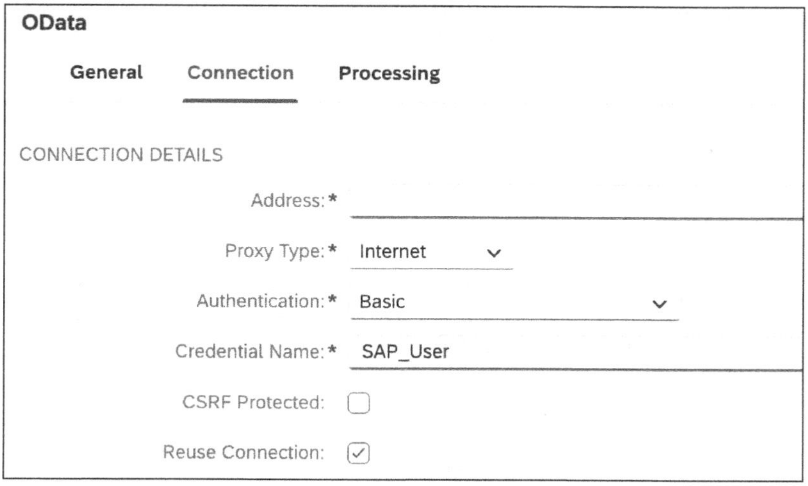

**OData**

**General**     **Connection**     **Processing**

CONNECTION DETAILS

| | |
|---|---|
| Address: * | |
| Proxy Type: * | Internet ⌄ |
| Authentication: * | Basic ⌄ |
| Credential Name: * | SAP_User |
| CSRF Protected: | ☐ |
| Reuse Connection: | ☑ |

**Figure 7.6** Inserting User Credential Name

### 7.1.3   OAuth

*OAuth 2.0* is a commonly used protocol for delegated authentication and authorization, and it was developed specifically for secure access to web resources. It has established itself as the standard for accessing APIs, not only on the public internet but also in enterprise applications and integration platforms such as Cloud Integration.

In contrast to basic authentication, where login data is transmitted directly, OAuth 2.0 works with time-limited access tokens that are issued by the authorization server. This process significantly increases security and at the same time allows fine-grained control of authorizations.

The central concept of OAuth 2.0 is that an application called the *client* accesses a resource on behalf of a user or system without the client having to know the user's password. Instead, access takes place via an access token that is provided by the authorization server (which is also the identity provider).

This token contains defined rights called *scopes*, it's only valid for a certain period of time, and it can be revoked at any time. OAuth is therefore not only an authentication procedure but also a secure authorization mechanism.

OAuth 2.0 is based on a clear role model with the following four actors:

- **Resource owner**
  This is the person or entity (e.g., a user, an organization) that has access to a protected resource.

- **Client**
  This is the application (e.g., an iFlow in the SAP Integration Suite) that wants to access the resource.

- **Authorization server**
  This is the entity (e.g., Identity Authentication, Microsoft Entra ID, third-party identity provider) that verifies the identity of the resource owner and issues access tokens.

- **Resource server**
  This is the server (e.g. an API or a system that provides data) on which the protected resource is located.

The process of OAuth 2.0–supported authentication can vary, depending on the flow. The flow that's most commonly used in integration scenarios is the *client credentials flow*, which is specifically designed for machine-to-machine communication. The typical process is as follows:

1. The client authenticates itself to the authorization server with a client ID and a secret.

2. The authorization server issues an access token if the check is successful.

3. The client sends this token to the resource server with every request (e.g., as an HTTP header):

   ```
   Authorization: Bearer eyJhbGciOi...).
   ```

4. The resource server checks the validity and scope of the token and, if it's valid, grants access to the requested resource.

The access token is valid for a limited time (often, 3,600 seconds) and can be renewed after expiration (ideally, automatically) via a refresh token or a new client login.

Within SAP Integration Suite, OAuth 2.0 is supported across various components. Some common use cases include the following:

- **HTTP receiver adapter**
  It's used when calling external REST or OData services that require OAuth-based authentication.

- **API Management**
  It's used to secure and control access to published APIs.

- **Open Connectors**
  It's used for secure access to third-party services that require authorization via OAuth.

Configuration takes place via security materials or BTP destinations in which the client ID, the client secret, and the token endpoint URL of the authorization server are stored.

For example, if an iFlow wants to call a third-party service via HTTP that uses OAuth, **OAuth2 Client Credentials** is selected as the authentication method in the HTTP adapter. The configured security material is then referenced, and SAP Integration Suite automatically retrieves an access token in the background and adds it to the request.

Some advantages of OAuth 2.0 are as follows:

- **Security**
  There are no static passwords in data traffic, just short-lived tokens.

- **Scalability**
  Decoupling of user identity and system access enables flexible authorization scenarios.

- **Delegation**
  Access can be granted on behalf of third parties, with a definable scope of rights.

- **Maintainability**
  Tokens can be centrally managed, revoked, and renewed.

A decisive advantage of OAuth 2.0 over basic auth is that access rights can be assigned depending on the context. For example, a token can only be valid for a specific API, at a specific time, and with precisely defined actions (e.g., read only, no writing).

However, despite all its advantages, OAuth 2.0 also brings the following challenges:

- The setup requires more configuration effort than simple user name/password combinations.

- You must thoroughly document the management of tokens, secrets, and authorizations.
- To integrate it with external authorization servers (e.g. Microsoft Entra ID), you need to have an understanding of the protocols used.

We therefore recommend that you do the following to implement productive use of OAuth 2.0:

- Use trusted authorization servers such as Identity Authentication or Microsoft Entra ID.
- Define scopes sensibly and restrictively.
- Manage tokens via central mechanisms (e.g., SAP Keystore).
- Don't make token runtimes too long (e.g., make them a maximum of 1–2 hours)

Next, here's an example of how to implement OAuth in a practical and accessible way. In it, you use an OData V2 adapter, but the same principles apply when you're using OAuth with other adapters. Start by creating a new HTTPS connection through a request-reply step that you connect to a receiver. From there, select the HTTP adapter as illustrated in Figure 7.7.

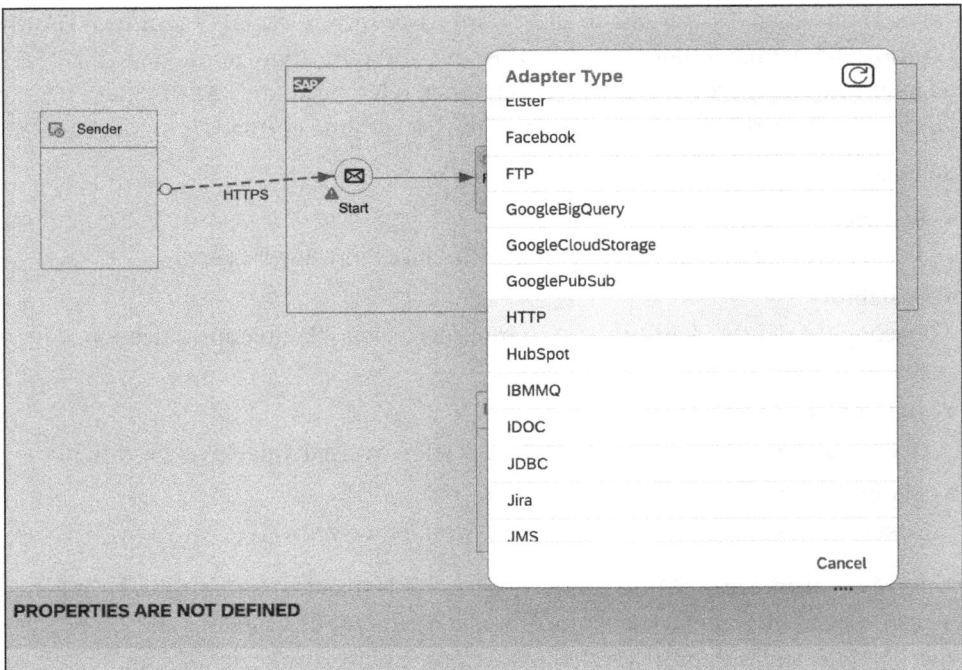

**Figure 7.7**  Creating OAuth Connection

Next, switch to the **Connection** tab and select **OAuth2 Client Credentials** for **Authentication** (see Figure 7.8).

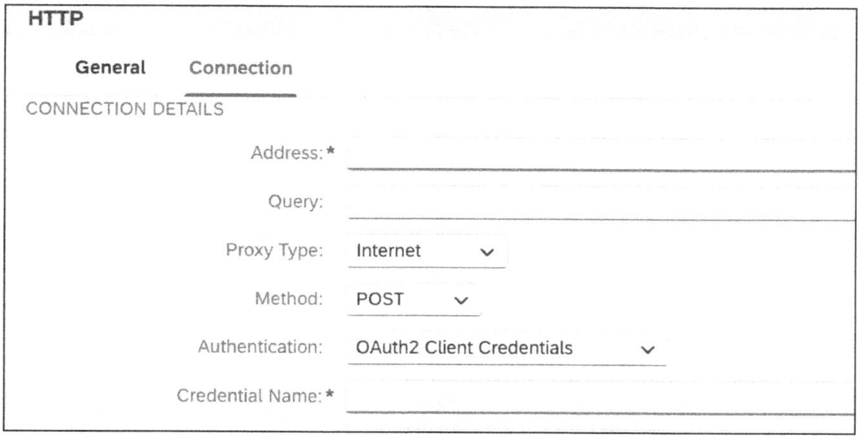

**HTTP**

**General**   **Connection**

CONNECTION DETAILS

Address: *

Query:

Proxy Type:   Internet ⌄

Method:   POST ⌄

Authentication:   OAuth2 Client Credentials ⌄

Credential Name: *

**Figure 7.8** Selecting OAuth as Authentication

Then, you'll be prompted to enter a **Credential Name**. To create it, switch to the **Security Material** section within Cloud Integration and follow the same procedure we described in Section 7.1.2. When you're in the **Security Material** menu, click **Create** at the upper right and select **OAuth2 Client Credentials** from the dropdown list (see Figure 7.9).

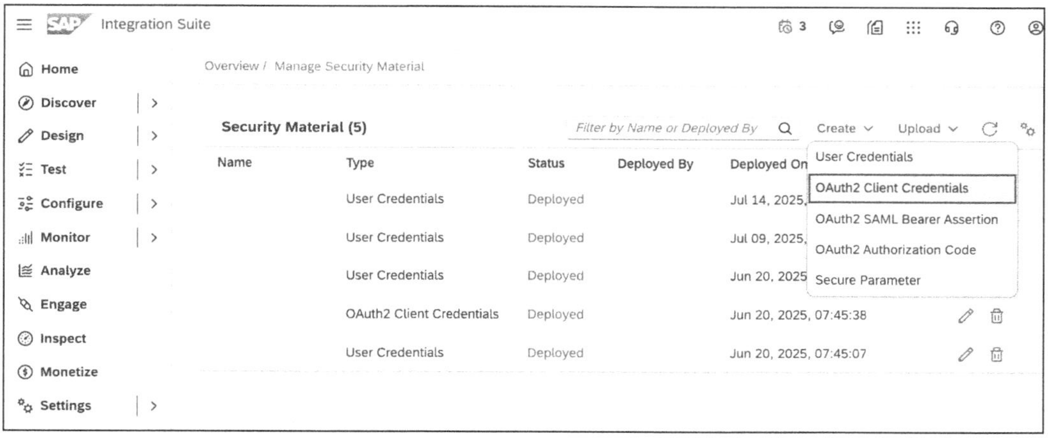

**Figure 7.9** Creating OAuth Credentials

A new window will open where you can enter the **Token Service URL** (which usually ends with "/oauth/token"), along with the **Client ID** and **Client Secret** (see Figure 7.10). You can also use **Client Authentication** field to choose whether to include these credentials in the header of the request or the body of the request, specifically in relation to the request that's sent to the token service endpoint. Furthermore, you can use the **Content Type** field to determine whether the request should be sent in JSON format or as x-www-form-urlencoded.

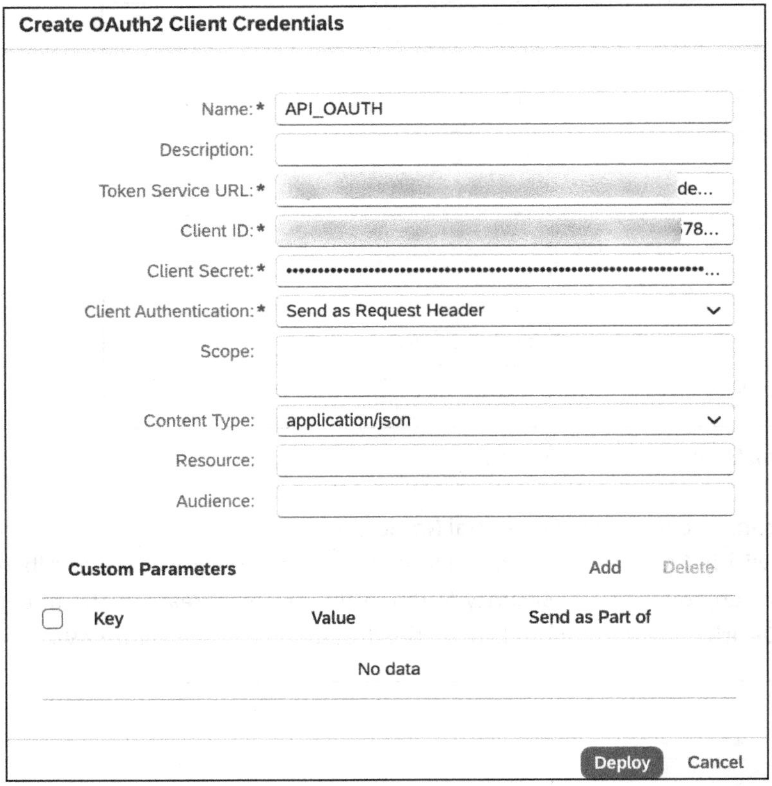

**Figure 7.10** Creating OAuth Credentials

Next, switch back to your iFlow and enter the name of the user credential that you've just created (as already described in Section 7.1.2). With that, you'll have successfully configured OAuth. Cloud Integration handles the token request and token renewal.

### 7.1.4   Client Certificate

*Client certificate authentication* (which is also known as X.509 client certificate authentication) is a highly secure authentication method that's based on digital certificates. In contrast to password-based and token-based methods, it uses an asymmetric cryptosystem in which a technical system authenticates itself to a server with a client certificate.

This method is most often used in integration scenarios with high security requirements, such as communication with on-premise systems, finance, and officially regulated infrastructures. It's frequently used in Cloud Integration to secure outgoing HTTPS connections to external systems or to securely authenticate incoming connections from trusted partners.

Client certificate authentication is based on the Transport Layer Security (TLS) protocol, which enables *mutual authentication* when a connection is established. This is also known as mutual TLS (mTLS). The following happens in client certificate authentication:

1. The client initiates an HTTPS connection with the target system (the resource server).
2. The server provides its server certificate to identify itself to the client.
3. The server requests a client certificate.
4. The client sends its own certificate to the server.
5. The server checks whether all of the following is true:
   - The certificate is valid.
   - It has been signed by a trustworthy certification authority (CA).
   - It hasn't expired or been revoked.
6. If the check is successful, the connection is established and the client is considered authenticated.

The actual exchange of data therefore only begins after the identity of the client has been cryptographically verified.

In Cloud Integration, certificate authentication can be used in two directions:

- **Outbound**
  An iFlow establishes a connection to an external HTTPS service, and a client certificate stored in the keystore is sent along.

- **Inbound**
  An external system calls an endpoint in the SAP Integration Suite. The HTTPS adapter can be configured so that only clients with a valid certificate are accepted.

Cloud Integration uses the central keystore of the appropriate subaccount in SAP BTP to handle certificates. You can import, manage, and reference certificates there.

Client certificates offer a number of significant advantages compared to other authentication methods:

- They provide very high security because no passwords are transmitted. Instead, authentication is based on cryptographic keys.

- They're not reusable because each certificate is uniquely assigned to a system and can't simply be copied or intercepted.

- They can be checked offline. The validity of a certificate can be checked locally without the involvement of an external identity provider.

- They are compliance-capable, which is important because many regulatory requirements demand certificate-based authentication mechanisms.

However, despite these advantages, the use of certificates also entails the following administrative and technical challenges:

- **Administrative burdens**
  Certificates need to be created, distributed, stored, and regularly renewed.

- **Trust chains**
  For a certificate to be recognized as valid, the entire certificate chain (the root CA and the intermediate CAs if necessary) must be stored in the system.

- **Dependence on infrastructure**
  Many companies have to operate their own CA or public key infrastructure (PKI) system or integrate certified third-party providers.

However, Cloud Integration provides suitable interfaces for these tasks, both for administration in the keystore and for dynamic use in iFlow. A typical example is an iFlow that transfers data to a third-party system, such as an API from a business partner. The HTTP receiver adapter is configured so that it sends a specific client certificate when the connection is established, and this certificate is stored in the keystore of the subaccount and provided with an alias. The target server verifies the certificate and only accepts requests that are made with a valid, known certificate. On the other side, the communication partner must store the Cloud Integration certificate in its own trust store, either as a single certificate or as part of a signed chain.

We recommend that you do the following to promote the secure and efficient use of client certificates in Cloud Integration:

- Give certificates a limited term (e.g., 1 year) and renew them before expiration.
- Check the keystore regularly for expired or revoked certificates.
- Only store sensitive keys in encrypted form and never export them in plain text.
- Subject access to the keystore and the management of certificates to clear roles and responsibilities.
- Plan and automate certificate renewal and rollout processes at an early stage, if possible.

You'll need to perform a few configurations to make Cloud Integration accept a client certificate sent by the sender system. First, you'll need to upload the client certificate of the sender system to SAP BTP within an instance of the SAP Process Integration runtime, using the integration-flow plan. Next, you'll need to add the certificate to the keystore of Cloud Integration. Finally, you'll need to reference the certificate within the iFlow itself.

We'll start by showing you how to configure a client certificate in SAP BTP. (If you already have one, you can use it and jump ahead to the screen shown in Figure 7.13.) Navigate to the subaccount where Cloud Integration is located, click on **Service Marketplace** in the left-hand menu, search for **SAP Process Integration Runtime** (as illustrated in Figure 7.11), and click the **Create** button in the upper right-hand corner.

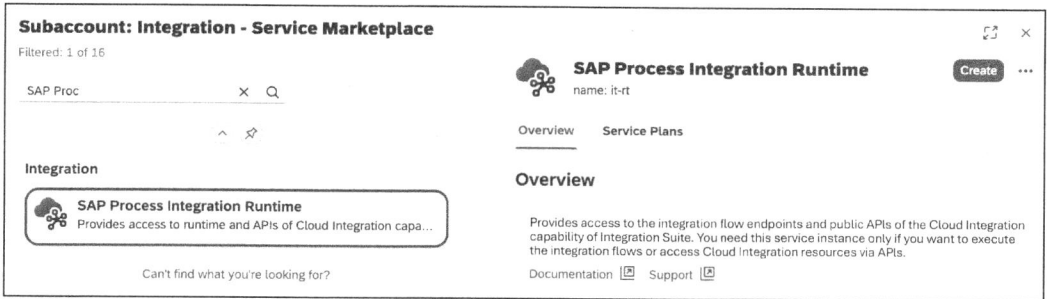

**Figure 7.11** Searching for SAP Process Integration Runtime

That will open a new window where you can configure the instance settings, as shown in Figure 7.12. Select **SAP Process Integration Runtime** as the **Service** and choose **integration-flow** as the **Plan**. Set the **Runtime Environment** to **Cloud Foundry**, select dev as the **Space**, and assign an **Instance Name** that's unique in that space—"pir-clientcert" is the one you should enter for this example. Then, click **Create** in the bottom right-hand corner.

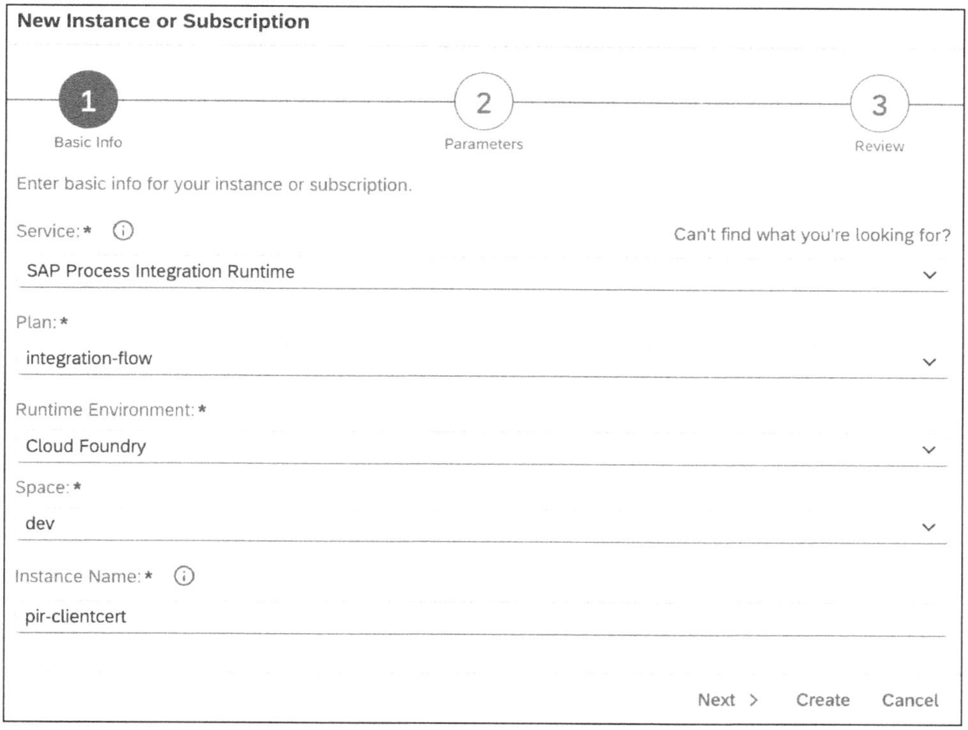

**Figure 7.12** Filled-Out New SAP Process Integration Runtime Screen

Next, navigate to the SAP Process Integration runtime instance that you just created. As shown in Figure 7.13, go to the **Service Keys** section and click **Create** to generate a new service key for this instance.

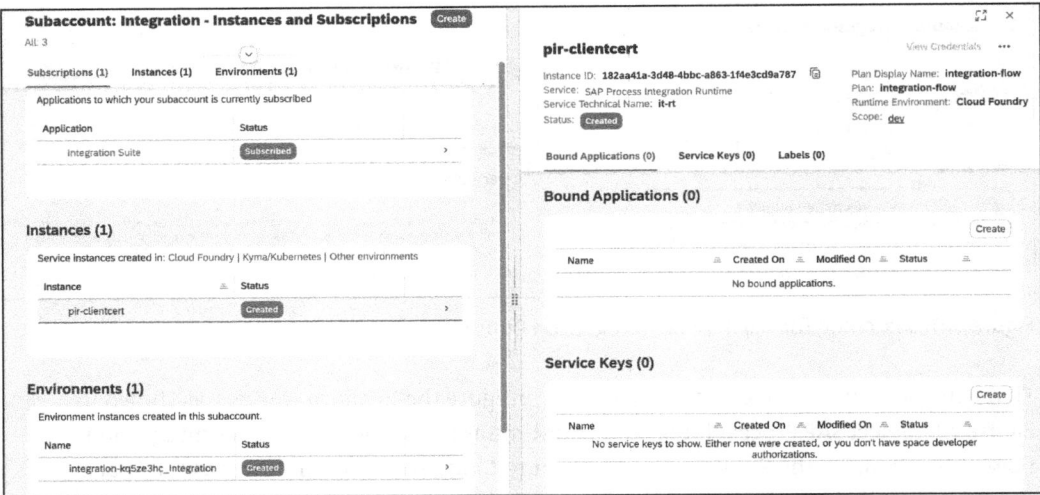

**Figure 7.13**  Switching to SAP Process Integration Runtime Instance

In the **New Service Key** dialog that's shown in Figure 7.14, define the settings for your new key. Enter a unique **Service Key Name** (e.g., "PIR_S4_KEY") and set the **Key Type** to **External Certificate**. Paste the contents of the client certificate into the **External Certificate** field, ensure that the **Pin Certificate** box is checked, and (optionally) adjust the **Validity** in days (e.g., 365) and the **Key Size** if you need to. Once you've configured everything, click **Create** to generate the service key.

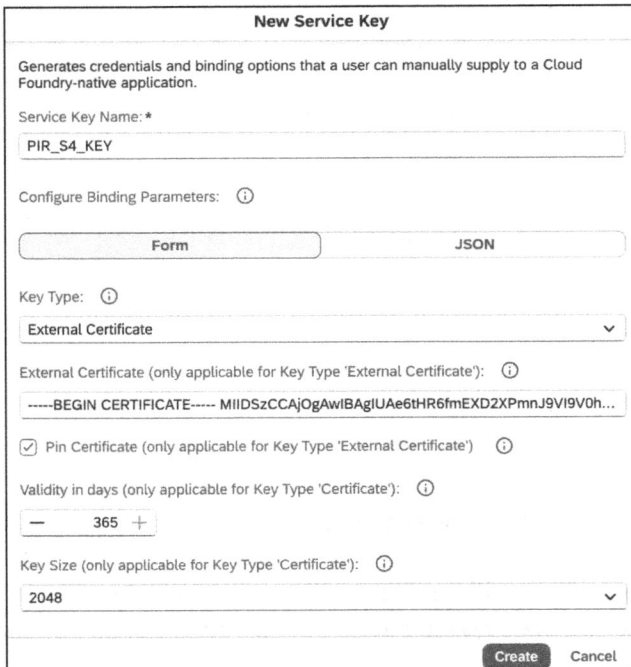

**Figure 7.14**  Filled-Out Service Key Creation Screen

Next, you need to upload the client certificate to the keystore of Cloud Integration. To do this, navigate to Cloud Integration, select **Monitor · Integrations and APIs**, and in the **Manage Security** section, select **Keystore**, as illustrated in Figure 7.15.

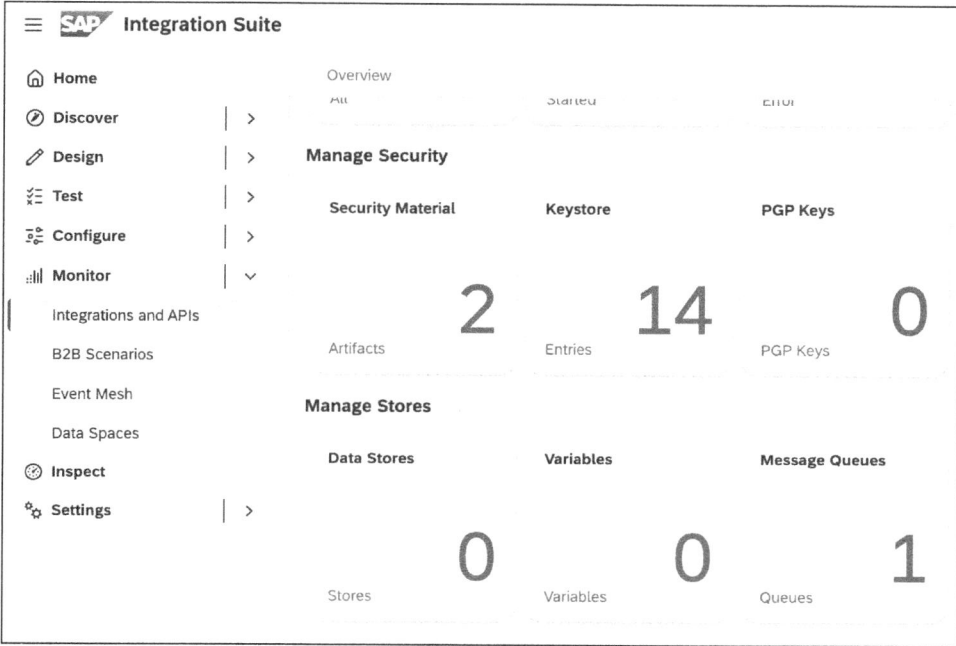

**Figure 7.15** Navigating to Keystore

In the **Keystore** section, as shown in Figure 7.16, click on **Add · Certificate** from the drop-down menu. This option allows you to upload the client certificate that's required for authentication.

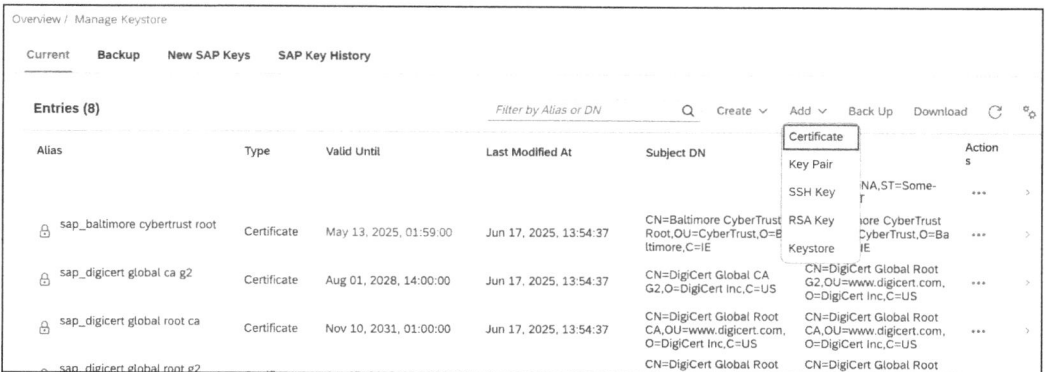

**Figure 7.16** Adding Certificate

In the **Add Certificate** dialog that's shown in Figure 7.17, enter a unique **Alias** for the certificate, upload the corresponding certificate **File**, and click **Add** to store the certificate in the keystore.

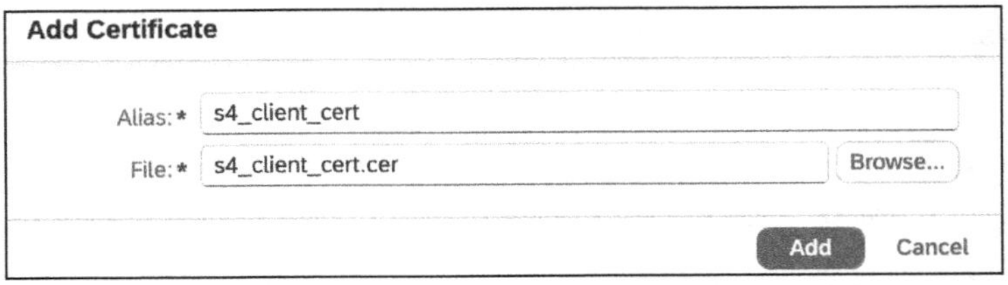

**Figure 7.17**  Uploading Client Certificate Keystore

In the final step, you configure the connection settings within the iFlow. As shown in Figure 7.18, set the **Authorization** method to **Client Certificate**, click on **Select**, and upload the client certificate of your sender system (e.g., "s4_client_cert") This step ensures that only sender systems that prove their identity, by providing the corresponding private key, are permitted to send messages to this iFlow.

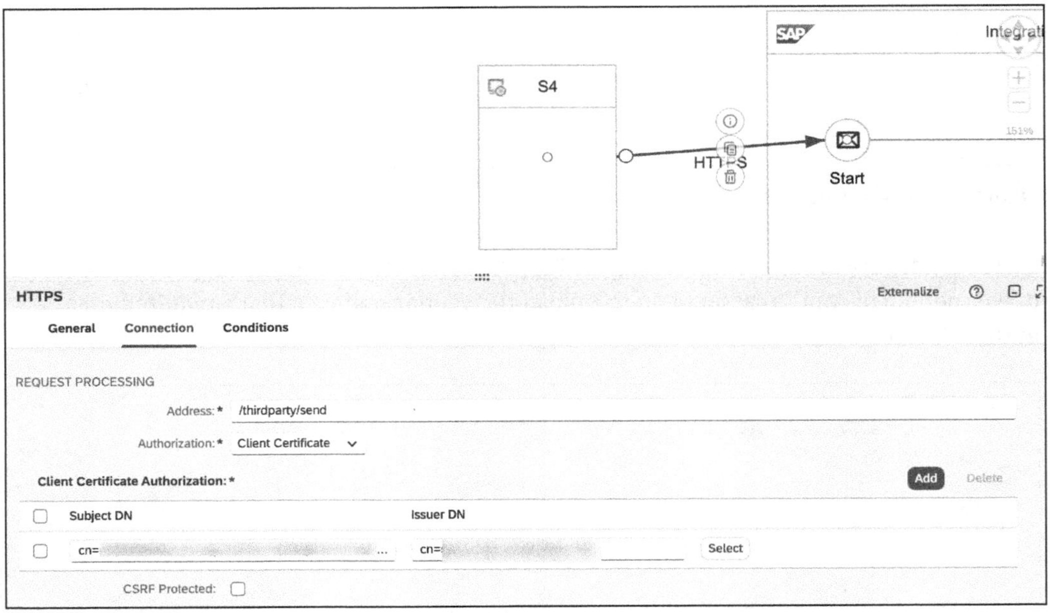

**Figure 7.18**  Inserting Certificate Alias into iFlow

## 7.2   Working with the Secure Store

In any integration landscape, the secure handling of sensitive data such as passwords, API keys, tokens, and certificates is of central importance. Cloud Integration provides a separate, specially protected area for this purpose: the *secure store*. It's the central repository for security-relevant information, and it ensures that confidential data is stored in encrypted form, is accessible on a role-based basis, and can be used programmatically without having to be stored in plain text within iFlows.

You can use the secure store to manage security-critical information on the system side and retrieve it at runtime within iFlows. The data in the secure store is stored in encrypted form, is only processed in decrypted form in the working memory, and is protected against unauthorized access by role and authorization concepts. Typical content that you'd store in the secure store is as follows:

- User name/password combinations for basic authentication
- Client IDs and client secrets for OAuth procedures
- Certificates and key pairs
- Private keys and signatures
- Passwords for target systems and databases
- User-specific tokens and API keys

In integration projects, you'll need to make sure this information isn't stored in plain text in iFlows or scripts. The secure store enables business logic and security data to be separated, which meets both security requirements and governance guidelines.

The contents of the secure store are managed in Cloud Integration under the term **Security Material**. You can maintain these in the SAP BTP cockpit or via the Cloud Integration web UI. You'll use different types of security materials, depending on the authentication procedure:

- **User credentials**
  This is the combination of user name and password.
- **OAuth2 client credentials**
  These are the client ID, secret, and token endpoint for OAuth 2.0.
- **Keystore entries**
  These are the certificates and keys, which are often referenced as aliases.
- **Secure parameters**
  These are any key-value pairs (e.g., for tokens and API keys).

You should give each security resource a uniquely name (e.g., "MyTargetSystem_Cred") and specify it as a reference in an iFlow via an adapter (e.g. HTTP, Mail, JDBC). The data itself is never displayed or saved in plain text in the flow.

Integration usually takes place via the configuration of an adapter in iFlow. When you're specifying the authentication method, you don't enter the user name that is entered directly—you enter a credential name that refers to a stored security artifact. The adapter resolves this name at runtime, retrieves the corresponding values from the security material and uses them for authentication. For example, if an HTTP receiver adapter transmits data to an external system that expects basic authentication, the password isn't entered in the configuration of the adapter, but the reference to the user credential object is stored in the security material. This ensures that passwords are never visible in the flow design.

The secure store is an integral part of secure integration architectures. It fulfills the following technical, regulatory, and operational requirements:

- **Centralized management**
  All sensitive access data is in one place and not scattered across iFlows.
- **Security standards**
  Storage is encrypted, and access is restricted to defined roles.
- **Maintainability**
  Changes to passwords or secrets don't require the iFlow to be redeployed.
- **Traceability**
  Changes in the secure store can be logged and versioned.
- **Auditability**
  Separation of logic and security facilitates audit security.

For productive use, we recommend that you follow these best practices:

- Make names of security materials descriptive and unique (e.g., "SAP_S4HANA_OAuth2")
- Before go-live, check all iFlows for hard access data.
- Assign roles for managing the secure store restrictively.
- Renew certificates and tokens should be regularly and check them for validity.

## 7.3    Working with Variables and Number Range Objects

In modern integration projects, you should maintain central values in one place and reuse them in various iFlows. Cloud Integration provides a powerful feature for this purpose: *variables* that you can define via a dedicated **Write Variables** step in the iFlow and save globally. This is supplemented by number *range objects* that generate consecutive numbers and are also configurable. Both functions support the aims of managing configuration values centrally, avoiding redundancy, and making iFlows dynamic and reusable.

The following sections introduce three core elements that support dynamic and maintainable iFlows. Section 7.3.1 covers *global variables*, which allow reusable values to be stored centrally and accessed across multiple flows. Section 7.3.2 presents *number range objects*, which are tools for generating sequential values (such as unique identifiers) directly within the runtime environment. Finally, Section 7.3.3 presents a practical example that illustrates how to combine these two features in a real-world scenario to create consistent, configurable, and scalable integration logic.

### 7.3.1   Global Variables

The **Write Variables** step allows you to define global variables that are stored centrally in Cloud Integration and are then available in any iFlow. This enables the separation of configuration data and flow logic, which is a significant contribution to the maintainability and scalability of integration scenarios.

One example could be the storage of a company number that is to be sent in several interfaces as part of a file name, as field content or in the header. Instead of hard-coding this value in each iFlow, you can define it once via the **Write Variables** step—for example with the "companyId" key and the "12345678" value. This entry will then be available centrally.

When this entry is accessed in another flow, it will be used via a content modifier where the variable can be saved in a header or exchange property. This allows the company number to be dynamically integrated into mappings, adapters, or Groovy scripts—without redundancy or the risk of inconsistent values. The use of global variables is particularly useful when the following are true:

- Recurring configuration values (e.g. IDs, URLs, technical parameters) are to be maintained centrally.
- A clear separation between flow logic and configuration data is required.
- Several iFlows require the same value (e.g. for client ID, language, country code, or organizational unit).
- Dynamic context control is required for deployments.

A major advantage is that a change to the variable's value doesn't affect the iFlow itself. The flow doesn't have to be redeployed—it's sufficient to update the value of the variable centrally.

We recommend that you adhere to these principles to promote the effective and secure use of global variables:

- Use unique, descriptive names (e.g., invoicePrefix, companyId, defaultCurrency).
- Separate technical and functional values by using naming conventions (e.g., tech_*, biz_*)
- Document all global variables centrally, ideally with where-used lists.
- Avoid making changes too frequently via productive iFlows; it's better to make them via dedicated admin flow.

### 7.3.2   Number Range Objects

*Number range objects* are more useful tools for process control. They are objects that make it possible to generate consecutive, unique numbers within an iFlow. This may be necessary, for example, to generate technical IDs, unique file names, or transaction-related references. The special feature of number range objects is that they're persistent and unique throughout the system. They store the last value assigned internally and increase it each time it's accessed—so they're comparable to counters in classic ERP systems.

In the iFlow, a number range object can be integrated into a content modifier as a header or exchange property. This step saves the current value of the number range object.

In practice, variables and number ranges can be used in combination. For example, an invoice number generated by the number range can be saved as a property, used in the message content (payload), and also output in logging. This information can in turn influence the flow of the iFlow via condition expressions—for example, when controlling whether it's a new or subsequent invoice.

The interaction of variables and unique identification (number ranges) is particularly valuable in scenarios with high throughput, differentiated message paths, or integrations with third-party systems that rely on consecutive references.

### 7.3.3   Practical Example

As shown in Figure 7.19, you can start by navigating to Cloud Integration and selecting **Monitor** and then **Integrations and APIs**. In the **Manage Stores** section, select **Number Ranges** to add a new number range.

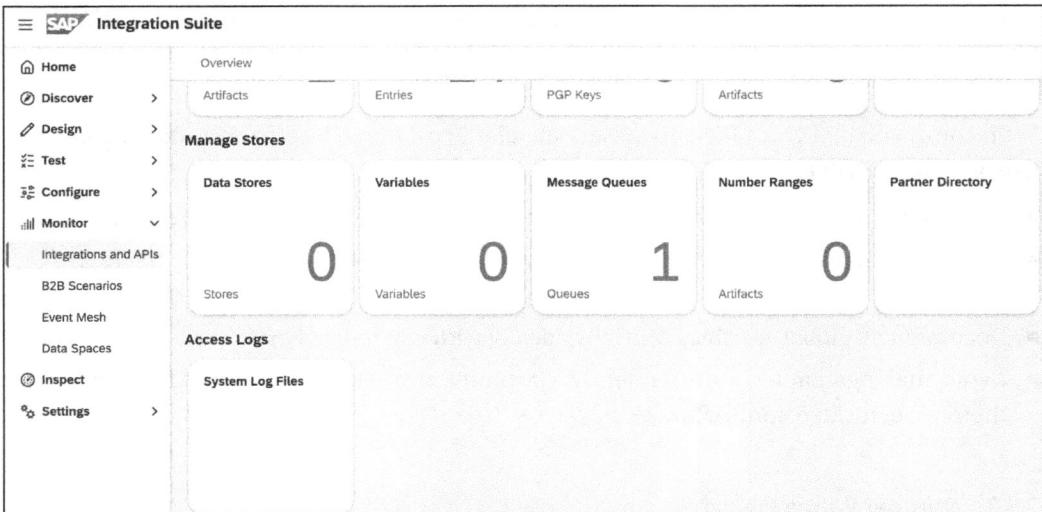

**Figure 7.19**  Switching to Number Range Objects

That will open a new window where you can add a new number range object by clicking **Add** at the top right (see Figure 7.20).

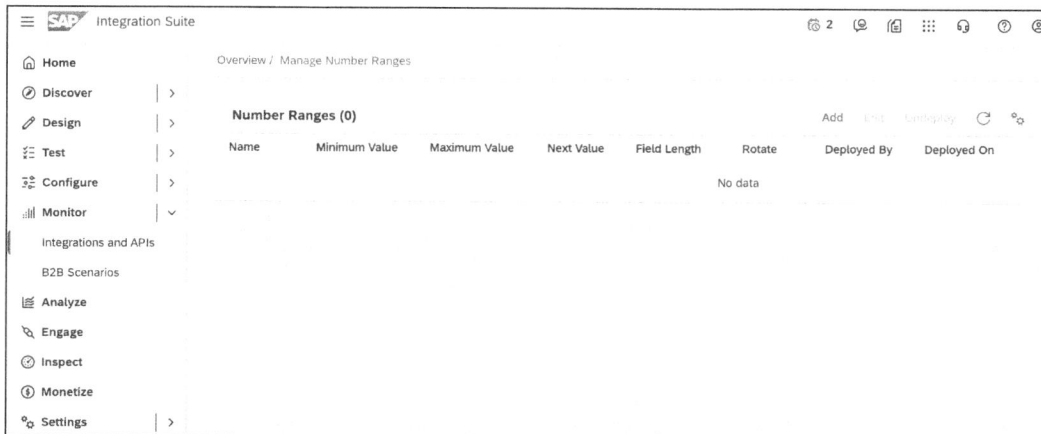

**Figure 7.20** Adding New Number Range Object

Next, define a number range object to provide sequential billing numbers. You can define the following parameters for each object:

- **Name**
  Enter a unique identifier for the number range.

- **Description**
  Enter a description of what this number range is intended for.

- **Minimum Value**
  Enter the minimum value.

- **Maximum Value**
  Enter the maximum value.

- **Field Length**
  This field displays the number of digits for the current value. If you entered "100" for the Maximum Value, then you should enter "3" in this field, if you entered "99999" for the Maximum Value, you should enter "5" in this field, and so on. If you enter "4" for the **Maximum Value** and "7" for the **Field Length**, then the number range will be reflected as 0007 and not just 7. This refers to the concept of padding zeroes into the value. If the Field Length value is 0, then the value of the current value attribute will be flashed because there's no padding of the number range. The maximum **Field Length** value is 99.

- **Rotate**
  If you check this box and the number range reaches the **Maximum Value** you specified, then the current value will reset to the **Minimum Value** you set.

As illustrated in Figure 7.21, the "numberRangeBilling" object is configured with a value range from 1 to 10,000, and it's set to rotate so the sequence can restart when it reaches the maximum.

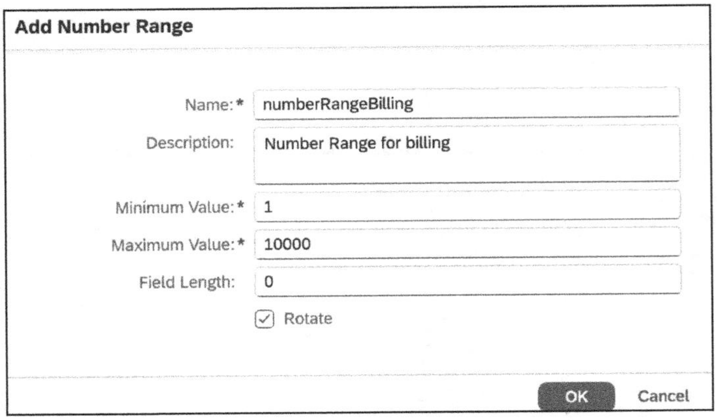

**Figure 7.21** Filled-Out New Number Range Object Screen

After this, you need to create a new iFlow. To define a constant value that can be reused throughout the iFlow, you create a global variable within the **Write Variables** step (see Figure 7.22) by assigning the value in the **Processing** tab and ensuring that the **Global Scope** box is checked.

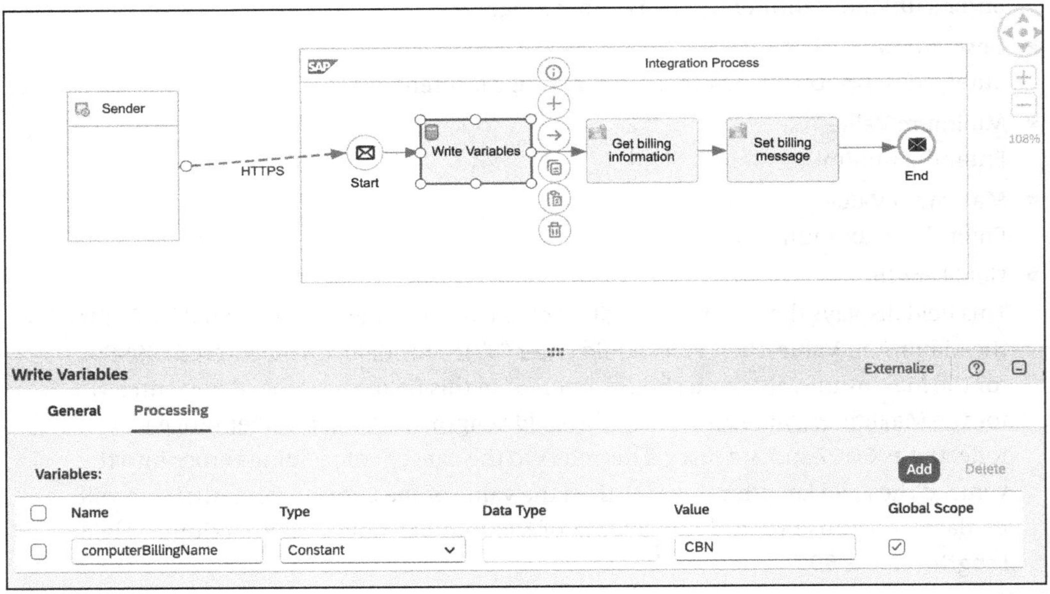

**Figure 7.22** Writing Variable

Once you've created both artifacts, you move to the **Content Modifier** step, where you use them to enrich the message. In the **Exchange Property** tab (see Figure 7.23), you assign the computerBillingName variable and the numberRangeBilling number range to new exchange properties: billingName and billingNumber.

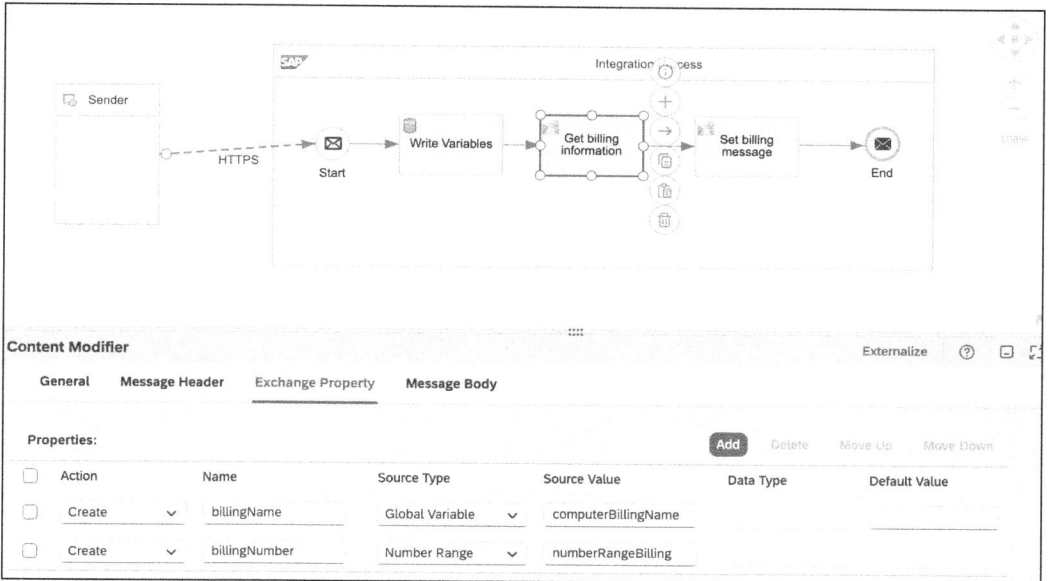

**Figure 7.23** Getting Variable and Number Range Object

Finally, as seen in Figure 7.24, you insert these properties into the payload in the **Message Body** tab.

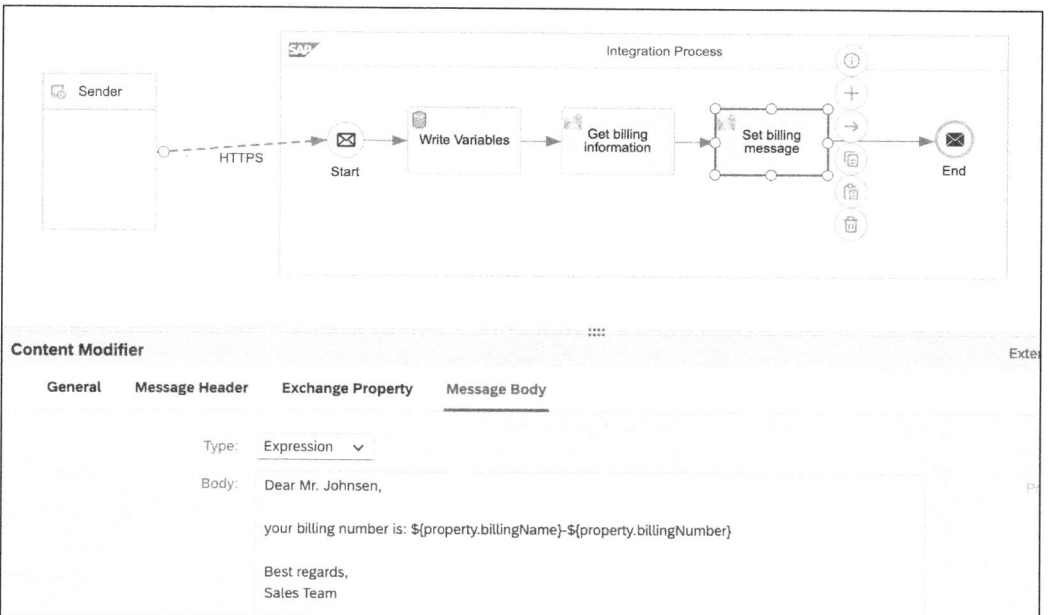

**Figure 7.24** Inserting Information into Payload

The result is a message that looks like this:

**Dear Mr. Johnsen,**

**Your billing number is CBN-1.**

**Best regards,**
**Sales Team**

A variable is also created, as shown in Figure 7.25.

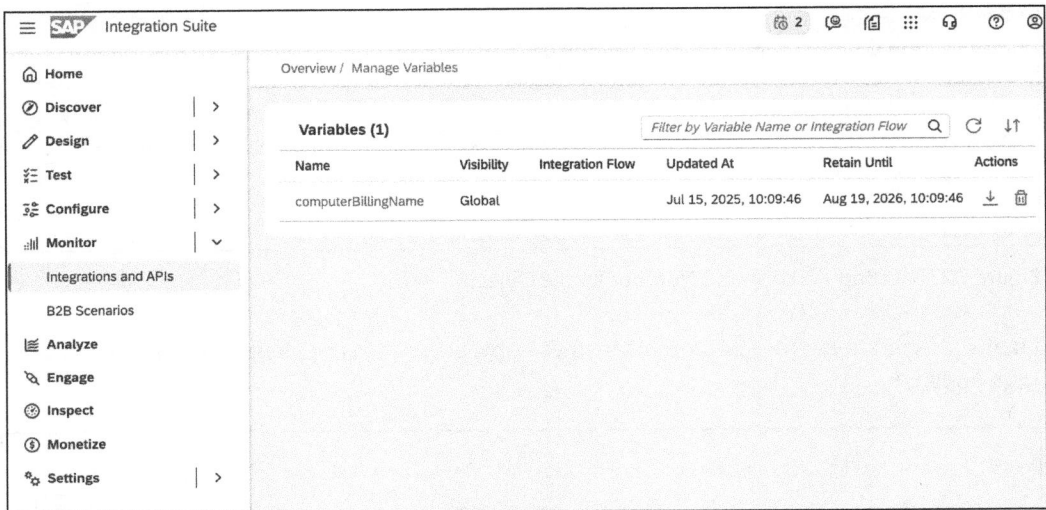

**Figure 7.25**  Managing Variables

## 7.4   Working with the Data Store

In integration scenarios, there are often requirements for temporary or persistent intermediate storage of data. Whether you need to resend messages in the event of an error, retain status information over a longer period of time, or manage system-independent test and comparison values, Cloud Integration offers a powerful means of persistent data storage within an iFlow in the form of the data store.

The *data store* is a configurable, cloud-based storage component where structured data objects can be stored permanently for a specific period of time, retrieved again in a targeted manner, processed automatically, and deleted. Unlike variables or header elements, which are only available within the runtime of a single iFlow instance, the data store is cross-process and state-preserving.

Typical application scenarios for the data store include the following:

- **Retry mechanisms**
  Messages that temporarily can't be delivered are temporarily stored in the data store and sent again later.

- **Auditability and traceability**
  This involves the backup of incoming or outgoing data over longer periods of time.

- **Status management**
  This involves the storage of check values (e.g., hashes, flags) to avoid duplicate processing.

- **Connectionless processes**
  These involve the storage of message fragments in asynchronous architectures.

Especially in B2B scenarios, in the context of exception handling and long-running processes, you need to use a data store to make integration processes robust and fail-safe. A data store in Cloud Integration consists of the following elements:

- **Data store name**
  This is the logical name of the store, and it's used for read and write access.

- **Data store entry**
  This is the actual data unit, and it consists of a key and the stored content (payload).

- **Key**
  This is a unique identifier per entry that can be generated statically or dynamically.

- **Retention time**
  This specifies how long the stored entry should remain in the store.

The data is stored internally in encrypted form and is only accessible via the respective iFlow in which it's used. Access takes place entirely within the graphical modeling, without the need to configure external databases or storage locations. Cloud Integration provides the following actions as part of the data store operations:

- **Write**
  This saves a new data record with an individual key.

- **Delete**
  This deletes a stored data record.

- **Select**
  This enables several entries to be read using search criteria (e.g., time, status).

- **Update**
  This overwrites an existing entry.

Developers can integrate these operations via dedicated steps in the iFlow, such as **Write to Data Store** or **Get from Data Store**, which they can flexibly combine with other process steps. Developers can therefore decide which data will be persisted or processed at which point in time.

You also need to know that data stores are client and tenant bound, which means that they're only visible within the subaccount and the respective iFlow in which they were created. Direct access from other flows or from the environment isn't provided, and that serves the purpose of security and data isolation.

Administrators can also access existing data store entries via the **Monitor** area of Cloud Integration. They can view, delete, and analyze entries there—for troubleshooting or runtime diagnostics, for example.

The following typical deployment patterns have become established in practice:

- **Retry queue**
  In the event of communication errors, a message is temporarily stored in the data store. Then, a separate, time-controlled iFlow processes the store regularly and attempts to send the messages again.

- **Idempotence check**
  An incoming data record is checked against a unique key in the store. If this already exists, processing is skipped.

- **Aggregation and split/join**
  In multistage processes, partial information is stored and merged into a complete message at a later point in time.

These patterns show that the data store not only serves to prevent errors but also forms the basis for state-oriented integration architectures.

Even though the data store is a powerful tool, it has certain restrictions:

- There is a maximum quota for storage space and entry size, depending on the Cloud Integration plan you select.
- The data structure is flat (i.e., there is no complex query logic, as in classic databases).
- The store isn't transactional (i.e., write and read processes are not linked atomically).

For these reasons, you should use the data store in a targeted and deliberate manner—for example, for technical control information but not for large user data or long-term archiving. We recommend that to use storage resources efficiently, you should perform regular cleansing via expiration rules or manual deletion handling.

Now, let's look at an example of data store operations. The iFlow in Listing 7.1 handles data store operations, and over HTTPS, it receives a customer's XML payload that consists of one or more Customer entries in the following format:

```
<Customers>
    <Customer>
        <CustomerID>1001</CustomerID>
        <FirstName>Anna</FirstName>
        <LastName>Schmidt</LastName>
        <Email>anna.schmidt@example.com</Email>
```

```
        <Country>Germany</Country>
        <Phone>+49-151-0000001</Phone>
    </Customer>
    <Customer>
...
</Customers>
```

**Listing 7.1** Incoming XML Customer's Payload

To begin working with data store operations in Cloud Integration, you count the number of customer entries in the incoming XML payload. As shown in Figure 7.26, the iFlow begins with a **Content Modifier** named **Count Customer**. In this step, you define a **Message Header** by entering the name "countCustomer" and using **XPath** as the **Source Type**. The **Source Value** is an XPath expression that counts the number of <Customer> nodes within the <Customers> element.

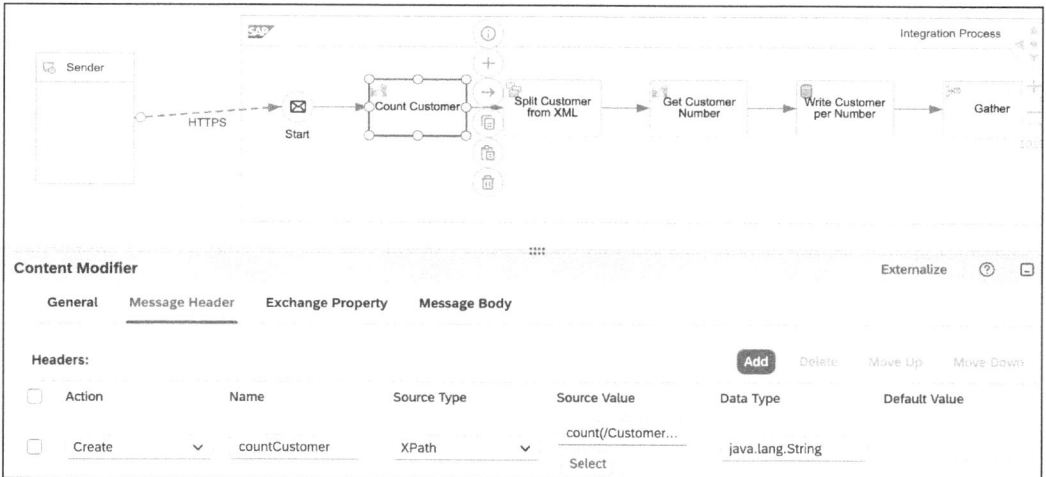

**Figure 7.26** Counting Customer Entries

Next, you need to split the XML payload to process each customer individually. As shown in Figure 7.27, you do this by using an **Iterating Splitter** step that's labeled **Split Customer from XML**. In the **Processing** tab of the splitter configuration, you define the **Expression Type** as **XPath** and set the **XPath Expression**.

The next step is to extract the unique identifier for each customer. As shown in Figure 7.28, you do this by using a **Content Modifier** that's labeled **Get Customer Number**. In this step, you switch to the **Exchange Property** tab to define a new property by entering "customerID" as its name. Choose **XPath for the Source Type**, enter "//CustomerID" for the **Source Value**, and enter "java.lang.String" for the **Data Type**.

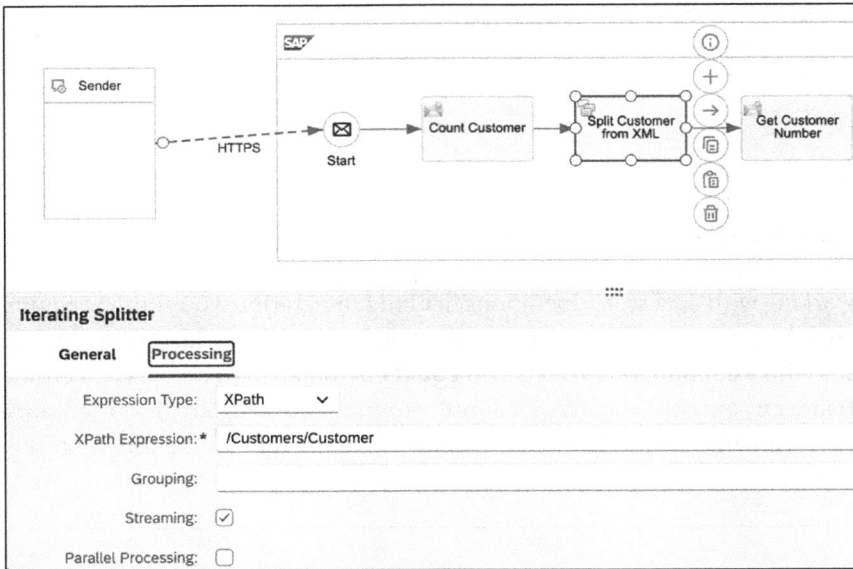

**Figure 7.27**  Split Customer XML

This configuration instructs the iFlow to extract the value of the <CustomerID> element from each split customer payload and store it as an exchange property. This ID will be used later for writing data to the data store and for referencing specific records during read and delete operations. Using exchange properties ensures that each customer's identifier is isolated and available for subsequent steps in the looped process.

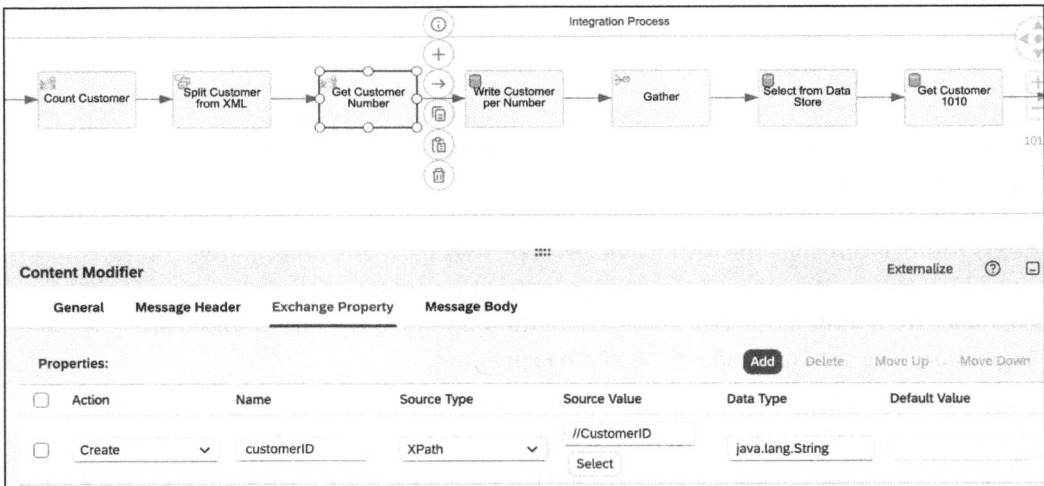

**Figure 7.28**  Getting Customer ID

Once you've extracted each individual customer ID, you can persist the customer data into a data store for further use. As illustrated in Figure 7.29, you do this by using a

**Write** step that's labeled **Write Customer per Number**. In the **Processing** tab, make the following configurations:

- **Data Store Name**
  Enter "Customers" as the logical name under which the messages will be stored.

- **Visibility**
  Choose **Integration Flow**, which limits access to this data store to the current iFlow only.

- **Entry ID**
  Enter "${property.customerID}" to dynamically assign the customer ID as the unique key for each stored entry.

- **Retention Threshold for Alerting (in d)**
  Enter "2" to define the number of days after which alerts should be raised for aging entries.

- **Expiration Period (in d)**
  Enter "30" to automatically remove entries after that many days.

- **Encrypt Stored Message**
  Check this box to ensure secure storage of sensitive data.

- **Overwrite Existing Message**
  Check this box to allow updates if an entry with the same ID already exists.

- **Include Message Headers**
  Optionally, you can check this box if you want headers to be preserved with the stored message.

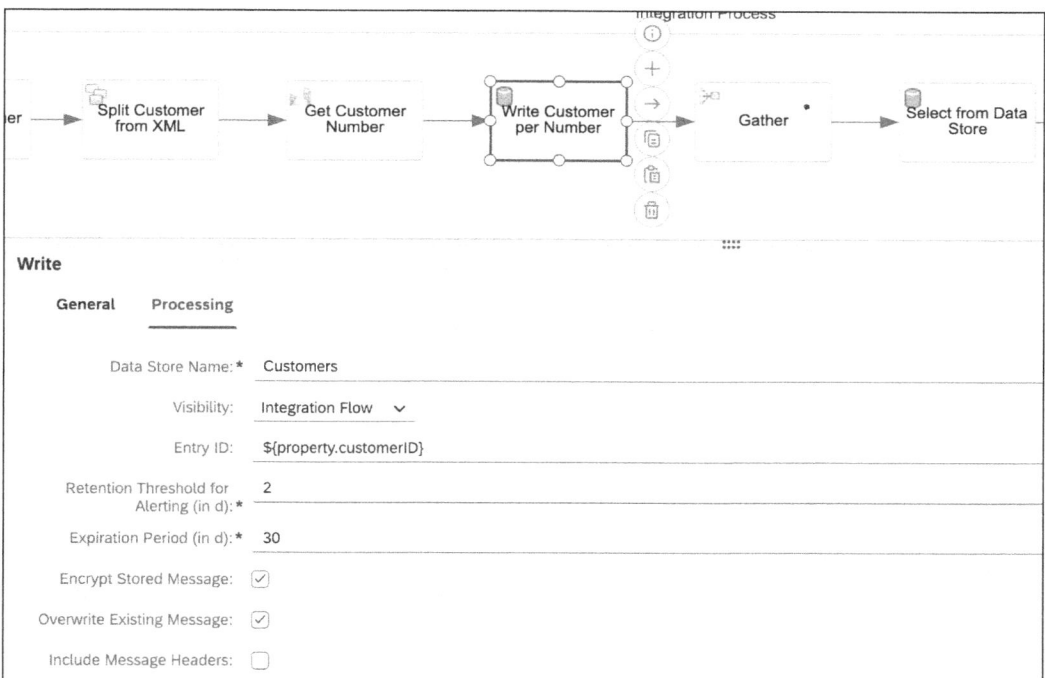

**Figure 7.29**  Writing in Data Store

This step ensures that each customer record will be securely written to the data store, uniquely identified by its customer ID, and ready to be retrieved or updated later in the process. You can see the results in Figure 7.30.

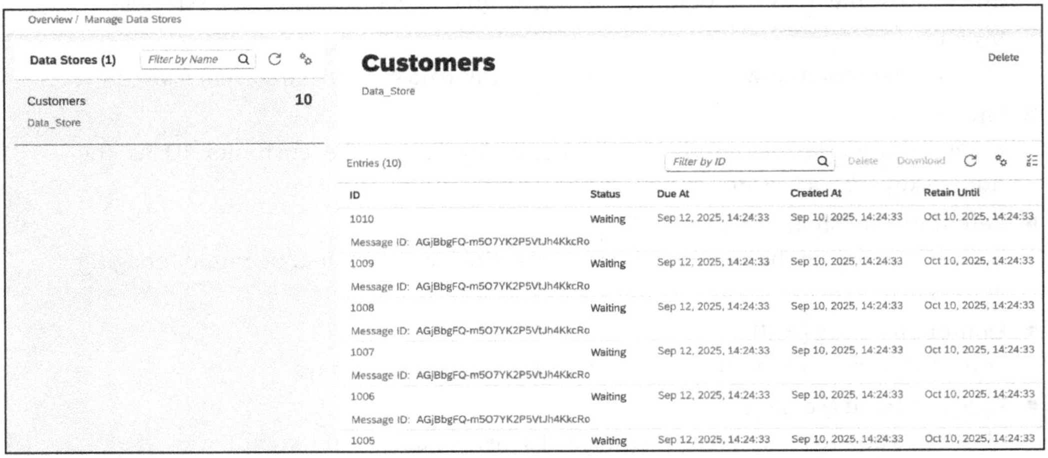

**Figure 7.30**  Data Store Entries

Each entry in the data store contains the header and payload of the messages in the state in which it was saved in the data store. The first one, for customer 1001, looks like Listing 7.2.

```
<?xml version='1.0' encoding='UTF-8'?><Customer>
        <CustomerID>1001</CustomerID>
        <FirstName>Anna</FirstName>
        <LastName>Schmidt</LastName>
        <Email>anna.schmidt@example.com</Email>
        <Country>Germany</Country>
        <Phone>+49-151-0000001</Phone>
    </Customer>
```

**Listing 7.2**  Customer Payload from Data Store

After you store each individual customer entry in the data store, the next step is to reassemble the message fragments before continuing with further processing. As shown in Figure 7.31, you do this by using a **Gather** step, the purpose of which is to aggregate the results from the previous loop iteration into a single, structured message. In the **Aggregation Strategy** tab, you should make the following configurations:

- **Incoming Format**
  Select **XML (Same Format)** to ensure that the payloads being combined share a common XML structure, which is necessary for proper aggregation.

- **Aggregation Algorithm**
  Select **Combine** to make this algorithm concatenate the incoming XML payloads into one single output.

This step is especially useful when a series of parallel or split steps produce multiple message fragments that must be recombined into a single payload for continued processing or final delivery. In this case, you use this to stop the splitting of the customers.

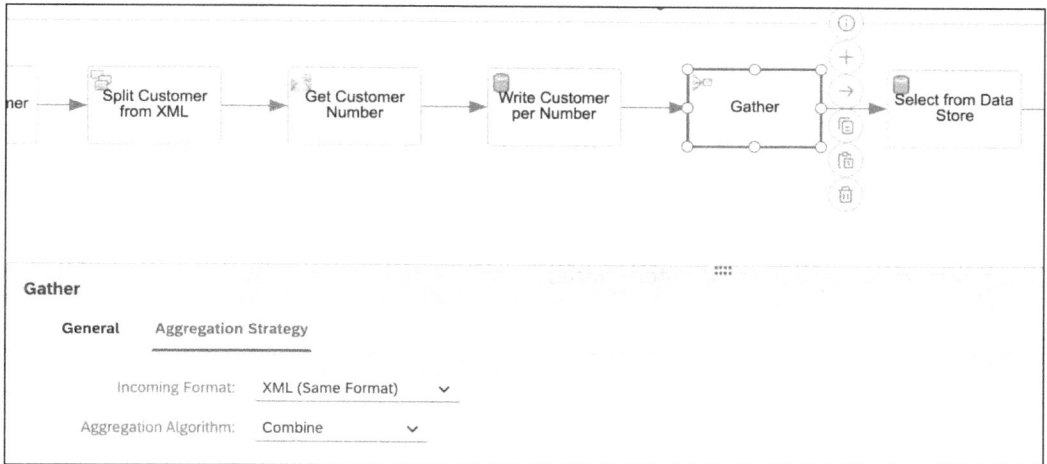

**Figure 7.31**  Gathering Information

After gathering and preparing the message data, you should read entries from the data store to let you retrieve and work with stored messages programmatically. As shown in Figure 7.32, you do this by using a **Select** step that's labeled **Select from Data Store**. In the **Processing** tab, configure the following parameters:

- **Data Store Name**
  Enter "Customers" as the name. It's the same name you used in the **Write** step, and it identifies which data store to access.

- **Visibility**
  Select **Integration Flow** to ensure that access is scoped only to the current iFlow.

- **Number of Polled Messages**
  Enter "${header.countCustomer}" to dynamically control how many entries to retrieve, based on the earlier message header that holds the total count of customers.

The **Delete On Completion** box isn't checked here. That means the entries will remain in the data store after they're read—which is ideal for auditing or follow-up operations.

This step enables batch retrieval of stored messages using the customer count calculated at the beginning. This ensures that only the expected number of records will be pulled.

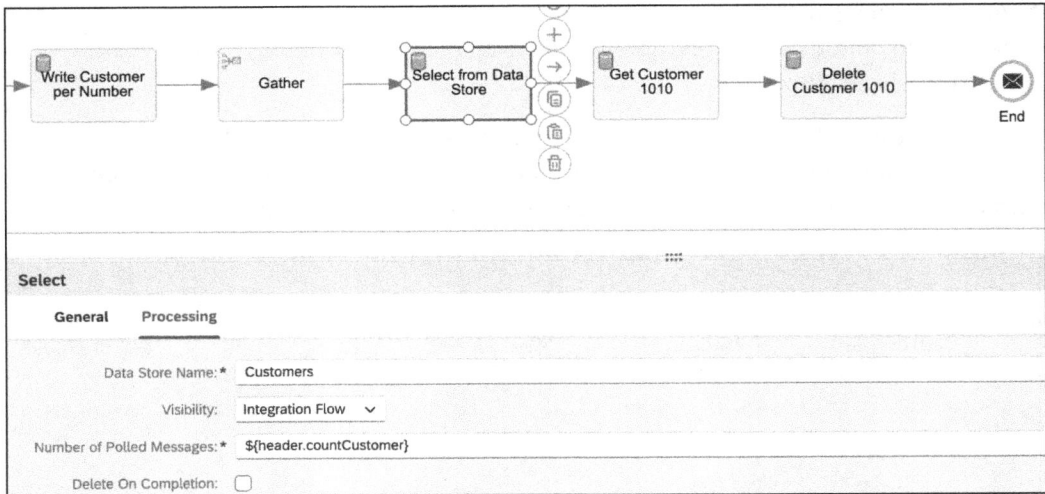

**Figure 7.32** Selecting Data Store Entries

When you use the **Select Data Store** operation, you generate a new payload with the data of the data store. In our example, the payload looks like Listing 7.3.

```
<?xml version='1.0' encoding='UTF-8'?>
<messages>
    <message id="1001">
        <Customer>
            <CustomerID>1001</CustomerID>
            <FirstName>Anna</FirstName>
            <LastName>Schmidt</LastName>
            <Email>anna.schmidt@example.com</Email>
            <Country>Germany</Country>
            <Phone>+49-151-0000001</Phone>
        </Customer>
    </message>
    <message id="1002">
        ...
</messages>
```

**Listing 7.3** Customer's Payload after Select Operation

After polling the available messages from the data store, you can retrieve a specific entry by using the **Get** step. As shown in Figure 7.33, you use the step labeled **Get Customer 1010** to read a message with a defined entry ID. In the **Processing** tab, you perform the following configuration:

- **Data Store Name**
  Enter "Customers" since it matches the name you used in previous write and select steps.

- **Visibility**
  Select **Integration Flow** to limit access to the current iFlow scope.
- **Entry ID**
  Enter "1010" because it statically references the entry to retrieve. (In real-world scenarios, you can parameterize this value to support dynamic lookups.)
- **Delete On Completion**
  Leave this box unchecked so that the entry will remain in the store after being retrieved.
- **Throw Exception on Missing Entry**
  Check this box to ensures that the integration will fail gracefully if the entry with the specified ID doesn't exist.

This step is useful for retrieving individual customer records on demand—for example, based on a specific customer ID passed via an API or message header.

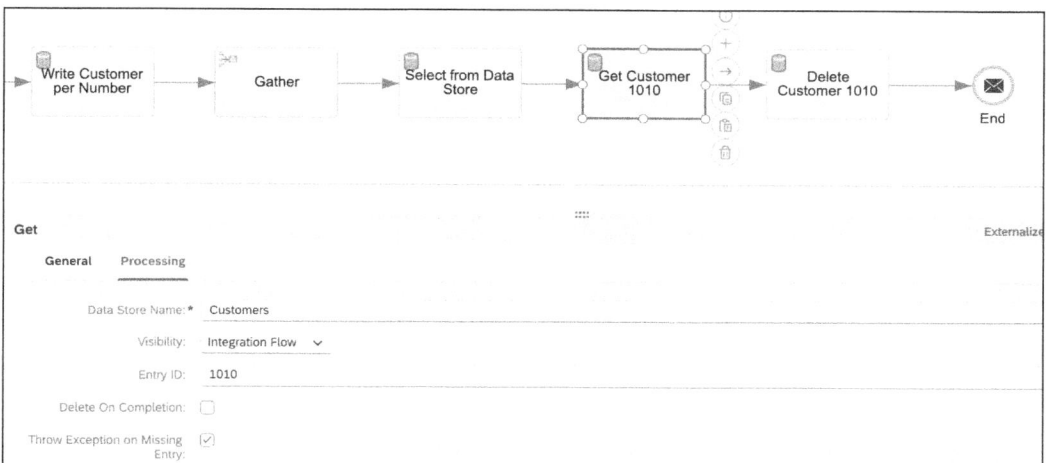

**Figure 7.33** Getting Specific Customer

The payload after the **Get** operation is the payload from the data store entry, and it looks like Listing 7.4.

```
<?xml version='1.0' encoding='UTF-8'?>
<Customer>
    <CustomerID>1010</CustomerID>
    <FirstName>Jonas</FirstName>
    <LastName>Schulz</LastName>
    <Email>jonas.schulz@example.com</Email>
    <Country>Austria</Country>
    <Phone>+43-699-1122334</Phone>
</Customer>
```

**Listing 7.4** Customer Payload after Get Operation

To complete the process, you can optionally remove the customer entry from the data store by using a **Delete** step. As shown in Figure 7.34, the step that's labeled **Delete Customer 1010** is responsible for deleting a specific entry by its unique identifier. In the **Processing** tab, perform the following configurations:

- **Data Store Name**
  Enter "Customers" because it refers to the same data store you've used throughout the flow.

- **Visibility**
  Select **Integration Flow** to ensure scoped access to the data store within the current iFlow.

- **Entry ID**
  Enter "1010" to identify the specific message entry to delete.

By targeting a known **Entry ID**, the iFlow can clean up customer data that is no longer needed, such as in cases of processed, archived, or invalidated records. This final step helps maintain clean and relevant data within the data store, especially in long-running and frequently executed iFlows.

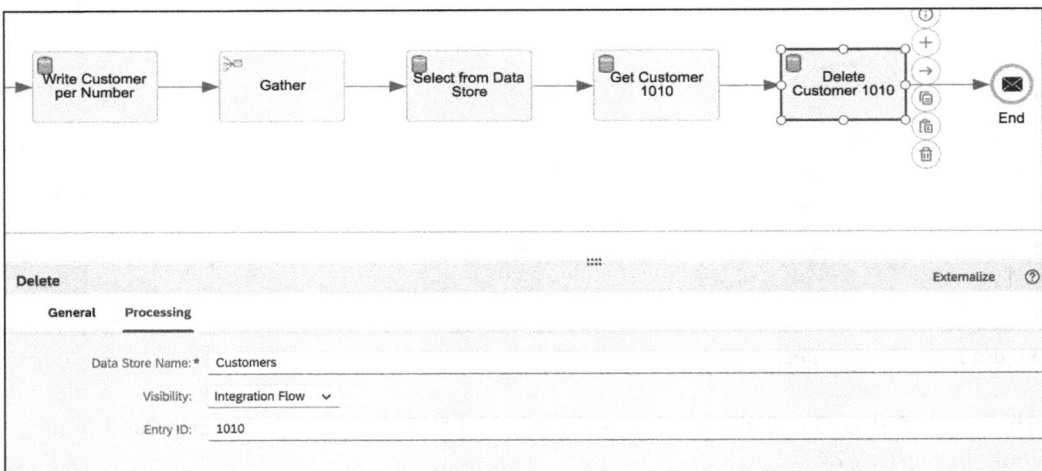

**Figure 7.34** Delete Specific Customer

After you complete the deletion step, you can observe the updated state of the data store in Cloud Integration. To do this, in Figure 7.35, navigate to **Monitor • Integrations and APIs** and open the **Manage Data Stores** view. Select the **Customers** data store to reveal a list of stored entries on the right-hand side. You'll see that each entry is listed with its unique ID, status, creation date, retention period, and message link.

You'll also notice that the previously deleted entry with an ID of 1010 is no longer present, and that will confirm that the **Delete** step successfully removed the specified message from the store. The remaining entries (with IDs from 1001 through 1009, excluding 1010) will still be in **Waiting** status, which means they're scheduled to be retained until the configured expiration date. This view provides a transparent and manageable way to monitor and maintain the data lifecycle in Cloud Integration and to ensure that temporary or obsolete records are efficiently cleaned up.

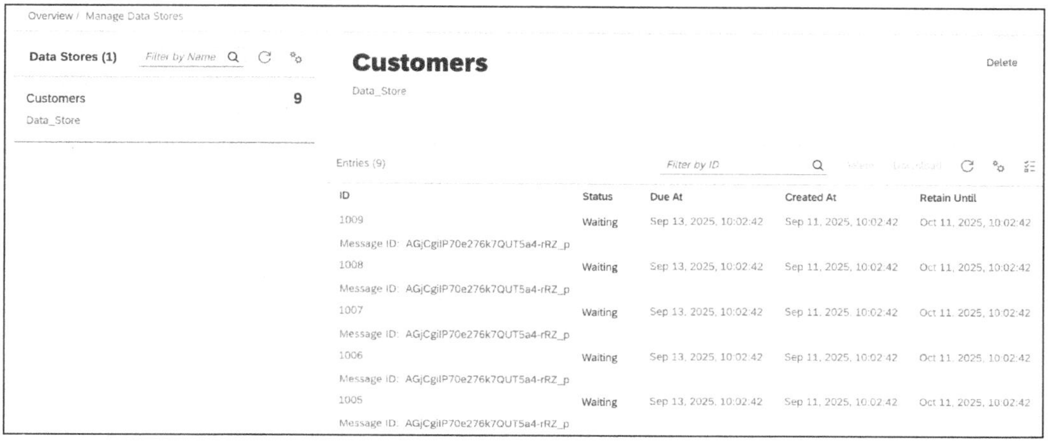

**Figure 7.35** Data Store with Deleted Entry

## 7.5   Summary

In this chapter, we systematically established the data management and security foundation for working with Cloud Integration. Our focus wasn't solely on feature demonstration or tool exploration, but rather on configuring a secure, reusable, and governance-compliant integration platform that fulfills both functional and nonfunctional requirements. We presented the configuration steps in a structured way to help you create an operational environment that supports high standards of system security, data governance, and operational robustness.

We started off by showing you how to configure the authentication and authorization mechanisms within the Integration Suite. After you established secure authentication patterns, we introduced you to the secure store and how to use it. This component plays a key role in the secure management of sensitive credentials and tokens. Another key component we introduced in this chapter is the use of global configuration variables, which are created and stored via the **Write Variables** operation. To support the generation of unique identifiers within process flows, we introduced you to number range objects and showed you how to configure them, retrieve values from them during runtime, and use those values in filenames, message properties, and audit trails. We also introduced the data store component, which serves as a persistent message store within the Integration Suite. We concluded by integrating all of the elements we described into a coherent example. It demonstrated not just how each tool functions in isolation but how to orchestrate them to create flows that are not only functional but auditable, maintainable, and ready for real-world deployment.

Overall, this chapter laid the technical groundwork for building iFlows that go far beyond simple data movement. The implementation of security mechanisms, configuration abstraction, state management, and controlled message persistence provides a framework for creating enterprise-grade, cloud-based integrations. By completing the steps described in this chapter, you have established a strong baseline for secure and maintainable integrations within the SAP Integration Suite. The approaches demonstrated here will serve as foundational patterns for all future flows and ensure that as business requirements grow in complexity, the integration layer will remain stable, secure, and scalable.

# Chapter 8

# Transport Management and Deployment

*In dynamic development environments, you must be able to transfer artifacts from development to test and ultimately to production to ensure consistency, compliance, and speed. However, this process involves more than just exporting and importing content—it requires structured version control, role-based governance, and ideally, automation via pipelines.*

Section 8.1 of this chapter explains how to use SAP Cloud Transport Management, which acts as the central foundation for structured and audit-compliant transports across subaccounts, tenants, and environments. This service allows you to model multistage landscapes, manage transport nodes, and monitor deployments centrally. Section 8.2 explores versioning and deployment strategies, and you'll learn how to manage artifact versions within Cloud Integration, how to deal with configuration parameters across environments, and how to establish deployment patterns that align with enterprise DevOps principles. Section 8.3 covers automated transports and CI/CD pipelines.

## 8.1    Transportation with SAP Cloud Transport Management

*SAP Cloud Transport Management* is a central transport tool within SAP BTP that enables software artifacts such as iFlows, value assignments, script collections, and UI components to be moved securely, traceably, and automatically among different subaccounts and system landscapes. It's not just a file transfer tool but a fully integrated transport service with governance, lifecycle, and monitoring functionalities that are based on the classic SAP transport system but designed specifically for cloud-native applications.

At its core, SAP Cloud Transport Management is based on the principle of *transport landscapes*, which are modeled as a sequence of transport nodes. Each of these nodes represents a destination for software artifacts, such as a development, test, and production systems. These can be different subaccounts within the same global account or even subaccounts in completely different regions or clients. You can graphically model transport landscapes in SAP Cloud Transport Management, version them, and flexibly adapt them to the needs of your organization.

A transport begins with the export of an artifact from a source, for example, an iFlow from a subaccount of Cloud Integration. This export is then stored as a transport request in SAP Cloud Transport Management and can then be "pushed" to the next node in the landscape. The transport requests are versioned, contain metadata, (e.g. author, timestamp, artifact type), and can be tracked.

A major advantage of SAP Cloud Transport Management is the central control and logging of all transport processes. Every transport is traceable, audit-proof, and linked to a clearly defined destination. You can control transport authorizations at a granular level, via roles and authorizations—for example, by specifying who can create, approve, or deploy transports. This is particularly important for firms in regulated industries, such as financial service providers and pharmaceutical companies.

You can also restrict access to certain nodes or provide transport requests with a digital signature mechanism. This can ensure that artifacts don't enter productive environments in an uncontrolled manner. In the context of Cloud Integration, SAP Cloud Transport Management a central role in the transport of the following:

- iFlows
- Value mappings
- Message mappings
- Script collections
- Adapter configurations

Cloud Integration is "transportable" in the sense of SAP's own lifecycle concept. Using transport mode deployment, you can export finished content from a source system, transfer it to SAP Cloud Transport Management as a transport object, and import it specifically into a target system. During this process, you can ensure that dependencies and versions are taken into account.

Before the import, you can adapt certain configuration parameters such as URLs, access data, and operating modes via environment configuration sets. You need to do this to cleanly separate environment dependencies and ensure reusability.

Using SAP Cloud Transport Management offers the following advantages for the operational control and quality assurance of integration solutions:

- Central control of all transport processes across subaccounts
- Transparency and traceability through versioned transport requests
- Granular assignment of rights to control and approve deployments
- The ability to integrate it into DevOps and CI/CD processes for automated transports
- Minimization of manual error sources through controlled and repeatable processes
- Standardization of deployment strategies across different SAP services

This section outlines the complete setup procedure you need to perform to enable SAP Cloud Transport Management as the central transport layer for Cloud Integration.

Section 8.1.1 explains how to create a dedicated subaccount and assign relevant service plans to isolate and manage transport operations independently. Section 8.1.2 covers how to define a structured landscape so artifacts can move between source and target systems using predefined transport routes. Section 8.1.3 and Section 8.1.4 show you how to create essential service instances, such as the SAP Content Agent and SAP Process Integration runtime, both of which are required for secure communication and artifact handling. Section 8.1.5 covers how to configure destinations by using service key credentials to ensure OAuth-secured integration between all components. Finally, Section 8.1.6 explains how to assign role collections to govern access and responsibilities for transporting and importing packages, and it walks you through a test transport to confirm that all components are correctly connected and the transport mechanism is fully operational within the SAP BTP landscape.

### 8.1.1   Creating a Subaccount and Assigning SAP Cloud Transport Management

To ensure a clean separation of the transport landscape and at the same time enable clear management of the transport processes, SAP recommends that you operate SAP Cloud Transport Management in a separate subaccount as a best practice. This separation allows centralized control of transports across subaccount boundaries without directly interfering with the respective development or production environments.

In this section, we'll show you step-by-step how to set up a dedicated subaccount for SAP Cloud Transport Management and which configurations you need to make to ensure a smooth and secure transport process. As shown in Figure 8.1, you'll start by creating a dedicated subaccount in which SAP Cloud Transport Management will be operated.

In your global account, switch to the **Account Explorer** and click the **Create** button in the upper right-hand corner to add a new subaccount. Then, in the **Create Subaccount** dialog box, enter, select, or verify the following:

- **Display Name**
  Enter a freely selectable display name for the subaccount (e.g., "Cloud Transport Manager").
- **Region**
  Choose the geographical region in which the subaccount is to be provided (e.g. **US East (VA)**).
- **Subdomain**
  This is an automatically generated (but customizable if required) subdomain that becomes part of the URL structure of the subaccount.
- **Parent**
  The parent global account ID is automatically displayed here.

After completing these fields, click on **Create** to create the subaccount. This step forms the basis for the separate management of transportation in SAP BTP, according to best practices.

375

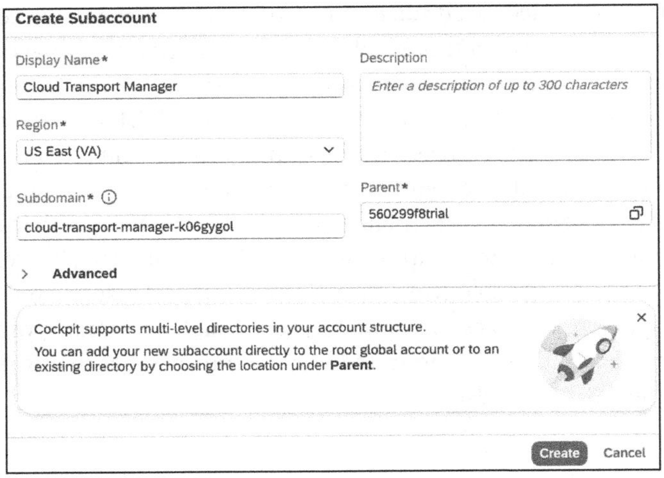

**Figure 8.1** Creating New Subaccount

Next, you must assign the required services via the entitlements so that they can be used in the subaccount. Figure 8.2 shows an overview of the **Entity Assignments** in the global account. The **Cloud Transport Manager** subaccount that you previously created will already be selected here, and you can use the **Edit** button to assign the services you want to the subaccount.

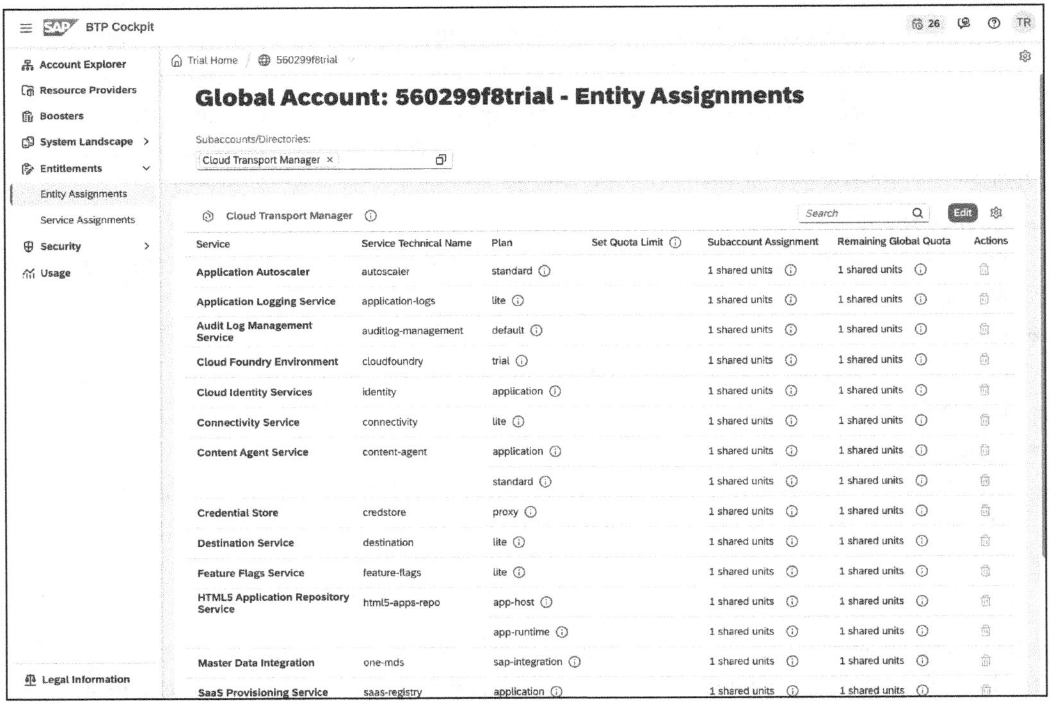

**Figure 8.2** Entity Assignment

To provide the subaccount with the necessary services, navigate to **Entitlements** • **Entity Assignments** in the SAP BTP cockpit and click on **Edit**. As shown in Figure 8.3, the Cloud Transport Manager subaccount will already be selected, and to add services, you click the **Add Service Plans** button on the right.

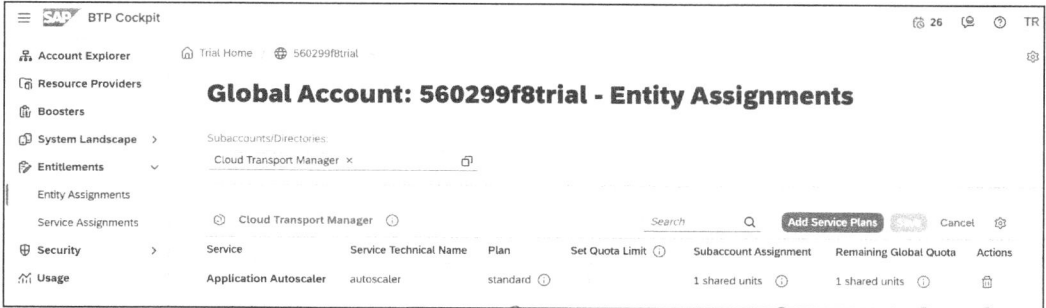

**Figure 8.3**  Adding Service Plans

That will open a dialog for assigning services to the subaccount, where you can search for the **Cloud Transport Management** and check the **standard** and **lite (Application)** boxes in the **Available Plans** section, as shown in Figure 8.4. In your case, the second box is also the standard plan for the application. To finish assigning service plans, you also need to assign the **Content Agent Service** with the **standard** plan.

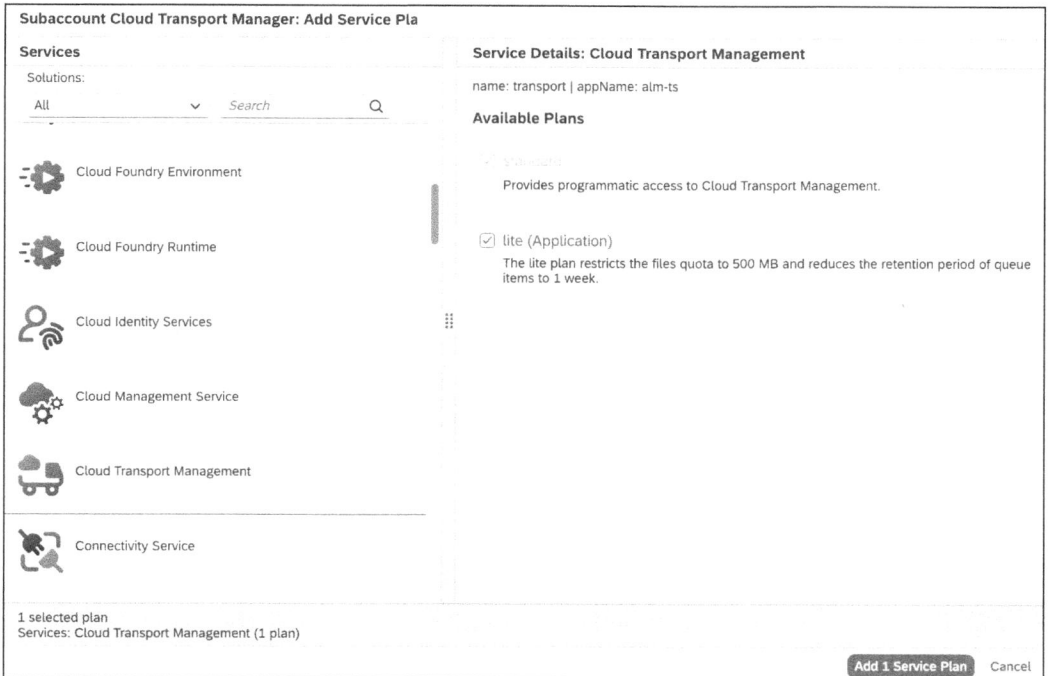

**Figure 8.4**  Assigning Cloud Transport Manager

Once you've selected and confirmed the service plan, the **Cloud Transport Management** service will appear in the list of assigned services in the **Entity Assignments** section of the subaccount (see Figure 8.5). There, you'll see that the service with the standard plan has been assigned to the Cloud Transport Manager subaccount. The **Subaccount Assignment** column will show that the assignment has been made with **1 shared units,** which means that the service is shared on the basis of the global quota.

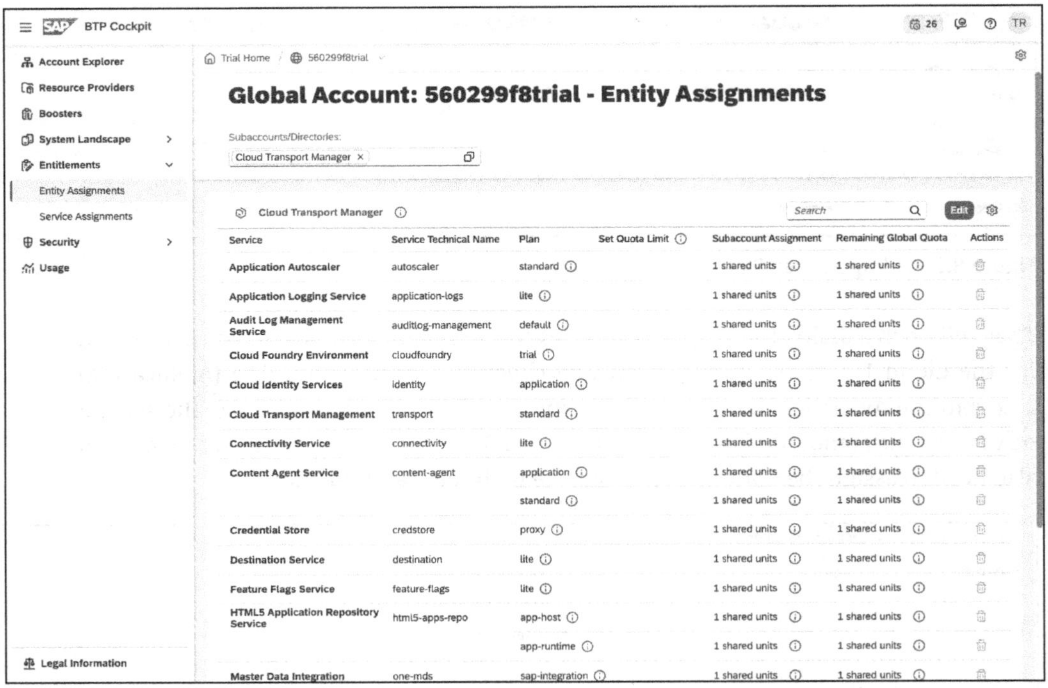

**Figure 8.5** Assigned Cloud Transport Manager

To be able to use SAP Cloud Transport Management, you must first activate the Cloud Foundry environment in the associated subaccount. Enter or verify the following information in the dialog box (see Figure 8.6):

- **Instance Name**
  Enter an individual name for the Cloud Foundry instance, which is derived here from the subdomain and service usage.

- **Org Name**
  This is automatically generated from the global account and the subdomain, and it identifies the organizational unit within Cloud Foundry.

Then, click **Create** to initialize the environment, which forms the basis for deploying and operating cloud-native applications such as SAP Cloud Transport Management.

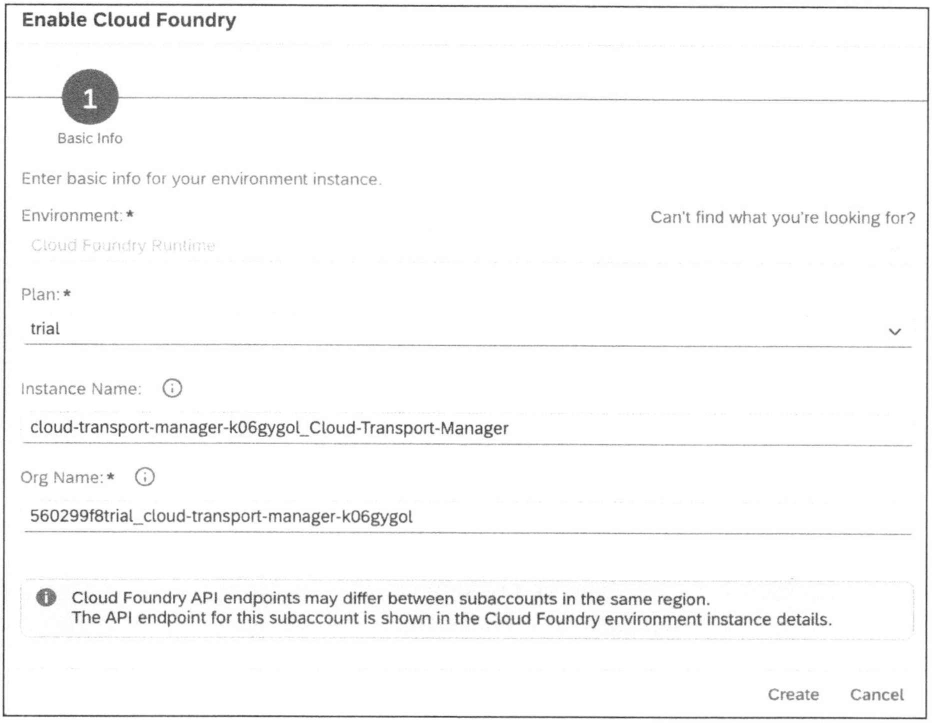

**Figure 8.6** Enabling Cloud Foundry

To make SAP Cloud Transport Management communicate correctly with the target subaccount, you need to obtain the **Region**, organization name (**Org Name**), and **Space** name from the target subaccount (see Figure 8.7).

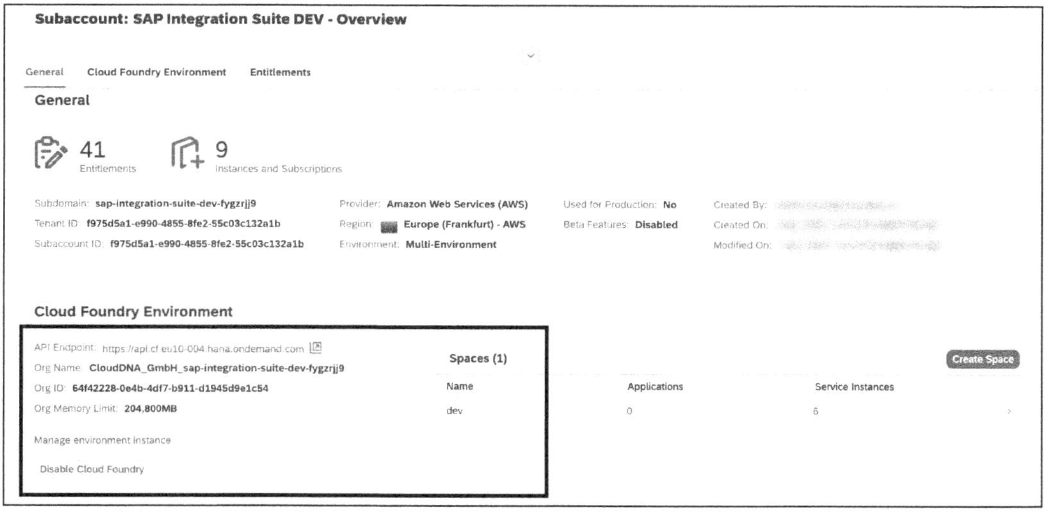

**Figure 8.7** Target Subaccount

Click on **Create Space**, give the space a name, and assign the necessary roles, as shown in Figure 8.8.

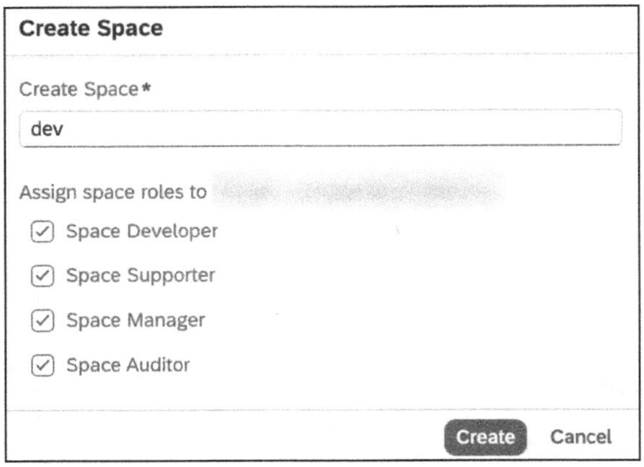

**Figure 8.8** Creating Space

In the next step, switch to the **Service Marketplace** of your **Cloud Transport Manager** subaccount to activate the actual SAP Cloud Transport Management there. Then, as shown in Figure 8.9, search for "Transport" and select **Cloud Transport Management** from the list. A brief description of the service will be displayed on the right-hand side; this service enables the management and execution of software transports between subaccounts and environments (such as Neo, Cloud Foundry, and Kubernetes). It supports the transport of application artifacts and their associated content across different landscapes. Click **Create** to start the creation process.

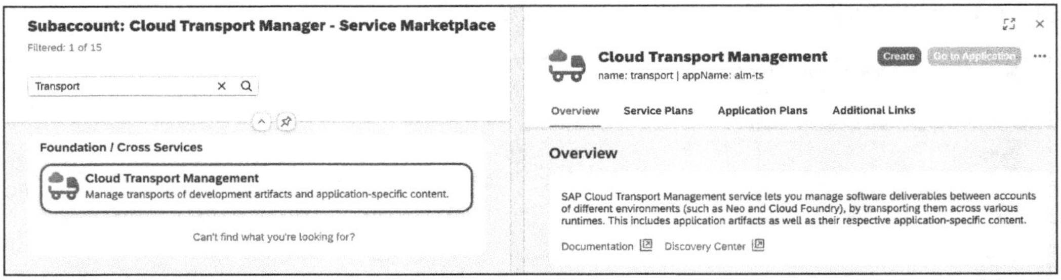

**Figure 8.9** Creating Cloud Transport Manager

In the next step, you create a new instance of the SAP Cloud Transport Management you previously selected. In the **New Instance or Subscription** dialog, select **Cloud Transport Management** as the **Service** again. For testing and development purposes, you'll opt for the **lite Plan** here. This offers basic functionality with restrictions: the maximum file size is limited to 500 MB, and artifacts are only stored for one week. Nevertheless, this plan is perfectly adequate for many use cases, especially for getting to know and trying out the service.

Click **Create** to create the instance and activate SAP Cloud Transport Management in the subaccount. You can then define transport nodes and configure your transport setup (see Figure 8.10).

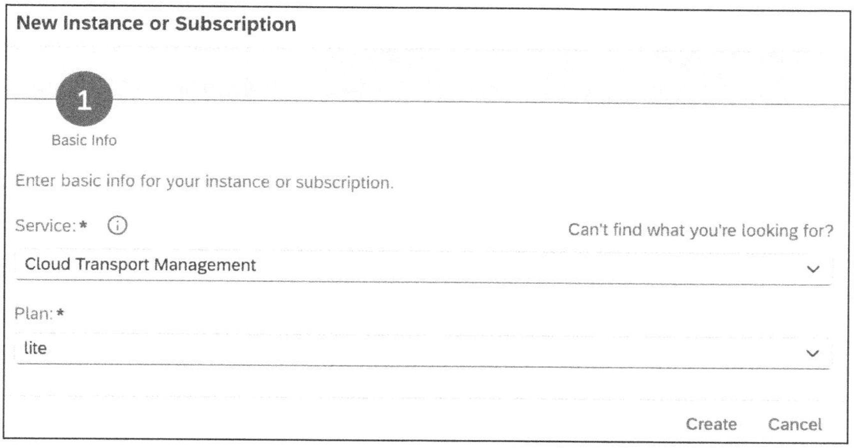

**New Instance or Subscription**

**1**

Basic Info

Enter basic info for your instance or subscription.

Service: *   ⓘ                                                          Can't find what you're looking for?

Cloud Transport Management                                                                    ∨

Plan: *

lite                                                                                          ∨

Create    Cancel

**Figure 8.10** Creating Cloud Transport Manager Application

Then, the subscription will appear under the **Instances and Subscriptions** menu item. As shown in Figure 8.11, the service will have been successfully subscribed and the status will be **Subscribed**. This will ensure that SAP Cloud Transport Management is active in the subaccount and can be used. On the right-hand side, you'll also find technical details such as the name of the application, the creation date, and further links to documentation and SAP Discovery Center.

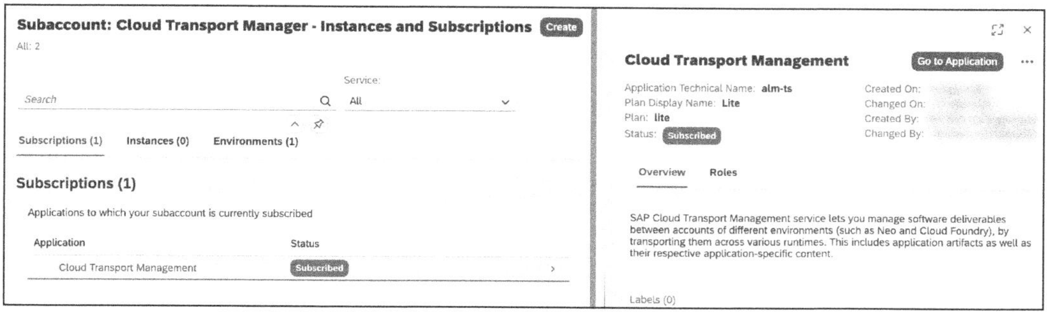

**Subcount: Cloud Transport Manager · Instances and Subscriptions**  Create

All: 2

Search                                           Q    Service:  All   ∨

Subscriptions (1)    Instances (0)    Environments (1)

**Subscriptions (1)**

Applications to which your subaccount is currently subscribed

Application                        Status

Cloud Transport Management          Subscribed                        >

**Cloud Transport Management**                            Go to Application   ···

Application Technical Name: **alm-ts**          Created On:
Plan Display Name: **Lite**                     Changed On:
Plan: **lite**                                  Created By:
Status: Subscribed                              Changed By:

Overview    Roles

SAP Cloud Transport Management service lets you manage software deliverables between accounts of different environments (such as Neo and Cloud Foundry), by transporting them across various runtimes. This includes application artifacts as well as their respective application-specific content.

Labels (0)

**Figure 8.11** Created Cloud Transport Manager

Next, you have to assign permissions for the Cloud Transport Manager subaccount. To do this, navigate to **Security • Users** in the left-hand menu, where you'll see all users who are assigned to the **Cloud Transport Manager** subaccount. As shown in Figure 8.12, select the user you want to make detailed information about that user appear on the right-hand side. In the lower section of the user information, you'll find the **Role Collections** area.

381

To assign the necessary permissions to the user, click on the three-dot menu button on the right in the **Role Collections** area and then select **Assign Role Collection**. This will open the dialog for assigning predefined permission groups that enable access to the functions of SAP Cloud Transport Management.

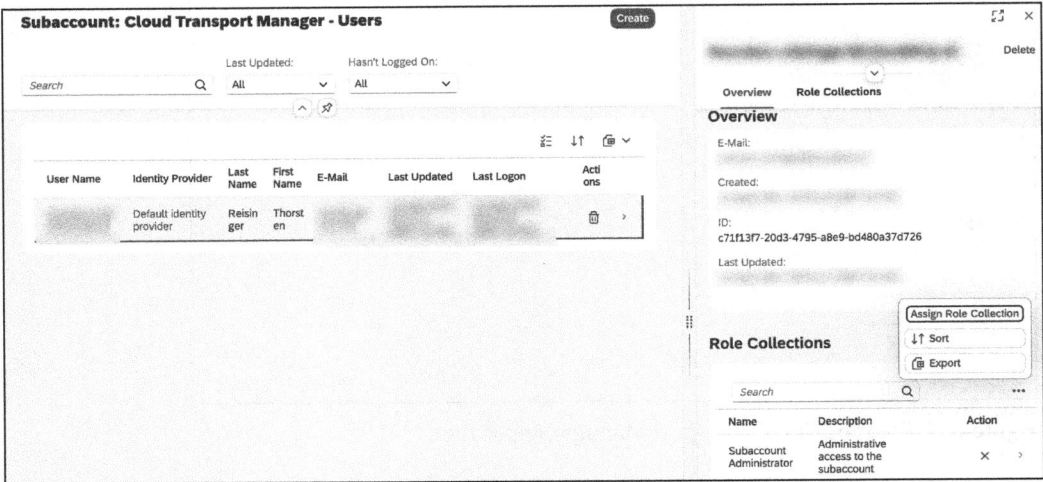

**Figure 8.12** Navigate Assigning Role Collections

Here, as shown in Figure 8.13, you check the box for the TMS_LandscapeOperator_RC role. It contains the TMS LandscapeOperator authorization, which allows the user to manage the transport landscape, define transport nodes, and configure target systems.

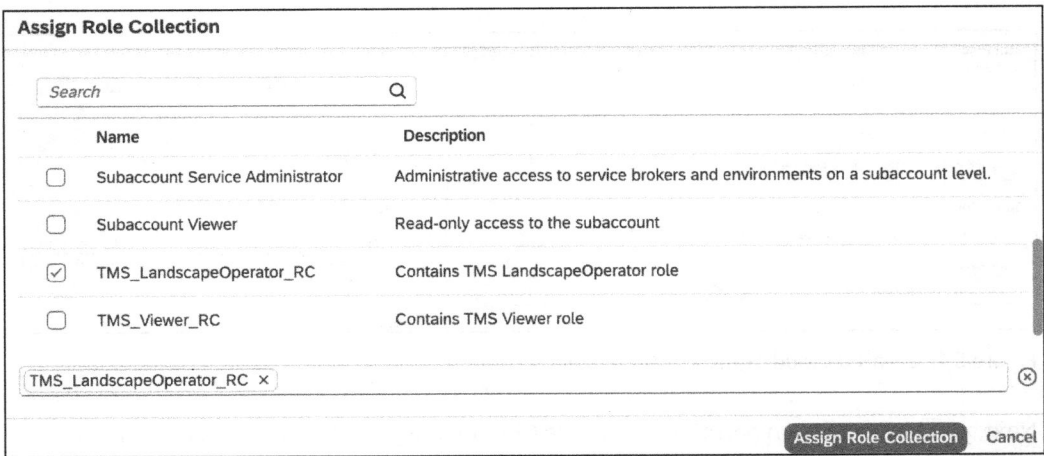

**Figure 8.13** Assing TMS_LandscapeOperator_RC

Next, you define an HTTP destination in the Cloud Transport Manager subaccount that points to your target system, which is the subaccount to which you want to transport software artifacts in the future. Go **to Connectivity • Destinations**, click on **Create Destination**, and choose a name for that destination (see Figure 8.14). Select **HTTP for the Type,** and in the **URL** field, you need to include the region, org name, and space name that you looked up in Figure 8.7. Structure the URL as follows: *https://deploy-service. cfapps.<region>.hana.ondemand.com/slprot/<org-name>/<Space>/slp*. Replace *<region>* with the region you previously determined (for example, "eu10"), *<org-name>* with the org name you previously determined, and *<Space>* with the space you previously determined.

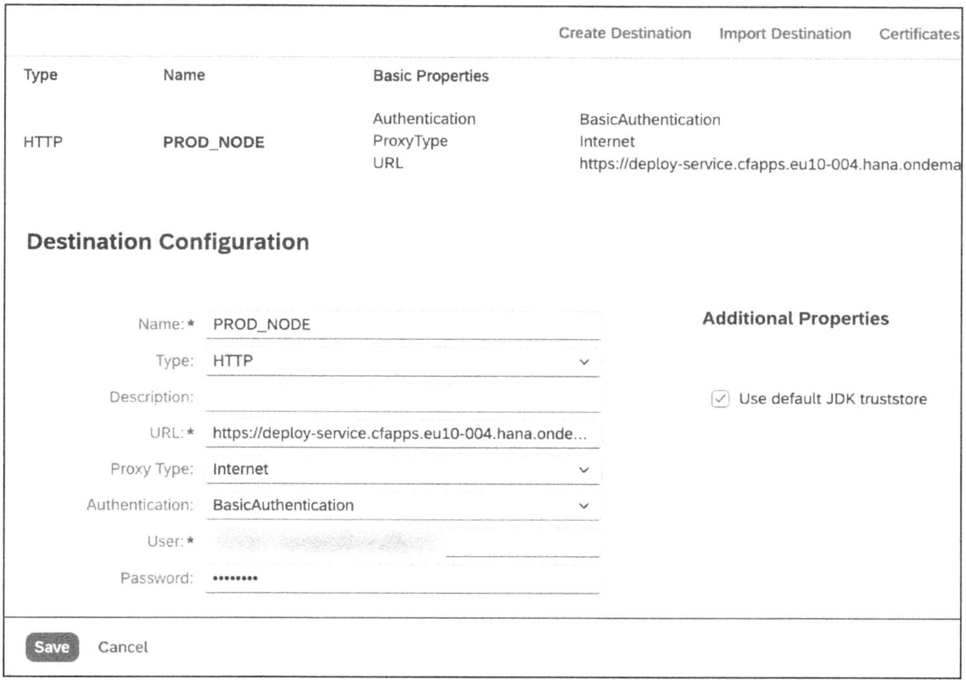

**Figure 8.14** Creating Target Subaccount Destination

Once you've fully configured and saved PROD_NODE as the HTTP destination, the next step is to test the connection (see Figure 8.15). To do this, click the **Check Connection** button, which should elicit a response with the status **200: OK.**

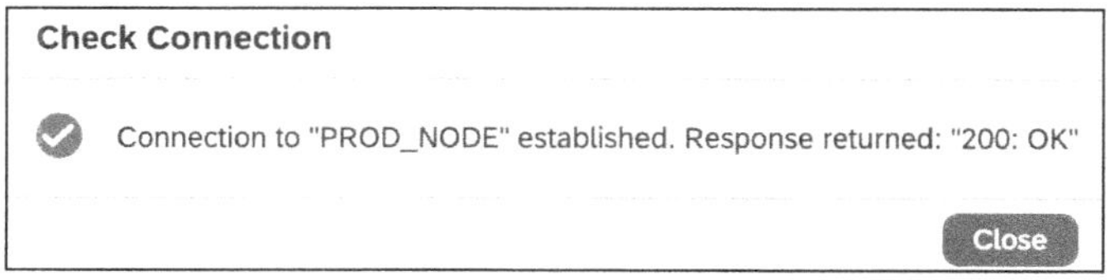

**Figure 8.15** Checking PROD_NODE Connection

## 8.1.2   Defining the Landscape

Next, open SAP Cloud Transport Management by clicking on **Instances and Subscriptions** on the left-hand side (see Figure 8.16) and then clicking on the **Cloud Transport Management** application.

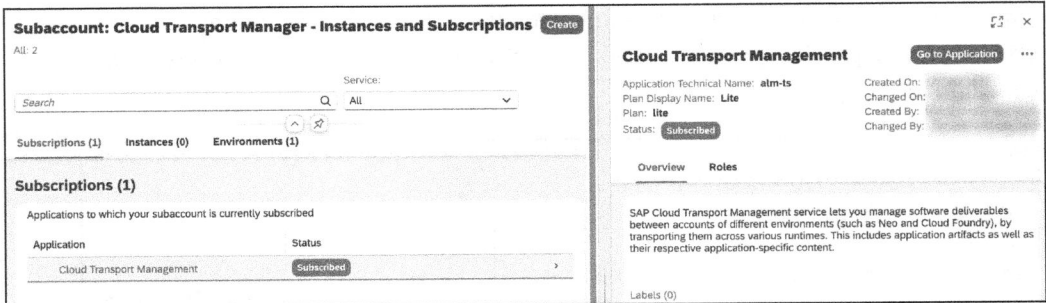

**Figure 8.16**  Opening Cloud Transport Management

Then, you'll be guided through the **Transport Landscape Wizard**, which simplifies the setup process (see Figure 8.17). On the first screen, you'll be asked to select a template for your landscape structure. Since your use case involves a simple development-to-production flow, you should choose the **Two-Node Landscape** option. This allows you to define two systems (or *nodes*) between which transports will flow: typically, a source (e.g., development) and a target (e.g., production) node. If your landscape requires more stages (e.g., QA, staging), you can opt for a three-, four-, or five-node template. For highly customized scenarios, the wizard also offers you the option to define nodes and routes manually via the navigation panel on the left.

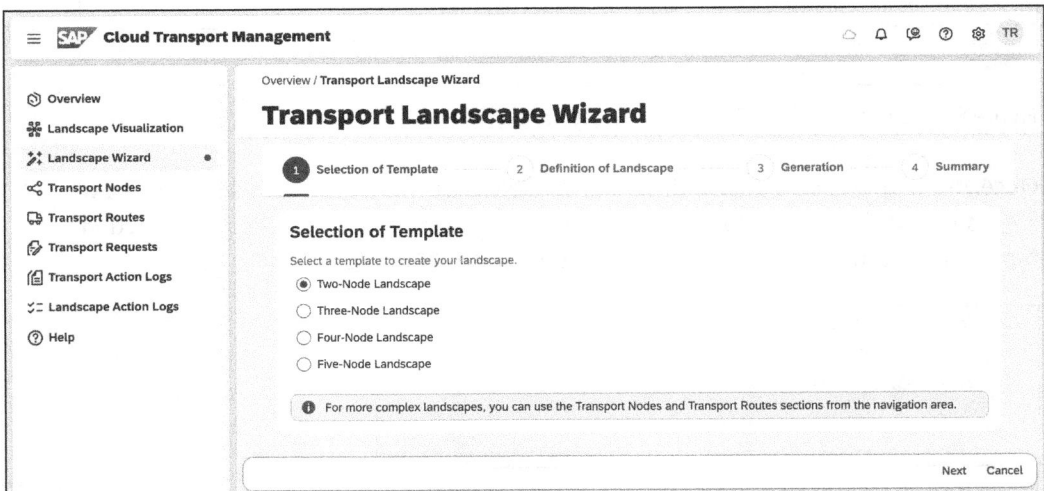

**Figure 8.17**  Landscape Wizard

After you select the landscape type, click on **Next** to proceed with the definition of the actual nodes.

Then, define the two transport nodes (see Figure 8.18), as follows:

- **Source: Node 1**
  - Name
    Enter "Dev" as the name of the source node.
  - Description
    Enter "Development" for the description of the source node.
  - **Allow Upload to Node**
    Check this box to allow content to be uploaded directly to this node.
  - Forward Mode
    Select **Auto** to ensure that transports are automatically forwarded to the next node when approved.
  - Content Type
    Leave this empty here because it's typically set on the target node.
  - Destination
    Leave this unassigned since the source node typically doesn't require a destination.
- **Target: Node 2**
  - Name
    Enter "Prod" as the name of the target node.
  - Description
    Enter "Production" as the description of the target node.
  - **Allow Upload to Node**
    Check this box if content needs to be uploaded directly, though this is less common for production.
  - Forward Mode
    Select **Auto.**
  - Content Type
    Select **Multi-Target Application,** which is important for transporting MTA artifacts.
  - Destination
    Select **PROD_NODE**, which is the HTTP destination you created earlier that points to the target subaccount.

Finally, assign a **Name** and **Description** to the transport route that connects the development and production nodes and then click **Next** to continue with landscape generation (see Figure 8.19).

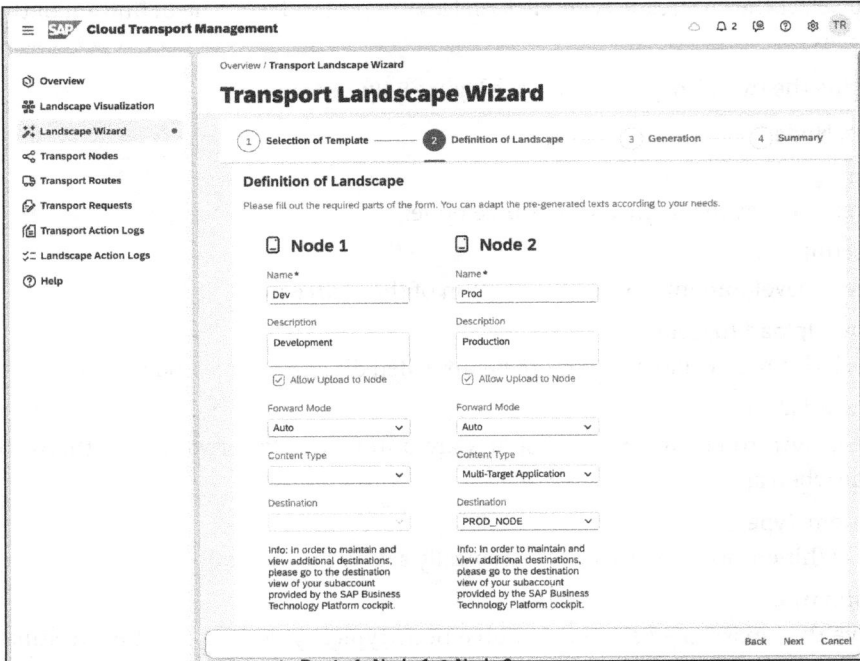

**Figure 8.18**  Defining Transport Nodes in Landscape Wizard

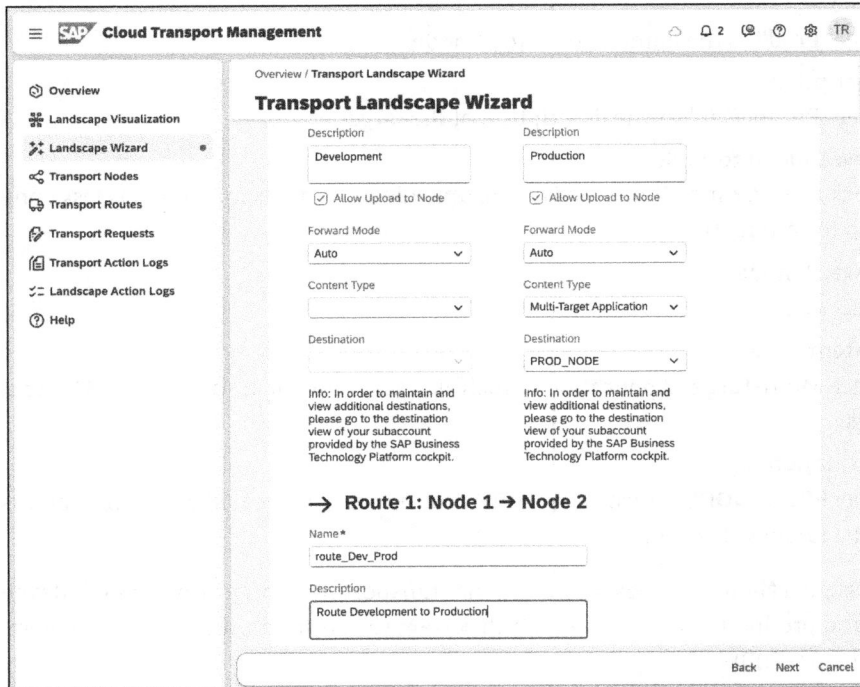

**Figure 8.19**  Assigning Route Name in Landscape Wizard

You can review details by expanding the **Show more** sections, and then, you can click **Finish** to complete the setup (see Figure 8.20).

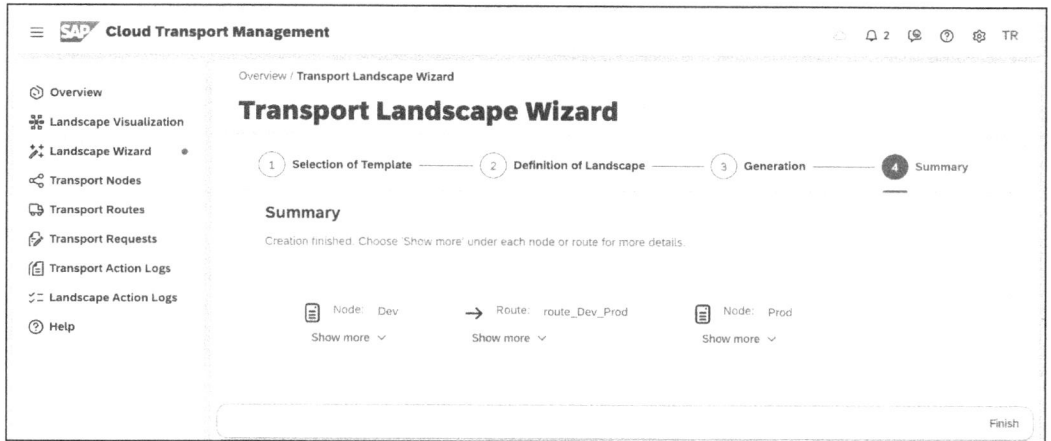

**Figure 8.20**  Created Landscape

### 8.1.3   Creating SAP Content Agent

In the next step, you have to create the SAP Content Agent service within the Cloud Transport Manager subaccount. To do this, navigate to the **Service Marketplace** of the subaccount and search for "Content Agent Service" (see Figure 8.21). This service facilitates generic content management operations such as viewing, exporting, and transporting content within the SAP Cloud Transport Management framework.

On the configuration screen, enter or select the required details:

- **Plan**
  Select **standard**.
- **Runtime Environment**
  Select **Cloud Foundry**.
- **Space**
  Select **dev**.
- **Instance Name**
  Enter "cas" as the instance name.

Then, navigate to the **Instances and Subscriptions** section of the Cloud Transport Manager subaccount, locate your newly created instance in the **Instances** tab, and select the instance to open its details view (see Figure 8.22). There, you can click the **Create button** in the **Service Keys** section.

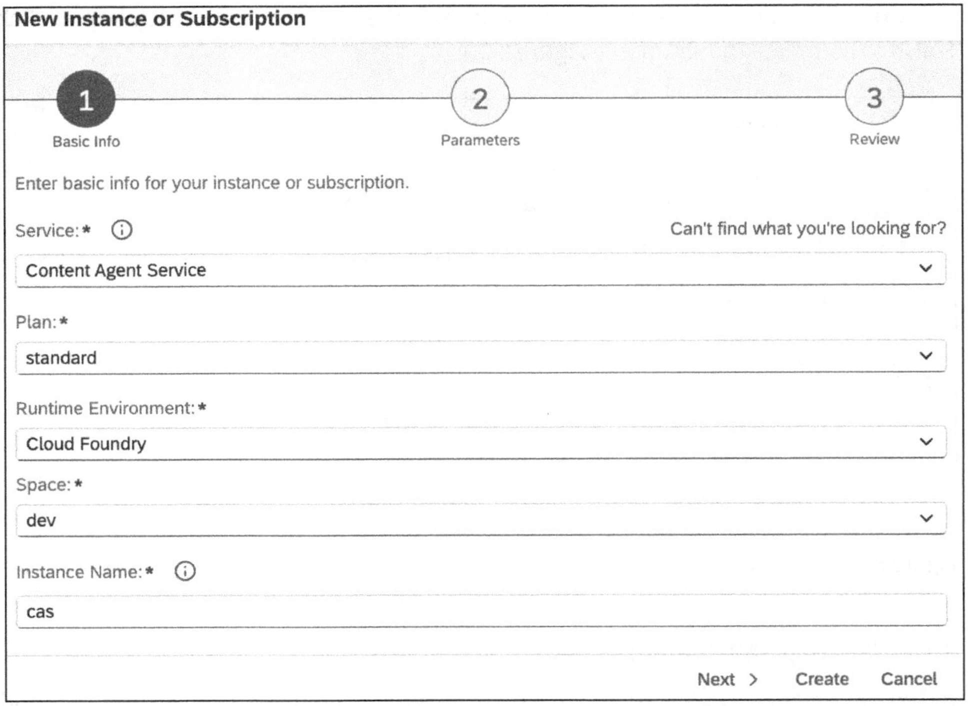

**New Instance or Subscription**

① Basic Info         ② Parameters         ③ Review

Enter basic info for your instance or subscription.

Service: *  ⓘ                                    Can't find what you're looking for?

Content Agent Service                                                  ⌄

Plan: *

standard                                                              ⌄

Runtime Environment: *

Cloud Foundry                                                        ⌄

Space: *

dev                                                                  ⌄

Instance Name: *  ⓘ

cas

Next  >      Create      Cancel

**Figure 8.21** Creating SAP Content Agent Service

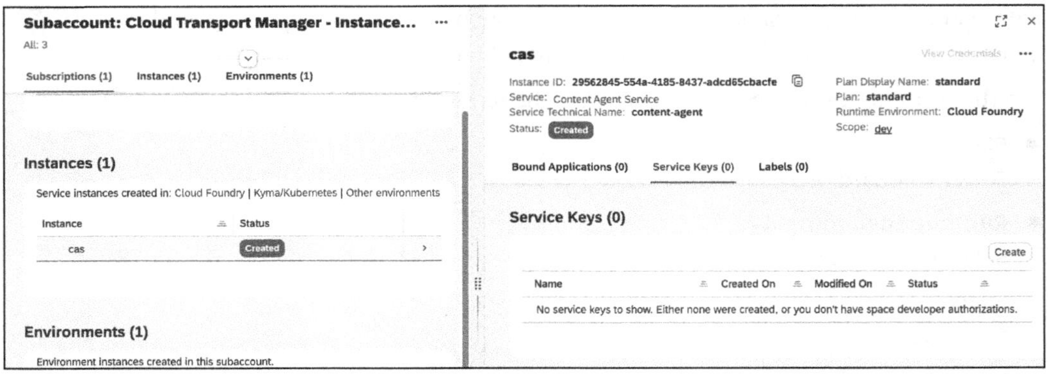

**Figure 8.22** Navigating to SAP Content Agent Service Instance

A *service key* provides the necessary credentials for external tools or services (like SAP Cloud Transport Management) to access SAP Content Agent securely. In the **New Service Key** dialog, you can enter a **Service Key Name (in this case, "cas-sk")**. Leave the configuration parameters empty (**{}**) because no additional settings are required in this case, and then, simply click **Create** to generate the service key (see Figure 8.23).

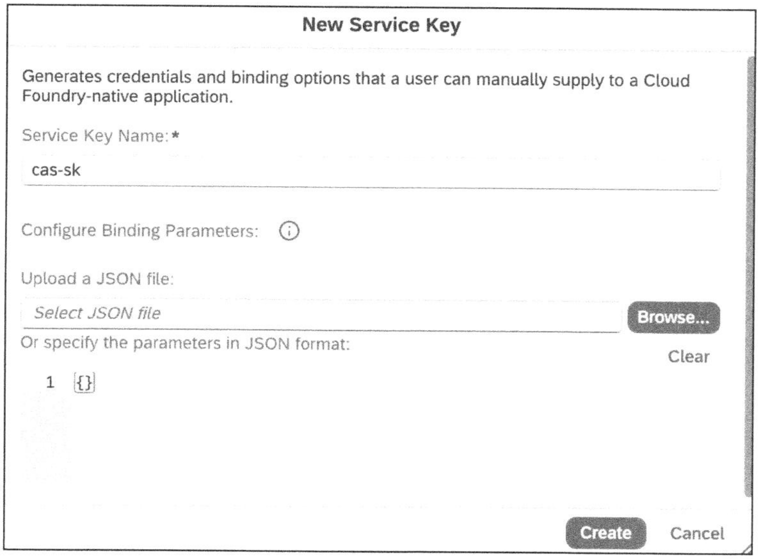

**Figure 8.23**  Creating SAP Content Agent Service Service Key

Once you've created the service key, you can click on it to view and copy the **Credentials**, which include important information such as the `clientid`, `clientsecret`, `url`, and other authentication parameters (see Figure 8.24). You'll need to use these credentials later when configuring a destination with SAP Cloud Transport Management.

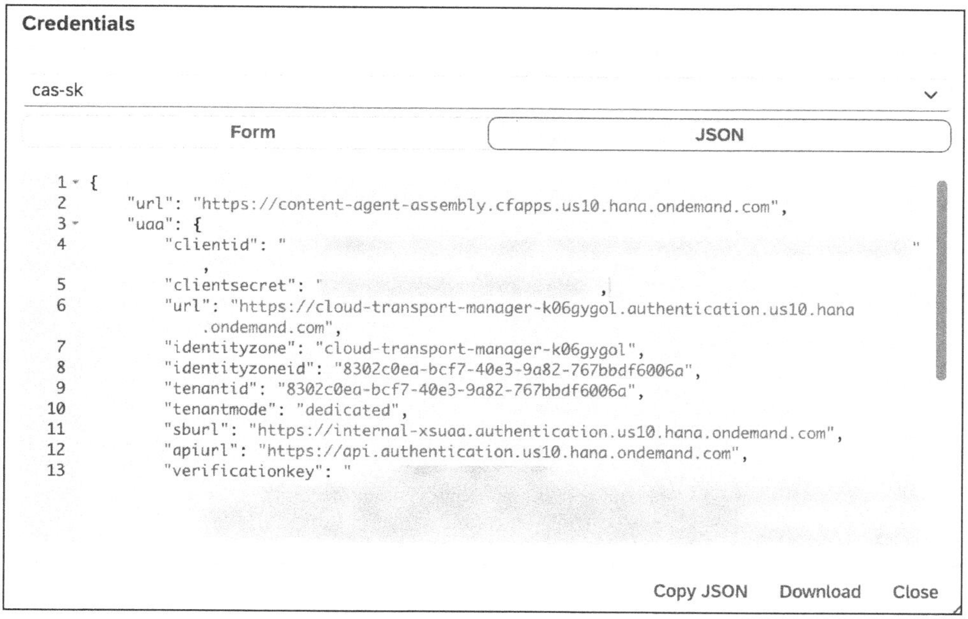

**Figure 8.24**  Copying SAP Content Agent Service Service Key with Client Secret ClientID

> **Important**
> Make sure to copy the entire JSON content and save it securely in a separate document. You'll need it in a later step to create a destination. Use the **Copy JSON** or **Download** button to easily export the data.

Next, you need to create a destination for the **ContentAssemblyService** in the subaccount where the development Integration Suite is located—that is, the one where you want to start your transportation. Begin by navigating **to Connectivity • Destinations** and clicking on **Create Destination**. Fill in the destination configuration by entering data and selecting options as follows (see Figure 8.25):

- **Name**
  Enter "ContentAssemblyService" (this *exact name* is required for SAP Cloud Transport Management to recognize and use the destination).

- **Type**
  Select **HTTP**.

- **Description**
  You can enter a description like "CI Content Assembly Service" (this is optional).

- **URL**
  Enter the main URL from the service key you saved earlier. This is the value in the top-level "url" field in the JSON (not from the **uaa** section).

- **Proxy Type**
  Select **Internet**.

- **Authentication**
  Select **OAuth2ClientCredentials**.

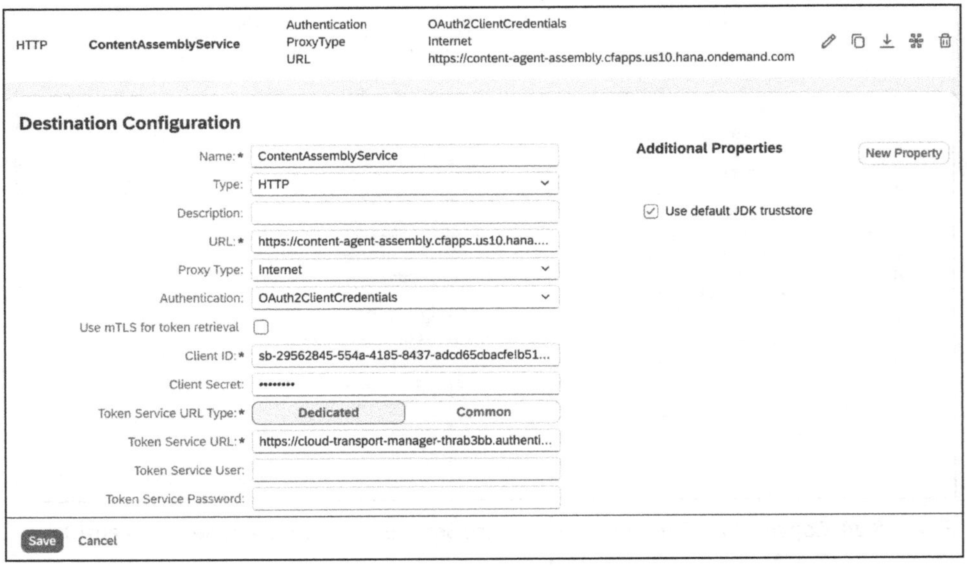

**Figure 8.25** Creating ContentAssemblyService Destination

Then, fill in the OAuth2 credentials with values from the **uaa** section of the saved service key:

- **Client ID** and **Client Secret**
Copy these values from the **clientid** and **clientsecret** fields in the **uaa** section of the service key.

- **Token Service URL**
Take the **url** value from the **uaa** section and append */oauth/token* to it.

Once you've filled everything in, click **Save** to create the **ContentAssemblyService** destination. Then, click on **Check Connection**. That should bring up a response of **401: Unauthorized** (see Figure 8.26).

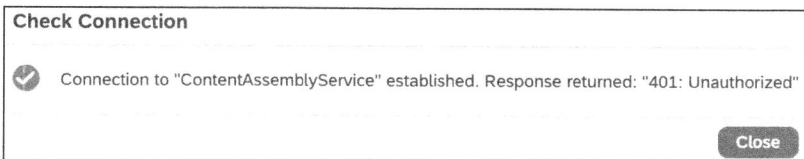

Figure 8.26  Checking ContentAssemblyService Connection

### 8.1.4   Adding SAP Process Integration Runtime

Staying in the Cloud Integration development subaccount, you now need to create an instance of the **SAP Process Integration Runtime** service with the **api** plan. To do this, go to **Service Marketplace**, select **SAP Process Integration Runtime**, and fill in the required fields, including **Instance Name** (see Figure 8.27). Then, click on **Next**.

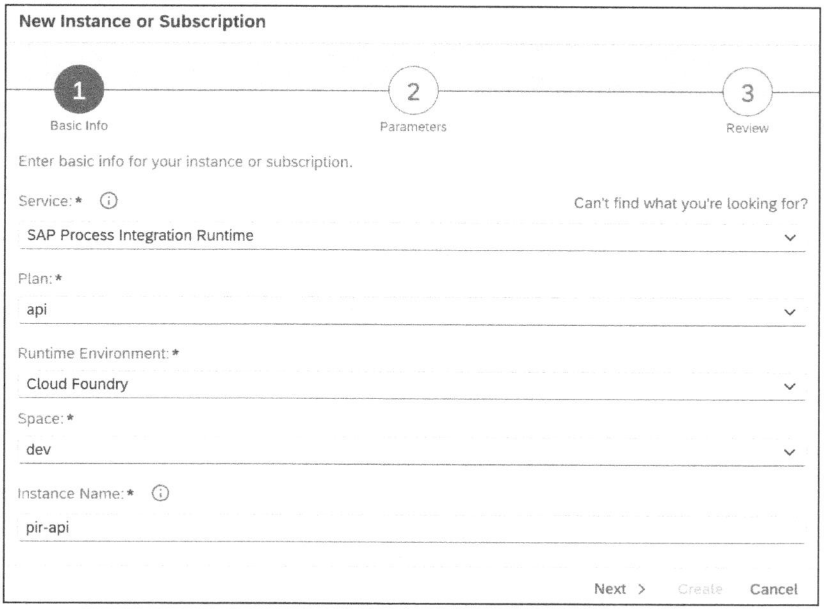

Figure 8.27  Creating SAP Process Integration Runtime API

Next, on the screen shown in Figure 8.28, you configure the parameters for the SAP Process Integration runtime instance. It's important to assign the correct role and authentication type, as follows:

- **Roles**
  Select the **WorkspacePackagesTransport** role to allow the service instance to handle package transport operations via SAP Cloud Transport Management.

- **Grant-types**
  Choose **Client Credentials** because the service will authenticate via OAuth using a client ID and secret.

- **Access Token Validity (in seconds)**
  You can typically leave the default value of 43,200 seconds (12 hours) unchanged, unless you need a shorter or longer validity period for your scenario.

After filling in these values, click **Next** or **Create** to finalize the instance setup.

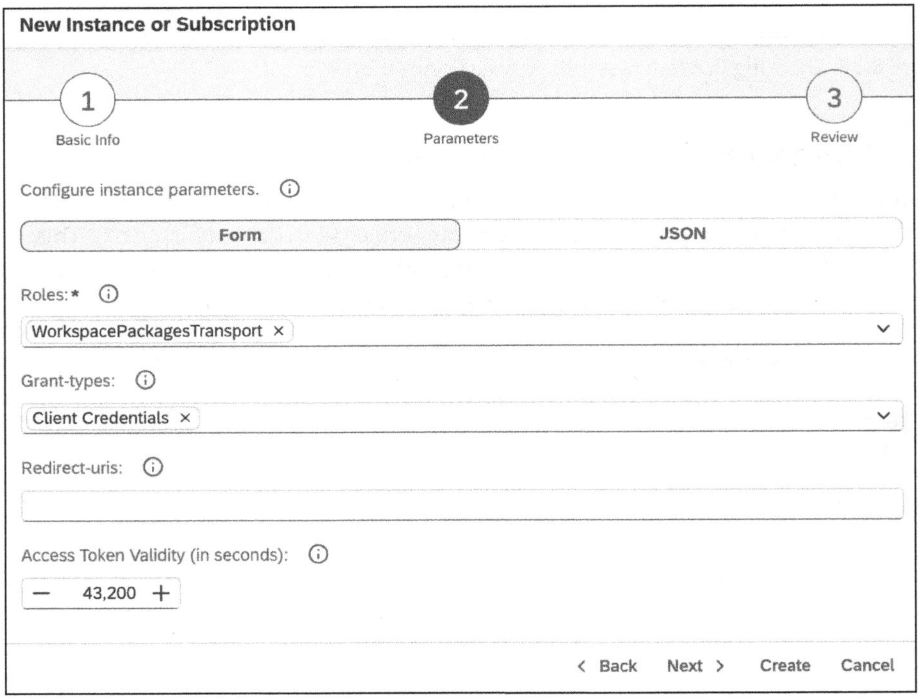

**Figure 8.28**  Assigning SAP Process Integration Runtime API Role

After you successfully create the SAP Process Integration runtime instance, you can open the instance in the **Instances and Subscriptions** section and click on **Create** in the **Service Keys** section (see Figure 8.29).

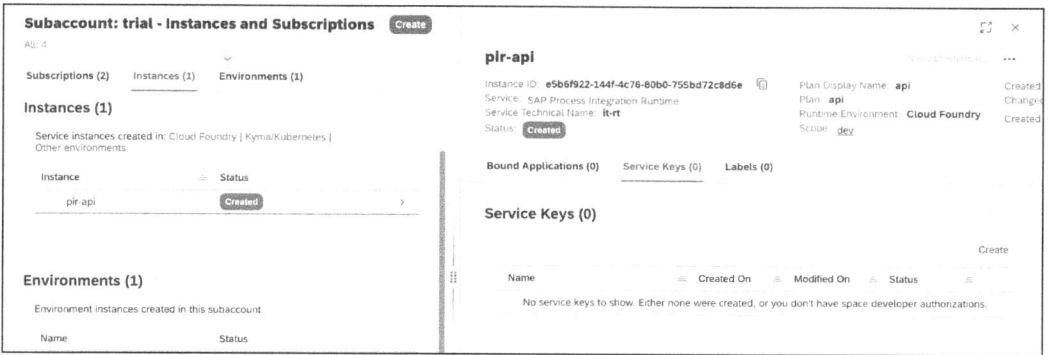

**Figure 8.29** Navigating SAP Process Integration Runtime API

Create the service key by making the following entries and selections (see Figure 8.30).

- **Service Key Name**

  Enter a unique name for the service key you want to create. You'll use this name for later identification within the subaccount, so ideally, it should reflect its intended use. In the example, we've entered the name "pir-api_sk" because it indicates that the key is intended for API communication with the SAP Process Integration runtime.

- **Configure Binding Parameters**

  This section allows you to configure additional parameters for the binding. You can choose between form-based and JSON-based input. Form input is usually sufficient for standard scenarios, while JSON offers additional flexibility for more complex configurations.

- **Key Type**

  Use this dropdown menu to select the type of authentication information to be generated. **ClientId/Secret** is selected by default, which means that client access is created with a client ID and secret.

- **External Certificate**

  You only fill in this field if you set the **Key Type** to **External Certificate**. If you did, you can upload a public certificate here to use it for authentication with the service.

- **Validity in days**

  If you chose **External Certificate** as the **Key Type**, specify how long the certificate should be valid. In this example, the default value is set to 365 days. This field isn't relevant for ClientId/Secret keys.

- **Key Size**

  If you chose **External Certificate** as the **Key Type**, define the key length you want to use when generating a certificate. Typical values are 2048 and 4096 bits—larger keys offer more security but also require more computing power.

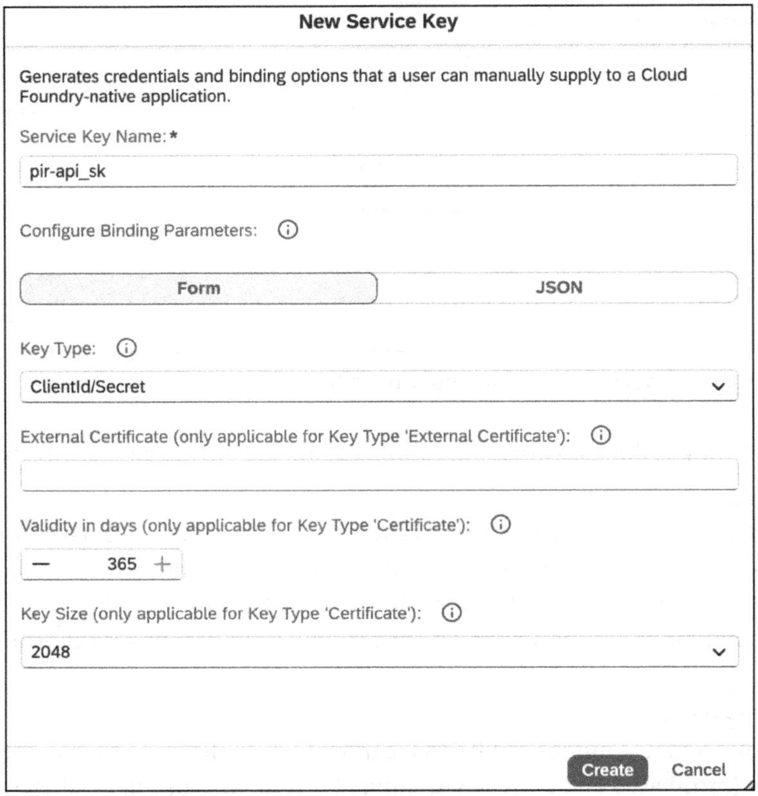

**Figure 8.30** Creating SAP Process Integration Runtime Service Key

Then, copy the **JSON** value of the service key (see Figure 8.31) and save it for later.

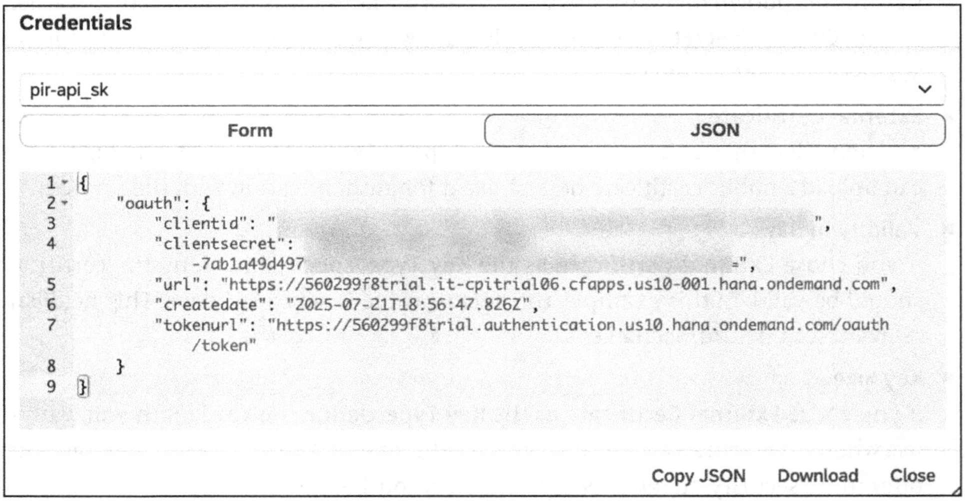

**Figure 8.31** Copying SAP Process Integration Runtime Service Key Credentials

Next, return to the Cloud Transport Manager subaccount and create a new destination by using the service key credentials from the SAP Process Integration runtime API instance. For this destination (see Figure 8.32), you'll enter and select the following:

- **Name**
  Enter "CloudIntegration" as the name of the destination.

- **Type**
  Select **HTTP**.

- **URL**
  Use the base URL from the service key and append "/api/1.0/transportmodule/ Transport" to it.

- **Authentication**
  Select **OAuth2ClientCredentials**.

- **Client ID/Client Secret**
  Copy it from the corresponding service key.

- **Token Service URL**
  Use the `tokenurl` from the service key.

Then, click **Save**.

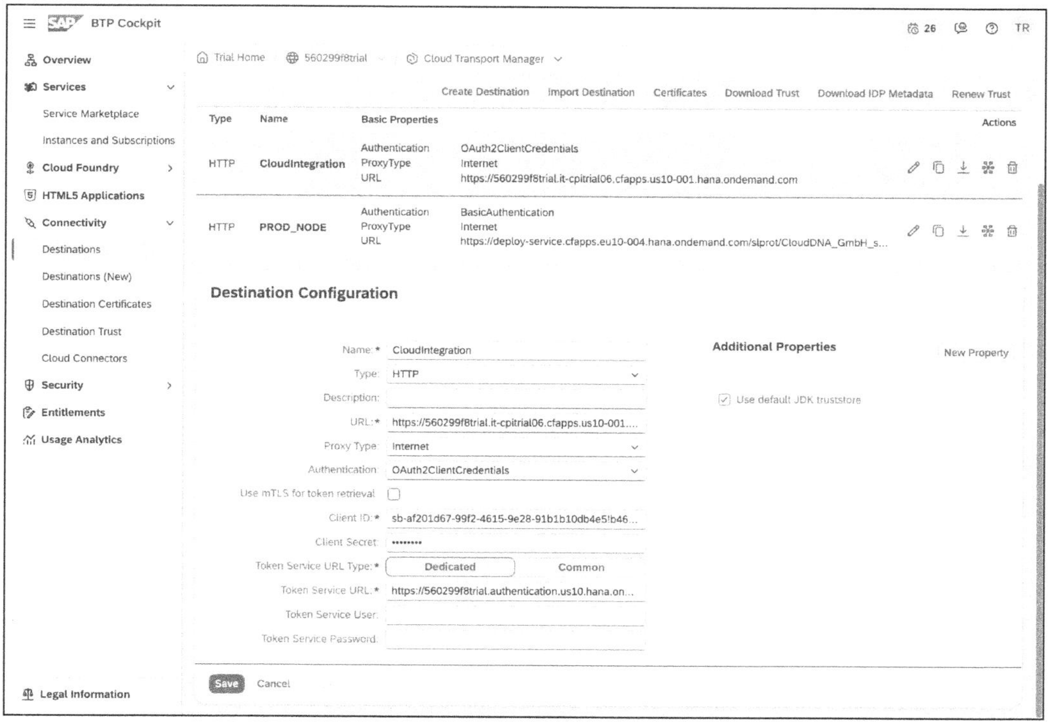

**Figure 8.32**  Creating CloudIntegration Destination

After that, click **Check Connection** to get the **401: Unauthorized** message, as shown in Figure 8.33.

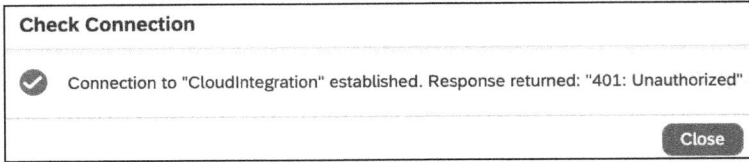

| Check Connection |
| --- |
| ✅ Connection to "CloudIntegration" established. Response returned: "401: Unauthorized" |
| Close |

**Figure 8.33** Checking CloudIntegration Connection

### 8.1.5   Setting Up an SAP Cloud Transport Management Instance and Destination

In the next step, you need to create a new instance of SAP Cloud Transport Management (see Figure 8.34) within the appropriate subaccount. To do this, navigate to the **Service Marketplace**, select **Cloud Transport Management**, choose the **standard** plan, set the runtime environment to **Cloud Foundry**, and select your space. For the instance name, enter a meaningful name (e.g., "ctm"). Then, click **Create** to create the instance.

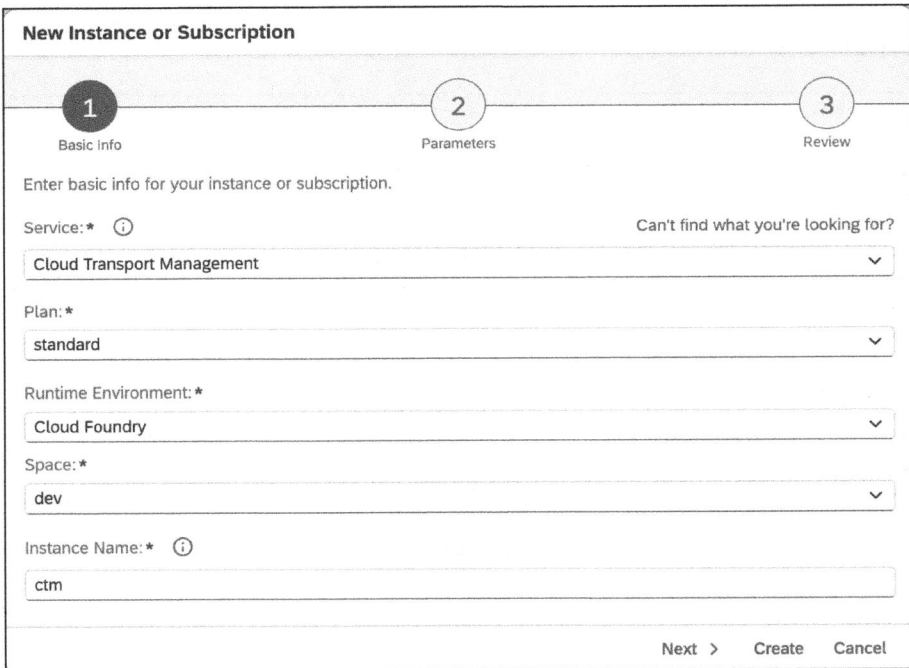

**New Instance or Subscription**

1 Basic Info     2 Parameters     3 Review

Enter basic info for your instance or subscription.

Service: *
Cloud Transport Management

Plan: *
standard

Runtime Environment: *
Cloud Foundry

Space: *
dev

Instance Name: *
ctm

Next >   Create   Cancel

**Figure 8.34** Creating Cloud Transport Manager Instance

Then, open the instance and create a service key for it as well. We previously covered how to create a service key in Section 8.1.3 (refer to Figure 8.23 in particular). Once you've done that, copy and save the service key (see Figure 8.35) so you can use it for your next destination.

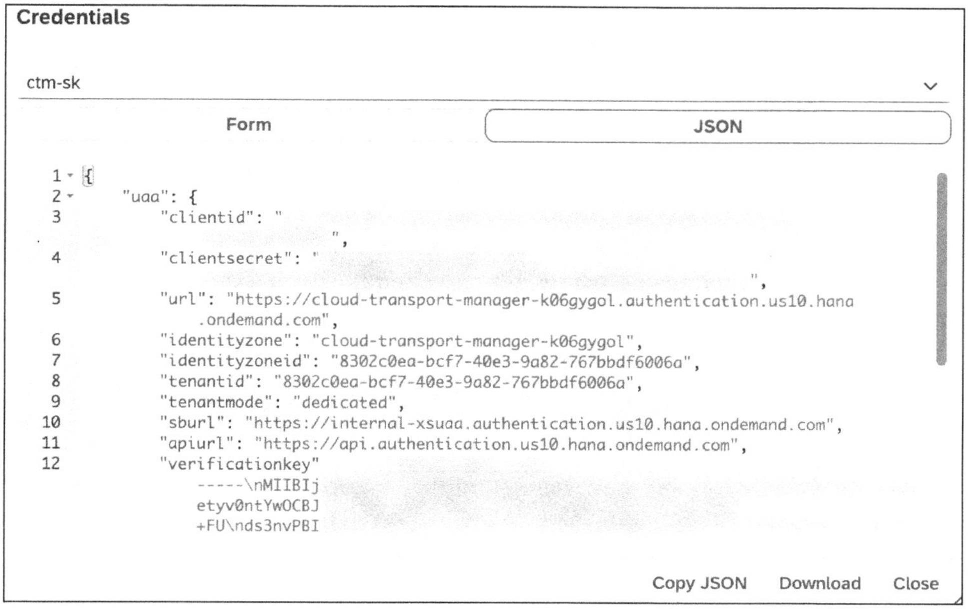

**Credentials**

ctm-sk

| Form | JSON |
|------|------|

```
1 ▾ {
2 ▾     "uaa": {
3           "clientid": "
                          ",
4           "clientsecret": '
                                                                ",
5           "url": "https://cloud-transport-manager-k06gygol.authentication.us10.hana
              .ondemand.com",
6           "identityzone": "cloud-transport-manager-k06gygol",
7           "identityzoneid": "8302c0ea-bcf7-40e3-9a82-767bbdf6006a",
8           "tenantid": "8302c0ea-bcf7-40e3-9a82-767bbdf6006a",
9           "tenantmode": "dedicated",
10          "sburl": "https://internal-xsuaa.authentication.us10.hana.ondemand.com",
11          "apiurl": "https://api.authentication.us10.hana.ondemand.com",
12          "verificationkey"
                -----\nMIIBIj
              etyv0ntYwOCBJ
              +FU\nds3nvPBI
```

Copy JSON     Download     Close

**Figure 8.35**  SAP Cloud Transport Management Service Key

Next, you need to create a new destination in the **Cloud Transport Manager** subaccount (see Figure 8.36). You must name the destination "TransportManagementService" *exactly* because that name is required for proper integration. You must also fill the **URL** field with the uri value from the SAP Cloud Transport Management service key (the one outside the uaa section).

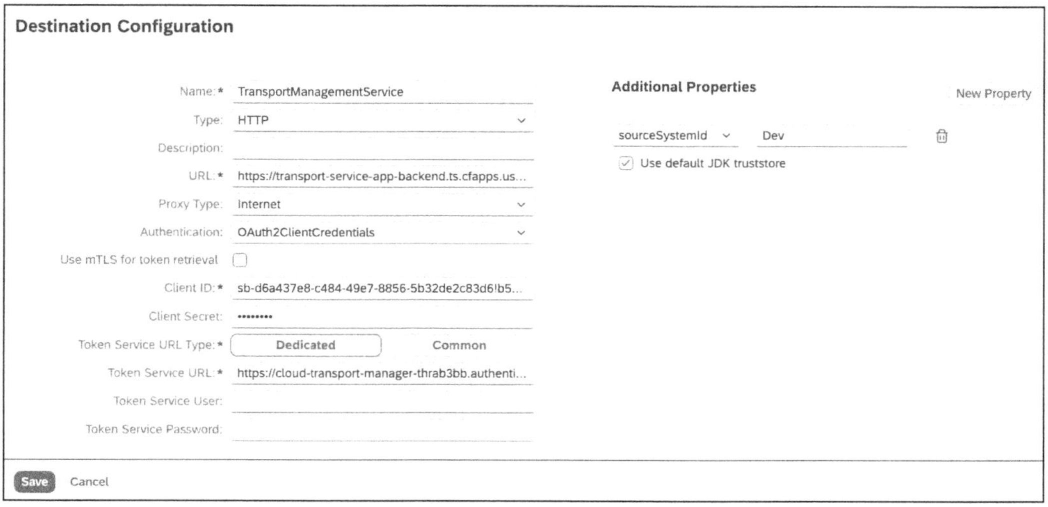

**Destination Configuration**

| | |
|---|---|
| Name: * | TransportManagementService |
| Type: | HTTP |
| Description: | |
| URL: * | https://transport-service-app-backend.ts.cfapps.us... |
| Proxy Type: | Internet |
| Authentication: | OAuth2ClientCredentials |
| Use mTLS for token retrieval | ☐ |
| Client ID: * | sb-d6a437e8-c484-49e7-8856-5b32de2c83d6!b5... |
| Client Secret: | •••••••• |
| Token Service URL Type: * | Dedicated    Common |
| Token Service URL: * | https://cloud-transport-manager-thrab3bb.authenti... |
| Token Service User: | |
| Token Service Password: | |

**Additional Properties**                                New Property

sourceSystemId ∨    Dev                    🗑

☑ Use default JDK truststore

**Save**   Cancel

**Figure 8.36**  Creating TransportManagementService Destination

For **Authentication**, select **OAuth2ClientCredentials**, and fill in the values for the **Client ID** and **Client Secret by taking them** from the same service key.

To fill in the **Token Service URL**, navigate to the uaa section of the service key, copy and paste the url value, and append "/oauth/token" to this URL to complete the token service endpoint. Then, set the **Token Service URL Type** to **Dedicated** and leave the **Token Service User** and **Token Service Password** fields blank.

Additionally, you must add a new property to the destination by clicking the **New Property** button and entering the name "sourceSystemId" for it *exactly*. Its value must match the name of the first node you defined in your Cloud Transport Manager landscape setup because that links the destination to the correct source system within the transport landscape. Then, click **Save** and **Check Connection** to get the **401: Unauthorized** message, as shown in Figure 8.37.

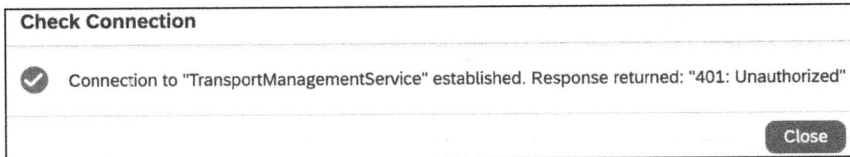

**Figure 8.37**  Checking TransportManagementService Connection

Once you've finished the setup, you need to verify that you've done the configuration correctly. To do this, navigate to your Cloud Integration development instance, go to **Settings** in the left-hand menu, select **Integrations**, and open the **Transport** tab. In the bottom right-hand corner, click **Edit** to enable the fields.

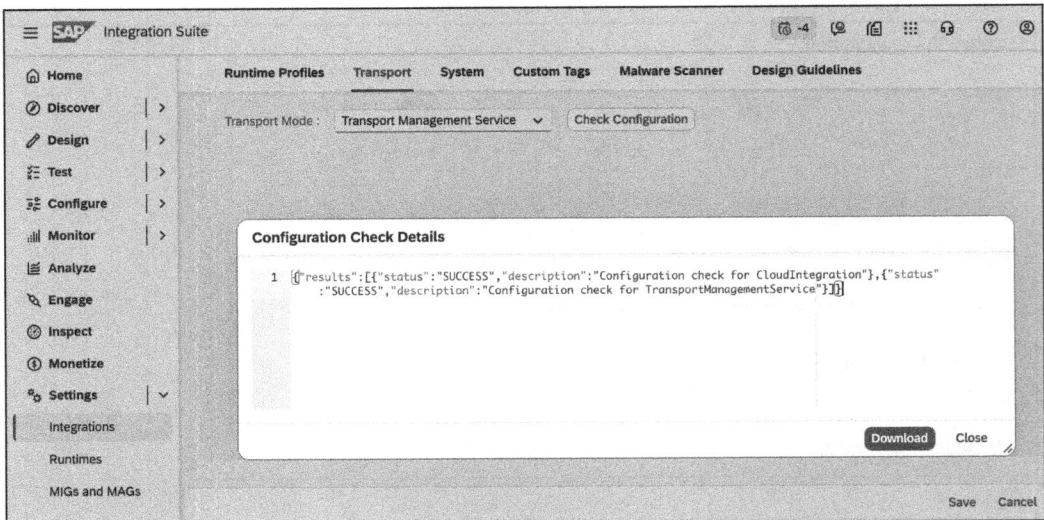

**Figure 8.38**  Testing

Then, choose **Transport Management Service** from the **Transport Mode** dropdown list and click the **Check Configuration** button. A popup will appear (see Figure 8.38) showing the results of the configuration check, and you should see "status": "SUCCESS" displayed for both CloudIntegration and TransportManagementService.

If both show success, click **Close** to exit the popup and then **Save** in the lower right-hand corner to finalize the configuration. This will confirm that the connection between Cloud Integration and SAP Cloud Transport Management is correctly established.

### 8.1.6   Assigning Role Collections and the Transport Integration Package

Next, you need to assign the appropriate roles to the user to ensure they have the necessary authorizations for transporting artifacts. Navigate to the **Role Collections** section in your development Cloud Integration subaccount, click on **Create** at the upper right to create a new role collection, and give it a meaningful name. Once you're inside, switch to **Edit** mode and click on the value help in the **Roles** section (see Figure 8.39).

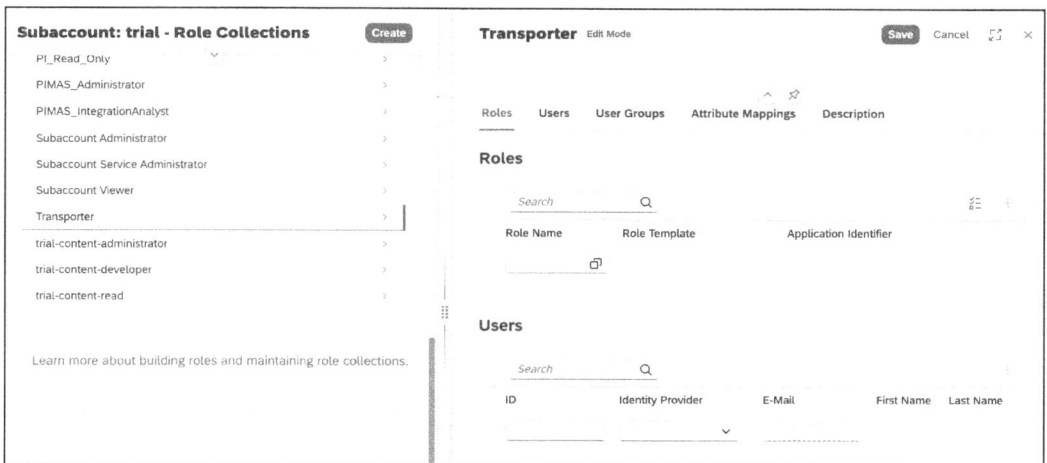

**Figure 8.39** Navigating Transporter Role Collection

Then, assign the **WorkspacePackagesTransport** role by searching for it and selecting it by clicking the box next to it. The role must be displayed on the left side of the **Selected Roles** section, and if it is, click on **Add** in the lower right-hand corner (see Figure 8.40).

Now that you've configured the **Transporter** role collection, you need to assign your user to it. In the **Users** section of the role collection, you click the **+** icon to add a new user, and after that, the user should appear in the list with their details filled in. Don't forget to click **Save** in the upper right-hand corner to apply the changes (see Figure 8.41).

**Figure 8.40** Assigning WorkspacePackagesTransport Role

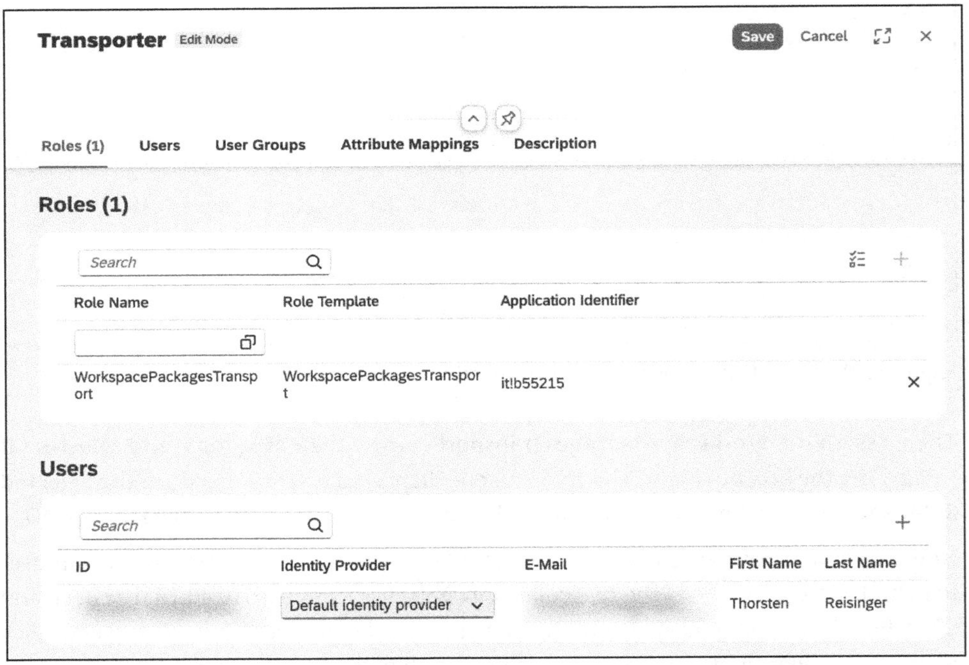

**Figure 8.41** Assigning User to Role Collection

Then, return to the Cloud Integration development subaccount and navigate to your Integration Package in the **Integration and APIs** section, where you should now see a

**Transport** button in the upper right-hand corner of the screen (see Figure 8.42). This will confirm that the transport functionality has been successfully activated for your user.

If the **Transport** button doesn't appear immediately, try accessing SAP Integration Suite in a private or incognito browser window. This often resolves caching or session-related issues. If the button still doesn't appear, make sure you've saved the Transport Manager configuration check as shown in Figure 8.38.

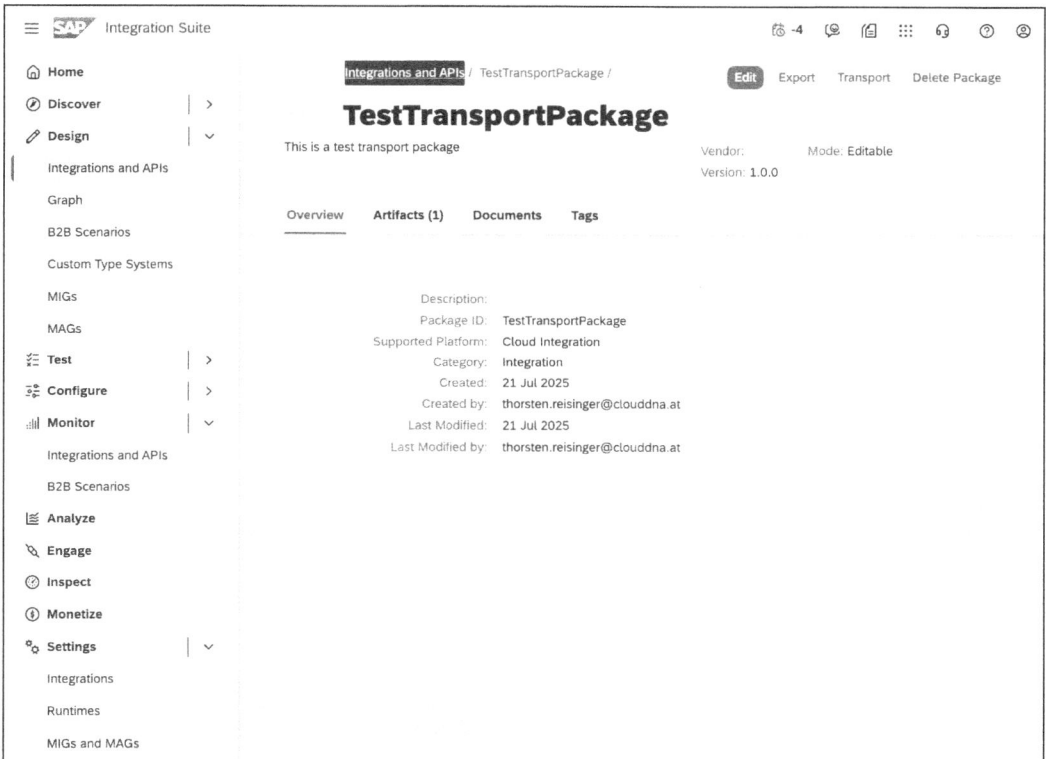

**Figure 8.42** Transport Integration Package

Next, click on the **Transport** button in the integration package view and the window shown in Figure 8.43 will appear. Here, you have to enter **Comments** describing the transport. The box to **Propagate logged-in user as transport owner** isn't checked by default, but if you check it, your user ID will appear as the initiator of the transport. Otherwise, a random GUID will be shown in SAP Cloud Transport Management.

To enable users to release transports, you need to create a new role collection and assign it the `ImportOperator` role. To do this, switch back to your Cloud Transport Manager subaccount, create a new role collection, add the `ImportOperator` role to it, and assign your user to it as well (see Figure 8.44).

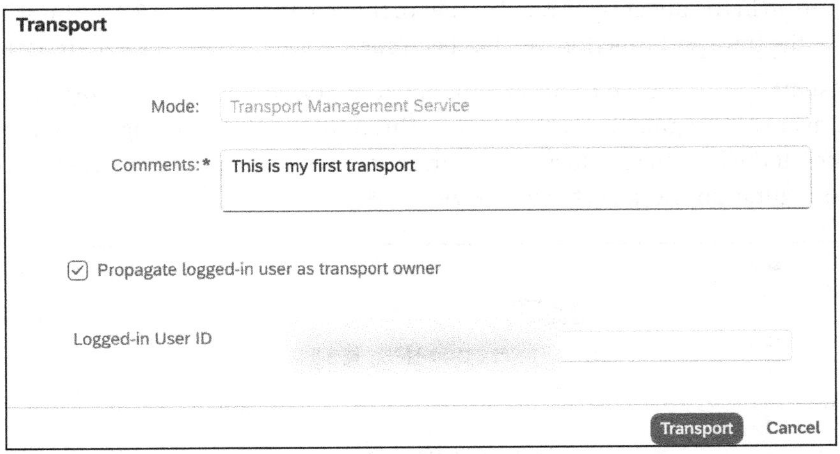

**Figure 8.43** Filled-Out Transport Menu

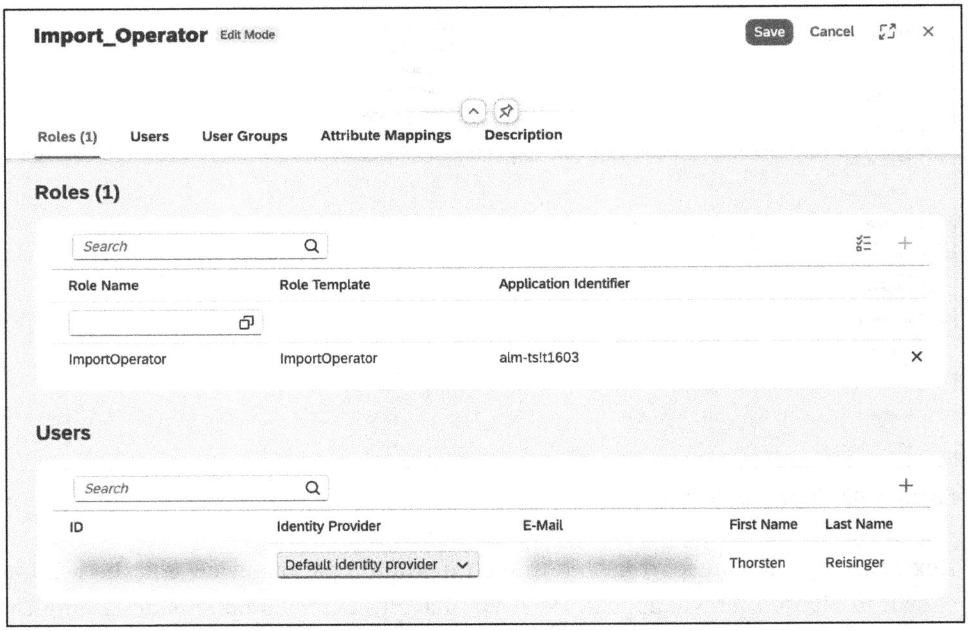

**Figure 8.44** Creating and Assigning SAP Cloud Transport Management Role Collection

Then, open SAP Cloud Transport Management from your subaccount. Once the application loads, you'll land on the overview screen, where you'll find a summary of the current transport activity. On the right-hand side, under **Top 1 Nodes**, you should see a list of nodes with pending transports. If your setup was completed correctly, you'll notice your target node (see Figure 8.45) displaying the number of transports waiting to be imported. This will confirm that the transport from your Cloud Integration has successfully reached SAP Cloud Transport Management.

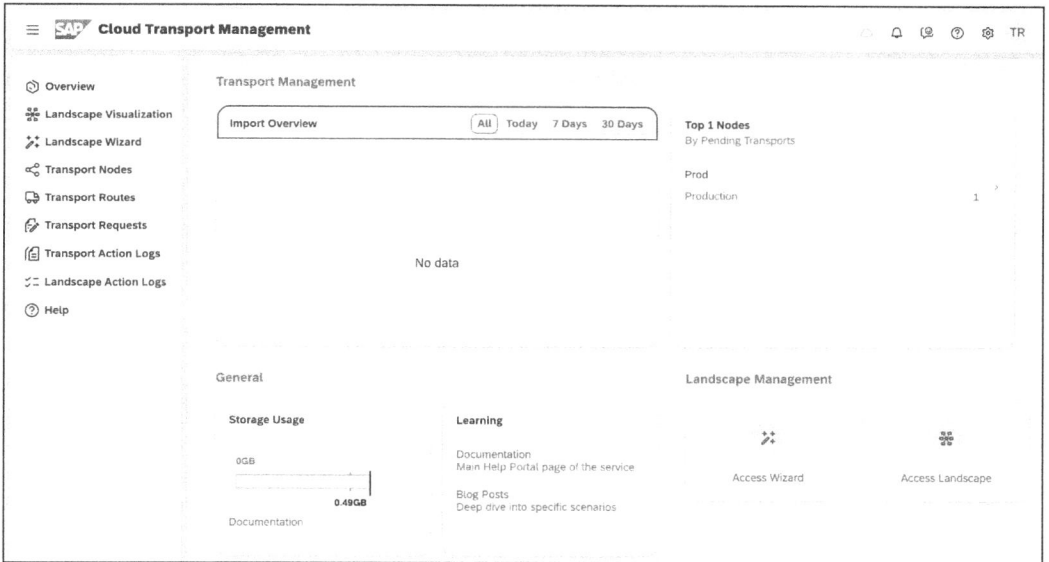

**Figure 8.45** Available Transport

Then, click on the target node and you'll be directed to the detailed view of that node. This screen will show the **Import Queue** with all transport requests that have been forwarded to this node but have not yet been imported (see Figure 8.46). There, you can see relevant details such as the transport request ID, description, owner, status, entry node, and timestamp. To import all listed transports at once, simply click on **Import All** on the right-hand side to trigger the import process for every request that is displayed in the node.

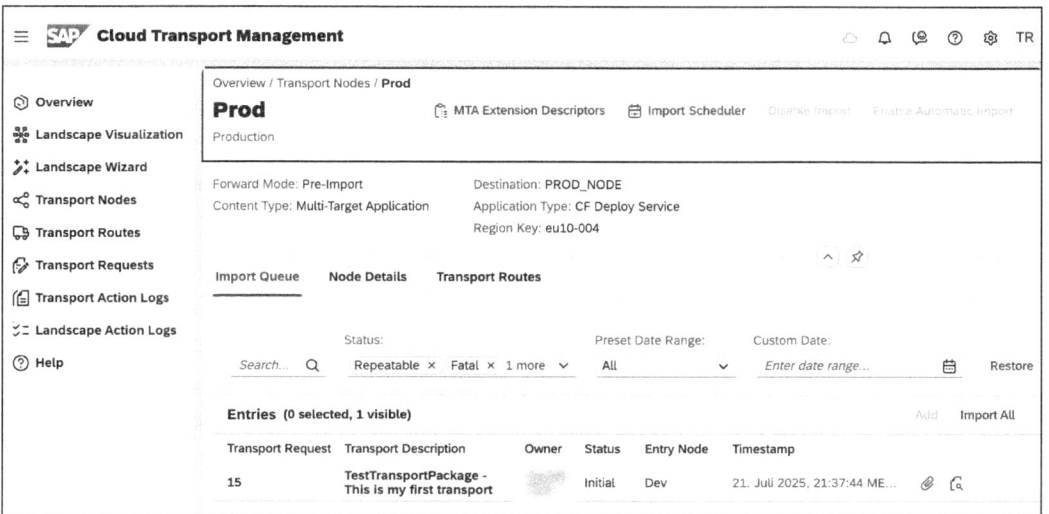

**Figure 8.46** Importing All Transport Requests

In Figure 8.47, you can see when you successfully process the transport, its status changes to **Succeeded to** indicate that everything was correctly configured and executed without errors. This confirms that the transport request was accepted and completed by the target node.

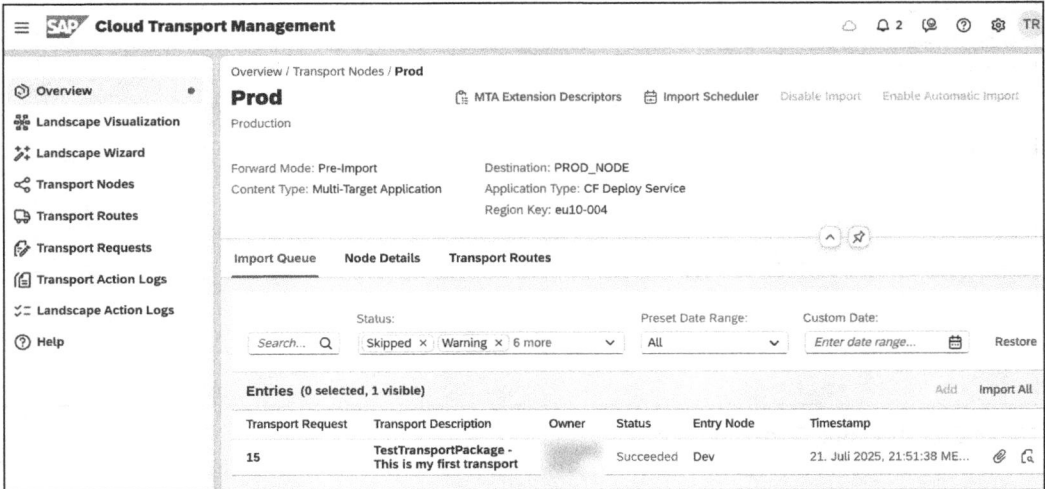

**Figure 8.47**  Transport Succeeded

Figure 8.48 shows the results of this process in the target Cloud Integration environment. The transported integration package appears in the design view, which means that the package has been fully delivered and is ready for use or further development in the destination system.

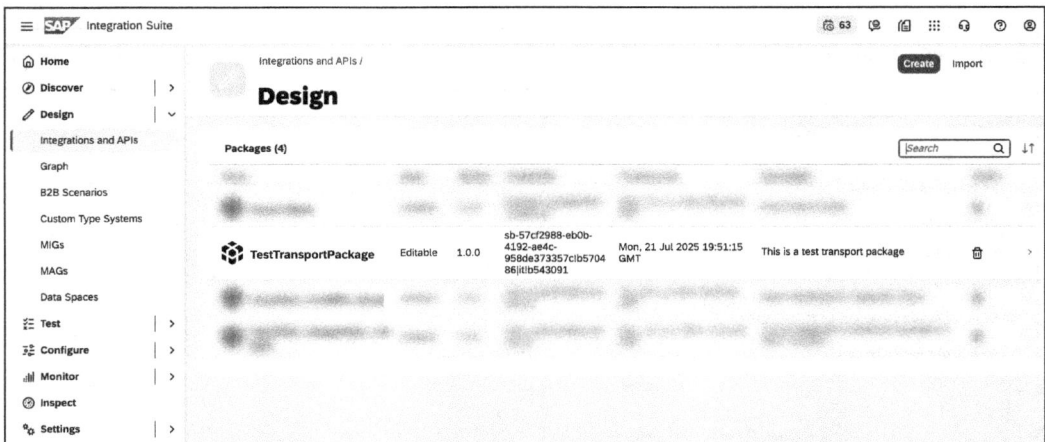

**Figure 8.48**  Transported Package

## 8.2   Version Controls and Deployment Strategies

Taking a robust and structured deployment approach is essential in any enterprise-grade integration project. When you're developing iFlows, value mappings, and other artifacts in Cloud Integration, it becomes increasingly important to establish a consistent and transparent versioning and deployment strategy. This not only ensures functional consistency across system landscapes but also supports team collaboration, lifecycle management, compliance, and rollback capability. This section introduces proven approaches to implementing version control and structured deployment in Cloud Integration projects.

Every development activity within Cloud Integration involves creating and modifying various types of artifacts—most notably, iFlows, message mappings, script collections, value mappings, and security artifacts. Each of these objects is saved and managed as a versioned entity within the tenant, and the versioning is performed automatically by the system whenever an artifact is saved or deployed.

Each version is timestamped and stored internally, and users can navigate through the version history, compare changes, and revert to previous versions if they need to. Teams can access this internal versioning mechanism directly within the web UI of Cloud Integration, and in practical terms, it means teams can roll back to a known-good configuration if a new version introduces errors or fails during runtime.

In addition to using the built-in version tracking, we encourage developers to adopt semantic versioning principles, which involve distinguishing among major, minor, and patch-level changes. Here are some examples of what the different changes do:

- A major version change indicates breaking changes in logic or structure.
- A minor version change reflects added functionality that remains backward compatible.
- A patch change resolves bugs or minor configuration updates without altering business logic.

By implementing semantic versioning as part of internal documentation and deployment policies, teams can maintain greater clarity across development and operations.

A key pillar of any deployment strategy is the separation of system environments, typically along the lines of development (DEV), quality assurance (QA), and production (PROD). All of these stages exist as separate subaccounts within SAP BTP, often with different configurations, user roles, connectivity options, and runtime limits.

To support this multienvironment setup, Cloud Integration allows artifacts to be exported from one environment and imported into another. This transportability becomes the technical foundation for managing lifecycle transitions. However, instead of relying on manual export-import actions, enterprises are strongly advised to utilize SAP Cloud Transport Management to automate, control, and audit these deployments (as described in Section 8.1).

An important consideration during deployment is how to handle environment-specific configurations such as URLs, credentials, endpoint selectors, and authentication logic. SAP addresses this requirement with the concept of parameterization via configuration sets, which allow developers to abstract environment-specific values from the iFlow logic and define them externally during deployment.

This decoupling of logic and environment ensures that the same artifact can be deployed across environments without code modification, thus reducing the risk of inconsistencies or human error.

Cloud Integration provides the following deployment modes for iFlows:

- **Direct deployment from the web UI**
  You'll use this mode during development and testing. It allows immediate activation of artifacts in the current tenant, but it isn't suitable for production.

- **Deployment via transport artifacts**
  In this mode, the iFlow is packaged and exported into a transport management node and then deployed into the target environment. This supports a formal review process and aligns with CI/CD principles.

- **Deployment via APIs**
  SAP provides APIs (such as `DeployIntegrationArtifact`) for automating deployments, typically in a scripted or pipeline-based context.

- **Deployment from Git**
  Some teams integrate external Git repositories to store version-controlled iFlow XML definitions, which are then imported and deployed manually or through APIs. This mode is only appropriate in limited use cases.

You can combine Each of these methods with automated testing, staging validation, and approval steps to ensure that deployment is smooth, traceable, and compliant.

Despite the internal versioning support within Cloud Integration, larger projects often benefit from externalizing their artifacts into source code repositories such as Git. This enables the following:

- Collaborative development with branching and merging
- Review and documentation of changes via pull requests
- Backup and traceability across projects and tenants
- Integration into CI/CD workflows

You can export artifacts such as iFlows, mappings, and Groovy scripts as XML or plain text files, and you can store them in Git. Although manual, this approach creates a single source of truth outside the platform and promotes discipline in version management. SAP's project "Piper" and related tooling offer scripts that can interact with Git repositories, automatically pull the latest changes, and trigger deployment workflows. By aligning SAP Integration Suite development with DevOps practices, organizations can improve code quality, reduce time to market, and simplify auditability.

Two main strategies are commonly used in integration deployment:

- **Push strategy**
  Artifacts are exported from the source (e.g., the DEV tenant) and pushed to the target (e.g., QA or PROD). This is the default in SAP's transport model and works well in tightly coupled teams and landscapes.

- **Pull strategy**
  The target system pulls the required version from a central repository (e.g., Git). This model provides more control to the receiving team and supports independent release cycles.

While SAP's tooling mainly supports the push model, hybrid approaches are possible. For instance, QA can approve a transport request while PROD waits for manual promotion or automated validation.

To summarize, effective versioning and deployment in Cloud Integration relies on a combination of platform features, development discipline, and lifecycle governance. Key recommendations include the following:

- Leverage the built-in versioning system for iFlows and artifacts.
- Adopt semantic versioning to make intent and change scope transparent.
- Use separate environments (subaccounts) for DEV, QA, and PROD.
- Externalize configurations with parameter sets to keep logic environment neutral.
- Manage transports by using SAP Cloud Transport Management for auditability.
- Integrate source control tools like Git for collaboration and traceability.
- Define a deployment policy that includes approvals, validation, and rollback mechanisms.
- Align with DevOps and CI/CD pipelines to scale deployment and reduce errors.

By following these strategies, organizations can ensure that integration artifacts are not only functionally correct but also operationally reliable, secure, and maintainable over time.

## 8.3   Automated Transports and Continuous Integration/Continuous Delivery Pipelines

The principles of continuous integration (CI) and continuous delivery (CD) have long been standard practice in modern software development. They enable the automated provision, testing, and delivery of software components, thereby accelerating development cycles and improving the quality of the delivered artifacts. In many development environments—especially for cloud-native applications—CI/CD pipelines are therefore deeply integrated into the daily workflow.

For SAP-specific technologies, however—especially Cloud Integration—implementing these concepts is more complex than in many other technology stacks. Although SAP offers a corresponding tool in SAP BTP in the form of SAP Continuous Integration and Delivery, its application in the context of the Cloud Integration isn't yet (as of August 2025) consistently practicable, isn't consistently productive, and doesn't currently offer clear added value for many companies.

SAP provides detailed instructions in the SAP Help Portal on how to integrate Cloud Integration artifacts into an automated pipeline by using SAP Continuous Integration and Delivery. This involves configuring a *job template* for Cloud Integration artifacts in which users can define which iFlows, value assignments, and other integration modules should be automatically packaged, versioned, and deployed to other subaccounts.

> **SAP Help Portal**
>
> If you want to use CI/CD anyway, please refer to this link: *https://help.sap.com/docs/continuous-integration-and-delivery/sap-continuous-integration-and-delivery/configure-sap-integration-suite-artifacts-job-in-job-editor?language=en-US*.

The theoretical idea behind this makes sense: developers check changes into a Git repository, an automated job builds a deployment artifact from them, ideally tests it, and transports it to other landscapes via transport management. So far, so DevOps.

However, the reality is different. Configuring these pipelines requires a number of manual steps, including the following:

- Setting up multiple SAP BTP service instances with sometimes unclear permissions
- Manually generating and managing service keys and OAuth tokens
- Establishing a correct and stable connection between Git repositories and job editors
- Defining artifact-related parameters, which are currently only documented or automatable to a very limited extent

These issues make setting up a functioning CI/CD pipeline for Cloud Integration not only technically challenging but also time-consuming and error-prone. Even minor deviations in the setup can cause jobs to fail or artifacts to be versioned incorrectly.

Many companies primarily expect a CI/CD service to increase productivity and improve quality assurance. In the case of the Cloud Integration, however, the opposite is true: the initial setup requires high effort, pipeline maintenance is complex, and the functional benefits are currently not commensurate with the effort involved.

Companies that already work with SAP Cloud Transport Management in particular can map many deployment steps much more easily, transparently, and auditably by using standardized transport orders. The additional benefits of a CI/CD pipeline are currently marginal—especially if fully automated testing of integration artifacts isn't possible.

Another problem is the lack of error diagnosis capabilities. When a deployment job fails, there are often no clear logs or clues as to what exactly was configured incorrectly. In addition, in many cases, there are SAP-specific restrictions (e.g., only certain artifact types are supported) and the transport of dependent objects (such as script collections) must be regulated separately.

From the developers' perspective, the integration of SAP Continuous Integration and Delivery into existing workflows is currently more of a hindrance than a help. The lack of integrated development environment (IDE) integration, the cumbersome setup in SAP's job editor, and the difficult-to-understand processes make it difficult to use such integration productively. Tools such as Jenkins and GitHub Actions offer more flexibility in theory, but they're only compatible to a limited extent in the SAP context and may require additional adaptation scripts, which in turn reduce maintainability.

The result is an inconsistent user experience that promotes neither developer-friendliness nor agility—which are two core objectives that CI/CD should actually fulfill. Against this backdrop and from today's perspective, we can't advise you to establish productive CI/CD pipelines based on SAP Continuous Integration and Delivery for Cloud Integration, as long as it's unclear what concrete advantages it will bring. The functional added value is low, the administrative effort is high, and the susceptibility to errors under real conditions is unacceptable.

Although tech-savvy teams with in-depth DevOps knowledge can take their first steps and gain experience with the toolset, it's currently more efficient for the vast majority of companies to focus on established mechanisms such as transport management (e.g., via SAP Cloud Transport Management), manual deployments, and semi-automated processes with clear auditing.

## 8.4   Summary

In this chapter, we helped you establish the strategic foundation for the controlled, traceable, and secure deployment of integration artifacts in Cloud Integration. Our objective wasn't merely to demonstrate transport capabilities but to have you configure a sustainable and governance-ready environment that aligns with enterprise-grade requirements for auditability, reliability, and scalability.

We began by introducing you to SAP Cloud Transport Management and walking you through the technical enablement of it as the core transport mechanism for Cloud Integration content. Once you had the landscape in place, you established technical connectivity by creating destinations, assigning roles, and deploying service instances such as SAP Content Agent and SAP Process Integration runtime. Governance was also a critical aspect of this chapter, and we had you assign role collections like TMS_LandscapeOperator_RC and ImportOperator to users to enforce role-based access control over transport creation and approval. We also demonstrated actual transport execution end-to-end.

Following this, we introduced version control and deployment strategies, and finally, we critically examined SAP Continuous Integration and Delivery. By completing the steps outlined in this chapter, you've created a fully configured and enterprise-ready deployment architecture for Cloud Integration artifacts. It provides a groundwork for robust integration lifecycle management—one that is secure, scalable, auditable, and ready to support the evolving needs of complex hybrid system landscapes.

# Chapter 9

# Logging and Monitoring

*A successful integration process doesn't end with the flawless execution of a flow—it only begins when it comes to operation, transparency, and error analysis.*

The development and productive operation of iFlows in Cloud Integration must prioritize stability, transparency, and responsiveness in the event of errors. This chapter gives you the complete picture of the available monitoring and error-handling functions, and it shows you how to use them to operate integrations reliably and analyze problems efficiently.

Section 9.1 introduces integrated monitoring, which is the central tool for monitoring message flows, adapter statuses, and system components. It will show you how to search for errors, analyze payloads, and evaluate runtime information. Section 9.2 is dedicated to debugging and error handling, and you'll learn how to quickly identify error causes, interpret logs correctly, and respond to typical problems in message processing.

Section 9.3 emphasizes the use of SAP Cloud ALM as the central tool for lifecycle management and monitoring of cloud integrations. This includes alerting, trace analysis, and business process monitoring. Section 9.4 introduces the SAP Alert Notification service for SAP BTP, which allows you to define notifications and automatic escalations. This ensures that you will be informed at an early stage if integrations go off track or defined thresholds are exceeded.

## 9.1 Integrated Monitoring

The **Monitor • Integrations and APIs** homepage in Cloud Integration serves as a central entry point for monitoring a tenant's integration operations (see Figure 9.1). The page is divided into several logically structured sections, each of which covers a specific area of responsibility. Each section contains tiles that display status information about messages or artifacts and link to a detailed view when you select them. The page automatically refreshes every five seconds, but this feature ends after five minutes.

The **Monitoring Status Overview** section provides a quick overview of the general status of message processing, which users can filter according to criteria they define.

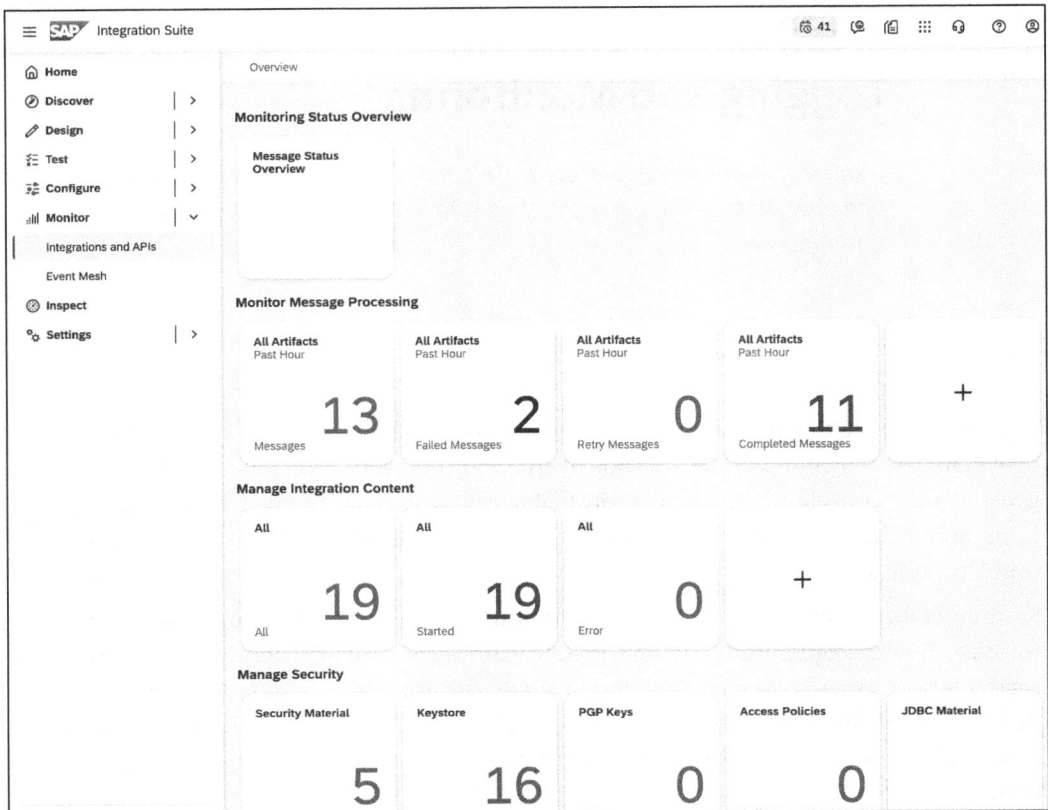

**Figure 9.1** Monitoring Overview

The **Monitor Message Processing** section shows how many messages have been processed within a specific time window, as well as their status. This section is the central point of contact for runtime monitoring of iFlows.

In the **Manage Integration Content** section, you can manage your tenant's integration content artifacts, such as iFlows. The tiles provide information about the number, status, and version of the deployed content.

The **Manage Security** section contains all security-related settings and information. Here, you can manage user credentials, keystore certificates, PGP keys, access policies, JDBC connections, user roles, and more. You can also test the connections to recipient systems.

The **Manage Stores** section lets you manage temporary storage, such as data stores for caching message content, runtime variables, and JMS message queues (if enterprise messaging is enabled). You can also manage number ranges for B2B scenarios here.

In the **Access Logs** area, you can view system logs and analyze errors that occurred during incoming HTTP requests, provided that SAP Cloud Logging is enabled.

Finally, the **Manage Locks** section allows you to view and manage lock entries. There's a distinction between *message locks*, which prevent the duplicate processing of running messages, and *design-time artifact locks*, which prevent the simultaneous editing of integration artifacts.

The **Message Status Overview** page provides a customizable overview of all messages that are processed within a defined period of time and their respective processing status (see Figure 9.2). It serves as the central starting point for quickly assessing the integration process and enabling navigation to the detailed message display in the **Message Monitor**. You can access the overview by clicking the **Message Status Overview** tile in the **Monitoring Status Overview** section on the **Monitor** home page. The table displays the data from the Cloud Integration runtime by default.

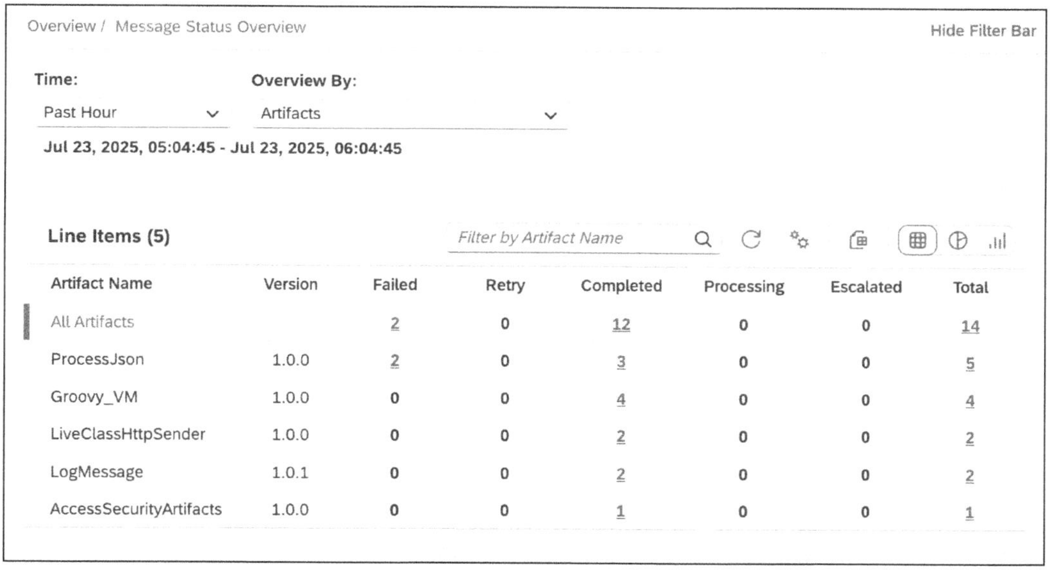

Overview / Message Status Overview                                                        Hide Filter Bar

**Time:**                    **Overview By:**

Past Hour           ⌄     Artifacts                        ⌄

**Jul 23, 2025, 05:04:45 - Jul 23, 2025, 06:04:45**

**Line Items (5)**                          *Filter by Artifact Name*   Q  C  ⚙  ⊞  ⊞  ⊕  .ıl|

| Artifact Name | Version | Failed | Retry | Completed | Processing | Escalated | Total |
|---|---|---|---|---|---|---|---|
| All Artifacts | | 2 | 0 | 12 | 0 | 0 | 14 |
| ProcessJson | 1.0.0 | 2 | 0 | 3 | 0 | 0 | 5 |
| Groovy_VM | 1.0.0 | 0 | 0 | 4 | 0 | 0 | 4 |
| LiveClassHttpSender | 1.0.0 | 0 | 0 | 2 | 0 | 0 | 2 |
| LogMessage | 1.0.1 | 0 | 0 | 2 | 0 | 0 | 2 |
| AccessSecurityArtifacts | 1.0.0 | 0 | 0 | 1 | 0 | 0 | 1 |

**Figure 9.2**  Message Status Overview

If multiple runtimes are available, you can select the environment you want from the dropdown menu. The table lists the processed elements—such as messages, artifacts, and iFlows—in groups, and it also displays The status filters that you have set. It also specifies the version number for certain object types, such as iFlows and artifacts. If the **Version** column is empty, there are a few possible reasons for it: the artifact may be in draft mode, the version data may not have been retrievable, or the corresponding row may have represented an aggregated overview (such as **All Artifacts** or **All Integration Flows**). The **Message Status Overview** provides a clear and systematic view of message processing statuses, so it offers a solid foundation for ongoing operational monitoring and error analysis.

You can adapt the Cloud Integration **Monitor** home page to your individual needs and working methods. Use drag and drop to add new tiles, remove existing ones, or change the arrangement of tiles within a section. For example, you can configure the home page to display only data for the iFlows that you have edited.

However, you can only customize the following sections of the page:

- **Monitor Message Processing**
- **Manage Integration Content**

To customize one of these sections, you can add a new tile by selecting an empty tile element (one with the + symbol) or edit or delete existing tiles by right-clicking them and selecting **Edit** (see Figure 9.3).

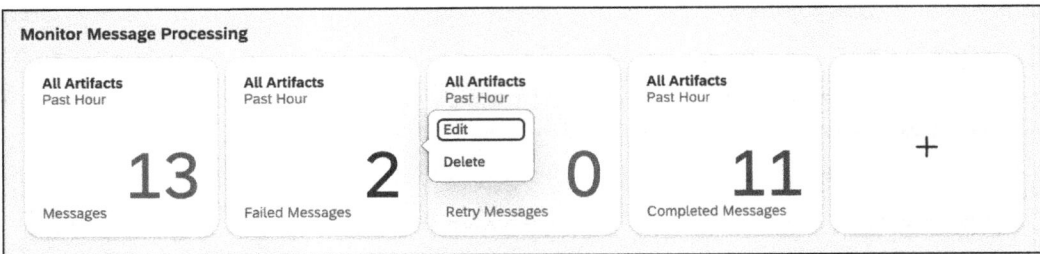

**Monitor Message Processing**

| All Artifacts<br>Past Hour | All Artifacts<br>Past Hour | All Artifacts<br>Past Hour | All Artifacts<br>Past Hour | + |
|---|---|---|---|---|
| **13**<br>Messages | **2**<br>Failed Messages | Edit<br>Delete<br>**0**<br>Retry Messages | **11**<br>Completed Messages | |

**Figure 9.3**  Customizing Monitor Message Processing Section

When adding or editing a tile, you can use specific filter categories:

- For **Monitor Message Processing** tiles, you can specify the status, time, and related integration artifact or package for the messages to be displayed (see Figure 9.4).
- For **Manage Integration Content** tiles, you can specify the status and type of the integration artifact to be displayed.

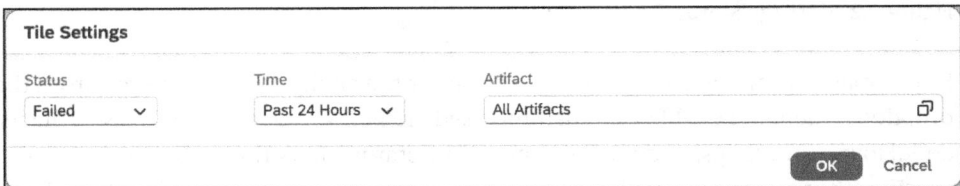

**Tile Settings**

| Status | Time | Artifact |
|---|---|---|
| Failed ∨ | Past 24 Hours ∨ | All Artifacts |

OK   Cancel

**Figure 9.4**  Monitoring Tile Filter Parameters

Clicking on the **All Artifacts** tile takes you to the monitoring screen. There, you'll see all messages that have passed through the filters you've selected, as shown in Figure 9.5. The message log details are divided into the **Status**, **Properties**, **Logs**, and **Artifact Details** tabs. In the **Logs** tab, you'll see that the **Log Level** is set to **Info**. By clicking on **Open Text View**, you can jump to the textual Camel message processing log. However, this is difficult to read, and a graphical log will only be available if the log level has been set to **Debug** or **Trace** or if the iFlow has encountered an error.

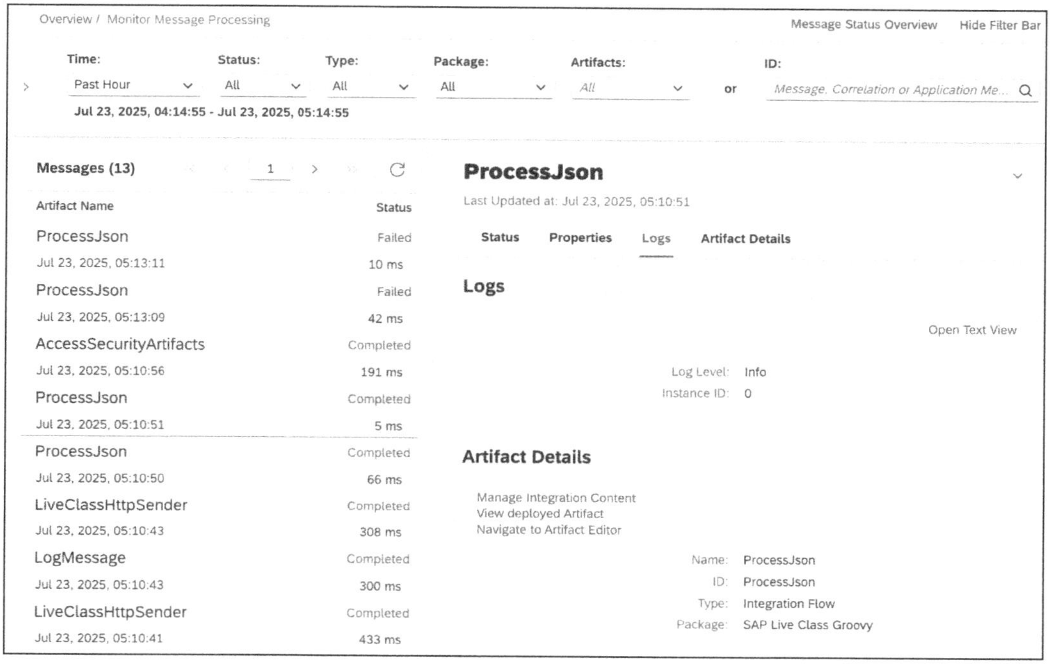

**Figure 9.5**  Log of Completed Message

The **Status** tab indicates the status of a message or how its processing was completed during runtime. The following status values are possible:

- **Completed**
  The message was successfully delivered to the recipient, and processing is complete.

- **Processing**
  The message is currently being processed. This status may change depending on the result.

- **Retry**
  An error occurred during processing, so the system automatically started a new processing attempt.

- **Escalated**
  An error occurred, but no retry was triggered. For synchronous messages, an error message is returned to the sender.

- **Failed**
  The message could not be processed or delivered, and no further processing attempts are possible. This is a final error status.

- **Cancelled**
  Processing was manually canceled (e.g., by deleting an entry from a JMS queue). Note that if the message is still being processed, this status can be overwritten by a final status, such as **Completed**.

- **Discarded**
  This status applies to iFlows with timer triggers. If a worker node fails (e.g., due to running out of memory), the flow can be restarted on another node. The original message is given the **Discarded** status while a new message with a new ID is generated.

- **Abandoned**
  Processing was interrupted, or there was no status update for a long time. This can happen during redeployment or undeployment of a flow or a controlled shutdown of a worker node. **Abandoned** isn't a final status—processing can continue if a retry is configured, for example. In the event of uncontrolled crashes, the system sets the status asynchronously.

If an error occurs while you're processing messages in iFlow, the message will display with the **Failed** status. The **Error Details** are displayed in the **Status** tab in the form of a Java stack trace (see Figure 9.6).

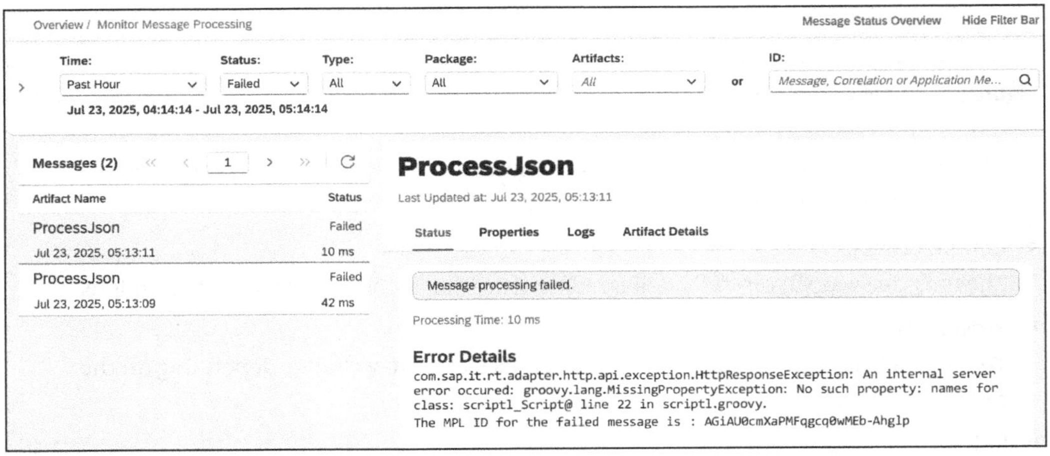

**Figure 9.6** Message Processing Error Message

As we mentioned earlier, you can jump to the graphical log, which displays errors as shown in Figure 9.7. In the **Integration Flow Model** area, you can see a graphical representation of the iFlow and the position where the error occurred.

In the **Log Content** area, you can find a textual description of the error by selecting the step in which the error occurred in the **Run Steps** table on the left side of the screen.

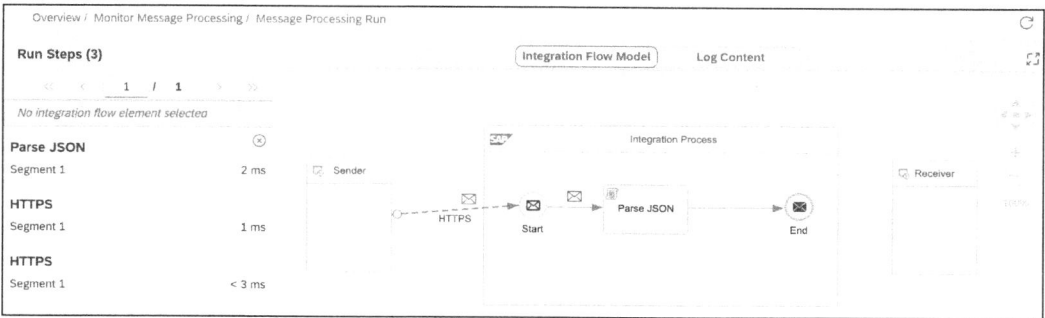

**Figure 9.7**  Error Message iFlow Model

**Figure 9.8**  Error Message Log Content

In the **Log Configuration** in the **Manage Integration Content** view (see Figure 9.9), you can set the log level you want individually for each iFlow. The following log levels are available:

- **None**
  At this level, no information is recorded or displayed in monitoring during message processing. Even if data archiving has been set up for the flow, no archiving takes place at log level **None**.

- **Info**
  At this level, basic information about message processing is recorded. The headers of the messages are always displayed, and in the event of faulty messages, detailed information about the last 50 processing steps will also be stored and displayed. This level is well suited for productive operation with moderate logging.

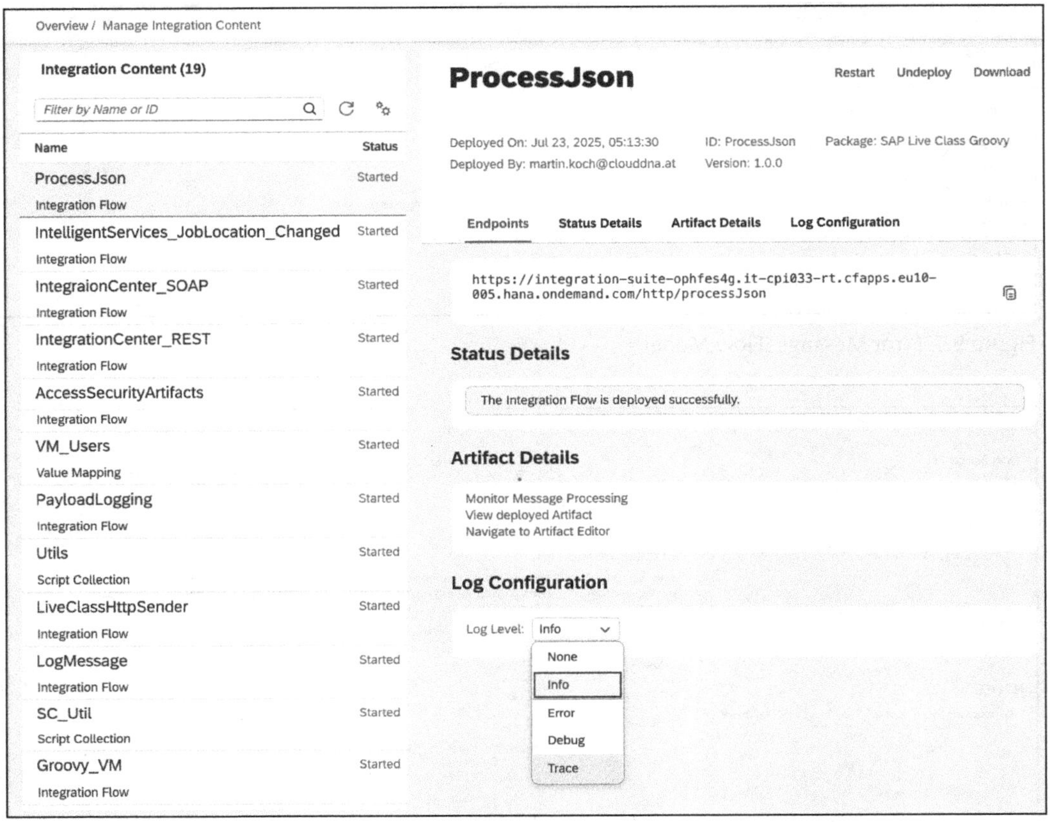

**Figure 9.9** Error Message Log Levels

- **Error**
  At this level, only failed message processing is logged. The last 50 processing steps are recorded in detail, but only for faulty messages. Successful messages are not recorded, and no log attachments are generated.

> **Important Note**
>
> If a retry scenario is configured and a subsequent delivery attempt is successful, those events aren't logged at the **Error** log level. The status of the message may remain **Retry** or **Failed**, even though the message was delivered correctly. Even if data archiving is enabled, no archiving takes place.

- **Debug**
  At this level, all details of message processing are recorded. These including all processing steps, header data, and other technical information. All steps are visible in the graphical user interface, but only the last 100 steps are displayed in the text view.

For performance reasons, this mode is limited to 24 hours—after that, the previously set log level is automatically reactivated. Due to the high resource consumption, you should only use the **Debug** level in development and test environments.

- **Trace**

  This level offers the most detailed logging: in addition to all processing steps, the entire message content is logged. This function is limited to 10 minutes, and then, it automatically returns to the previous log level. The logged content (payload) is stored for one hour by default.

Changing the log level to **Trace** will bring up a popup window like the one shown in Figure 9.10. This window will inform you that the **Trace** log level will revert to the previous log level after 10 minutes, and you'll also see that the trace data will be deleted after an hour.

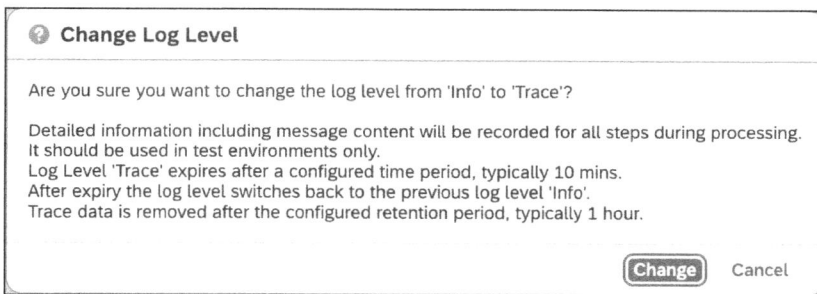

> **? Change Log Level**
>
> Are you sure you want to change the log level from 'Info' to 'Trace'?
>
> Detailed information including message content will be recorded for all steps during processing. It should be used in test environments only.
> Log Level 'Trace' expires after a configured time period, typically 10 mins.
> After expiry the log level switches back to the previous log level 'Info'.
> Trace data is removed after the configured retention period, typically 1 hour.
>
> [ **Change** ]  Cancel

**Figure 9.10**  Log Level Confirmation

In the **Trace** log level, you'll see the **Message Content** area on the **Monitor** home page, as shown in Figure 9.11. There, you can view the **Header**, **Exchange Properties**, and **Payload** tabs.

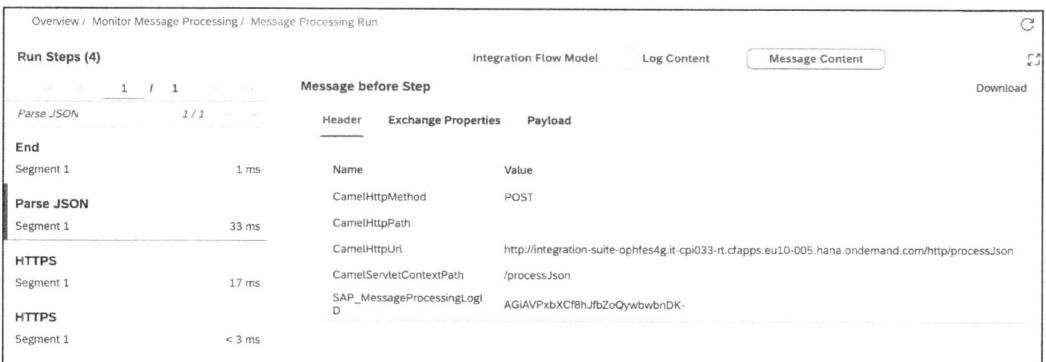

**Figure 9.11**  Message Content Monitoring

You can also use custom header properties to search messages by content. These properties are written to a Groovy script, as shown in Figure 9.12.

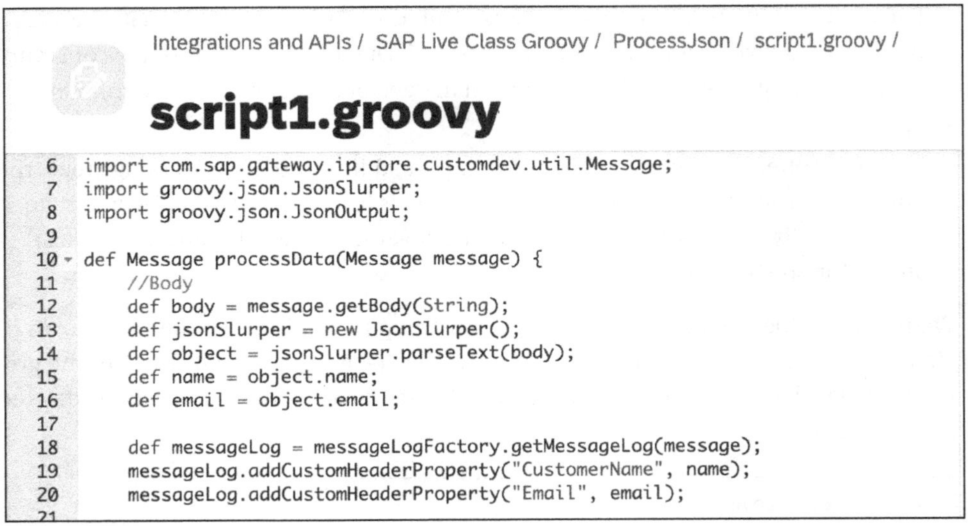

Integrations and APIs / SAP Live Class Groovy / ProcessJson / script1.groovy /

## script1.groovy

```
6   import com.sap.gateway.ip.core.customdev.util.Message;
7   import groovy.json.JsonSlurper;
8   import groovy.json.JsonOutput;
9
10  def Message processData(Message message) {
11      //Body
12      def body = message.getBody(String);
13      def jsonSlurper = new JsonSlurper();
14      def object = jsonSlurper.parseText(body);
15      def name = object.name;
16      def email = object.email;
17
18      def messageLog = messageLogFactory.getMessageLog(message);
19      messageLog.addCustomHeaderProperty("CustomerName", name);
20      messageLog.addCustomHeaderProperty("Email", email);
21
```

**Figure 9.12** Custom Header Property in Groovy Script

You can view the custom header properties in the **Monitor Message Processing** log, in the **Properties** tab (see Figure 9.13).

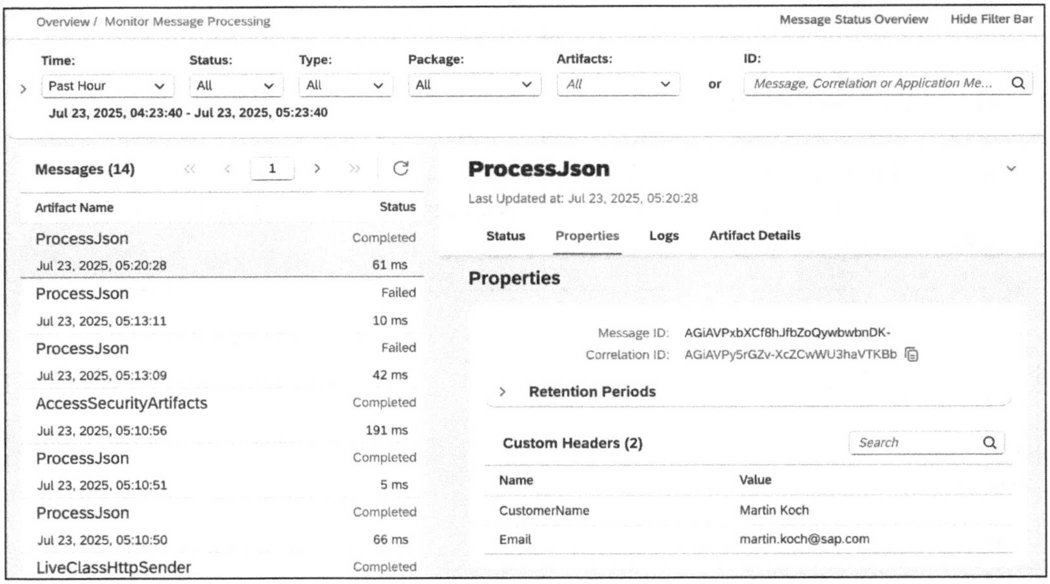

**Figure 9.13** Custom Header Property Monitoring

You can also search for custom header properties by expanding the filter options. Then, you'll see the **Custom Header** input field, as shown in Figure 9.14. You can search for properties by entering the "CustomerHeaderPropertyName=ValueOfCustomHeader-Property" syntax.

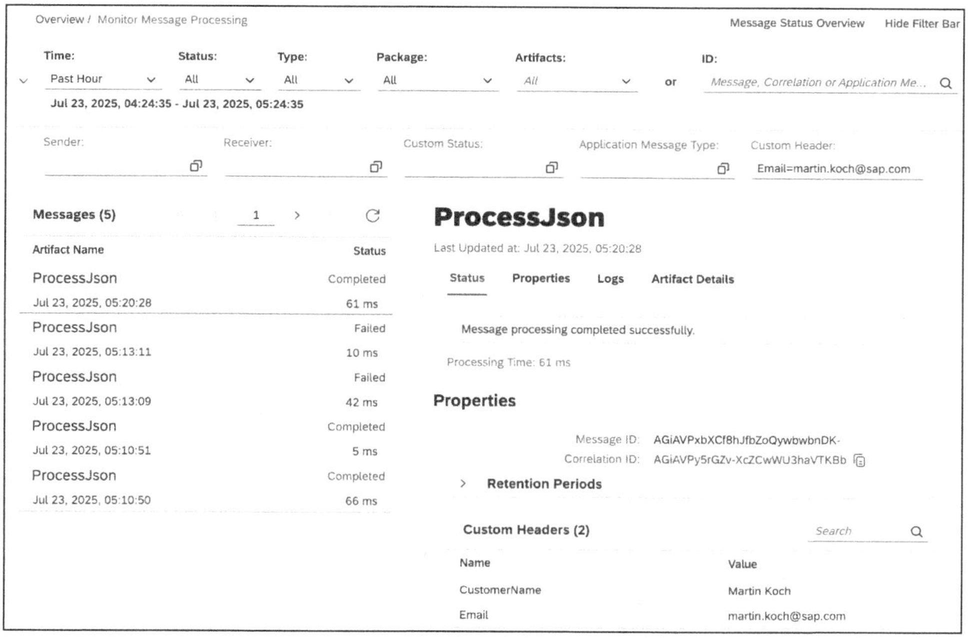

**Figure 9.14**  Custom Header Property Search

Monitoring is somewhat more complex if a message passes through several iFlows in Cloud Integration, because you need to identify the related iFlows. So, when messages are processed, a **CorrelationID** is created to facilitate this identification. You can view this ID in the **Properties** area of the **Status** tab of the message processing log, as shown in Figure 9.15. You can also click on the **Correlation ID** to see all related messages.

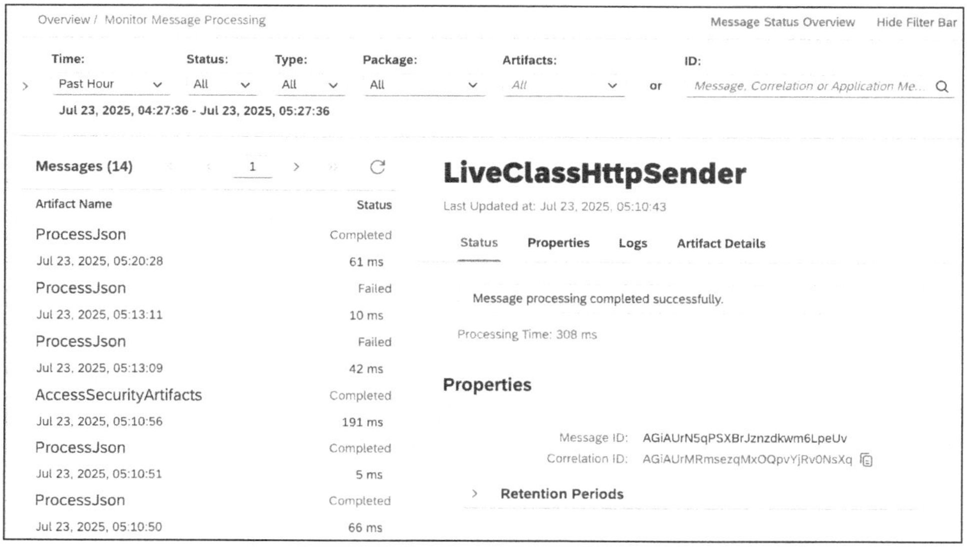

**Figure 9.15**  Correlation ID

As shown in Figure 9.16, you can also search for a correlation ID by entering the filter into the **ID** field.

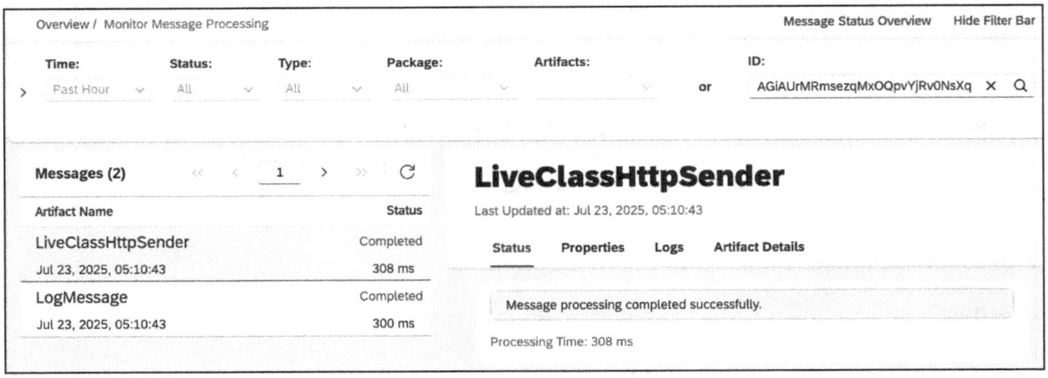

**Figure 9.16** Correlated Messages

## 9.2   Debugging and Error Handling

Efficient error handling involves identifying potential sources of error at an early stage and developing clear strategies for dealing with problems that arise. For example, error messages should be precise and understandable so that developers can respond quickly. Automated monitoring solutions also help to detect and analyze malfunctions in real time.

Structured debugging involves the use of specialized tools such as log analysis tools and cloud-specific debuggers that enable targeted troubleshooting. Best practices include systematic testing of integration processes, the use of version control, and the documentation of changes and known issues. The combination of proactive error handling and structured debugging minimizes downtime and increases the overall reliability of cloud systems.

Clean error handling is one of the basic principles. It ensures that exceptions—for example, mapping errors, routing problems, and technical malfunctions—are caught during message processing. In Cloud Integration, you can handle errors flexibly at various levels, from adapters to iFlows to the message level.

A proven pattern is to use an exception subprocess in iFlow. As soon as an exception occurs in a main process, this subprocess is activated to respond specifically to the error. Information can be logged, notifications can be triggered, and structured error messages can be sent back to the sender. In Figure 9.17, you can see a simple **Exception Subprocess** that just handles exceptions that are thrown during the HTTP call to the receiver. In the **Exception Subprocess**, you can get more details on the exception by using ${exception.message} or ${exception.stacktrace}.

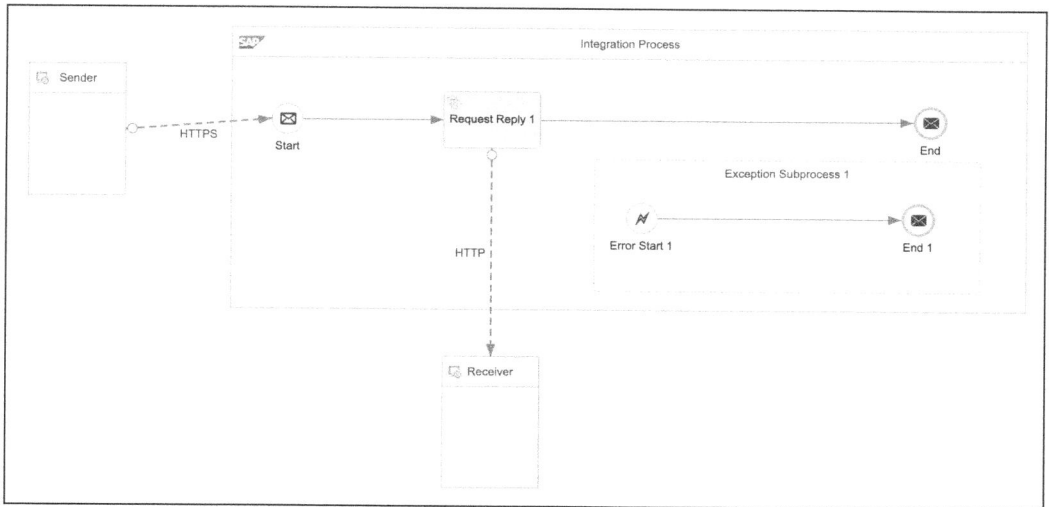

**Figure 9.17** Exception Handling

You can add other flow elements between the start and end events. For example, you can select **Add Service Call** from the context menu of a connection within the pool, and that will allow you to contact another system to handle the exception. Note that the following elements are not supported within an **Exception Subprocess**:

- Another Exception Subprocess
- Integration Process
- Local Integration Process
- Sender
- Receiver
- Start Message
- Terminate Message
- Timer Start
- Start Event
- End Event
- Router
- Aggregator

Please be aware that an **Exception Subprocess** can't handle an exception that occurs when a **Data Store Write** step fails because the entry already exists (a duplicate key exception). This is because the database transaction is rolled back, regardless of whether an **Exception Subprocess** is used.

However, an exception doesn't always have to be thrown when an error occurs. For example, the HTTP receiver adapter offers the option of not throwing an exception when an error occurs. To do this, uncheck the box at the bottom of the screen shown in Figure 9.18. This allows you to respond to the HTTP status code in subsequent steps.

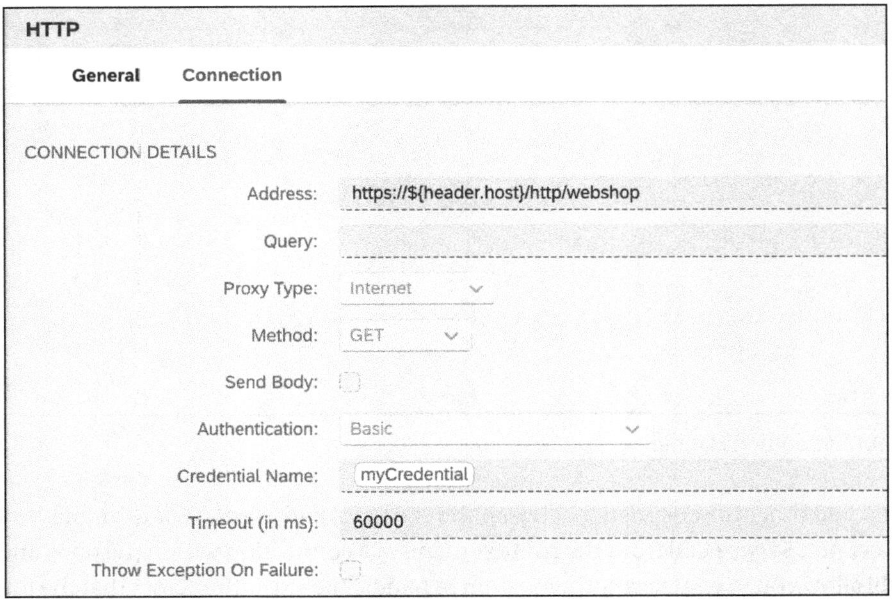

| HTTP | |
| --- | --- |
| **General** | Connection |

CONNECTION DETAILS

| | |
| --- | --- |
| Address: | https://${header.host}/http/webshop |
| Query: | |
| Proxy Type: | Internet ∨ |
| Method: | GET ∨ |
| Send Body: | ☐ |
| Authentication: | Basic ∨ |
| Credential Name: | myCredential |
| Timeout (in ms): | 60000 |
| Throw Exception On Failure: | ☐ |

**Figure 9.18**  HTTP Receiver Adapter

As shown in the example for the HTTP adapter, you can access the response code in a subsequent router by using the Camel expression language, as illustrated in Figure 9.19. This enables you to briefly intercept and react to individual responses.

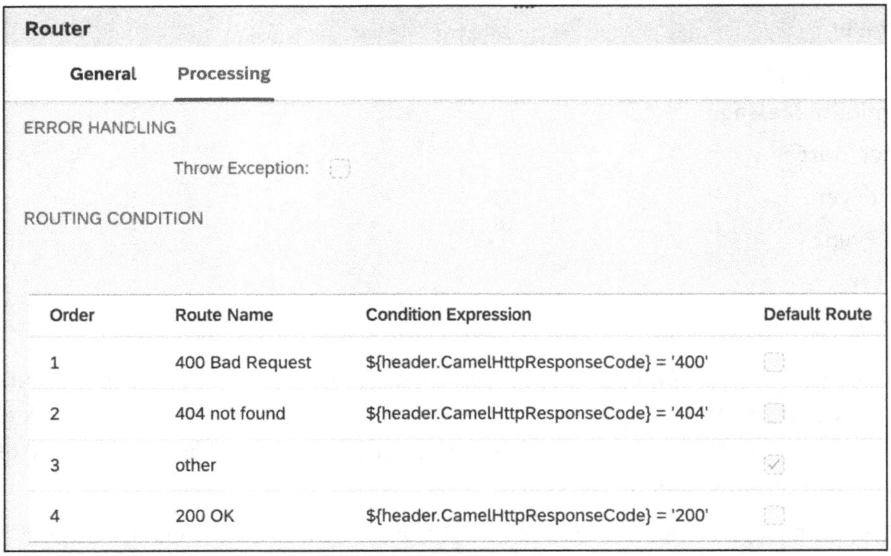

| Router | |
| --- | --- |
| **General** | Processing |

ERROR HANDLING

Throw Exception:  ☐

ROUTING CONDITION

| Order | Route Name | Condition Expression | Default Route |
| --- | --- | --- | --- |
| 1 | 400 Bad Request | ${header.CamelHttpResponseCode} = '400' | ☐ |
| 2 | 404 not found | ${header.CamelHttpResponseCode} = '404' | ☐ |
| 3 | other | | ☑ |
| 4 | 200 OK | ${header.CamelHttpResponseCode} = '200' | ☐ |

**Figure 9.19**  Router HTTP Status Code Conditions

Another option is to use a Groovy script to write information to the message processing log (see Listing 9.1). The addAttachmentAsString method adds an attachment to the
message processing log.

```
def Message processData(Message message) {
    def body = message.getBody(java.lang.String)
    def messageLog = messageLogFactory.getMessageLog(message)
    if (messageLog != null) {
        messageLog.addAttachmentAsString('Attachment Name', body, 'application/json')
    }
    return message
}
```

**Listing 9.1**  Writing Attachment to MPL

The attachment will be visible in the message processing log, in the **Attachments** section (see Figure 9.20).

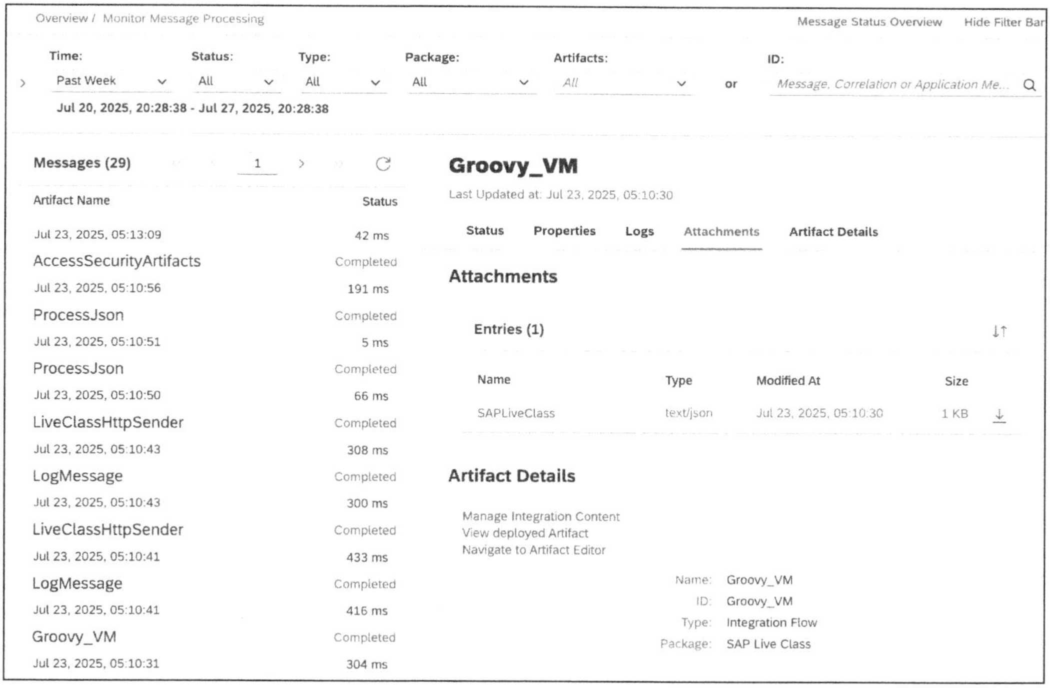

**Figure 9.20**  Message Processing Log Attachments

You can't set breakpoints or debug Groovy scripts in Cloud Integration, and that
applies to both Groovy script and iFlows in general. However, you have the option of
simulating iFlows. With the simulation feature, you can test an iFlow or a subset of it to
see if you can achieve the desired outcome before deploying the iFlow. Then, based on

the results of the simulation, you can decide whether to deploy the iFlow or resolve any errors. When you're simulating an iFlow, not all flow steps will behave exactly as they would in a live execution. Here's an overview of the supported elements by category:

- Message Transformers
  - Content Modifier
  - Converter
  - Decoder
  - Encoder
  - Filter
  - Message Digest
  - Script
    HTTP calls inside **Script** steps are not supported during simulation.
- Call
  - External Call/Request Reply/Content Enricher/Send
    These steps are supported, but the actual call to the receiver system isn't made. You can mock the response by selecting the connected **Receiver Adapter**, clicking **Add Simulation Response**, and entering a custom payload or headers. If no mock is defined, the previous payload will be used.
  - Local Call/Process Call/Looping Process Call
    These are fully supported in simulation.
- Message Routing
  - Join
  - Router
  - Splitter
  - Multicast
  - Gather
  - Aggregator
    This isn't supported in simulation mode.
- Security Elements
  - Decryptor
  - Encryptor
  - Signer
  - Verifier
- Persistence
  - **Write Variables** and **Data Store Operations**
    Only read operations can be simulated.
  - XML Validator
    This is fully supported.

The following simple example illustrates how the simulation works. First, you use the speed buttons to set a simulation start point, as shown in Figure 9.21. Then, click on ⏵ to set the simulation input.

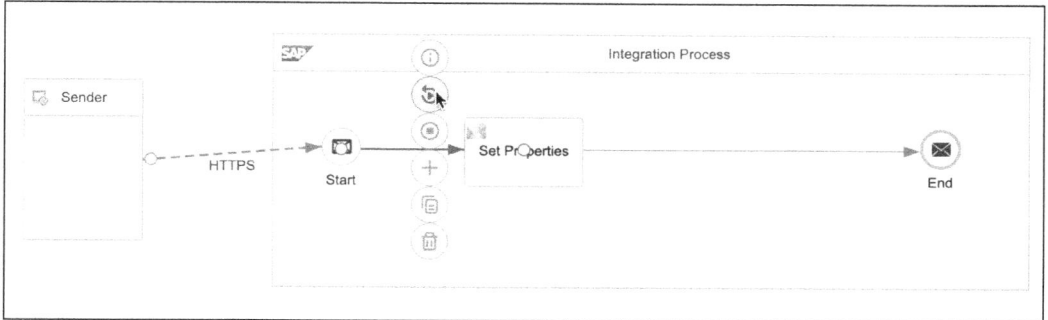

**Figure 9.21** Setting Simulation Start Point

As shown in Figure 9.22, you can set the **Header**, **Properties**, and **Body**.

**Figure 9.22** Setting Simulation Input

427

Then, as shown in Figure 9.23, set the simulation and point at the location you want by clicking on the icon. You can place any number of components between the start and end points. Click ▶ to start the simulation.

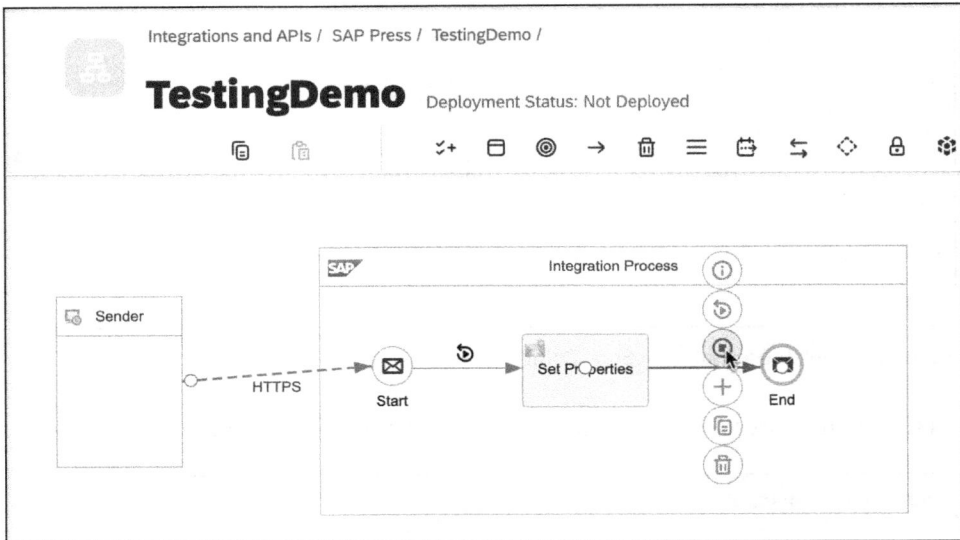

**Figure 9.23** Setting Simulation Endpoint

Then, you'll see the popup shown in Figure 9.24. Please note that depending on the complexity of the iFlow, the simulation may take longer.

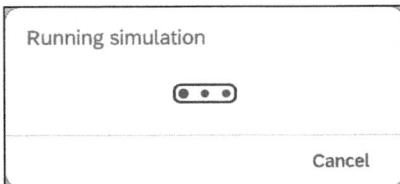

**Figure 9.24** Running Simulation

The status and results will be displayed immediately after the simulation has run (see Figure 9.25).

Click on ✉ to view the message content (see Figure 9.26), which will be displayed in the lower area. You can view the **Headers**, **Properties**, and **Body**.

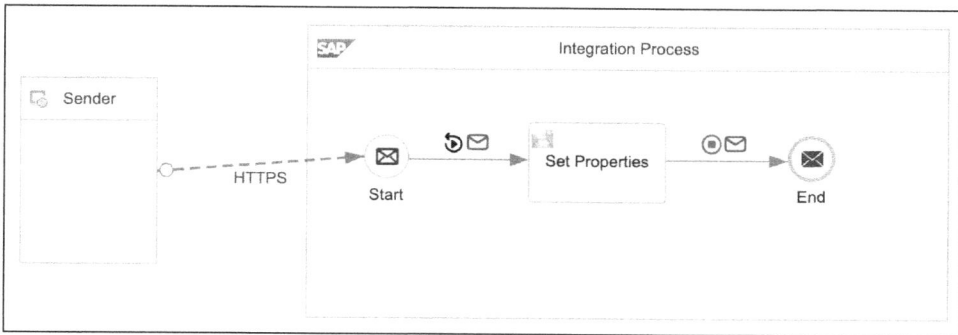

**Figure 9.25** Simulation Results

9

**Figure 9.26** Simulated Message Content

## 9.3   SAP Cloud ALM

*SAP Cloud ALM* is a central, cloud-based monitoring tool for SAP Cloud systems and services. You can also use it to monitor Cloud Integration and perform comprehensive monitoring of integration processes in a business context. SAP Cloud ALM supports integration and exception monitoring that provide integration experts with a central overview of exceptions and failed messages across the entire integration landscape. SAP Cloud ALM allows users to monitor end-to-end integration scenarios and drill down to individual messages, thus enabling them to quickly identify and address faults. Additionally, the **Health Monitoring** page in SAP Cloud ALM offers a centralized platform for monitoring resources, such as certificates and JMS queues, within Cloud Integration. This is especially useful for tracking the depletion and expiration of such resources. You can define KPIs and configure automatic notifications to help you address critical conditions early on.

When you create a Cloud Integration instance, it's automatically registered in SAP Cloud ALM. All you need to do is establish a connection to read the monitoring data. To do this, open **SAP Integration Suite (Cloud Integration)** in the **Services & Systems** area, open the **Endpoints** tab, and click **Add** (see Figure 9.27).

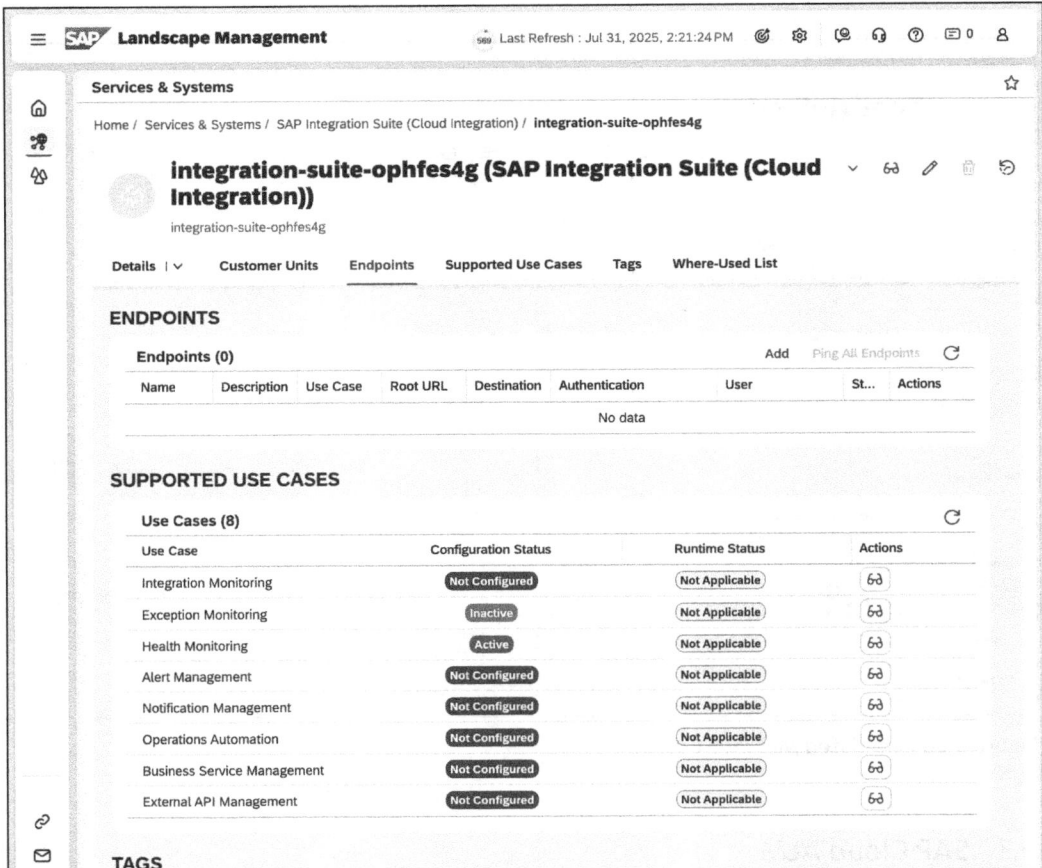

**Figure 9.27** System Endpoint

Before you can perform the configuration, you must create a service instance of the SAP Process Integration runtime and then create a service key for this instance. The service key contains all the necessary connection information, and you should assign a meaningful endpoint name. As shown in Figure 9.28, you should also select the following use cases: **Exception Monitoring**, **Health Monitoring**, **Integration Monitoring**, and (alternatively) **External API Management**. Then, copy the URL from the service key into the **Root URL** field; copy the **Client ID**, **Client Secret**, and **Token Service URL** from the service key; and click **Check Connection**.

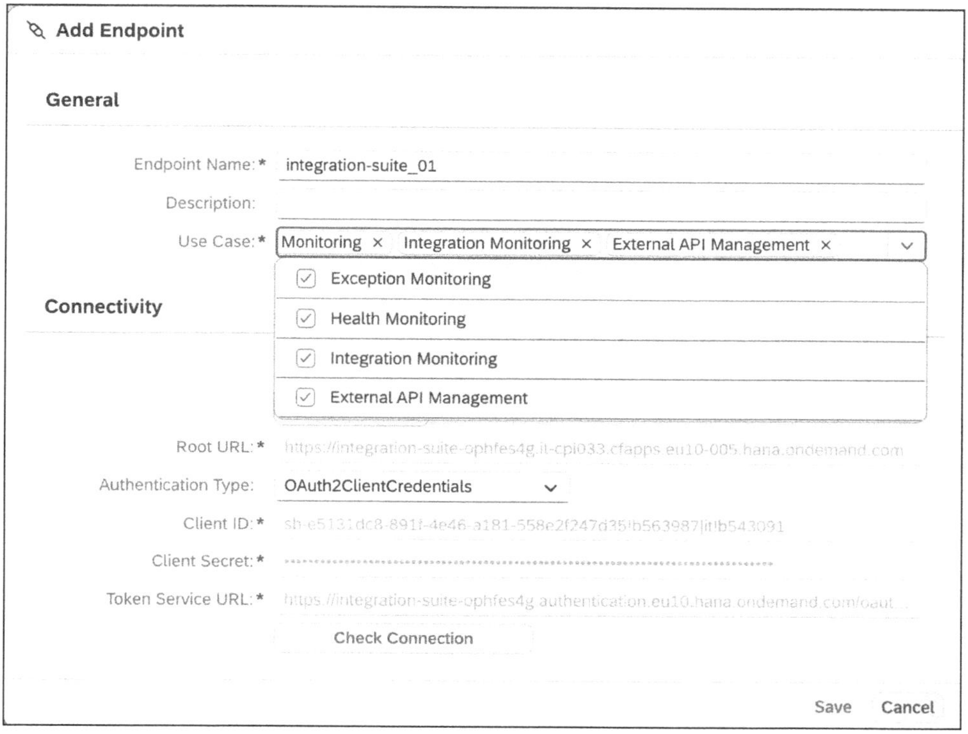

Figure 9.28 Adding Endpoint

If the connection check returns a **Success** status for all use cases, as shown in Figure 9.29, click **Close** and save the endpoint.

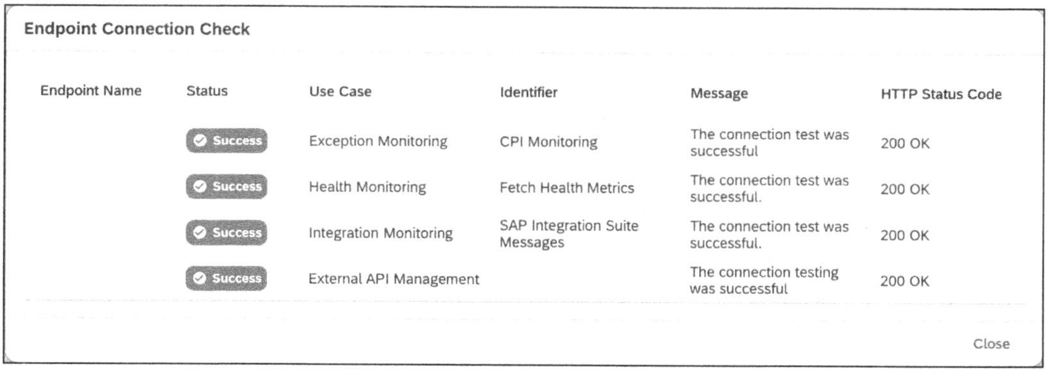

Figure 9.29 Connection Check

All you need to do now is perform a final configuration by navigating to **Integration & Exception Monitoring**, looking for the **SAP Integration Suite (Cloud Integration)** tile, and clicking on **Edit Configuration** (see Figure 9.30).

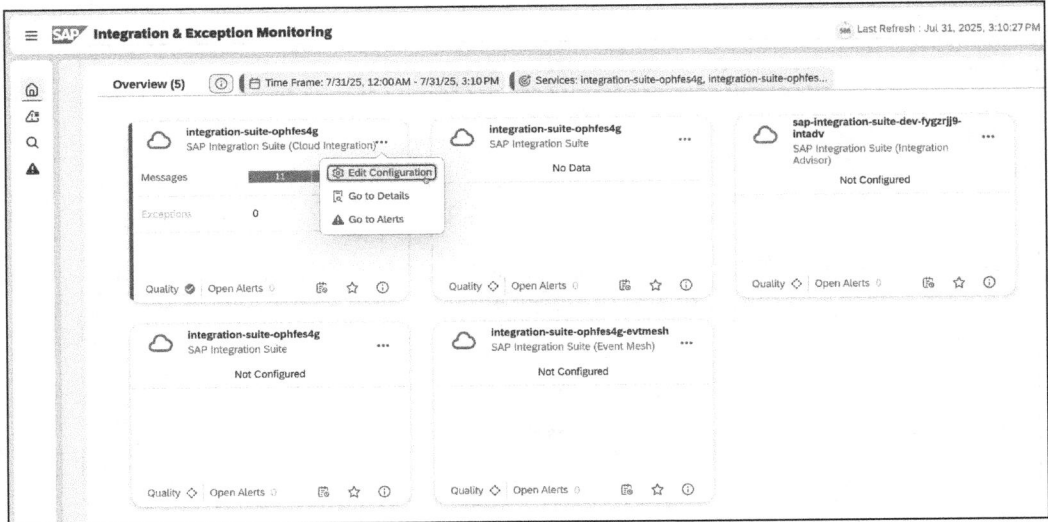

**Figure 9.30** Performing Final Configuration

Then, activate **Data Collection** for SAP Integration Suite by sliding the switch to **ON**, as shown in Figure 9.31.

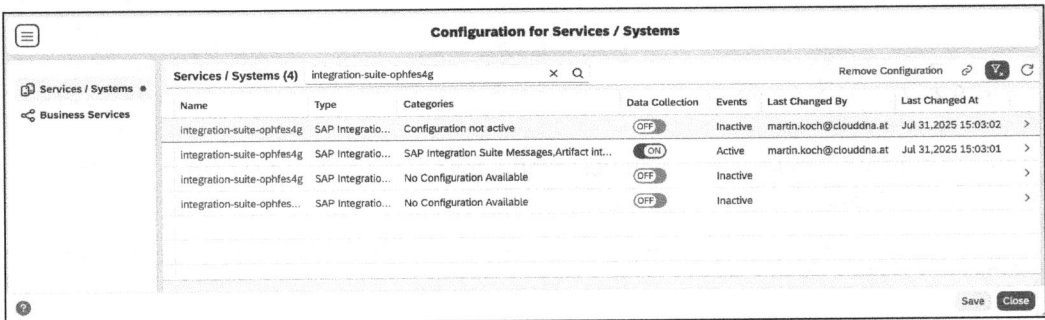

**Figure 9.31** Enabling Data Collection

Next, navigate to the details of the SAP Integration Suite entry and check whether the desired categories have been activated in the **Monitoring** tab (see Figure 9.32).

You have now completed the configuration, and from this point on, the monitoring data will be loaded from Cloud Integration and displayed in SAP Cloud ALM.

Next, open the **Operations** tab in SAP Cloud ALM and click on **Integration & Exception Monitoring** to access the monitoring screen (see Figure 9.33).

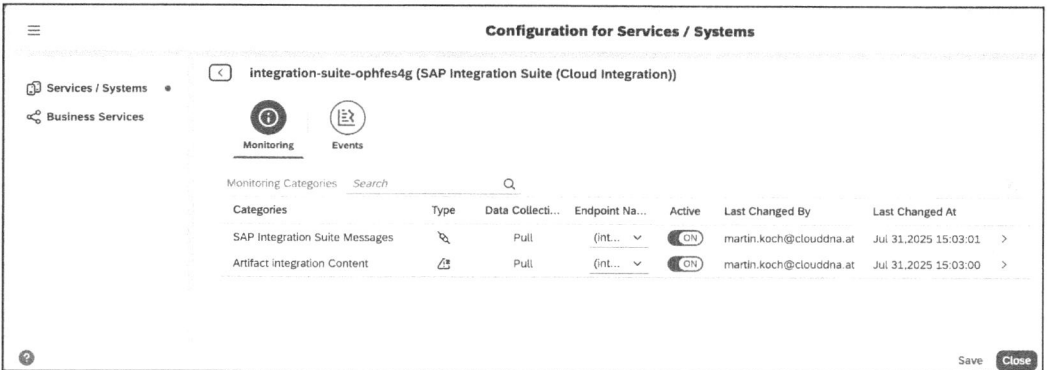

**Figure 9.32** Activating Categories

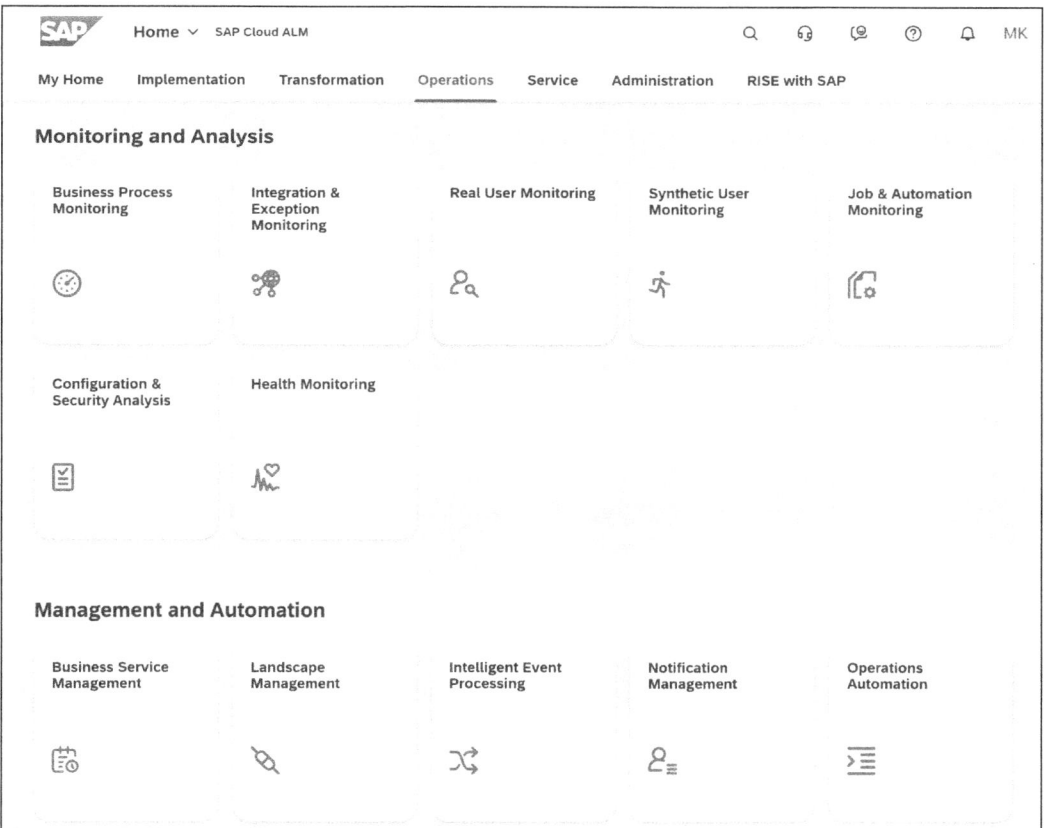

**Figure 9.33** Monitoring Launchpad

Figure 9.34 shows an example of a Cloud Integration instance with both failed and successfully processed messages. You can click on the corresponding tile to jump to the details.

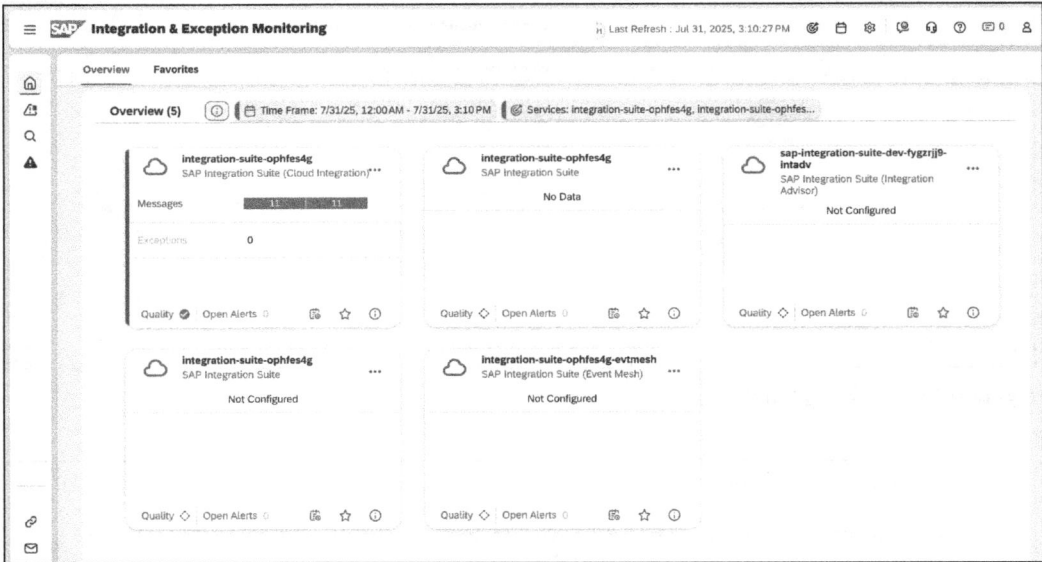

**Figure 9.34** Monitoring Cloud Integration

The monitoring overview shows aggregated information about error statuses, as shown in Figure 9.35. Below that, the **Messages** table shows messages that contain errors. Click on a message to jump to its details.

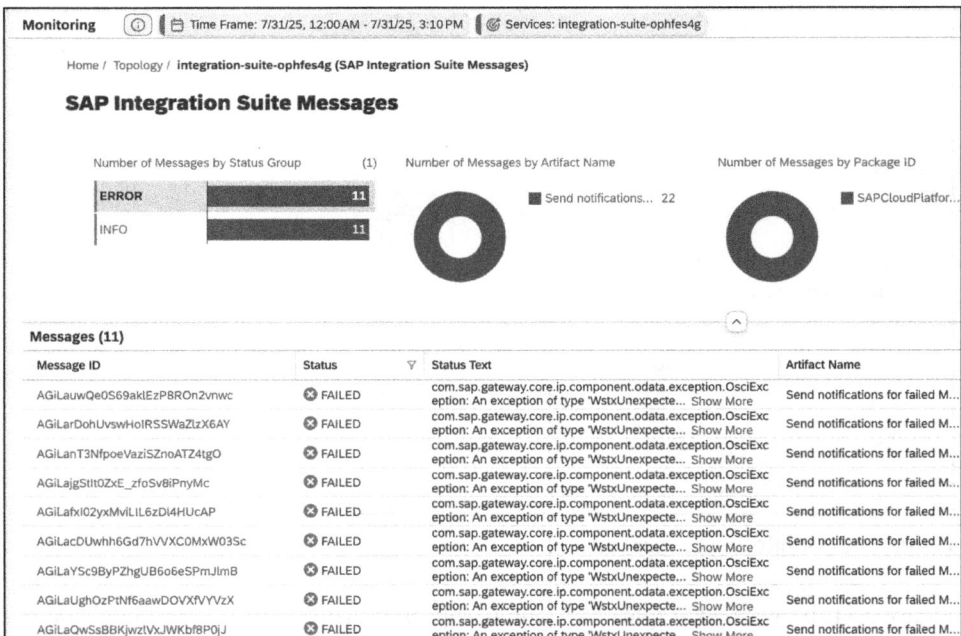

**Figure 9.35** Monitoring Message Overview

In the **Message Details**, you can see detailed information about the cause of the error, as shown in Figure 9.36.

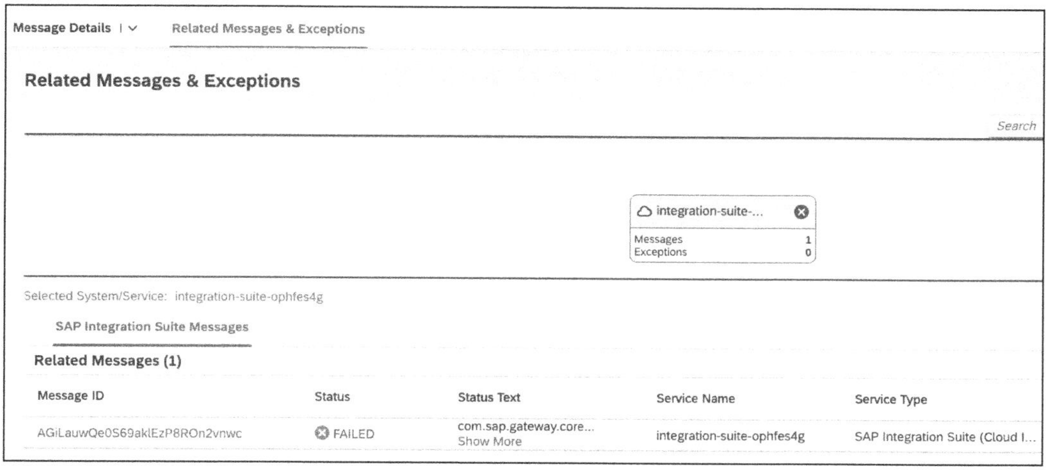

**Figure 9.36** Related Messages

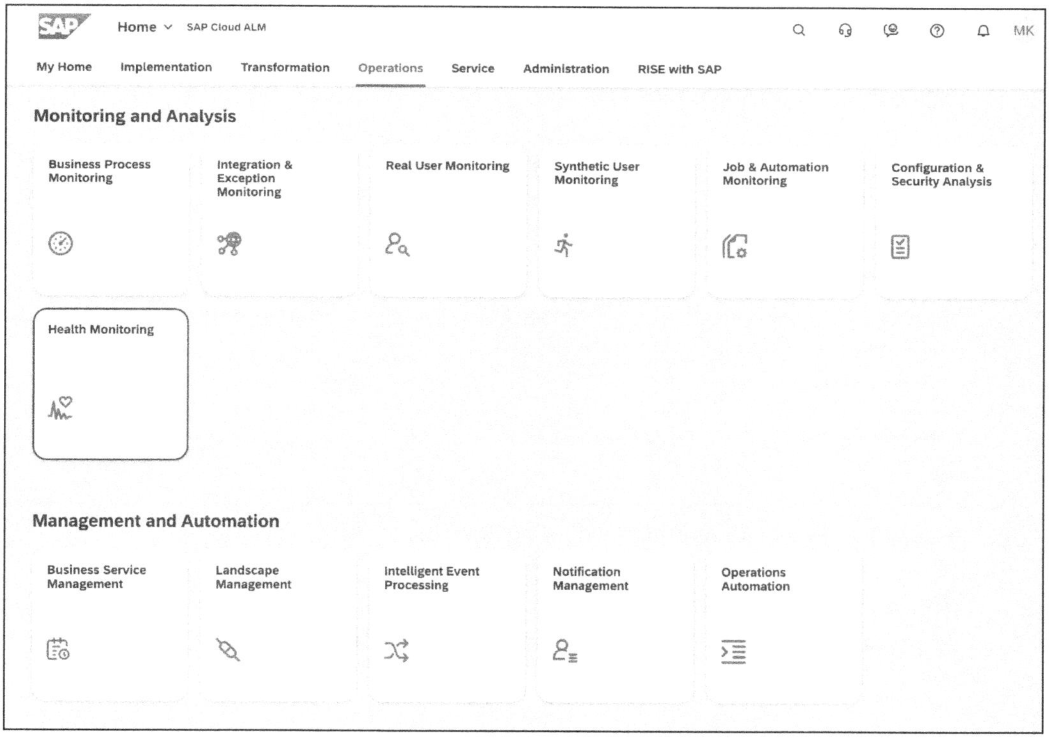

**Figure 9.37** Operations Overview

In SAP Cloud ALM, you also have the option of performing health monitoring. To do this, open SAP Cloud ALM as shown in Figure 9.37, select the **Operations** tab, and click the **Health Monitoring** tile.

Figure 9.38 shows the monitor displaying an overview of **SAP Integration Suite (Cloud Integration)**. Click on the tile to view the details.

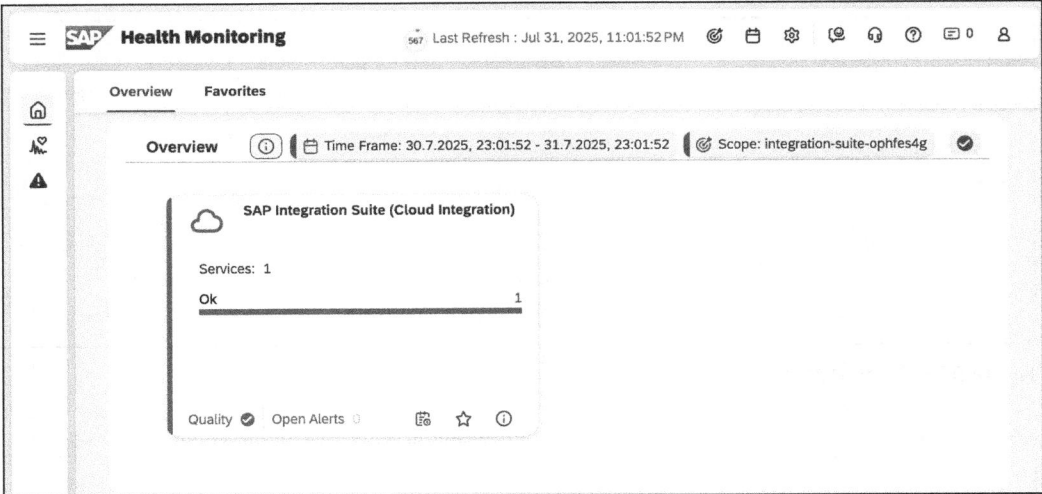

**Figure 9.38**  Health Monitoring Overview

There, you'll see an overview of the health status, as well as messages in tabular form, as shown in Figure 9.39. Clicking on the status icon will take you to the health status.

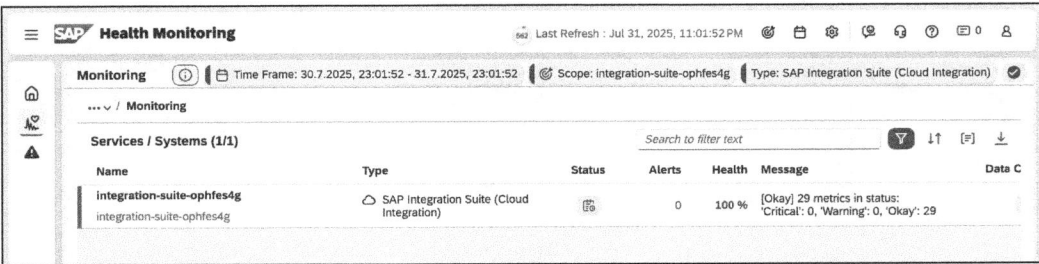

**Figure 9.39**  Health Monitoring Services and Systems

This will bring up the screen shown in Figure 9.40, where you can view the most important metrics for your **SAP Integration Suite (Cloud Integration)** instance.

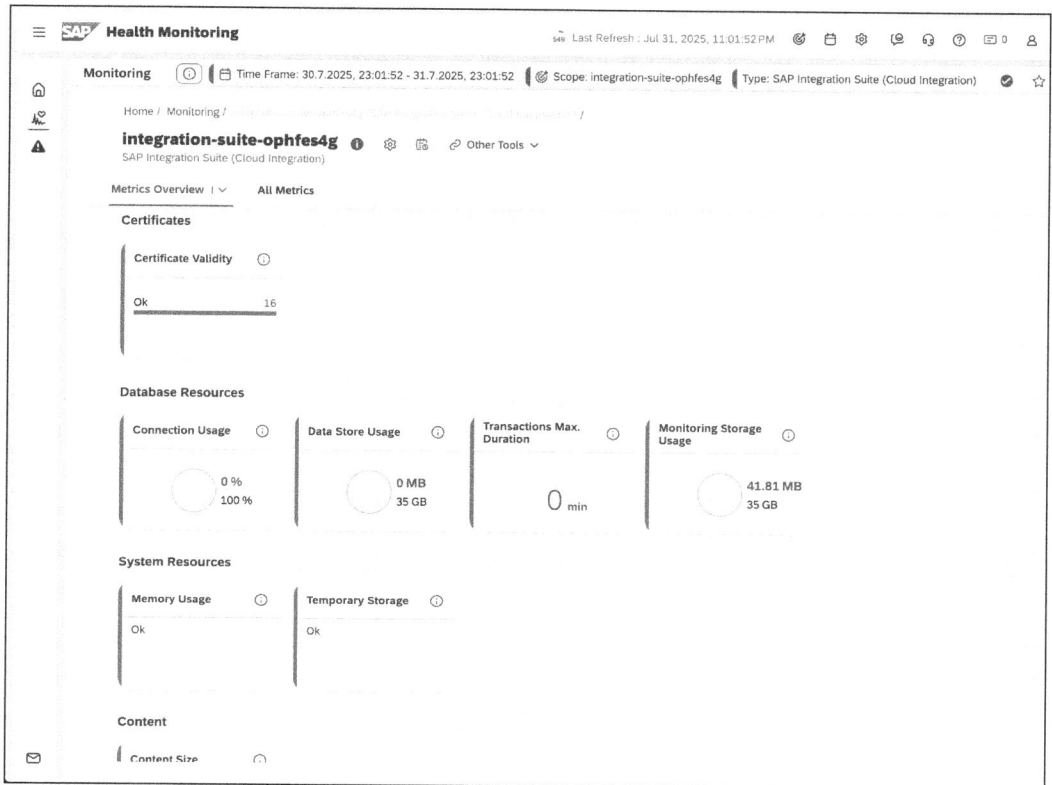

**Figure 9.40**  Health Monitoring Results

## 9.4    SAP Alert Notification Service for SAP BTP

SAP Alert Notification service for SAP BTP is a key component of the SAP BTP DevOps portfolio, and it's designed to provide operational teams with real-time, actionable insights across cloud-native and hybrid landscapes. The service provides a centralized API that allows service providers to publish alerts and enables consumers—such as administrators, developers, and external systems—to subscribe to and receive these alerts. The service's core functionality is the automated notification of technical and operational events that are relevant to business continuity and system health. The service can send these alerts via multiple channels, including email, Microsoft Teams, Slack, webhooks, and incident management tools. A major strength of the service is its standardized, environment-agnostic model, which ensures consistent configuration across all SAP BTP services. This allows organizations to apply unified monitoring and alerting strategies, regardless of the underlying architecture or runtime environment.

SAP Alert Notification service for SAP BTP primarily collects critical technical information from SAP BTP services, such Cloud Integration, Kyma, and Cloud Foundry. However, you can also extend it to support custom scenarios within your own applications.

Before you can use SAP Alert Notification, you must add it to the subaccount you want.
To do this, open the service assignment in the SAP BTP cockpit of the global account,
search for "Alert" under **Solutions**, and select **Alert Notification**, as shown in Figure 9.41.
In addition to the paid **standard** plan, you can test the entire service in the **free** tier. Add
the service plan you want by selecting it and clicking **Add 1 Service Plan**.

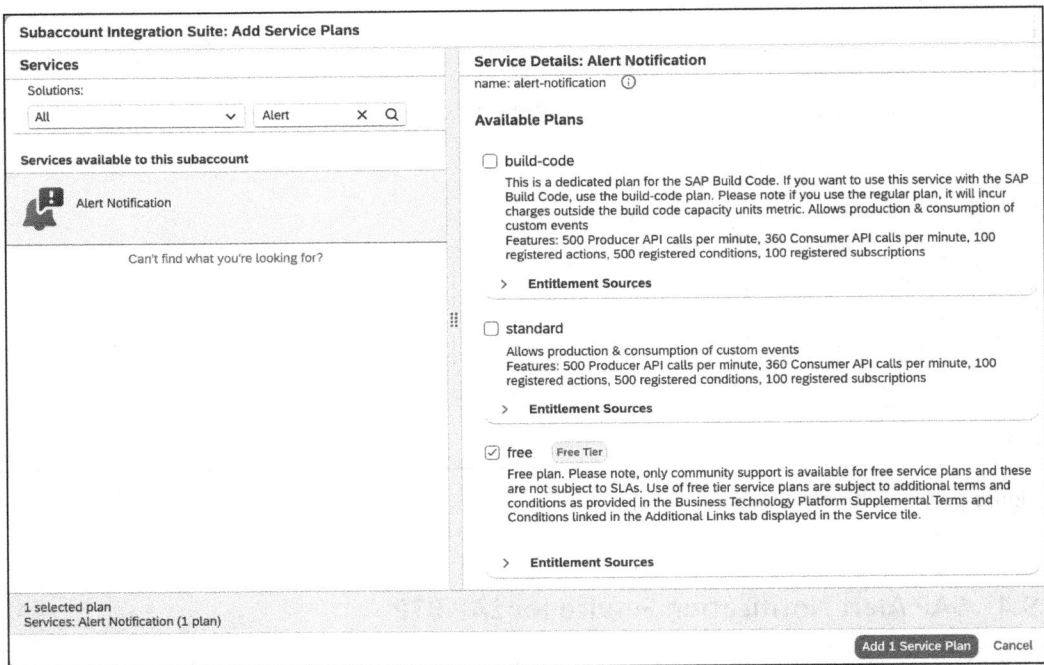

**Figure 9.41** Adding Service Plan

Then, switch to the subaccount and create a new service instance as usual. Select **Alert
Notification** as the **Service** and **Plan** you previously added, select **Cloud Foundry** as the
**Runtime Environment,** and assign an existing **Space** (see Figure 9.42). Enter a meaning-
ful name in the **Instance Name** field and then click **Next**.

If you don't want to change any of the parameters, leave them as they are and click
**Next** (see Figure 9.43). Then, in the **Review** section, check the settings you've made and
create the service instance by clicking **Create**.

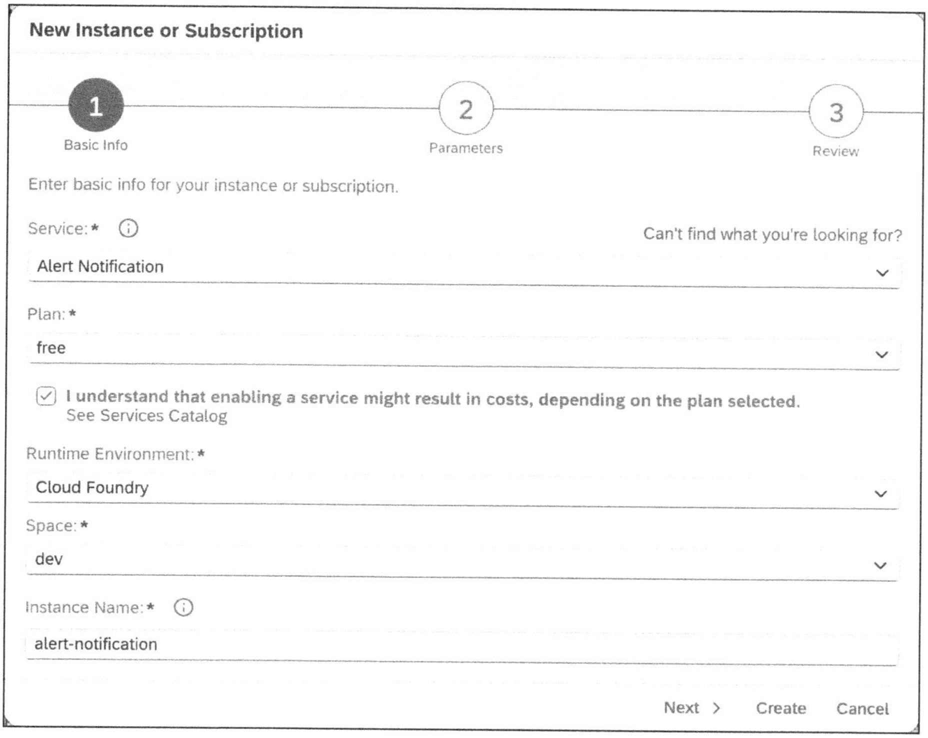

**Figure 9.42**  Creating Service Instance

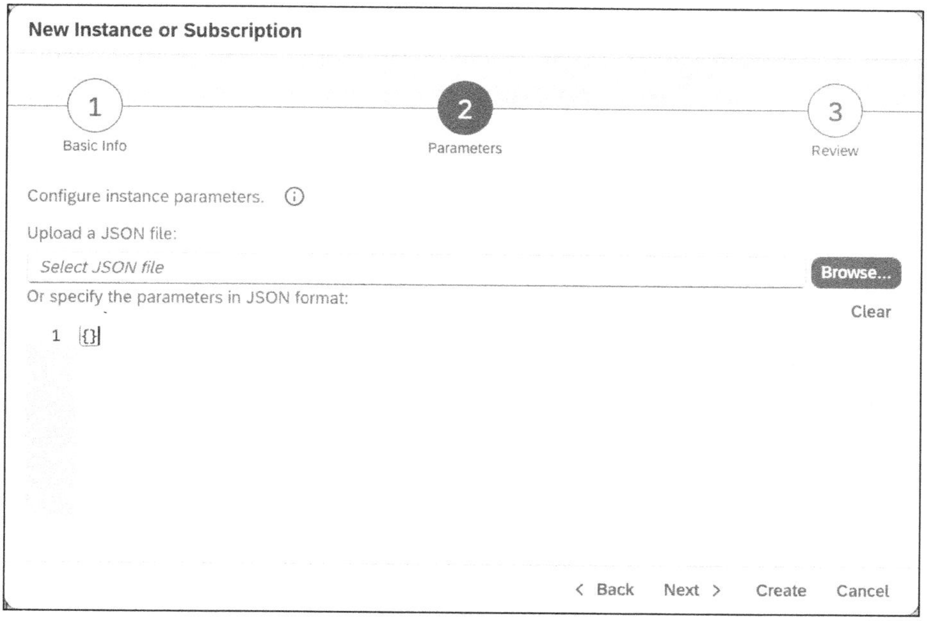

**Figure 9.43**  Configuring Parameters

Next, create a service key for this service as shown in Figure 9.44. This contains the credentials required to call the service from the iFlow.

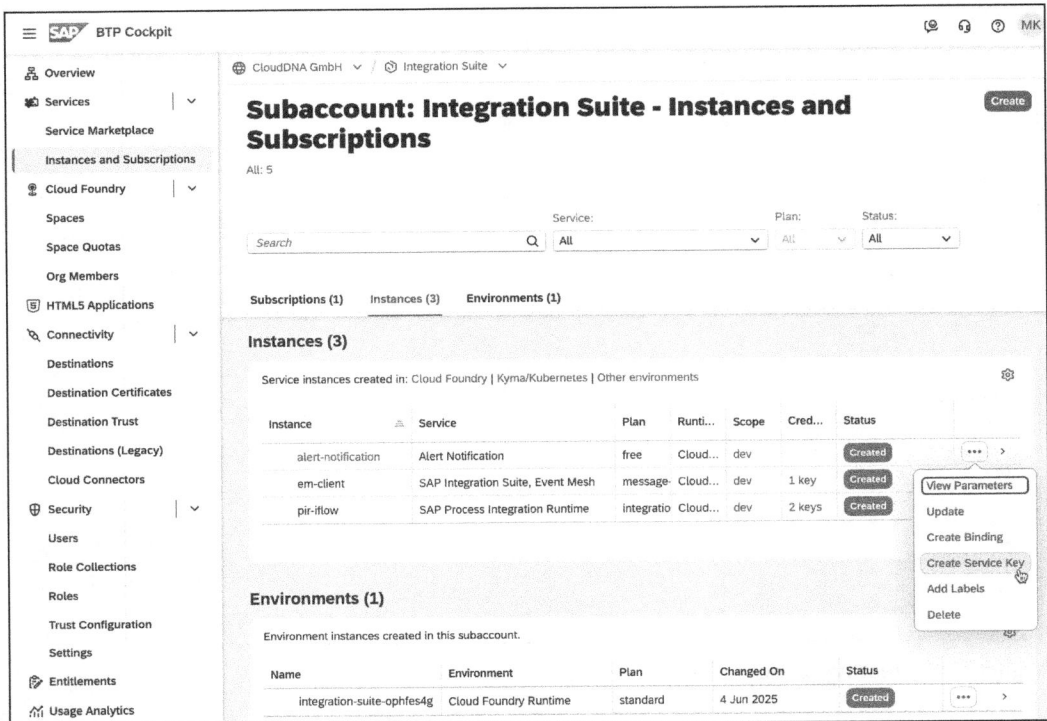

**Figure 9.44**  Creating Service Key

Then, assign a suitable **Service Key Name**, as shown in Figure 9.45, and adjust the parameters if necessary. (We'll leave the parameters unchanged in our example.) Then, click **Create**.

You can control the SAP Alert Notification service via an iFlow, which reads the message processing log via the Cloud Integration OData API and transfers it (either filtered or unfiltered) to SAP Alert Notification. To do this, you must create a service instance of the **SAP Process Integration Runtime** with the API service plan (see Figure 9.46).

Next, create a service key for this service instance, as shown in Figure 9.47.

Then, assign an appropriate **Service Key Name** as shown in Figure 9.48, select **ClientId/ Secret** as the **Key Type**, and click **Create**. At that point, you'll have created the conditions that are necessary for integration between Cloud Integration and SAP Alert Notification.

Next, you need to create the security artifacts for the service keys you previously created in Cloud Integration. Start by creating the user credentials for OData API access, as shown in Figure 9.49. Enter any **Name** you like, select **User Credentials** as the **Type,** and copy the ClientId of the service key into the **User** field and the ClientSecret into the **Password** field. Finally, click **Deploy**.

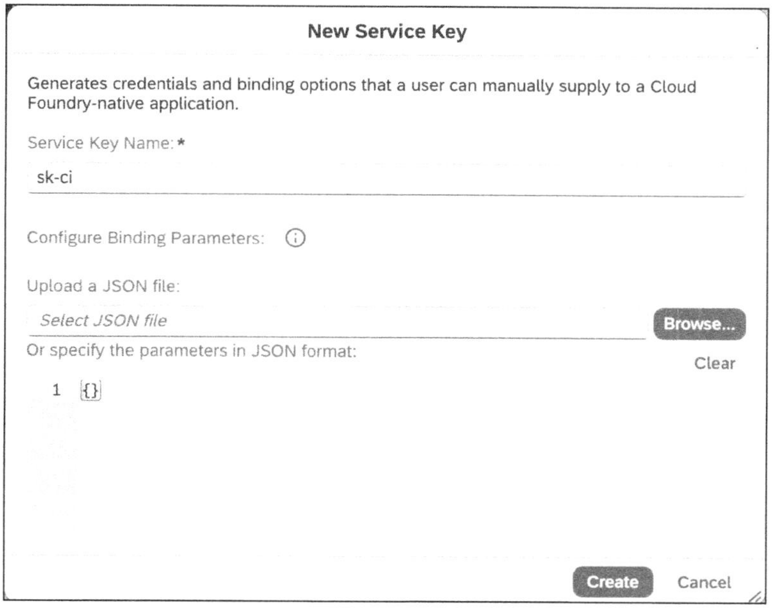

**Figure 9.45** Service Key Details

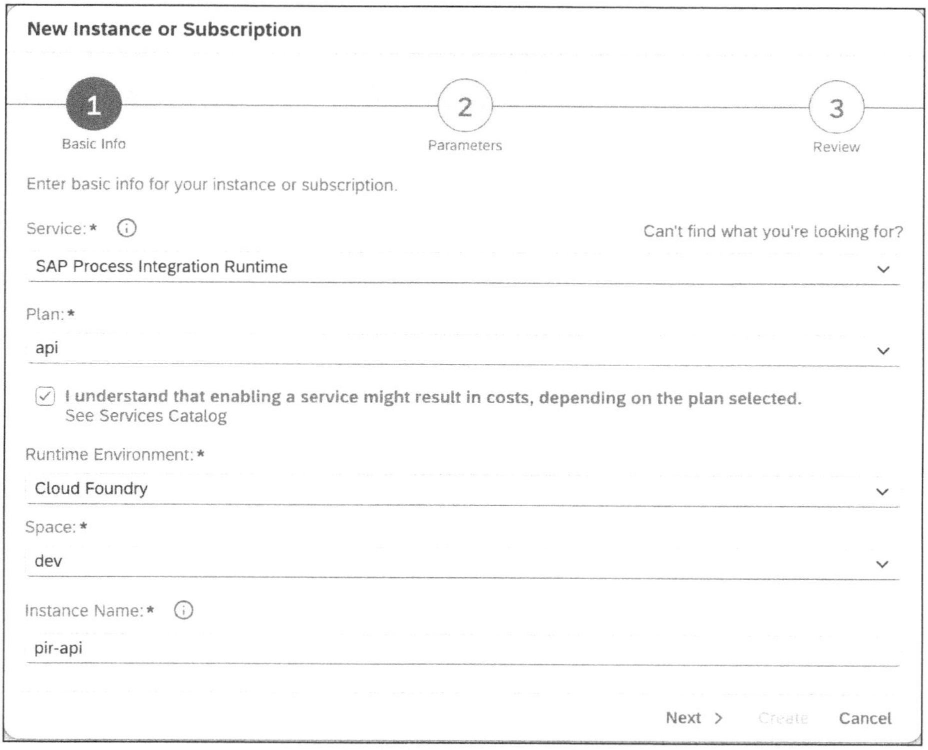

**Figure 9.46** Creating SAP Process Integration Runtime API

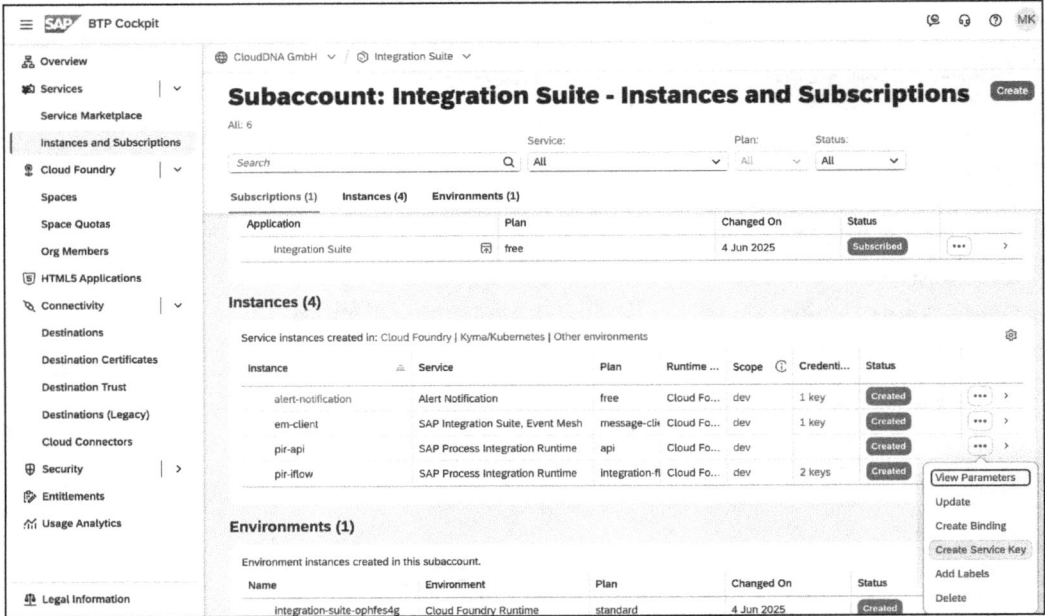

**Figure 9.47**  Creating Service Key

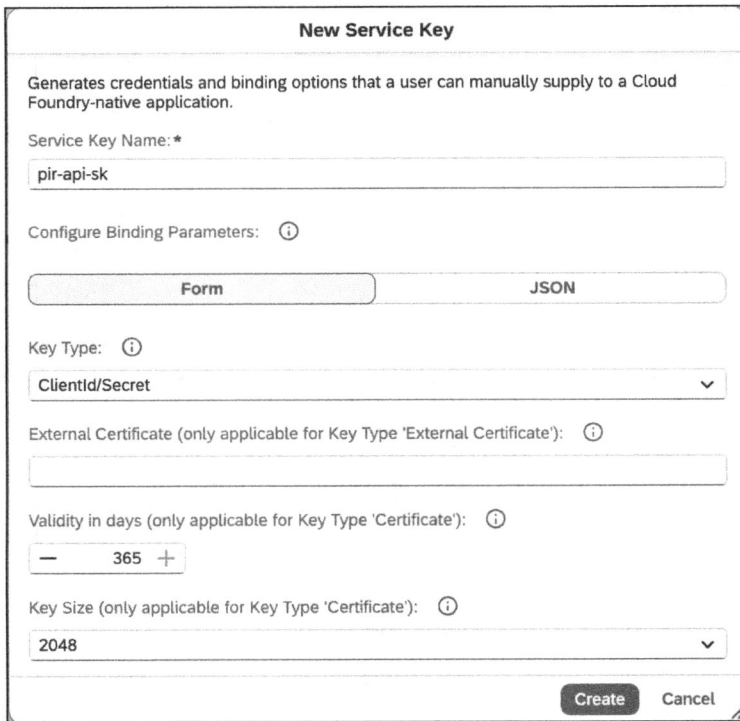

**Figure 9.48**  Service Key Details

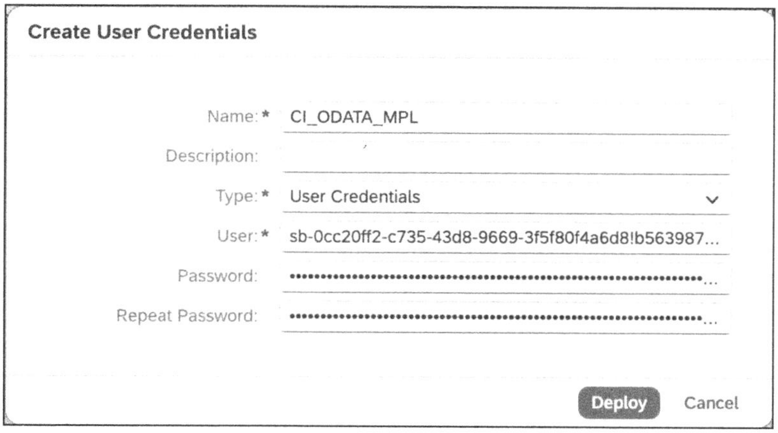

**Create User Credentials**

| | |
|---|---|
| Name:* | CI_ODATA_MPL |
| Description: | |
| Type:* | User Credentials ∨ |
| User:* | sb-0cc20ff2-c735-43d8-9669-3f5f80f4a6d8!b563987... |
| Password: | ●●●●●●●●●●●●●●●●●●●●●●●●●●●●●●●●●●●●●●●●●●●●●●●... |
| Repeat Password: | ●●●●●●●●●●●●●●●●●●●●●●●●●●●●●●●●●●●●●●●●●●●●●●●... |

**Deploy**   Cancel

**Figure 9.49**  Creating User Credentials for OData APIs

Then, you need to create another security artifact. Select the **OAuth2 Client Credentials** type, enter "SERVICE_TECHNICAL_CLIENT" as the **Name** (which will also be used in the iFlow), and copy the **Token Service URL**, **Client ID**, and **Client Secret** from the service key you created for SAP Alert Notification. Finally, click **Deploy** (see Figure 9.50).

**Create OAuth2 Client Credentials**

| | |
|---|---|
| Name:* | SERVICE_TECHNICAL_CLIENT |
| Description: | |
| Token Service URL:* | https://integration-suite-ophfes4g.authentication.eu1... |
| Client ID:* | sb-ef72973b-df23-4846-a35e-923d5d4dc2bb!b5639... |
| Client Secret:* | ●●●●●●●●●●●●●●●●●●●●●●●●●●●●●●●●●●●●●●●●●●●●●●●●... |
| Client Authentication:* | Send as Request Header ∨ |
| Scope: | |
| Content Type: | application/json ∨ |
| Resource: | |
| Audience: | |

**Custom Parameters**                                    Add    Delete

☐   Key                    Value                    Send as Part of

No data

**Deploy**   Cancel

**Figure 9.50**  Creating Client Credentials for SAP Alert Notification Service

SAP provides an integration package for integration with SAP Alert Notification that you simply need to copy and configure. In the side menu, navigate to **Discover • Integrations,** search for the "SAP Cloud Integration with SAP Alert Notification Service for SAP BTP" package, go to the package details, and click **Copy** (see Figure 9.51) to transfer the integration package to your Cloud Integration instance.

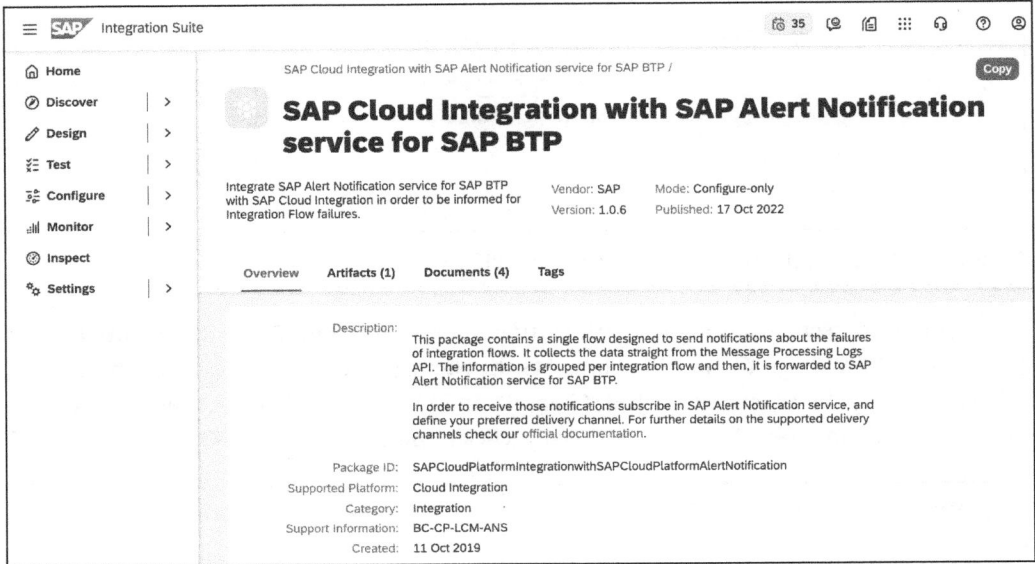

**Figure 9.51**  Copying Integration Package

In the side menu, navigate to the **Design • Integrations and APIs** area, open the integration package you copied earlier, and go to the **Artifacts** tab (see Figure 9.52). There, you'll find an iFlow named **Send notifications for failed Message Processing Logs**. Click on **Configure** under **Actions**.

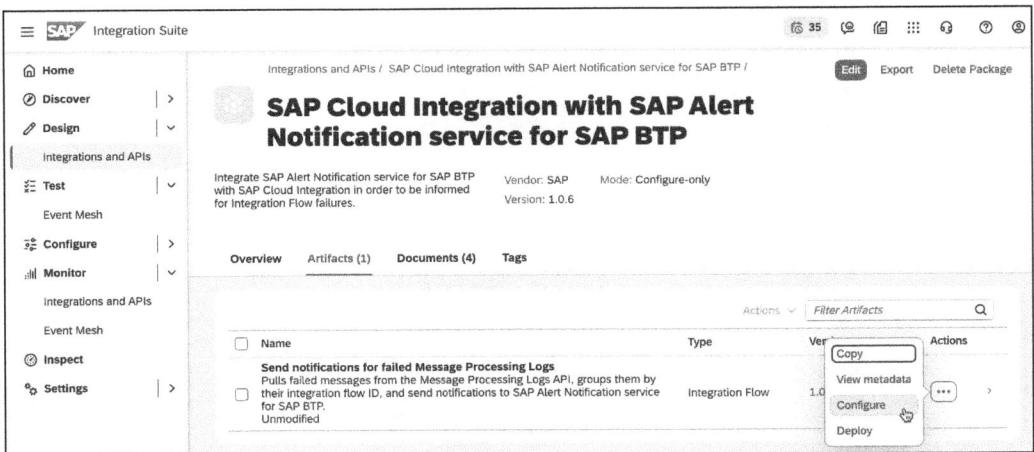

**Figure 9.52**  Configuring iFlow

Next, open the **Receiver** tab (see Figure 9.53) and copy the URL of the Cloud Integration tenant management node into the **CI Tenant Base URL** field. Then, enter the name of the user credentials you created earlier for the Process Integration API service instance into the **Credential Name** field.

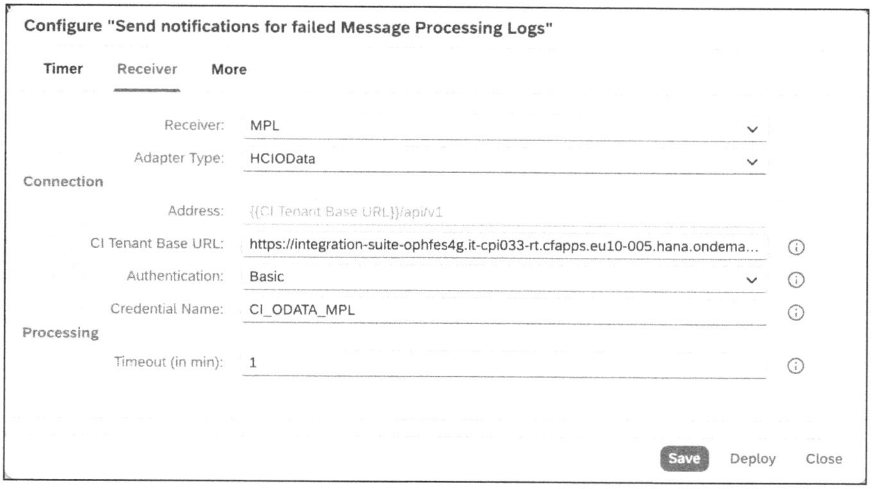

Configure "Send notifications for failed Message Processing Logs"

| Timer | Receiver | More |

Receiver: MPL
Adapter Type: HCIOData

Connection

Address: {{CI Tenant Base URL}}/api/v1
CI Tenant Base URL: https://integration-suite-ophfes4g.it-cpi033-rt.cfapps.eu10-005.hana.ondema...
Authentication: Basic
Credential Name: CI_ODATA_MPL

Processing

Timeout (in min): 1

Save    Deploy    Close

**Figure 9.53** Configuring OData Adapter

Then, open the **More** tab (see Figure 9.54), which is where the connection to SAP Alert Notification is configured. The **CI Tenant Base URL** is taken from the **Receiver** tab, but you'll need to copy the URL from the service key of SAP Alert Notification that you previously created into the **Service Resource Events URL** field. Add the "/cf/producer/v1/resource-events" path to this, and finally, click **Save** and then **Deploy**.

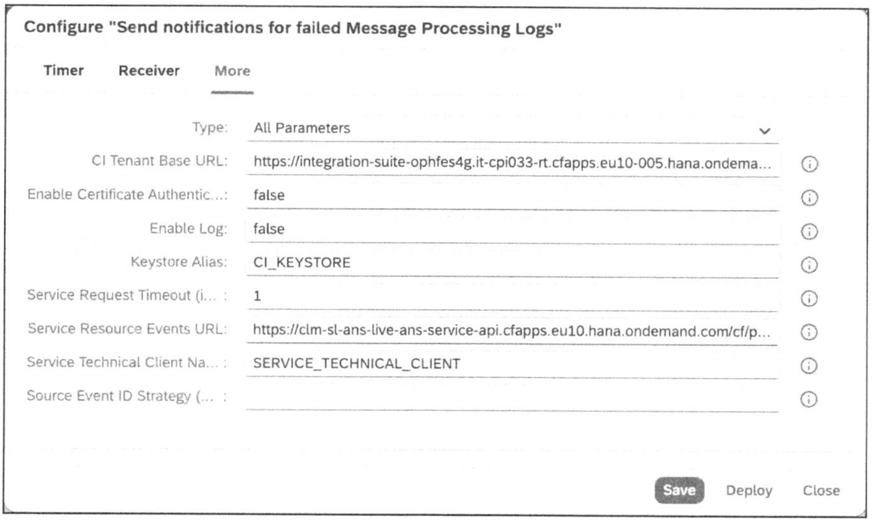

Configure "Send notifications for failed Message Processing Logs"

| Timer | Receiver | More |

Type: All Parameters
CI Tenant Base URL: https://integration-suite-ophfes4g.it-cpi033-rt.cfapps.eu10-005.hana.ondema...
Enable Certificate Authentic...: false
Enable Log: false
Keystore Alias: CI_KEYSTORE
Service Request Timeout (i... : 1
Service Resource Events URL: https://clm-sl-ans-live-ans-service-api.cfapps.eu10.hana.ondemand.com/cf/p...
Service Technical Client Na... : SERVICE_TECHNICAL_CLIENT
Source Event ID Strategy (... :

Save    Deploy    Close

**Figure 9.54** Configure SAP Alert Notification Connection

At this point, you've learned all the steps required to ensure that incorrect messages are transferred to SAP Alert Notification. However, depending on the implementation of the iFlows, messages may still be saved with the status **Completed** even if an error occurred during processing due to improper use of exception handling.

## 9.5   Summary

Stable and traceable monitoring is essential for successfully operating iFlows in Cloud Integration. In this chapter, you learned about the key tools and concepts that help you maintain an overview and respond quickly and effectively to errors.

Section 9.1 showed you how to use integrated monitoring to analyze message flows, evaluate runtime data, and search for errors in a targeted manner using the payload view and trace functions. Section 9.2 presented specific debugging and error-handling strategies, such as dealing with exception subprocesses, user-defined logging, and typical adapter configuration pitfalls.

Section 9.3 introduced SAP Cloud ALM, which is a central tool that provides comprehensive insights into cloud integrations, including status monitoring and automated alerts at the business process level. Section 9.4 introduced SAP Alert Notification service for SAP BTP, which helps you proactively identify critical events and automatically notify relevant people or systems. It's an important building block for operational management.

PART II

# Practical Interface Scenarios

# Chapter 10

# Integration with SAP SuccessFactors Integration Center

*The SAP SuccessFactors integration center is a tool within the SAP SuccessFactors suite that enables you to create, execute, plan, and monitor simple integrations between SAP SuccessFactors and other systems, including both SAP and third-party solutions. It's designed specifically for business and nontechnical users (such as HR analysts) who want to implement integration scenarios but don't have in-depth programming knowledge.*

The integration center in SAP SuccessFactors connects simple reporting tools and comprehensive middleware solutions, such as Cloud Integration. It allows business users and administrators to create integrations via an easy-to-use, step-by-step process that doesn't require any technical knowledge. In addition to a wide range of predefined templates, you can create and reuse your own templates—which makes it much faster to implement standardized and customized integration scenarios. The integration center is great for regular, organized data transfers (like monthly exports to payroll accounting and reports for insurance companies). It's based on the SAP SuccessFactors OData APIs, which enable access to data from various modules such as SAP SuccessFactors Employee Central, Recruiting, Performance & Goals, and Onboarding. CSV, TXT, XML, and EDI files are often created, and they can be stored on an SFTP server and retrieved by target systems. The integration center also supports SOAP- and REST-based outbound integrations to cover a wide range of target architectures. Users can select and customize fields, define filter rules, perform field transformations and calculations, use lookup tables to enrich data, and individually configure headers and footers.

You can plan and automate the process of integrating these systems by using a scheduler. Built-in monitoring functions allow you to check and analyze status, logs, and errors any time, and that supports smooth operation and quick troubleshooting. With the integration center, SAP SuccessFactors provides a flexible, user-friendly tool that enables companies to efficiently design data flows while bridging the gap between independent departments and IT-driven middleware approaches.

This chapter will give you all the information you need about the integration center. Section 10.1 provides a basic overview of the architecture, its main functions, and common ways in which it's used. It will also discuss how the integration center is different from traditional middleware approaches and show in which cases it can be an ideal addition or alternative.

Section 10.2 explains the technical basics of the underlying interfaces, especially REST. You'll learn how to use a REST architecture to reliably get data from modules like SAP SuccessFactors Employee Central, Recruiting, and Performance & Goals and transfer it to target systems (in our case, Cloud Integration). This makes it easy to connect to different system landscapes, and it supports both on-premise and cloud scenarios.

Section 10.3 covers the technical basics of SOAP-based integrations, and it will teach you how to use a SOAP for outbound integrations. Section 10.4 talks about how to handle errors correctly so that users can trust integrations and integrations continue to work in the future. It will explain how you can detect common errors early on, check logs, set up notifications, and implement measures to ensure high operational reliability.

All these sections together make a complete guide that will help you get the most out of the SAP SuccessFactors integration center. It covers how to set up the integration center, how to manage it professionally, and how to improve your integration scenarios. You'll gain technical expertise and the methodological tools you need to implement integrations in a flexible and efficient manner.

*Intelligent services* provide another way to integrate with SAP SuccessFactors, and they're designed for integration during specific events. They recognize certain events in the business world—like when an employee joins, a person is promoted, or a location changes—and they automatically trigger actions in other systems or processes. Intelligent services are more focused on real-time and process integration, while the integration center is traditionally used for batch-based data exchange. We'll talk more about the intelligent services in Chapter 11.

## 10.1   Overview of the Integration Center

The integration center is part of the SAP SuccessFactors HCM suite package, and it works even without Employee Central. It supports building integrations with OData APIs for the following modules:

- SAP SuccessFactors Employee Central
- SAP SuccessFactors Recruiting
- SAP SuccessFactors Performance & Goals
- SAP SuccessFactors Succession & Development

In the integration center, you can create and edit integrations (see Figure 10.1), monitor existing integrations, and access the data model navigator. You'll learn how to create integrations later in this section. In the **Monitor Integrations** section, you can see how the different integrations are doing.

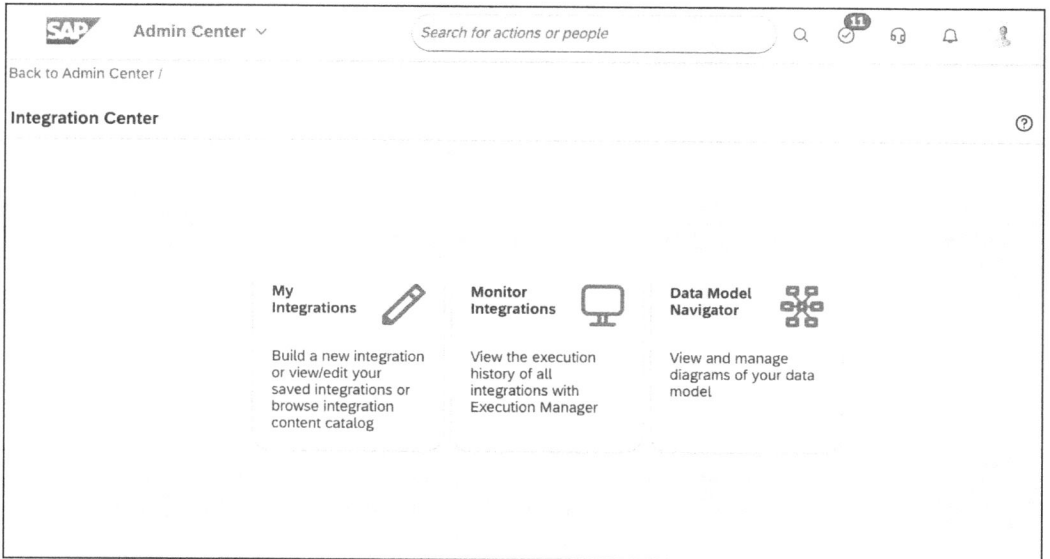

**Figure 10.1** Integration Center Overview

Figure 10.2 shows a summary of the last 24 hours with successful and failed integrations.

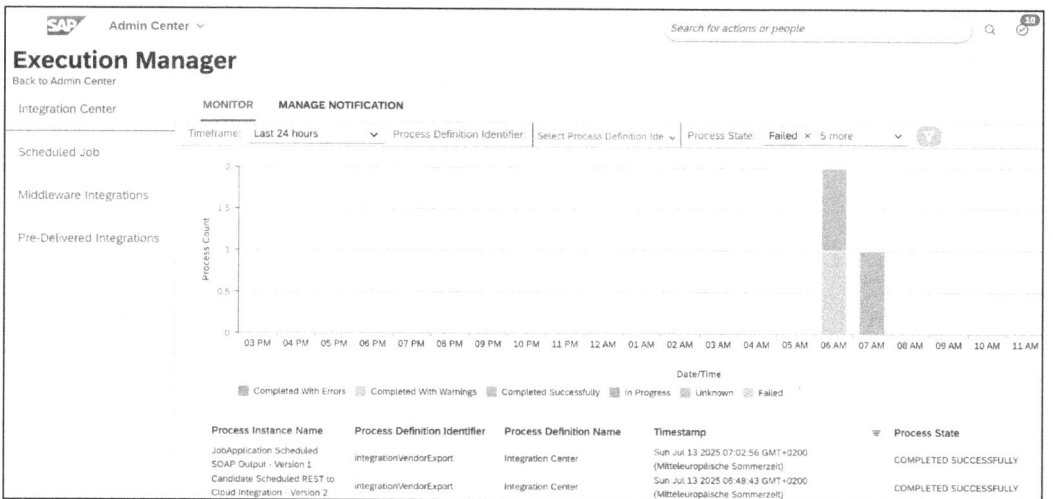

**Figure 10.2** Integration Center Monitoring

A key reason you'll want to use the integration center is because it lets you move data into SAP SuccessFactors and move it out again from SAP SuccessFactors. This lets you use the data in different ways in the business. Although SAP SuccessFactors is *the* human experience management (HXM) solution, it's rarely used as a standalone application. Instead, HR data must be regularly exchanged with SAP S/4HANA, SAP S/4HANA

Cloud, and even third-party applications like payroll systems, time tracking, and analytics solutions. The integration center can send data directly to Cloud Integration, in the following ways:

- By using SFTP files with a structured format (like CSV or XML) to send data automatically to an SFTP server and then retrieve it by an integration flow from Cloud Integration
- By using SOAP-based services to send data to a specific SOAP endpoint of Cloud Integration
- By using REST APIs, which provide a quick, modern, and adaptable way to send data to Cloud Integration right away

You can also use the integration center for different types of integration. The first step you take when creating an integration is to select a **Trigger Type**, such as by using a scheduler. However, the Intelligence Services can also start the integration process. You can also start the integration directly from the application or user interface in SAP SuccessFactors (see Figure 10.3). We discuss the intelligent services in Chapter 11.

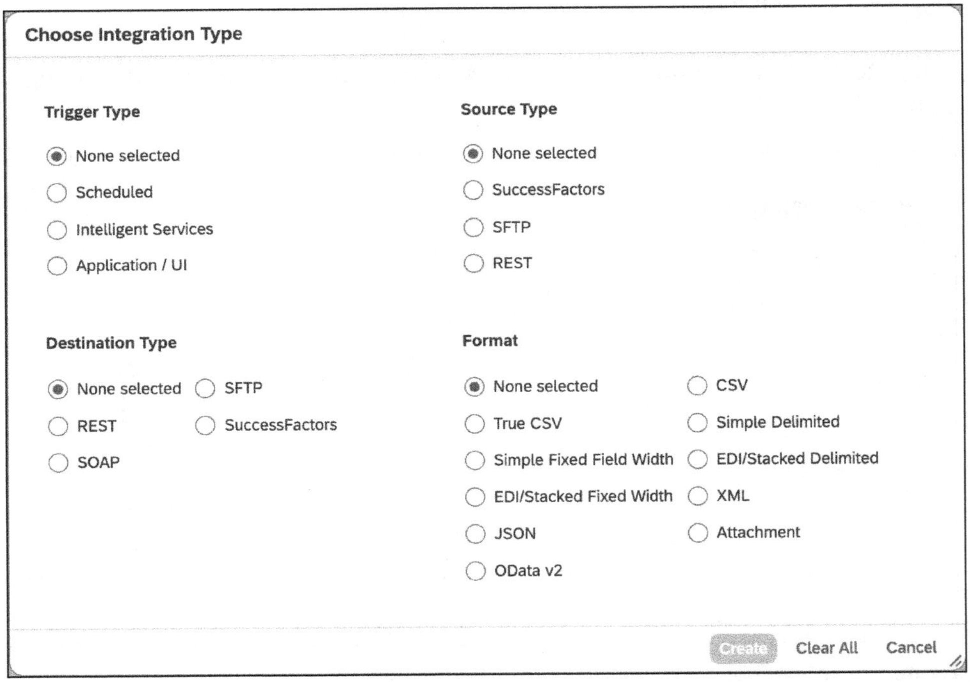

**Choose Integration Type**

| Trigger Type | Source Type |
|---|---|
| ● None selected | ● None selected |
| ○ Scheduled | ○ SuccessFactors |
| ○ Intelligent Services | ○ SFTP |
| ○ Application / UI | ○ REST |

| Destination Type | | Format | |
|---|---|---|---|
| ● None selected  ○ SFTP | | ● None selected | ○ CSV |
| ○ REST  ○ SuccessFactors | | ○ True CSV | ○ Simple Delimited |
| ○ SOAP | | ○ Simple Fixed Field Width | ○ EDI/Stacked Delimited |
| | | ○ EDI/Stacked Fixed Width | ○ XML |
| | | ○ JSON | ○ Attachment |
| | | ○ OData v2 | |

Create   Clear All   Cancel

**Figure 10.3**  Integration Types

The next step is to choose the **Source Type** (i.e., where the data come from), and you can choose **SuccessFactors**, **SFTP**, or **REST**. If you want to transfer data from SAP SuccessFactors to another system, such as Cloud Integration, you'll use **SuccessFactors**. This is called outbound integration. **SFTP** and **REST** are for integrating data that has been sent from another system to SuccessFactors. Depending on the source type you choose, you'll have different choice of **Destination Type** (i.e., where the data goes to). Choose **REST, SOAP, SFTP**, or **SuccessFactors**.

Depending on what choices you've made so far, you'll be able to choose one **Format** or another. If you've chosen the **REST** destination type, you can choose between a **JSON** format and an **XML** format. If you've chosen **SOAP**, **XML** will be the default format, and if you've chosen the **SFTP** destination type, you can select all format options except **OData v2**.

## 10.2    Using REST Interfaces

In this section, we'll look at how to achieve integration by using REST API. For starters, you must make its endpoint available by creating an iFlow (see Figure 10.4). We've chosen a simple iFlow that doesn't have any functionality except for the HTTP sender adapter, which provides our REST endpoint. As you can see in the figure, all you need to do is define an **Address**, the **Authorization**, and the associated **User Role**. You can also uncheck the **CSRF Protected** box so you don't have to request a token before the post request that transfers the data. This simplifies the process. Finally, you save the iFlow and deploy it.

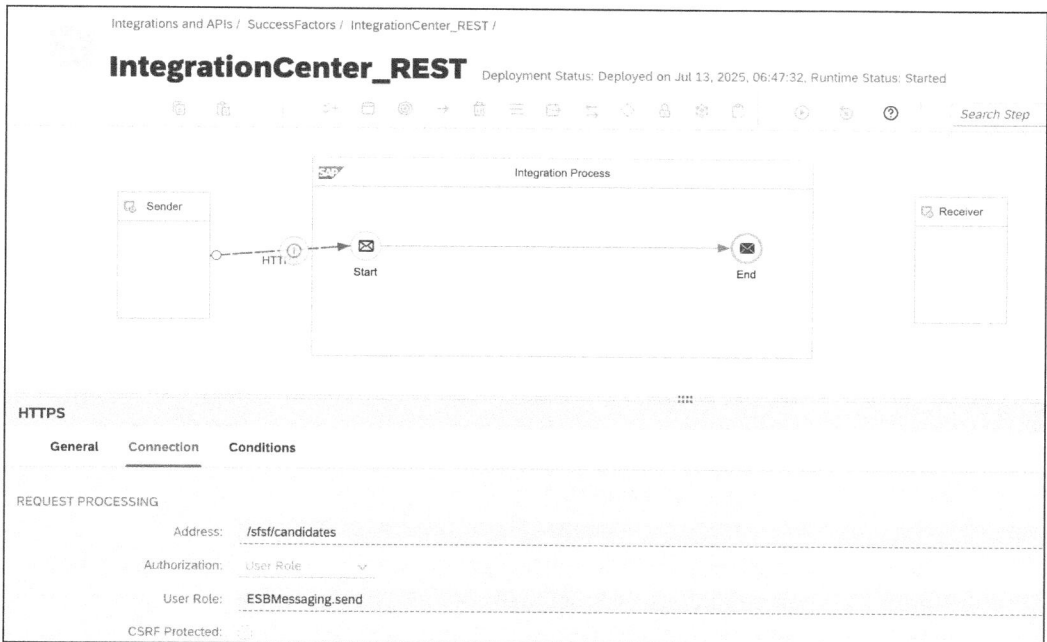

**Figure 10.4**  REST API iFlow

Then, you can jump to the iFlow you just deployed in **Monitor Integrations** on the screen shown in Figure 10.5, where you'll see the HTTP **Endpoint**. You'll need this later in SAP SuccessFactors for configuration.

**Figure 10.5**  REST Endpoint

Next, open the integration center in SAP SuccessFactors and click on **Create · More Integration Types** from the context menu there (see Figure 10.6).

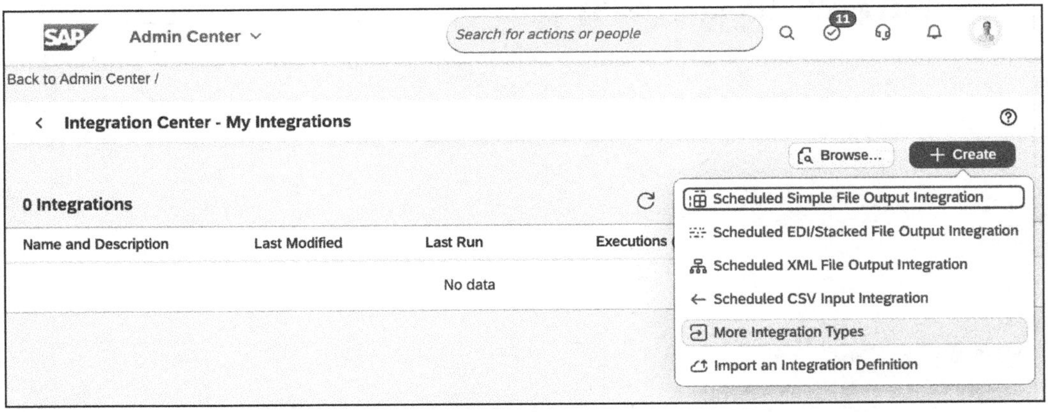

**Figure 10.6**  Creating New Integration

In our example, the export of data is triggered by a scheduler. Therefore, as shown in Figure 10.7, you should select **Scheduled** for the **Trigger Type SuccessFactors** for the **Source Type**, **REST** for the **Destination Type**, and **JSON** for the **Format**. Finally, click the **Create** button.

Then, you must choose the starting entity. We've chosen the **Candidate**, and you should click on the **Select** button (see Figure 10.8).

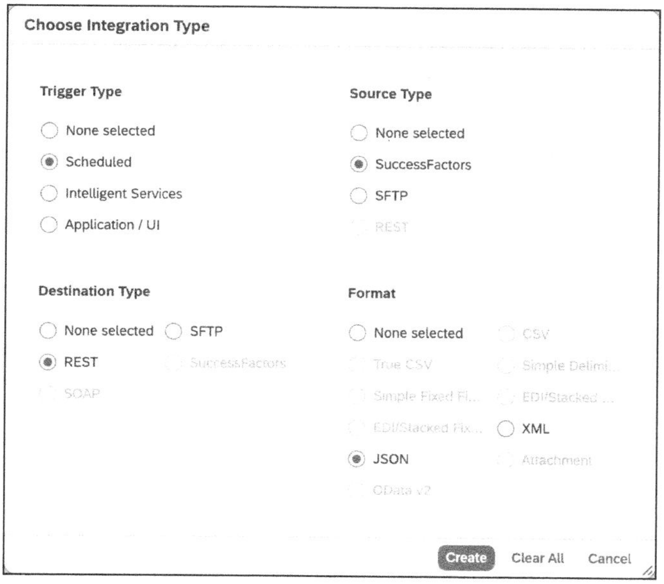

**Figure 10.7** Configuring Integration Type

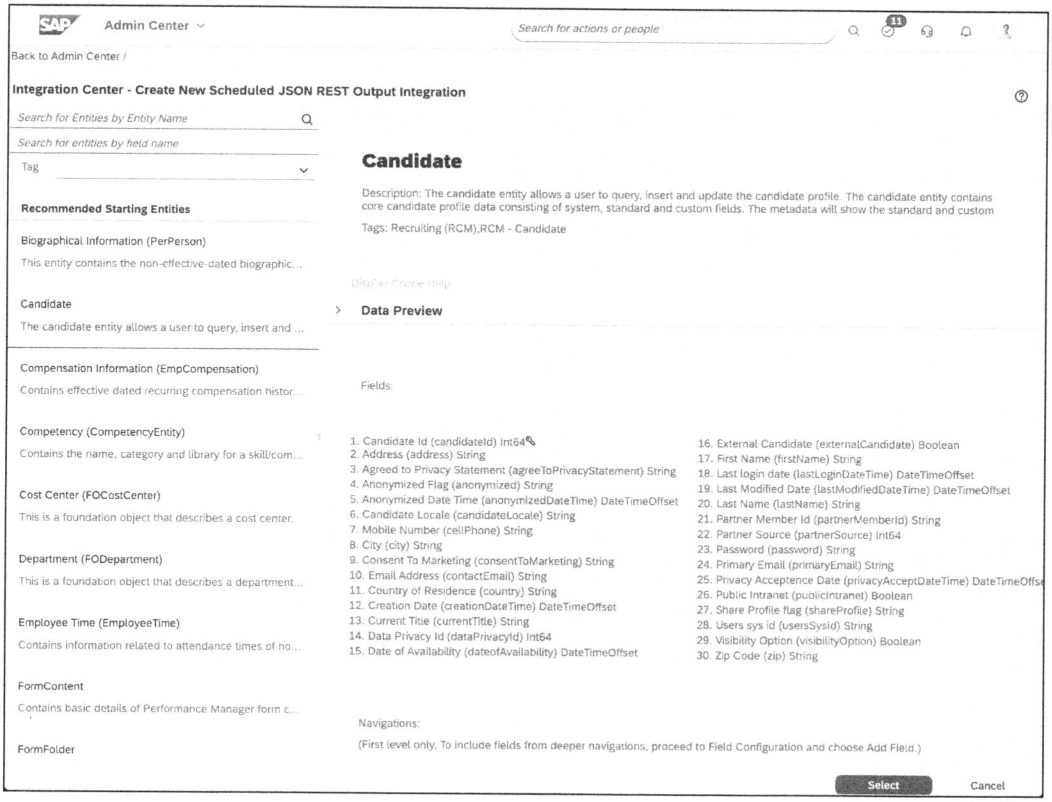

**Figure 10.8** Selecting Starting Entity

Now, you need to click through the wizard step by step. In the first step, you configure the options as shown in Figure 10.9. You assign an appropriate **Integration Name** and **Description**, and you can also specify the **Destination Page Size**, which controls how many messages are bundled and sent to the destination. In our example, we've chosen the value **1**, which means that each candidate is transferred individually and triggers its own iFlow.

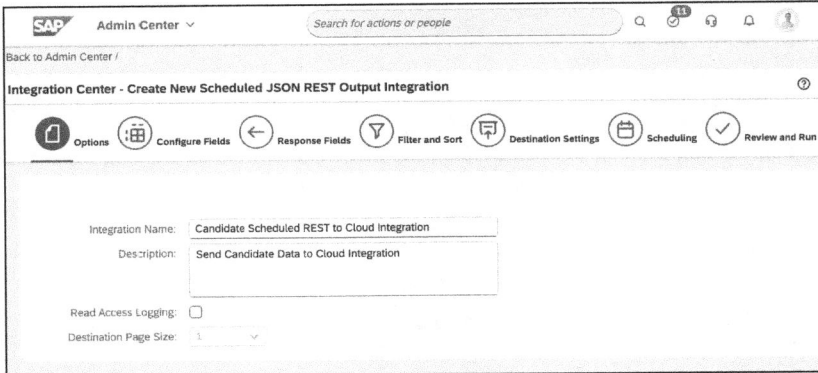

**Figure 10.9**  Defining Options

In the second step, which is called **Configure Fields**, you need to build the structure that will be sent to the iFlow. You do this manually, as shown in Figure 10.10. Click on the **+** button and select **Insert Sibling Element** from the popover menu.

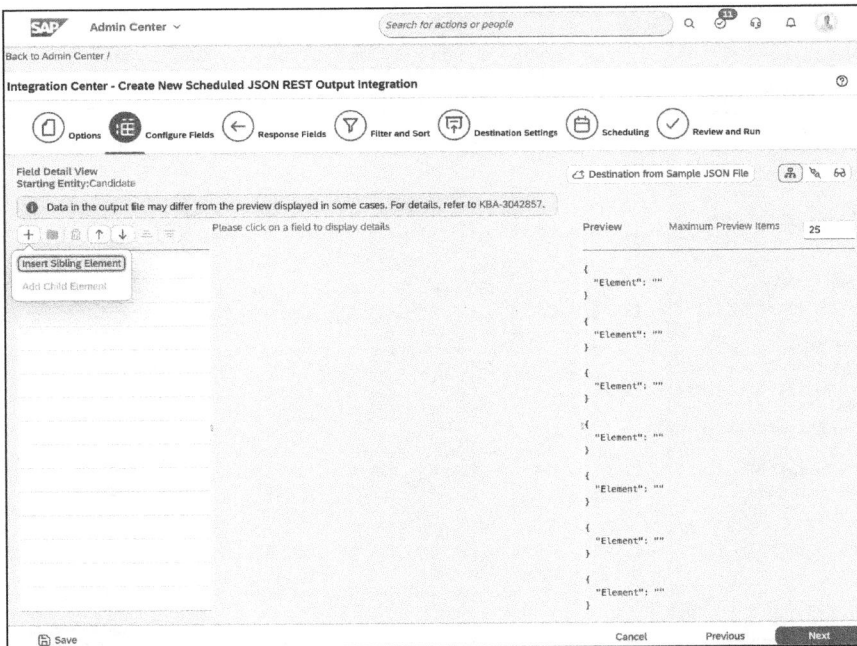

**Figure 10.10**  Configuring Fields and Defining Structure

Then, select the inserted element to navigate to the details as shown in Figure 10.11. There, you can change the **Label** to a meaningful name and also assign a **Description**. As you can see, the element is defined as the **Edm.String** data type by default.

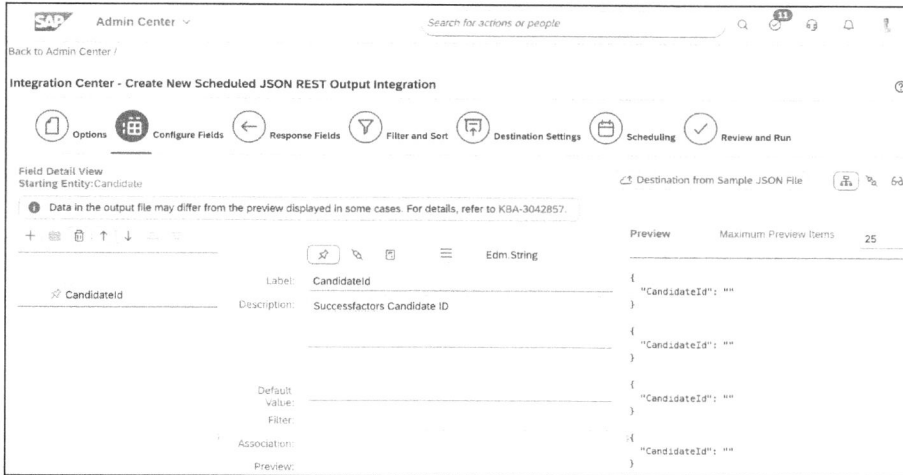

**Figure 10.11** Configuring Label

The next step is to map the SuccessFactors entity candidate to the target field you just created, as shown in Figure 10.12. To do this, click the ⬚ button and drag and drop the field from the left side (i.e., from the **Candidate** entity) to the right side of your target structure.

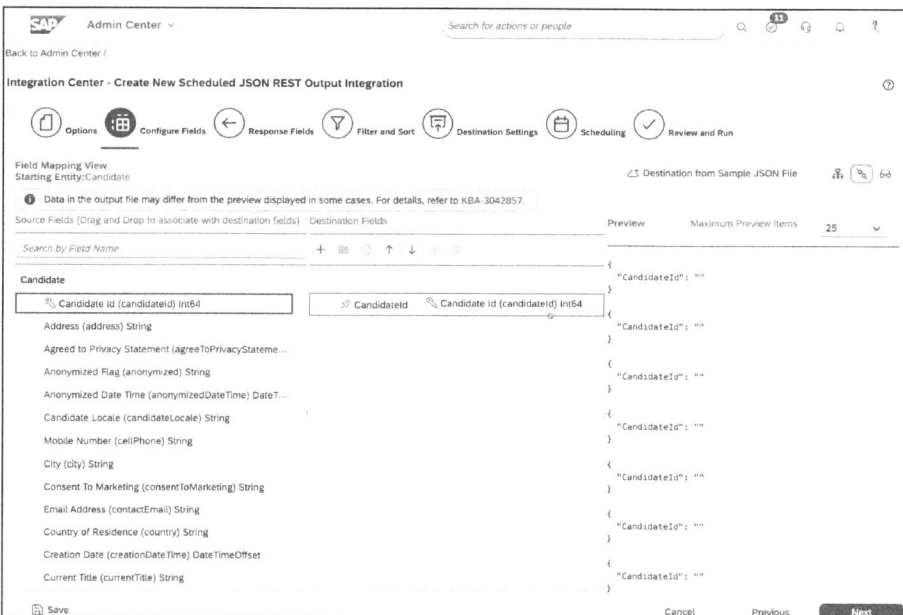

**Figure 10.12** Mapping Fields from Source Entity

Next, you create the other elements of the structure. In Figure 10.13, you can see that we've added the **Phone**, **Email**, and **CreatedAt** fields and embedded a structure named **Name**, which has the **Firstname** and **Lastname** child elements.

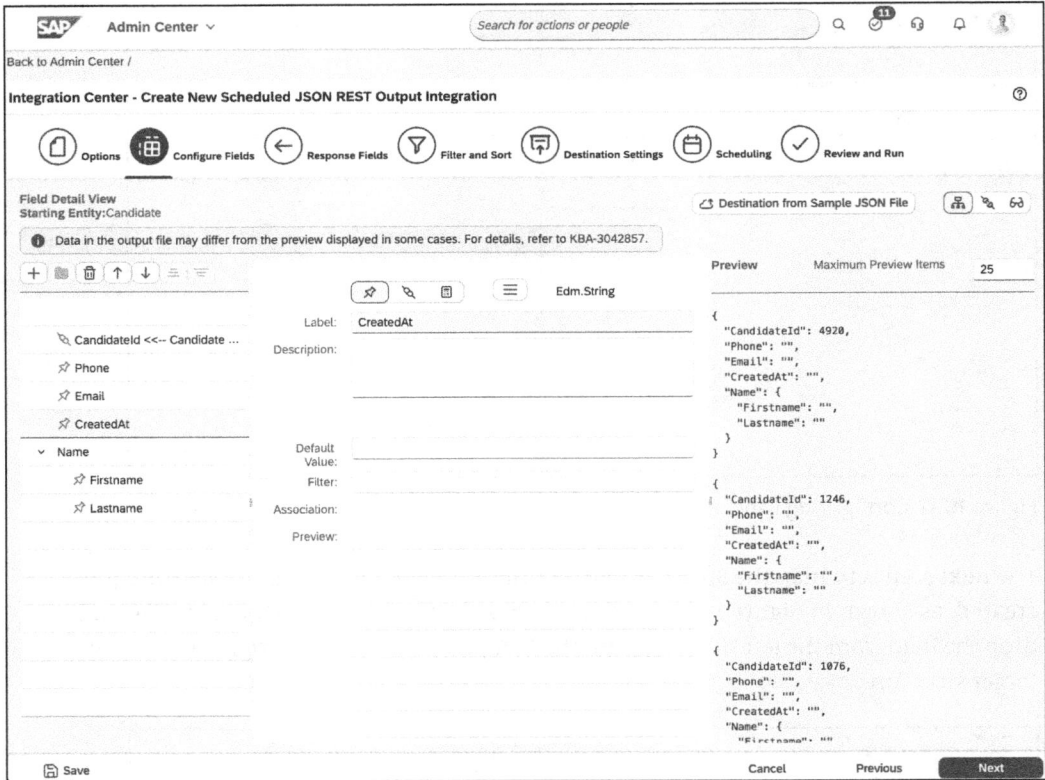

**Figure 10.13** Adding Another Field

If you look at the details of the **CreatedAt** field as shown in Figure 10.14, you'll notice that the **Edm.String** data type has been selected here as well. That's incorrect because this is a date or timestamp, so you can adjust the data type by clicking on ☰ to bring up a menu and selecting the **More Field Options** option from it.

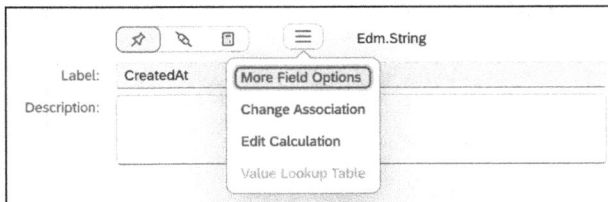

**Figure 10.14** Opening Field Options Menu

Next, select **DateTime** for the **Data Type** (see Figure 10.15) and click **OK** to close the dialog box.

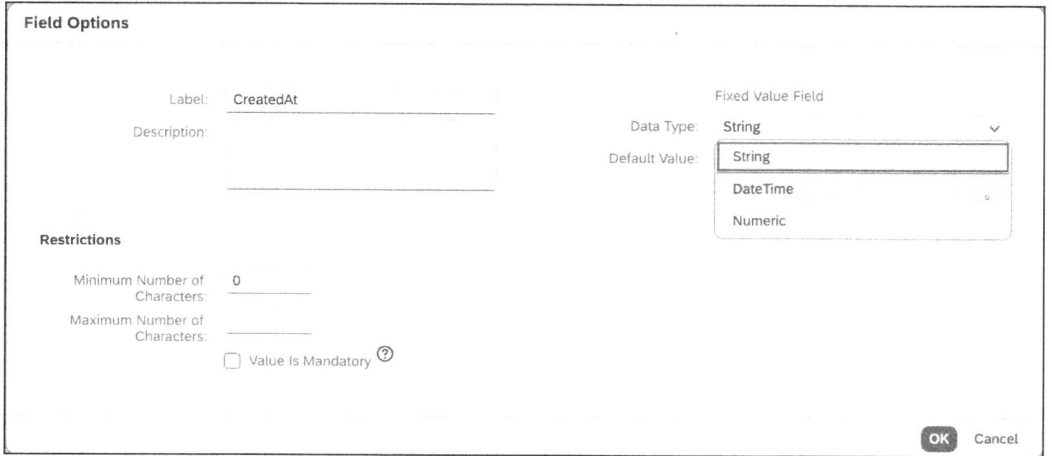

**Figure 10.15** Changing Data Type

Then, check again that all fields in the structure have been mapped, as shown in Figure 10.16. Also look at the preview on the right-hand side of the screen and check that all elements have been filled with the correct values. (These values are taken from the data that's available in the SAP SuccessFactors system.) Then, click on the **Next** button.

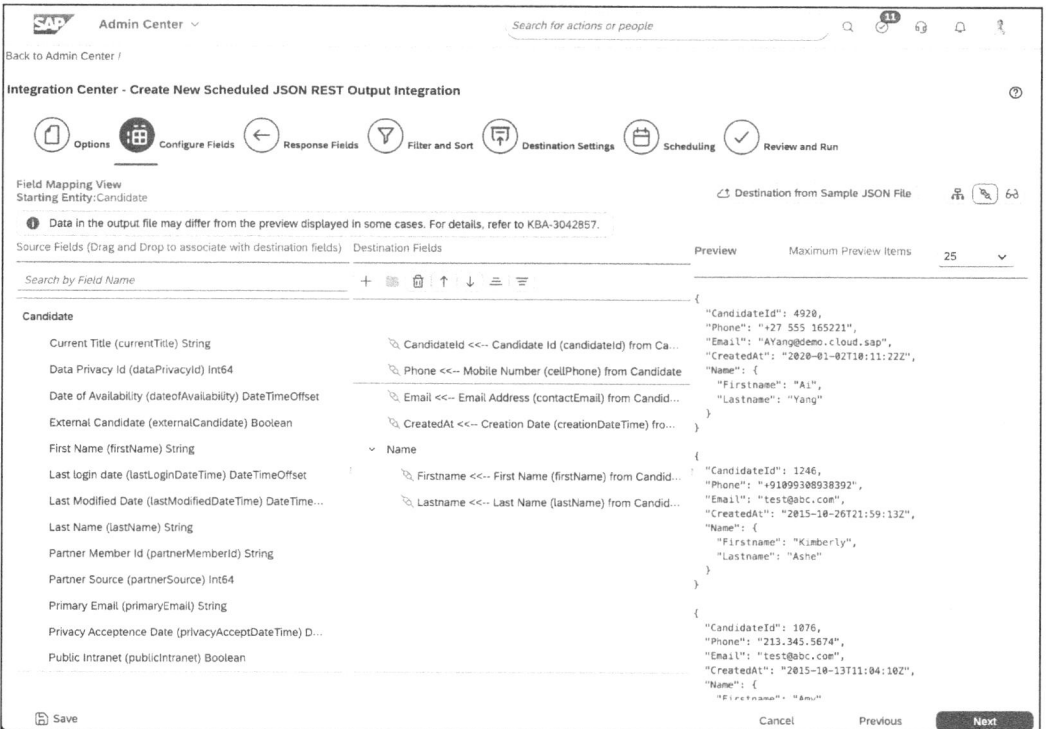

**Figure 10.16** Final Request Structure and Data Preview

At this point, you can define the response fields as shown in Figure 10.17. This step is optional, and in the example, we uploaded a sample JSON by clicking on **Upload Sample JSON**, which also displays the sample in the **Sample Data View** area.

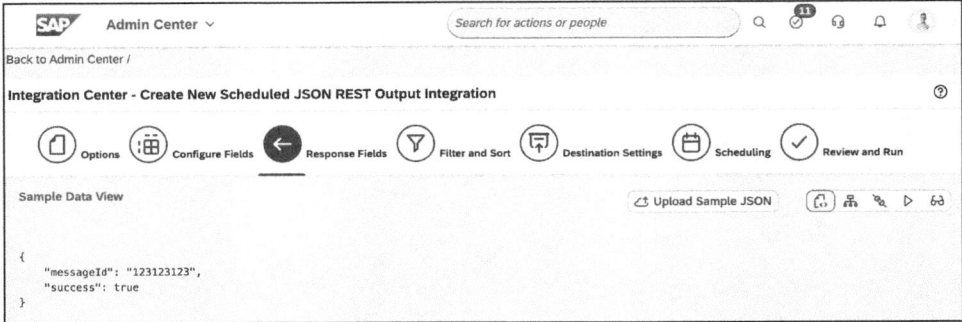

**Figure 10.17** Configuring Response Fields

In the next step, **Filter and Sort**, you can define filters and configure the sorting options. This step is also optional. In our example, we've opted for one of the **Time-Based Filters**: the **Last Run Time** (see Figure 10.18), which means that only those entities that have changed since the last run of the integration are loaded. Then, click the **Next** button.

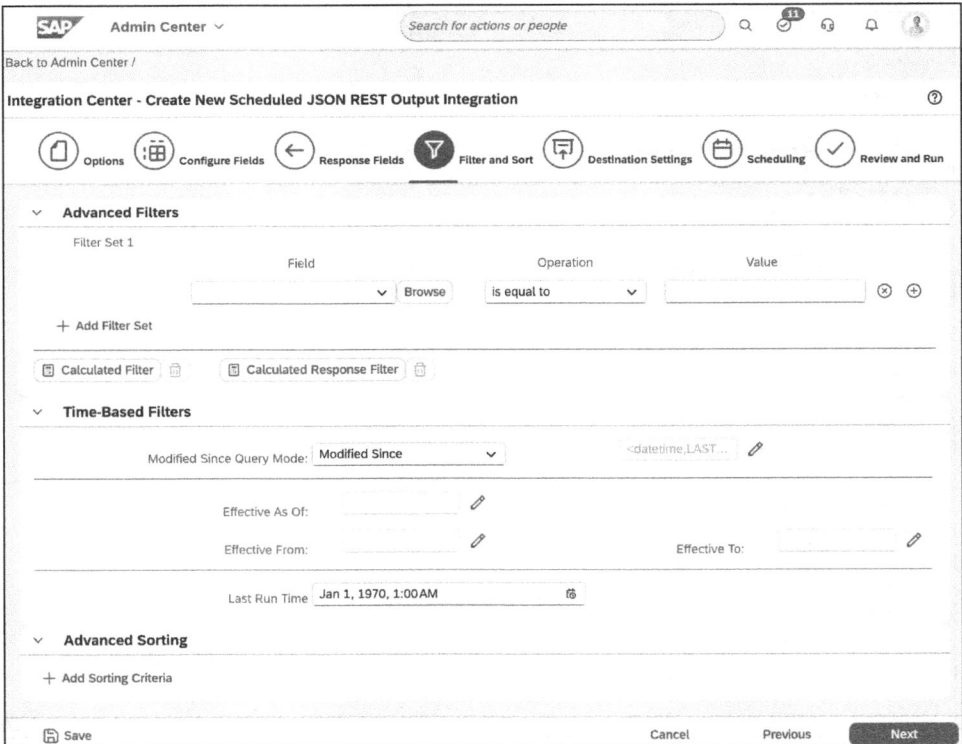

**Figure 10.18** Adding Filters and Sorting Options

In the next step, **Destination Settings** (see Figure 10.19), you configure the connection to the iFlow. To do this, select the **REST Server Settings** option, assign a meaningful name in the **Connection Name** field, copy the HTTP endpoint of the iFlow that you previously determined into the **REST API URL** field, and select **OAuth** for the **Authentication Type**. As you'll see, no OAuth configurations will be available at that point, so you'll need to create them in the next step. To do this, click on the **Click to manage OAuth Configuration** link.

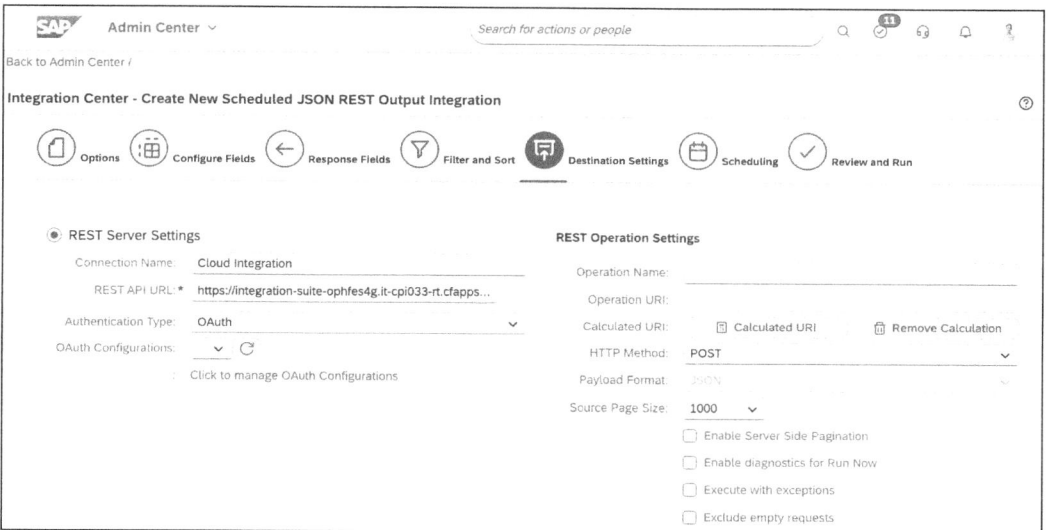

**Figure 10.19** Destination Settings

That will take you to the **Security Center** (see Figure 10.20), where you can click **Add** on the left side of the table next to the **OAuth Configurations** header to create a new configuration. Enter a name in the **Configuration Name** field, select **OAuth 2.0** in the **OAuth Type** field, select **Client_Credentials** for the **Grant Type**, copy the client ID from the Cloud Integration service key into the **Client ID** field, and copy the corresponding client secret into the **Client Secret** field. You must also specify the endpoint used to generate the token in the **Token URL** field. (You can also find this in the service key.) Then, define **POST** as the **Token Method** and click the **Save** button.

That will take you back to the integration center in the **Destination Settings** (see Figure 10.21). In the **OAuth Configurations** field, select the configuration you just created. Then, click the **Next** button.

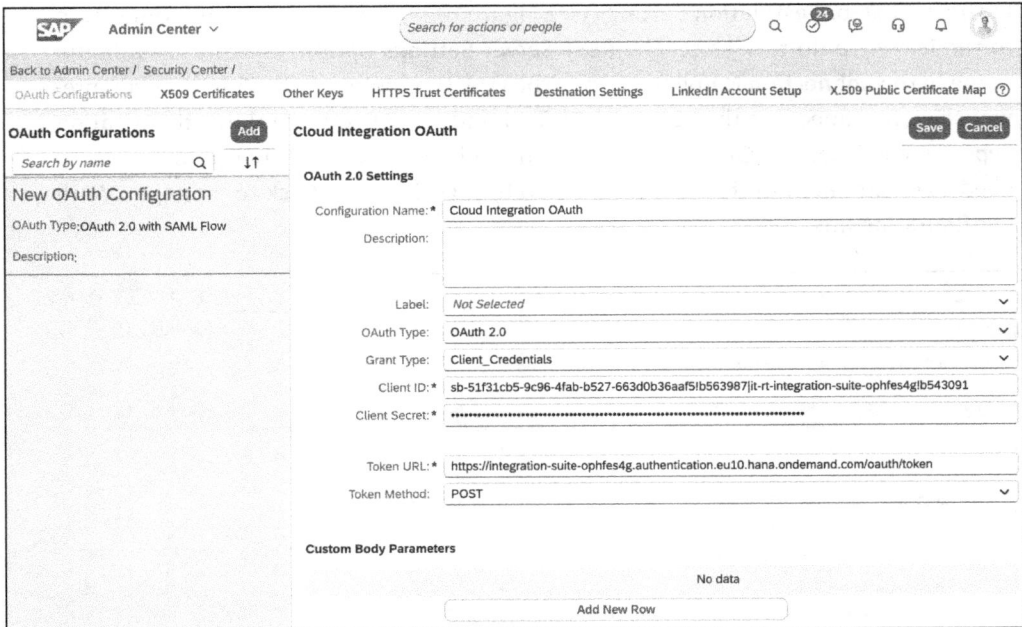

**Figure 10.20**  Adding OAuth Configuration

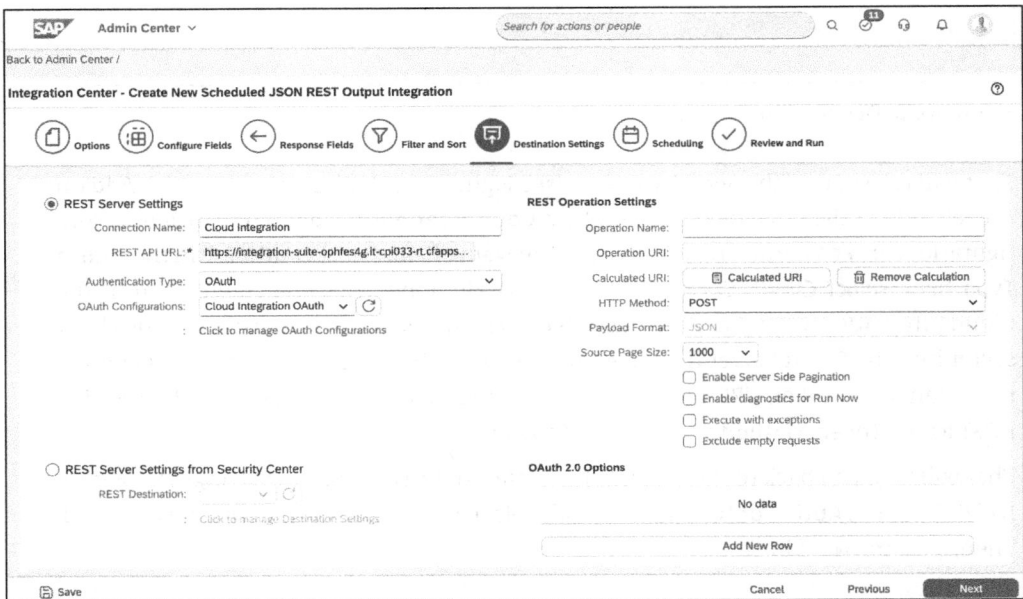

**Figure 10.21**  Finalizing Configuration

The next step is to set up **Scheduling**, as shown in Figure 10.22. Choose from among the different options and decide how often the integration should run. You can make it run once, daily, weekly, or several times a day.

In the last step, **Review and Run**, you'll see a summary of the integration that you configured earlier (see Figure 10.23). Click the **Save** button and then click **Run Now** to start
the integration process right away.

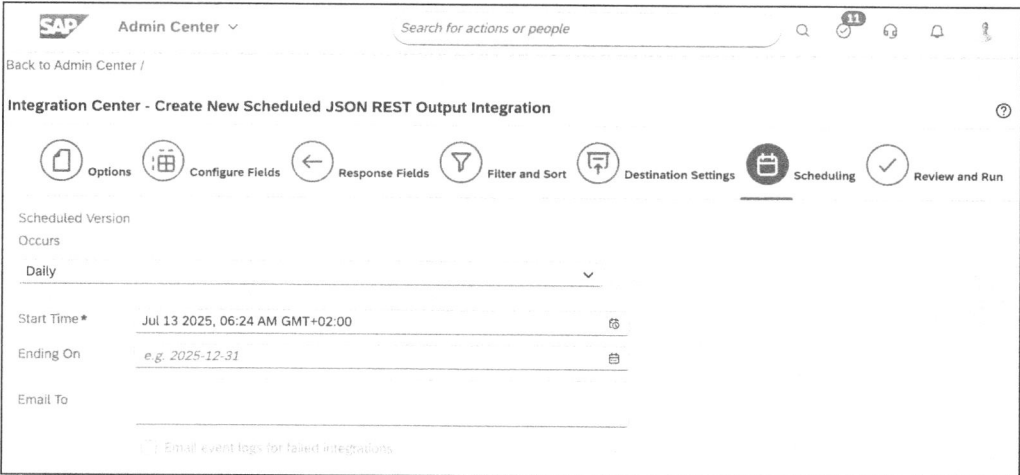

**Figure 10.22**  Configuring Scheduler

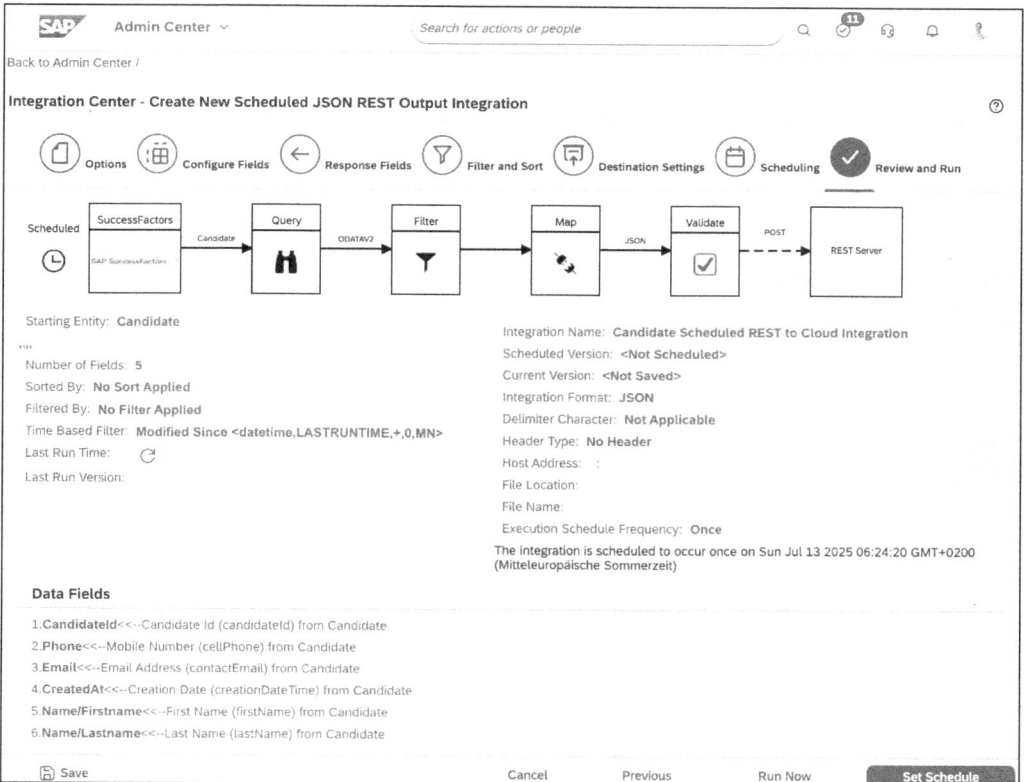

**Figure 10.23**  Review

Then, navigate to **Monitor • Integrations and APIs • Processed Messages** and check whether the integration was executed successfully. The example shown in Figure 10.24 shows that the integration ran once with an error and once successfully. Then, by clicking on the date in the **Last Run** column, you can jump to the job log.

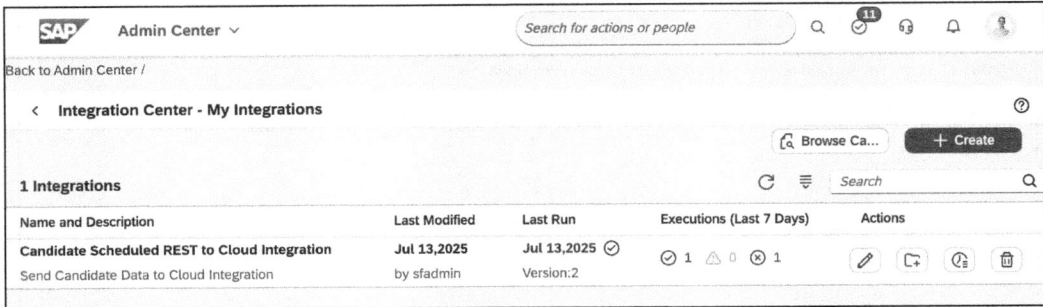

**Figure 10.24** Integration Monitoring

There, you'll see the event details (i.e., the individual steps that were executed), as shown in Figure 10.25.

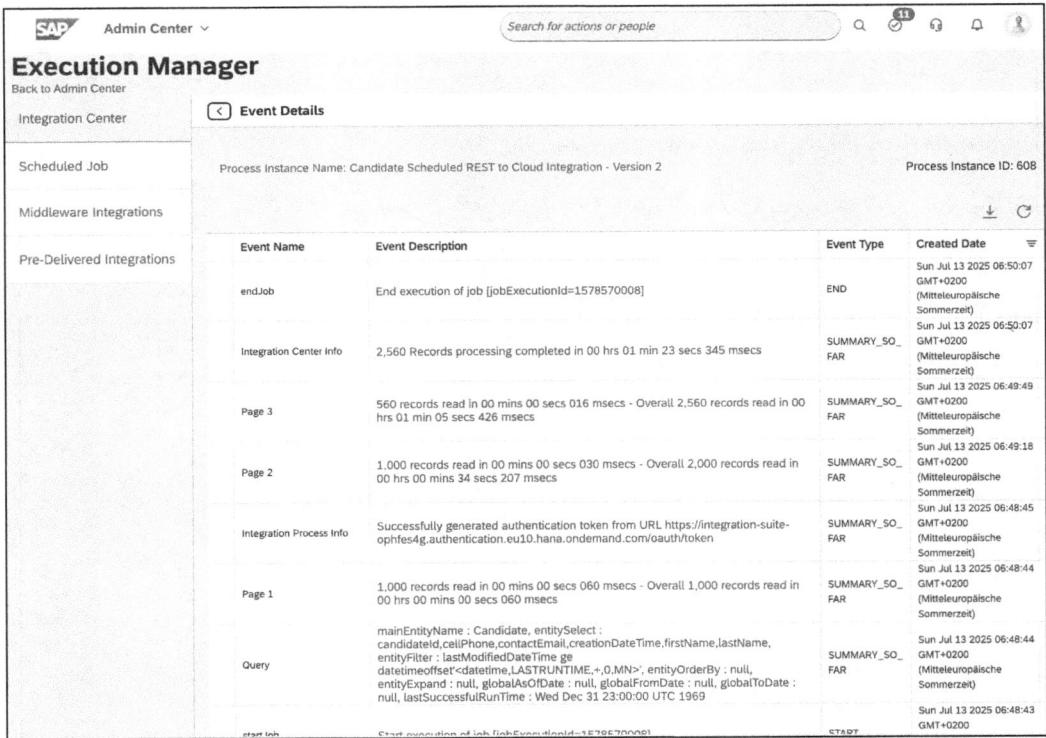

**Figure 10.25** Event Details

This section showed how you can use REST to send data from the integration center to Cloud Integration or a deployed iFlow. As you've seen, this process is very simple and intuitive. We always recommend that you start by creating an iFlow, because then, you'll have the required endpoint in hand and won't need to enter a dummy value for the URL.

## 10.3   Using SOAP Interfaces

In this section, we'll look at how you can send messages from the integration center to a SOAP endpoint (i.e., a classic web service endpoint). In many respects, the configuration will overlap with the REST interfaces.

Here, too, we recommend that you create an iFlow in the first step (see Figure 10.26). Select a **SOAP** adapter of type SOAP 1.x as the sender adapter. Then, in the **Address** field, configure the path you want—which must be unique across the entire SAP Integration Suite instance. Then, select **Manual** for **Service Definition** because you haven't prepared a WSDL. Alternatively, you can upload a WSDL at this point to specify the structure— but we've deliberately decided not to do that in our example because it allows you to be more flexible.

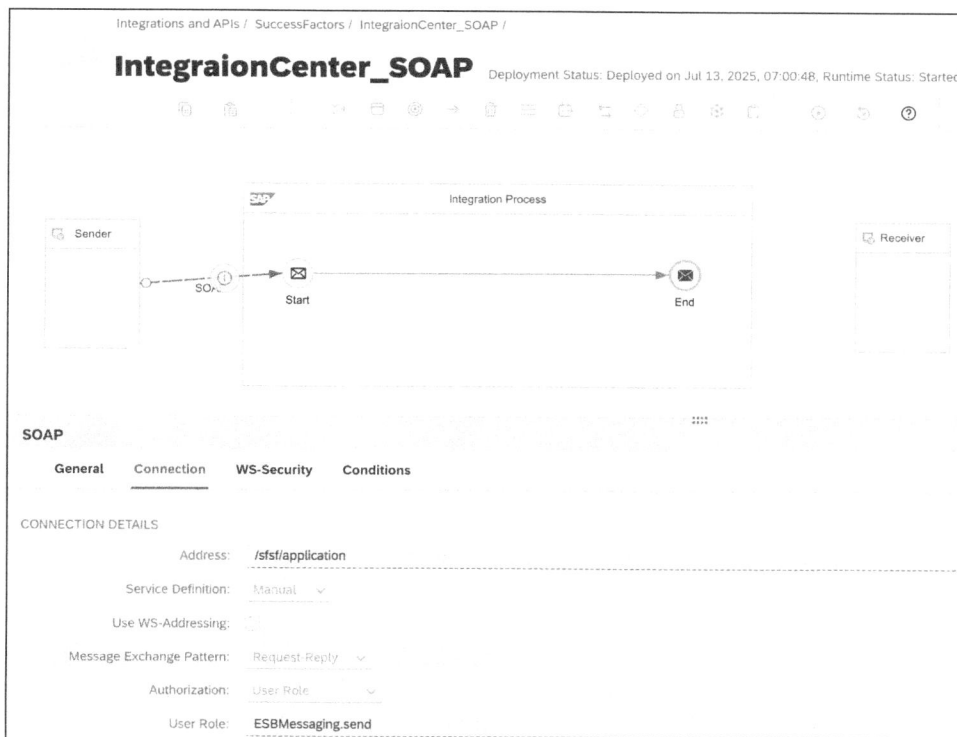

Figure 10.26  SOAP iFlow

Then, you select **Request-Reply** as the **Message Exchange Pattern**, which means that the call will take place synchronously (i.e., a response will be expected). Then, select **User Role** as the **Authorization** type, and we recommend that you enter "ESBMessaging.send" in the **User Role** field. (You can adjust this if you need to). Save and deploy to complete the iFlow.

Then, navigate to **Monitor • Integrations and APIs • Processed Messages** and determine the endpoint for the SOAP call, as shown in Figure 10.27. You'll need this later in SAP SuccessFactors in the integration center.

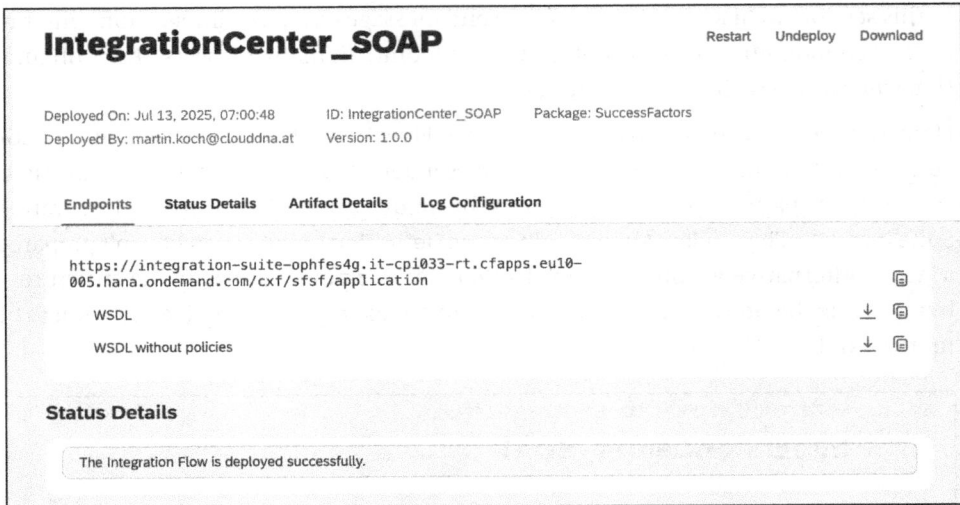

**Figure 10.27** SOAP iFlow Endpoint

Next, open the integration center in SAP SuccessFactors and click on the **Create • More Integration Types** from the context menu (see Figure 10.28).

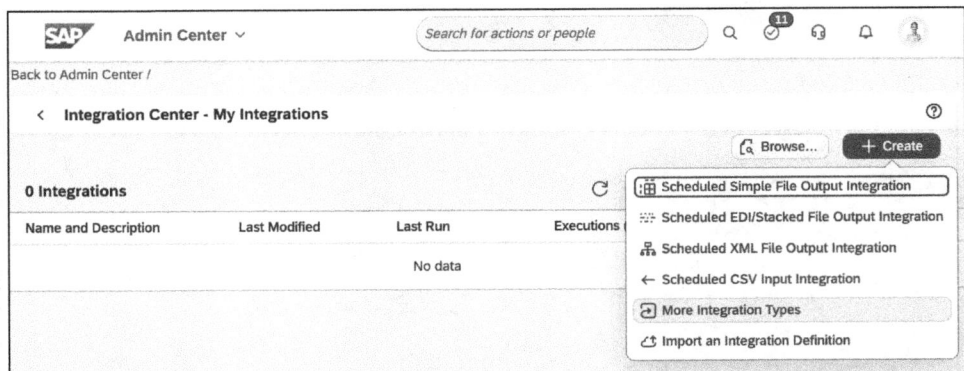

**Figure 10.28** Creating Integration

In our example, the export of data is triggered by a scheduler, so as shown in Figure 10.29, you select **Scheduled** for the **Trigger Type**, **SuccessFactors** for the **Source Type**, **SOAP** for the **Destination Type**, and **XML** for the **Format**. Finally, click the **Create** button.

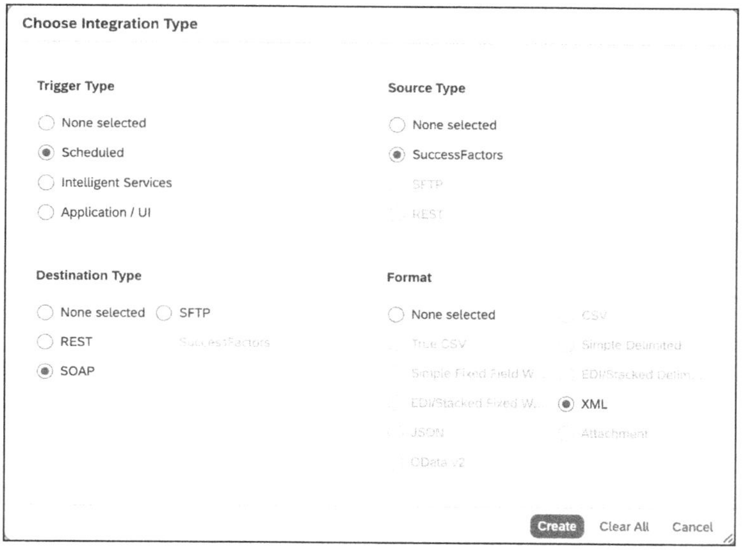

**Figure 10.29**  Configuring Integration Type

Next, you must choose the starting entity. In our example, we've chosen the **JobAppli-cation**. Then, click on the **Select** button (see Figure 10.30).

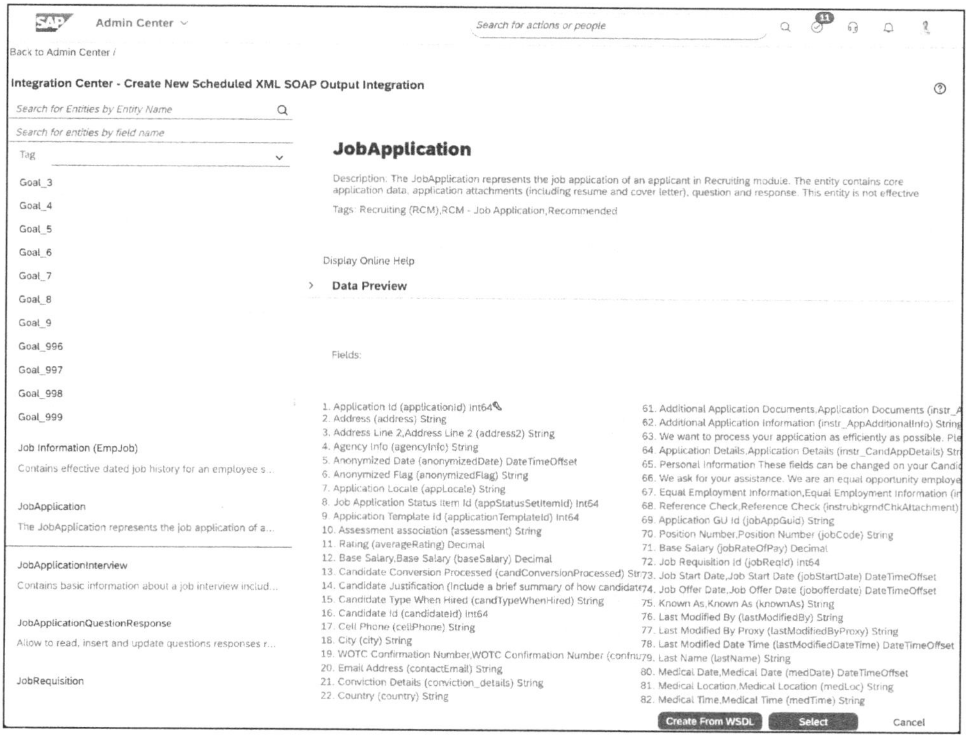

**Figure 10.30**  Selecting Source Entity

Then, you need to click through the wizard step by step. In the first step, configure the options as shown in Figure 10.31. You need to assign an appropriate **Integration Name** and **Description**. You can also specify the **Destination Page Size**, which controls how many messages are bundled and sent to the destination. In our example, we've chosen **1** for the page size, which means that each candidate will be transferred individually and will trigger its own iFlow.

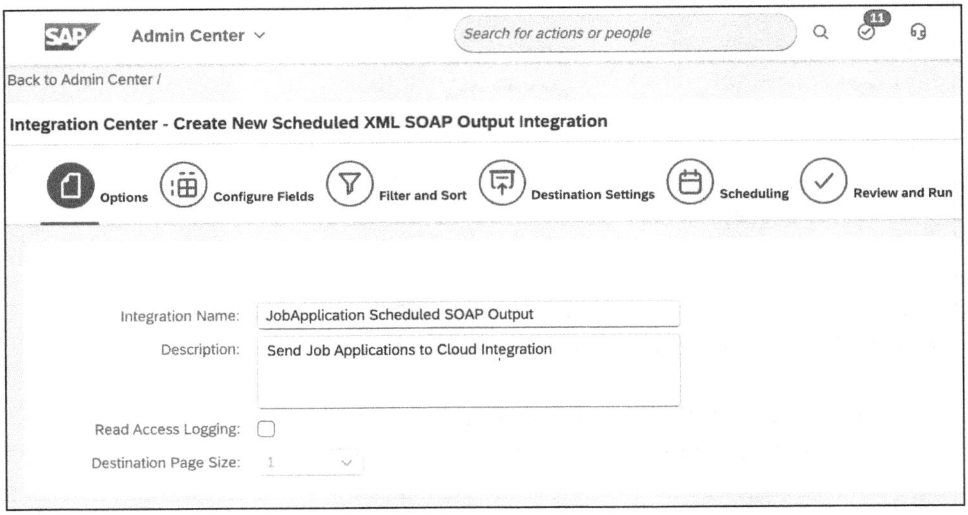

**Figure 10.31** Configuring Options

As with REST integration, you must also build the request structure manually here. Since a SOAP request always uses a SOAP envelope, which in turn can contain headers and a body, you must build the structure within the body (see Figure 10.32). Once again, you map the individual fields from the source entity to the destination fields by using drag and drop. Then, click on the **Next** button.

In the next step, **Filter and Sort**, you can define filters and configure the sorting options. This step is optional, and in our example, we've opted for a time-based filter and have chosen **Last Run Time** (see Figure 10.33). This means that only those entities that have changed since the last run of the integration will be loaded. Then, click on the **Next** button.

In the next step, **Destination Settings** (see Figure 10.34), you configure the connection to the iFlow. To do this, assign a meaningful name in the **Connection Name** field, copy the HTTP endpoint of the iFlow that you previously determined into the **SOAP API URL** field, select **Basic Authentication** as the **Authentication Type**, and enter a **User Name** and a **Password**. You can use the client ID from your service key as the **User Name** and the client secret as the **Password**, and you can also use them for both **OAuth 2.0 Client Credentials** and **Basic Authentication**.

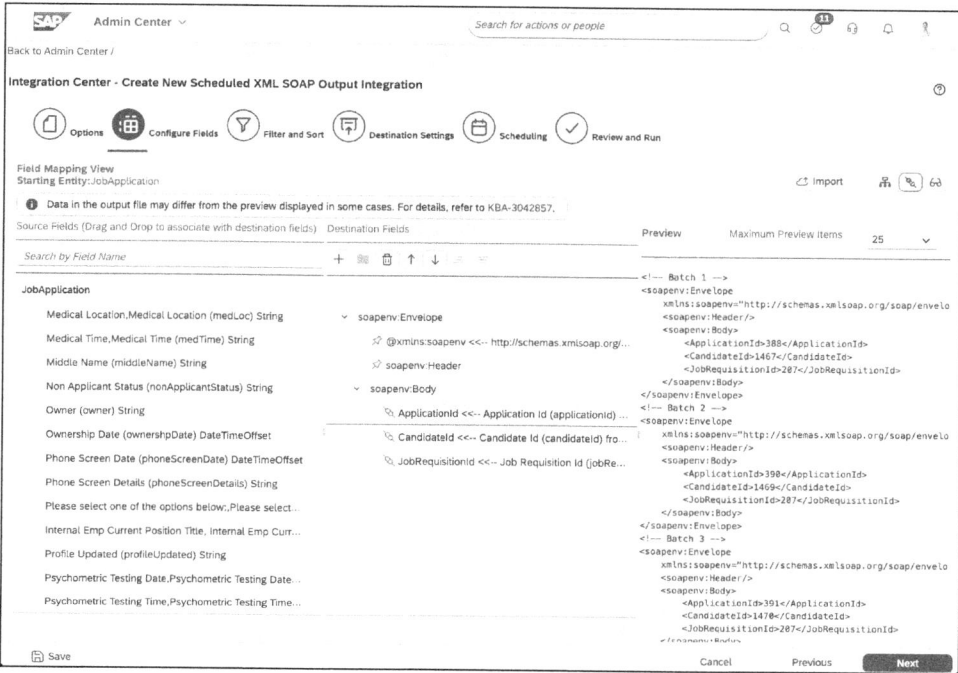

**Figure 10.32** Configuring Request Structure

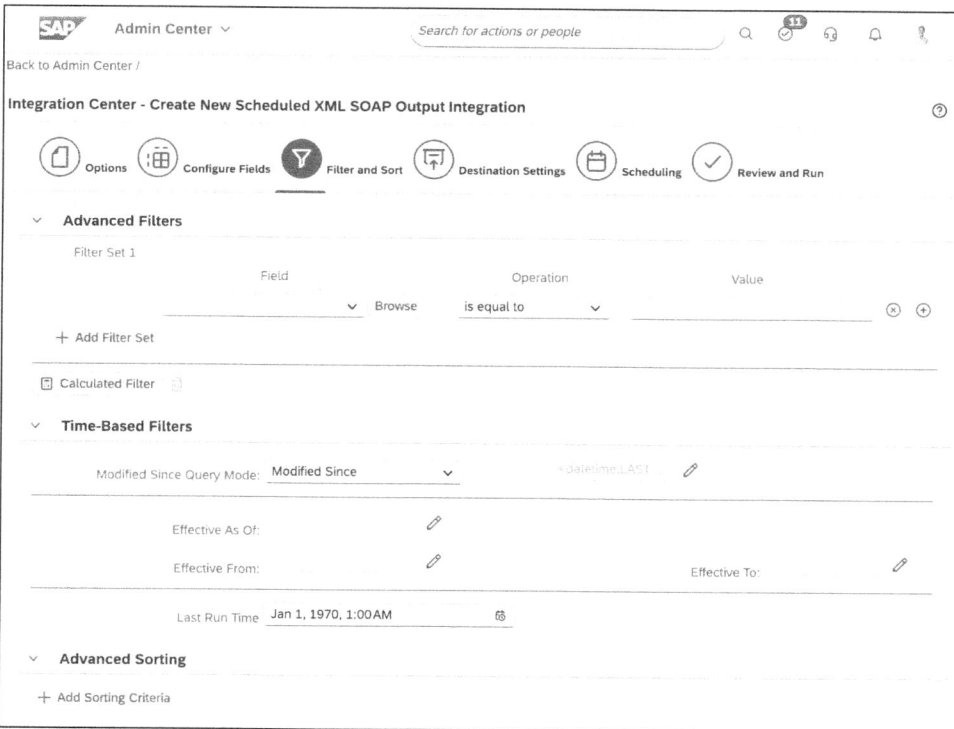

**Figure 10.33** Defining Filters and Configuring Sorting

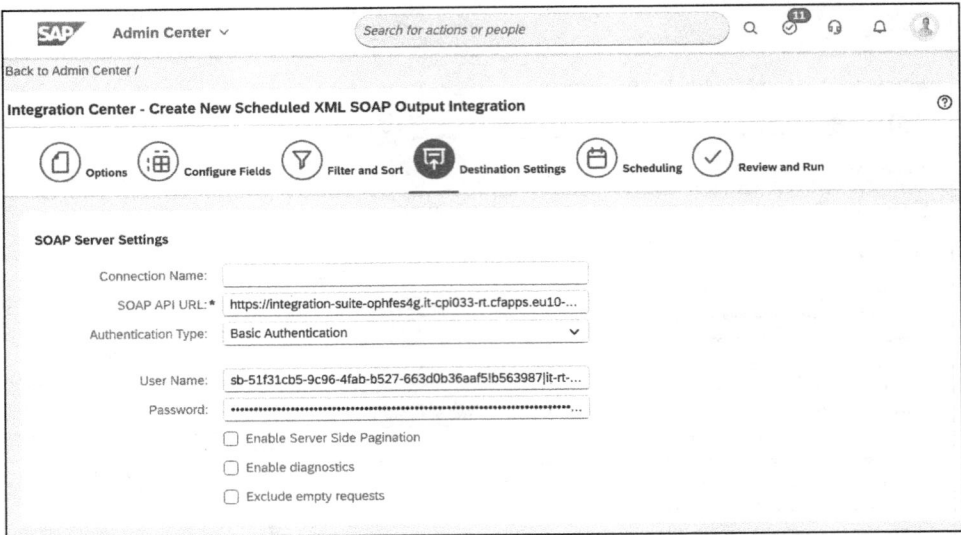

**Figure 10.34** Maintaining Destination Settings

The next step is to set up **Scheduling** as shown in Figure 10.35. You can choose between different options and decide how often the integration should run: once, daily, weekly, or several times a day.

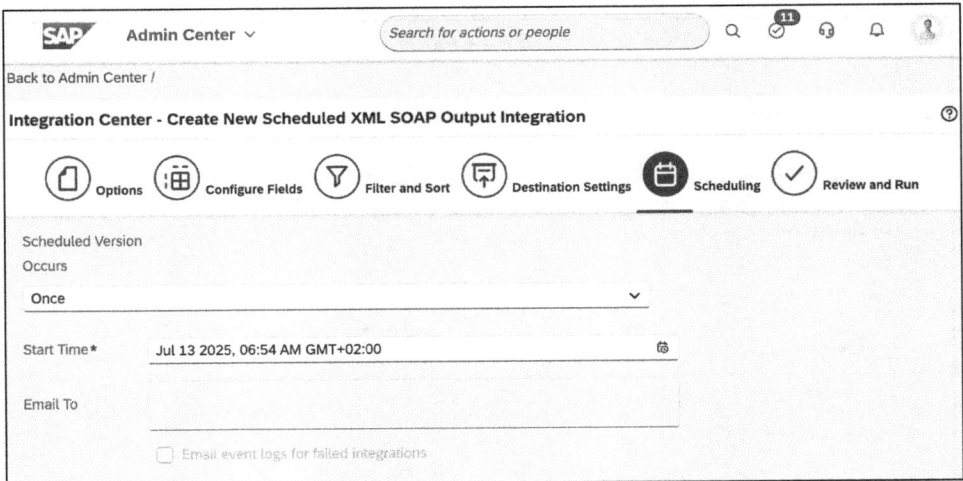

**Figure 10.35** Configuring Scheduling

In the last step, **Review and Run**, you'll see a summary of the integration that you configured earlier (see Figure 10.36). Click the **Save** button and then click **Run Now** to start the integration process right away.

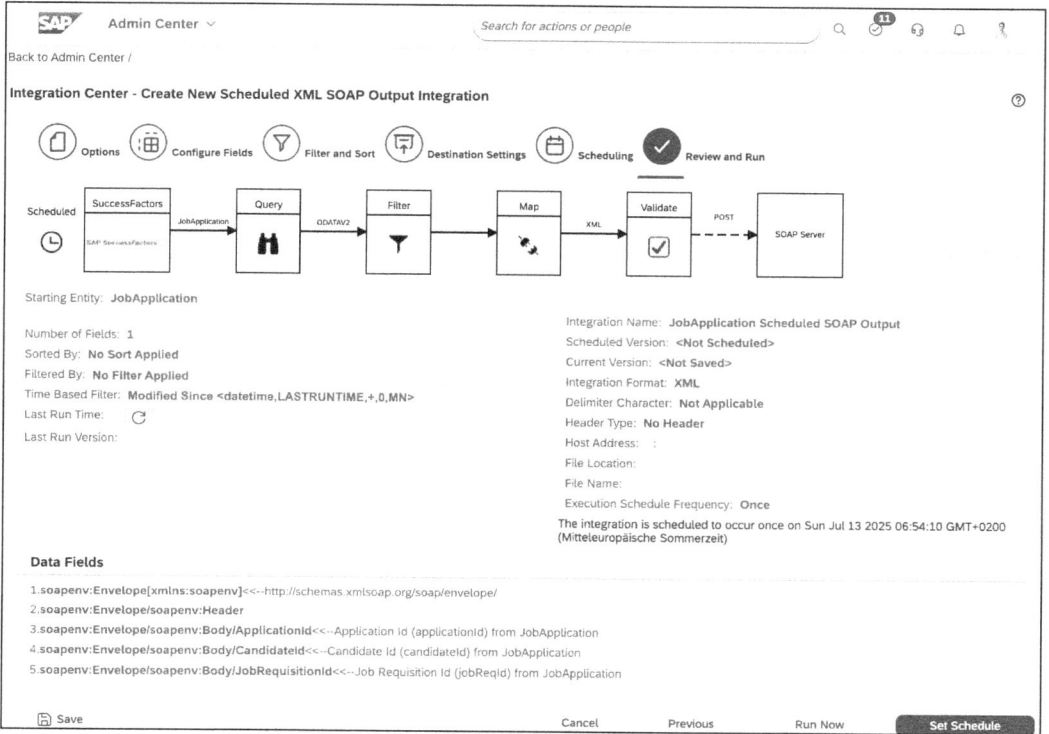

**Figure 10.36**  Reviewing and Running Integration

Then, navigate to **Monitor • Integrations and APIs** and check whether the integration was executed successfully. In the example in Figure 10.37, the integration ran once with an error and once successfully. Then, click on the date in the **Last Run** column to jump to the job log.

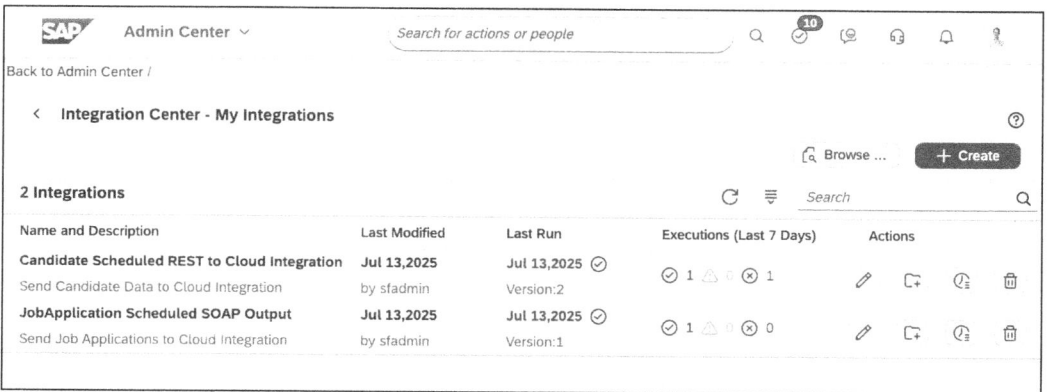

**Figure 10.37**  Integration Job Overview

There, you'll see the **Event Details** (i.e., the individual steps that were executed), as shown in Figure 10.38.

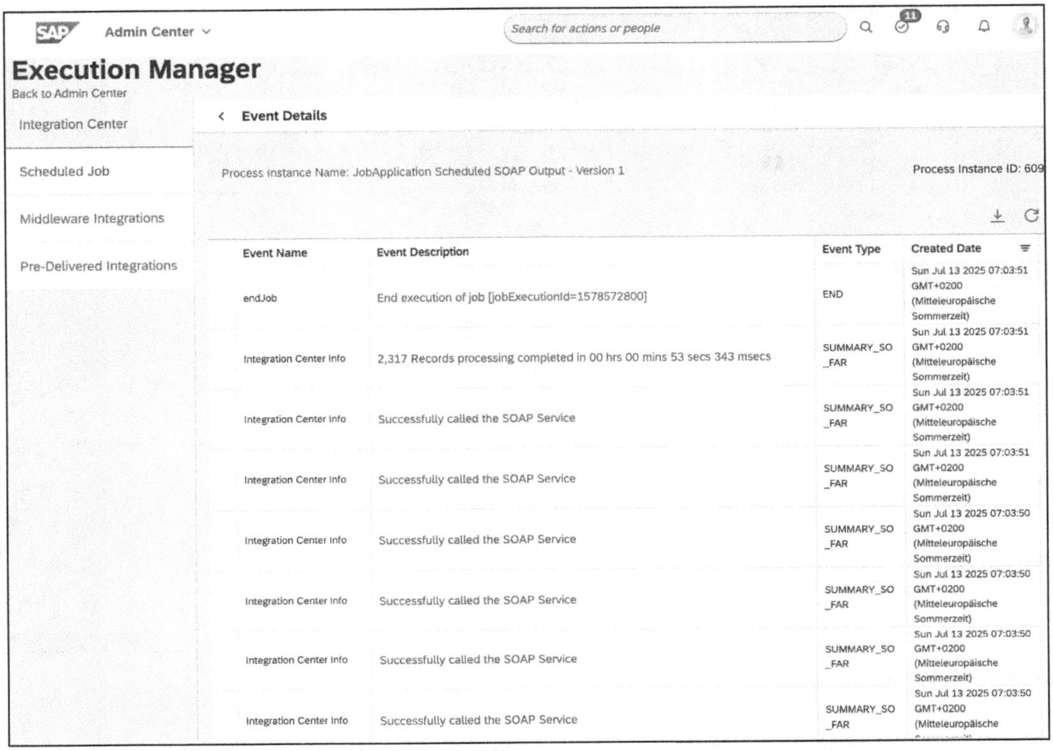

**Figure 10.38** Monitoring Integrations

This section showed how you can use SOAP to send data from the integration center to Cloud Integration or a deployed iFlow. As you've seen, this process is very simple and intuitive. We always recommend that you start by creating an iFlow, because then, you'll have the required endpoint in hand and won't need to enter a dummy value for the URL.

## 10.4   Error Handling and Monitoring

Monitoring is very simple. There are two places you need to monitor: you can check the SAP SuccessFactors side in the integration center to make sure everything is running smoothly, and you should also check the Cloud Integration monitor in SAP Integration Suite to make sure everything is running successfully there too.

SAP SuccessFactors uses the **Execution Manager** app to monitor integrations, and you can use various filters there as well. Figure 10.39 shows the messages from the last seven days for all statuses. You can also jump to the event details from the list below the graph by clicking on the arrow icon that's displayed in the last column on the right.

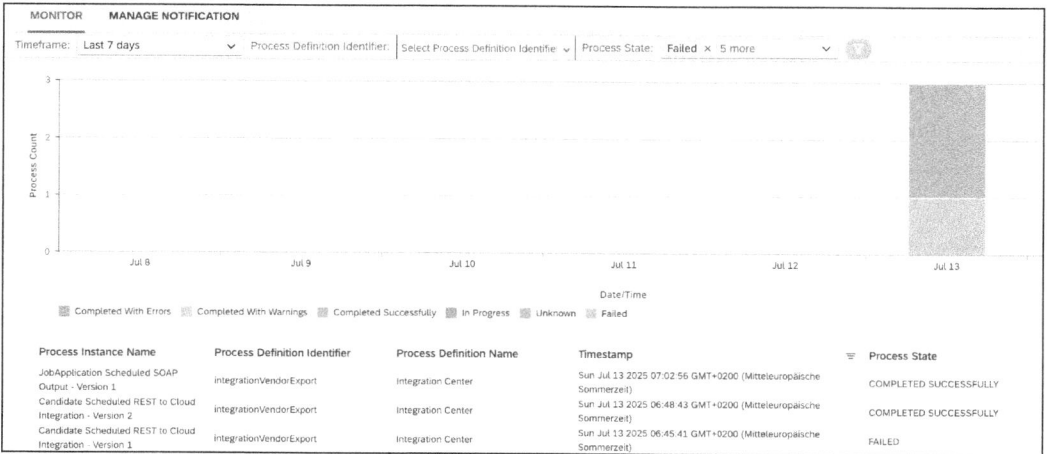

**Figure 10.39** Monitoring in Integration Center

In the **Event Details** section, you can find a comprehensive list of all actions taken in SAP SuccessFactors for an underlying integration. Figure 10.40 shows the event details for a failed integration, and as you can see from the log, an HTTP 403 error occurred when Cloud Integration was called. This means that the authentication worked but there were problems accessing the iFlow. This may be because the user doesn't have the right roles or because they didn't ask for a CSRF token.

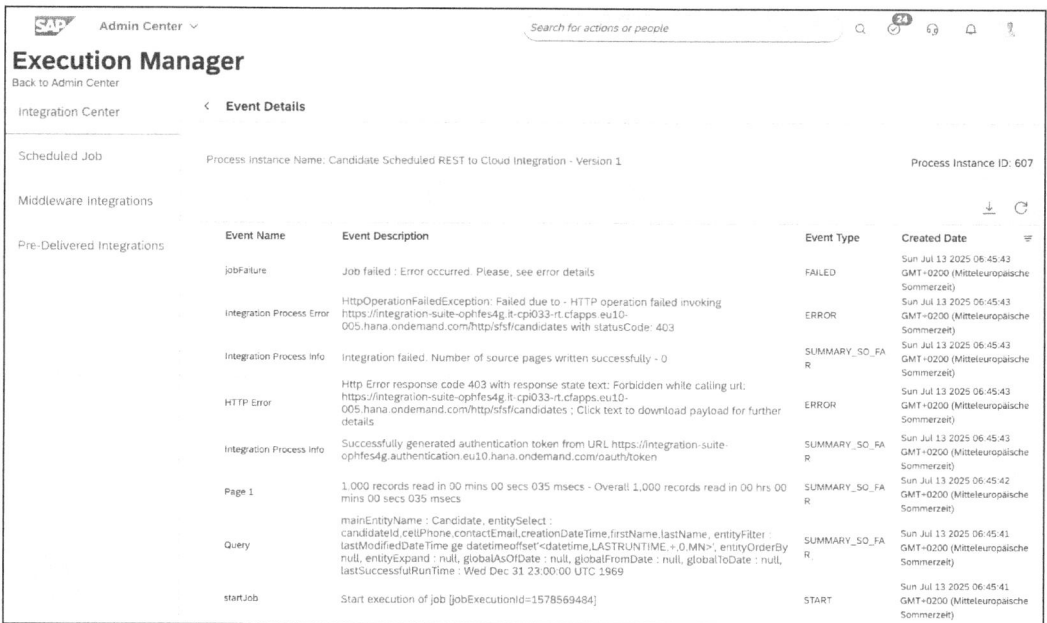

**Figure 10.40** Event Details

You can set up notifications to avoid having to monitor continuously. To do this, click on the **MANAGE NOTIFICATION** tab, as shown in Figure 10.41. In the first line, **Process Type**, you can select the types of error messages you want to receive, and in the **Email** field, you can also enter the address where you want these messages to be sent.

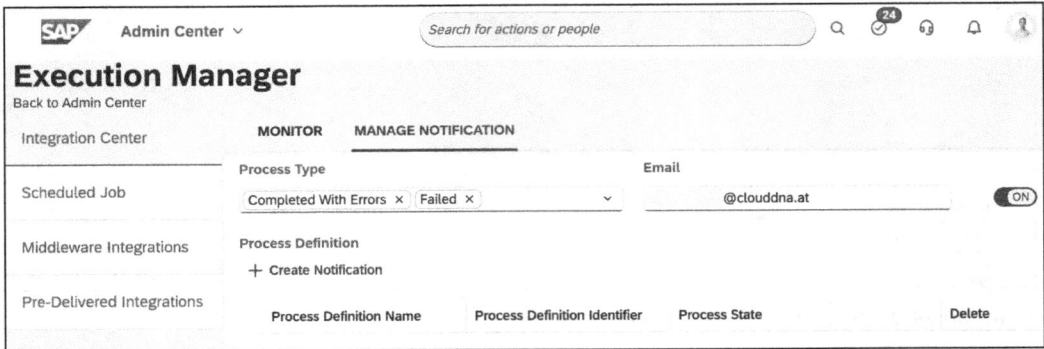

**Figure 10.41** Managing Notifications

By clicking on the **Create Notifications** button, you can choose which process definitions you would like to receive notifications for. As shown in Figure 10.42, you must again choose the **Process State** and the **Process Definition Name** you want. Finally, click **Save**.

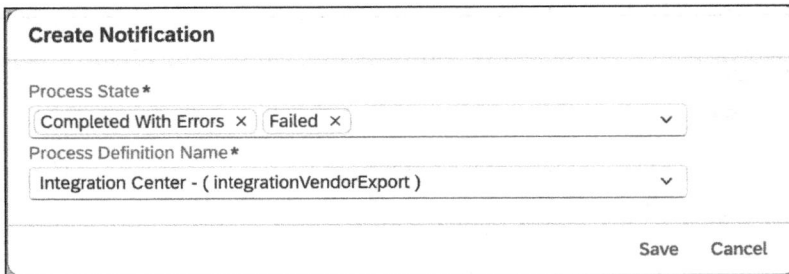

**Figure 10.42** Creating Notification

After saving everything, you should see the configured notifications, as shown in Figure 10.43.

In addition to monitoring integrations in SAP SuccessFactors, you can monitor the execution of iFlows in SAP Integration Suite, as we mentioned earlier. To do this, open **Monitor • Integrations and APIs • Processed Messages** (see Figure 10.44), where you can also use various filters. Please note that messages are deleted from the log after a certain period of time. If you've completed an iFlow successfully, you'll only see the information shown in the figure. There's no way to view further details unless the developer of the iFlow writes something in the message processing log, for example.

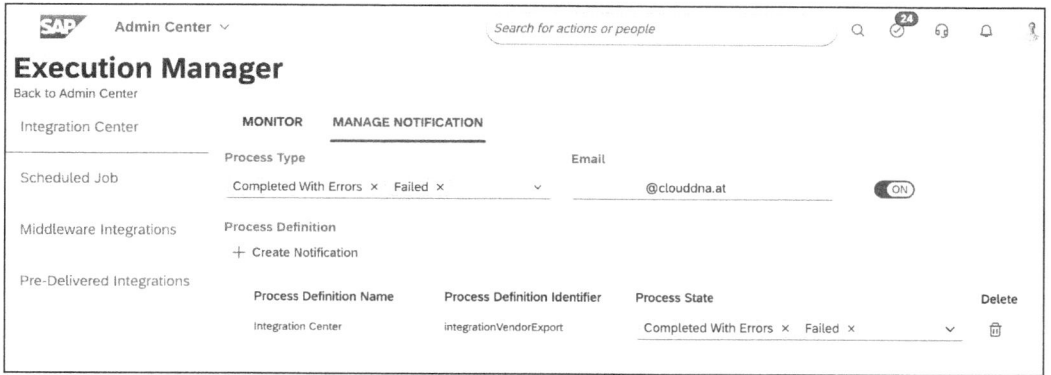

**Figure 10.43**  Configured Notifications

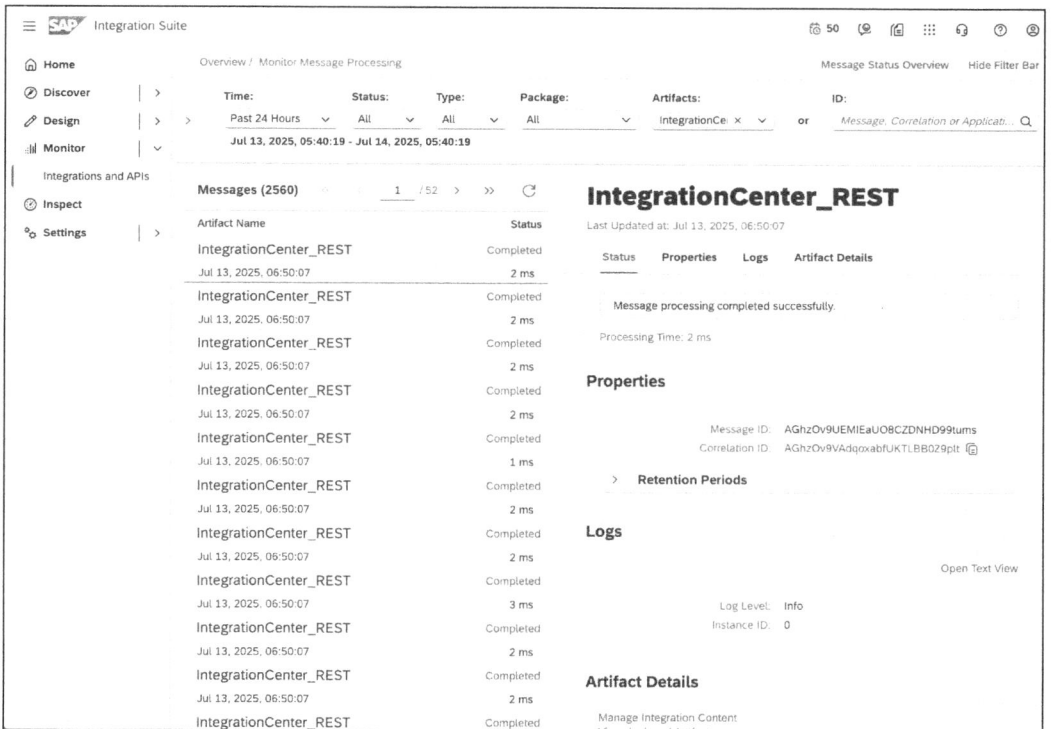

**Figure 10.44**  Monitoring Message Processing

You can also access the system log files, in addition to the message processing log. To do this, go to the **Access Logs** section in the **Monitor** area, as shown in Figure 10.45, and click on the **System Log Files** tile.

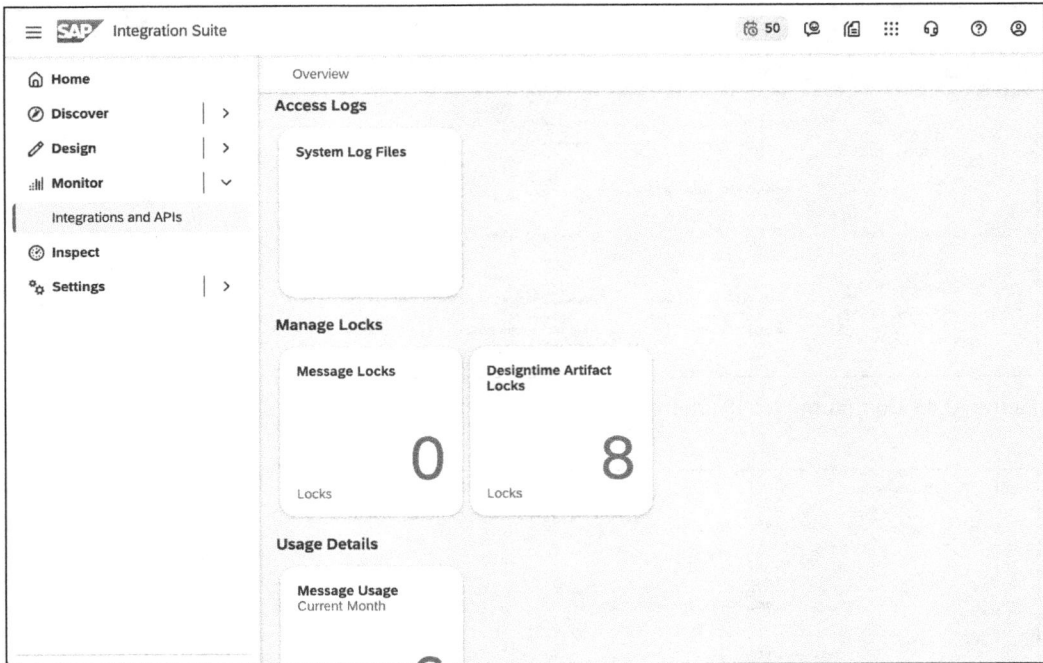

**Figure 10.45** System Log Files

That will take you to a screen where you can click on the **Collections** tab and find the **HTTP Access Log** and the **Trace Log** (see Figure 10.46). To download the log files to your computer, click the **Download** icon.

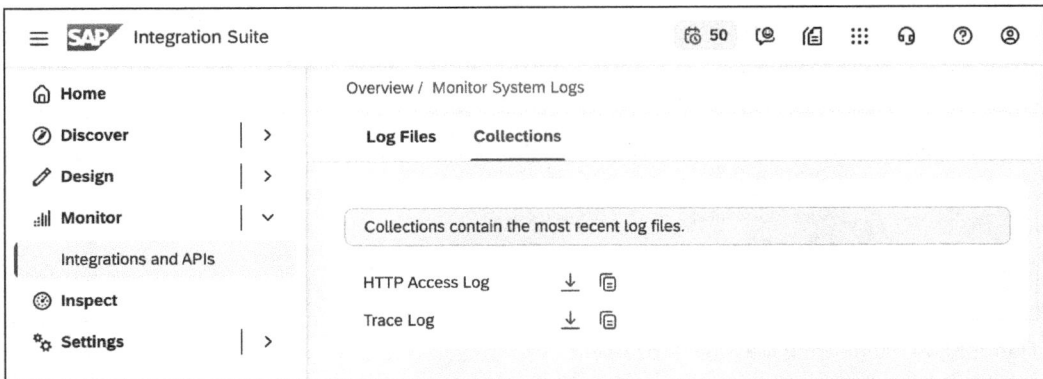

**Figure 10.46** Downloading Log Files

As shown in Figure 10.47, you can view the HTTP access log for access to the previously created integration. The first two lines show the HTTP 403 error that was previously shown in **Monitor • Integrations and APIs.**

Figure 10.47 Access Log

If you can't fix the error, you can use SAP's troubleshooting guide to help you find the cause of the error. You can access this guide by clicking on the user icon in the upper right corner. A popup menu will open, and you can click on the link to the **Troubleshooting Guide**. There, you'll be guided through a series of questions about possible causes of the error and their solutions (see Figure 10.48).

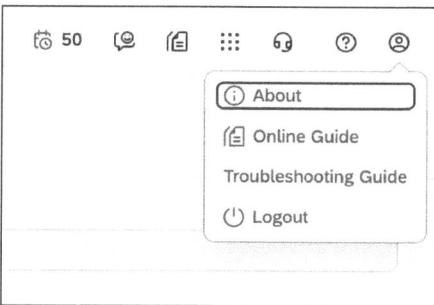

**10**

Figure 10.48 Troubleshooting Guide

## 10.5 Summary

This chapter has given you all the essential knowledge you need to understand and confidently work with the integration center. Section 10.1 provided a thorough review of the architecture, main functions, and common uses of the integration center. It also explained how the integration center is different from traditional middleware methods and when it can be a good addition or even a standalone solution.

Section 10.2 covered the technical basics of the underlying interfaces, with a special focus on integrations based on the REST framework. You learned how to reliably extract data from key modules like SAP SuccessFactors Employee Central, Recruiting, and Performance & Goals, and transfer it seamlessly to target systems such as Cloud Integration.

In Section 10.3, we looked at the technical parts of SOAP-based integrations and talked about how you can use SOAP for outbound scenarios. This provides a strong and consistent option for connecting SuccessFactors to various systems.

Finally, in Section 10.4, we talked about how to deal with and watch for errors. We reviewed how to detect and analyze errors early, monitor logs, set up automated notifications, and implement measures to ensure high operational reliability and data consistency.

These sections have given you a solid foundation for using the integration center—not just as a tool for simple data exports but also as a strategic component for building secure, scalable, and future-proof integrations. These integrations connect SAP SuccessFactors with both SAP and non-SAP systems.

# Chapter 11

# Using Intelligent Services for Event Triggers in SAP SuccessFactors

*SAP SuccessFactors intelligent services are powerful tools for automating, connecting, and making HR processes smarter.*

SAP SuccessFactors intelligent services are key parts of the SAP SuccessFactors HCM Suite. They're designed to improve HR processes through automation, seamless integration, and intelligent, predictive support. These services fundamentally change how HR transactions are handled by shifting from isolated activities to interconnected, end-to-end process chains that proactively engage systems across the organization.

At the heart of intelligent services is a process orchestration model that is driven by events. In the *intelligent services center*, administrators can define what will trigger the system—for example, they can enter "Employee Hired" or "Employee Transferred to Another Location" as triggering events. When one of these events occurs, the system automatically starts one or more follow-up actions like tasks, notifications, and integrations with other modules and third-party systems. This allows HR processes to be extended beyond SAP SuccessFactors itself.

This automation is managed through a set of rules that enable organizations to adapt follow-up actions precisely to their business processes. For example, when an employee is transferred, not only are internal SAP SuccessFactors modules updated, but external service providers (such as relocation agencies) are automatically informed. This ensures a consistent, fully automated experience.

Another great feature of intelligent services is that it lets you connect SAP SuccessFactors to other applications. The *event notification subscription* lets external systems—like Dell Boomi, SAP Integration Suite (Cloud Integration), and custom endpoints—subscribe to specific events. When an event is triggered, these systems receive notifications and can react immediately to enable real-time, cross-system workflows.

Here's an example: if an employee moves from the United Kingdom (UK) to Germany, this single event can automatically create a new job opening in SAP SuccessFactors Recruiting for the UK, adjust salary details in SAP SuccessFactors Employee Central Payroll, and notify a relocation partner to organize the move. This chain reaction shows how intelligent services can automate and manage complex HR processes across different departments and systems with minimal manual intervention.

Intelligent services do the following:

- They integrate all processes from start to finish. This means HR transactions are no longer separate. They're part of a continuous workflow that connects all the right systems and people.
- They reduce manual work. Such automation can help reduce the workload for HR departments, shared service centers, and external partners by automatically starting and checking on follow-up activities.
- They reduce errors and improve compliance. Automated workflows reduce the incidence of human error and help make sure processes and deadlines are followed.
- They provide support that takes action before problems arise. This helps the system predict what the next steps are likely to be, and it can either suggest such steps or automatically start them.

From a technical point of view, intelligent services can be set up in many ways. HR administrators can define events, rules, and actions directly in the intelligent services center, and they don't need to use custom code. This works with other systems through standard interfaces and notification mechanisms, such as webhooks and REST APIs. To integrate events into Cloud Integration, you can either call an HTTP endpoint of an iFlow directly via REST or use Event Mesh.

In Section 11.1 of this chapter, we show you how to create and configure event triggers in SAP SuccessFactors that call an HTTP endpoint of an iFlow. Section 11.2 demonstrates the more elegant option for calling Cloud Integration by using Event Mesh.

## 11.1   Configuring Event Triggers in SAP SuccessFactors

Events are linked to the **Intelligence Services Center** via Cloud Integration. In the **Intelligence Services Center**, you can see all events that are triggered within SAP SuccessFactors (see Figure 11.1).

After navigating to the details of an event, you'll see whether flows have already been configured for this event, as shown in Figure 11.2. You can click **New Flow** to create a new flow, and you can also create and use rules to define filters that are used to trigger the event. In our example, you want to send a message to Cloud Integration when an event is triggered, and you do it by clicking on the **Integration** entry in the **Activities** column on the right-hand side.

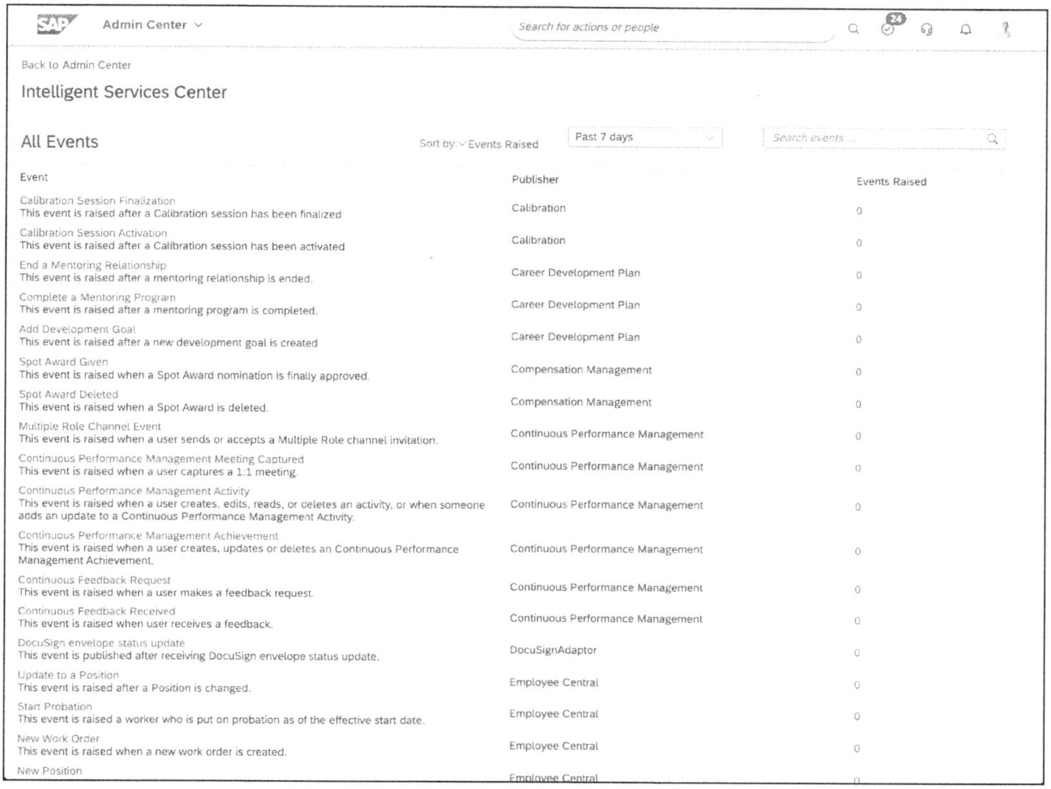

**Figure 11.1** Intelligent Service Center Events

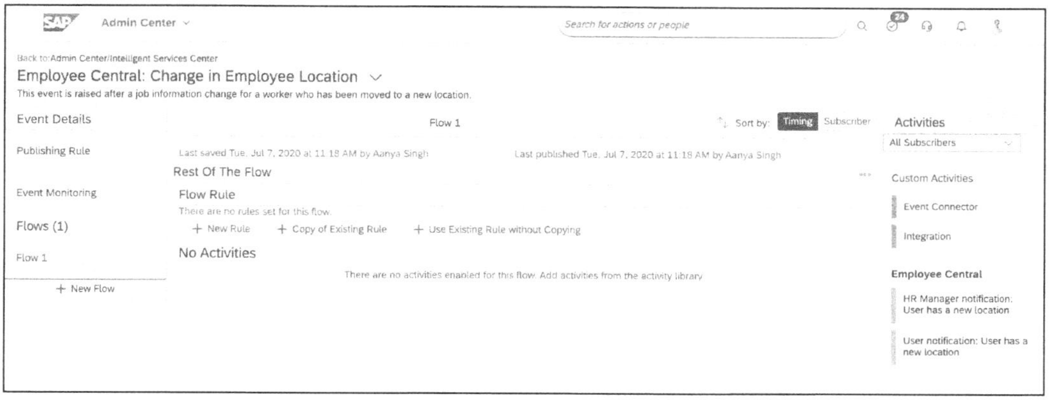

**Figure 11.2** Configuring Event Flow

You'll then see a pop-up window that will indicate no integration has been defined for this event (see Figure 11.3). Confirm this by clicking **OK**.

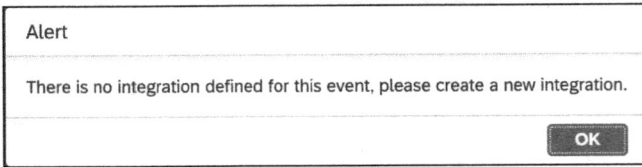

**Figure 11.3**  Alert for Missing Integrations

If an integration already exists, a pop-up window will open where you can select it (see Figure 11.4). Otherwise, you can create a new integration by clicking **Create New Integration**, which will take you to the integration center.

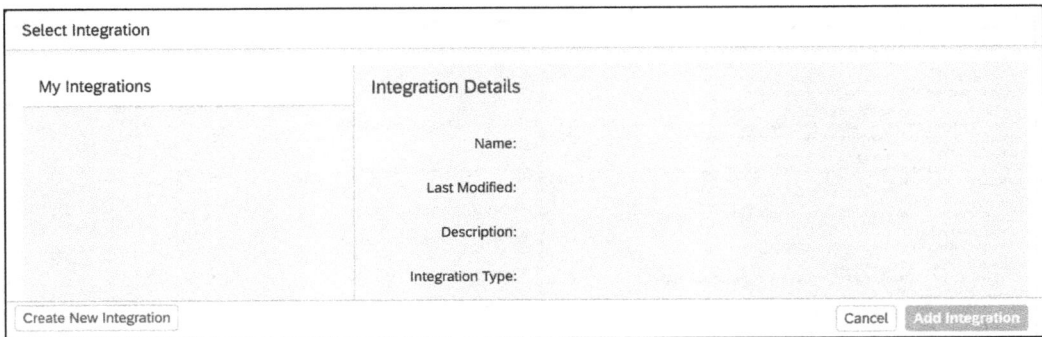

**Figure 11.4**  Dialog for Selecting Integration

Select **REST** as the **Destination Type** and **JSON** or XML as the file **Format**. Then, click **Create** (see Figure 11.5).

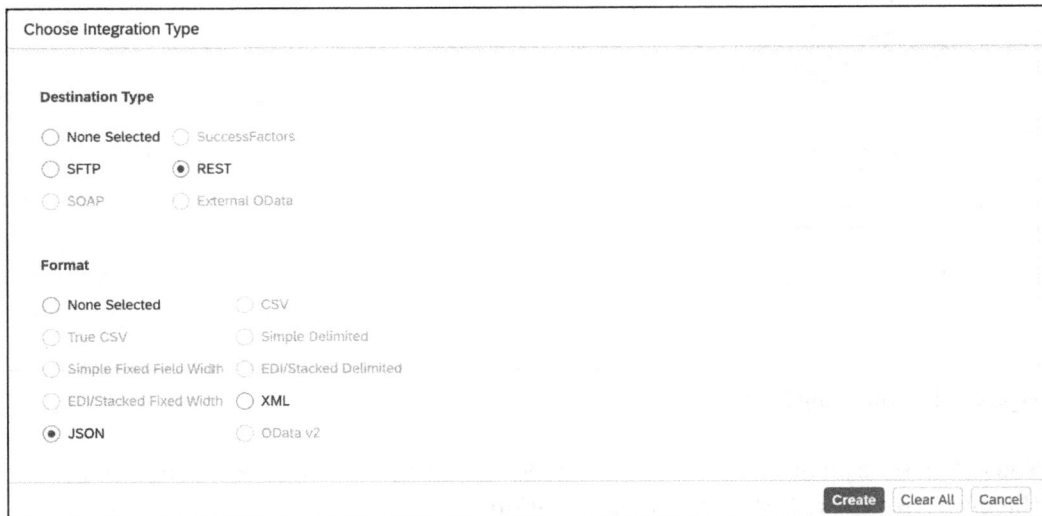

**Figure 11.5**  Configuring Integration Type

As shown in Figure 11.6, the **Integration Name** field is prepopulated. You can change the name if you need to be entering an appropriate name, and you can also enter a description in the **Description** field. The **Destination Page Size** allows you to control the number of messages that can be sent to the target system at the same time.

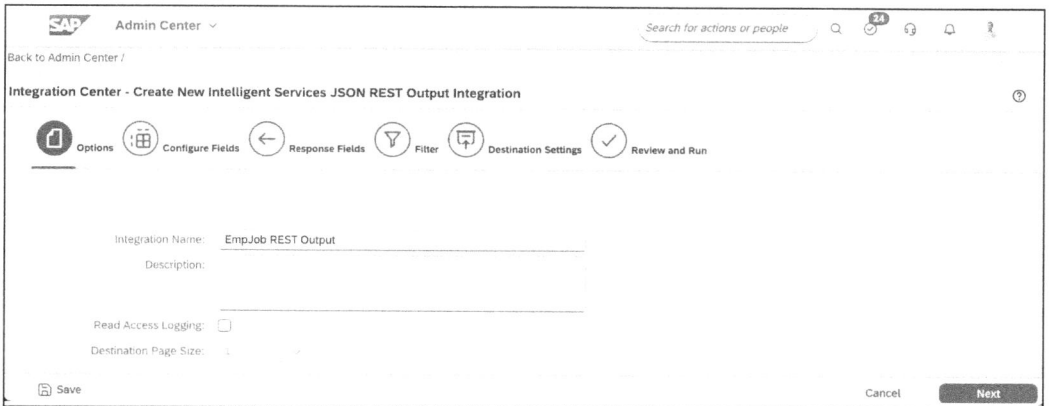

**Figure 11.6** Integration Options

In the next step, called **Configure Fields**, you need to set up the target structure (see Figure 11.7). Alternatively, you can upload the target structure by clicking on **Destination from Sample JSON File**.

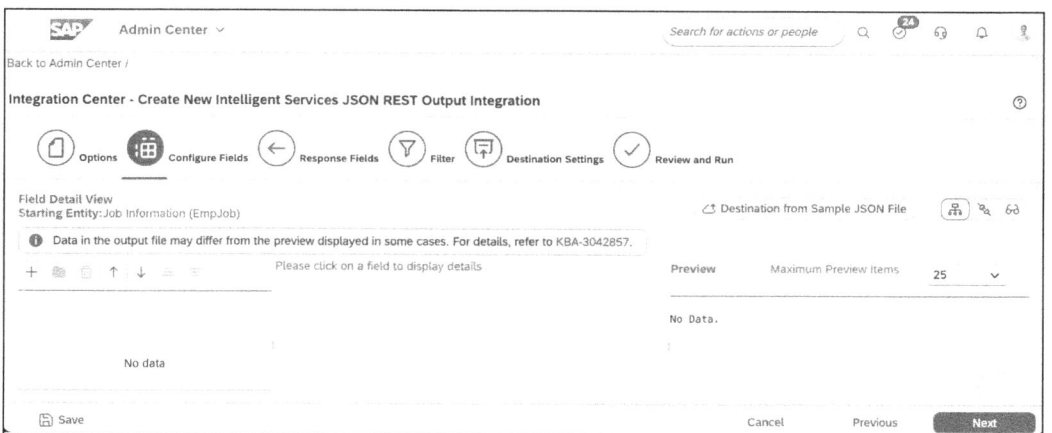

**Figure 11.7** Configuring Integration Fields

Click the **+** button as shown in Figure 11.8 and then select **Insert Sibling Element** from the popover menu.

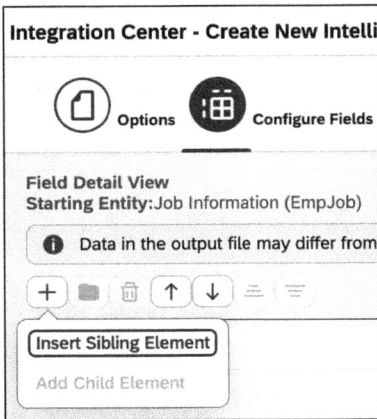

**Figure 11.8**  Creating Request Structure

Then, click on the field you just created, and enter the name you want to use for the element in the target structure in the **Label** field, and enter a description in the **Description** field. You also have the option of defining a **Default Value** (see Figure 11.9) that is used if no corresponding value is found via the mapping. This ensures that the field is never empty.

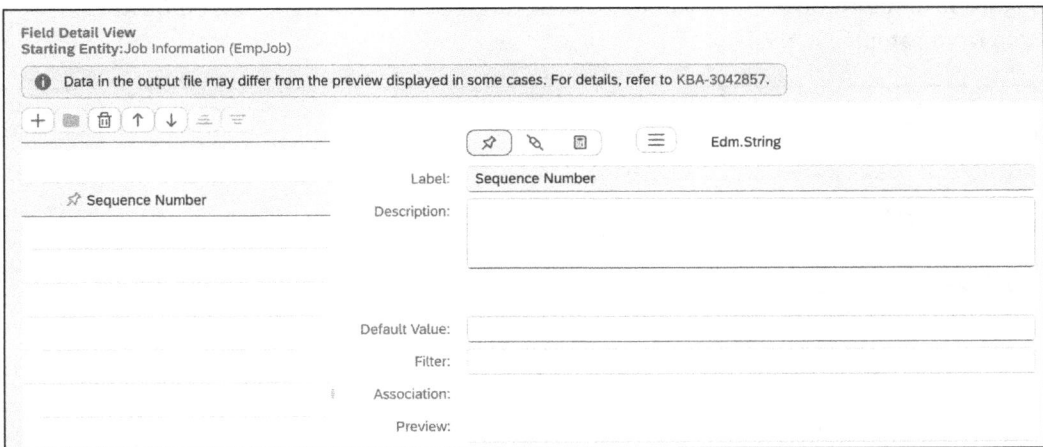

**Figure 11.9**  Changing Element Name

There are only three data types supported, and that simplifies development and modeling. Change the data type by clicking ☰ and then selecting **More Field Options** from the popover menu (see Figure 11.10).

As you can see in Figure 11.11, three data types are available: **String**, **DateTime**, and **Numeric**. Select the type you want in the **Data Type** field and then click **OK** to close the dialog.

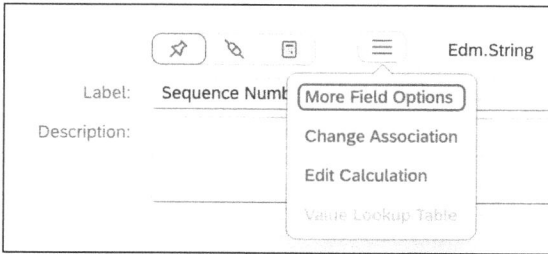

**Figure 11.10** Opening More Field Options Menu

You also have the option of creating nested structures. To do this, start by creating the parent element, which in Figure 11.12 is the **Location** field. Then, click on **+** and select **Add Child Element** from the popover menu.

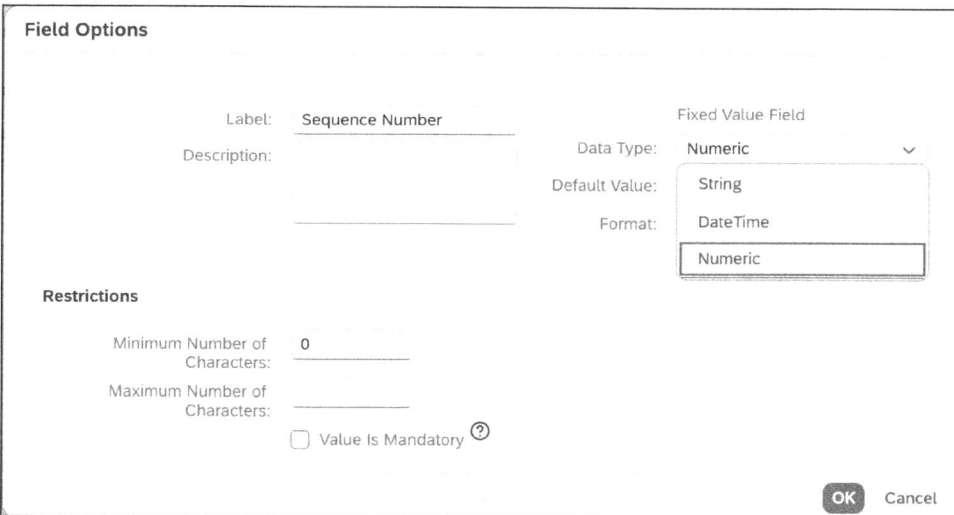

**Figure 11.11** Changing Data Type

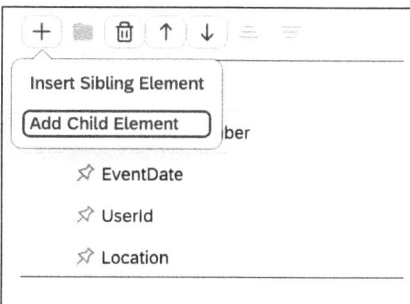

**Figure 11.12** Adding Nested Structure

Then, create the child elements below this structure as shown previously. In our case, you can see the final result in Figure 11.13.

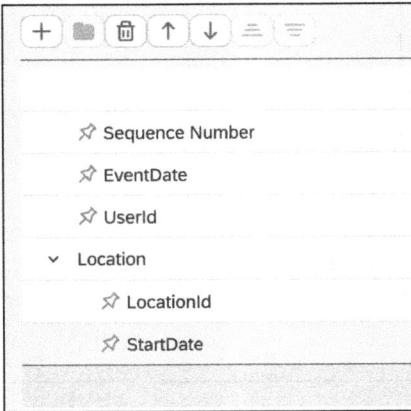

**Figure 11.13** Adding Child Elements

After creating the structure, you must map the individual fields to the SAP Success-Factors entities. Open the mapping editor by clicking ⬚, where you'll see the **Job Information** entity as shown in Figure 11.14. Drag and drop the elements of the SAP SuccessFactors entity into the destination field you want on the right-hand side.

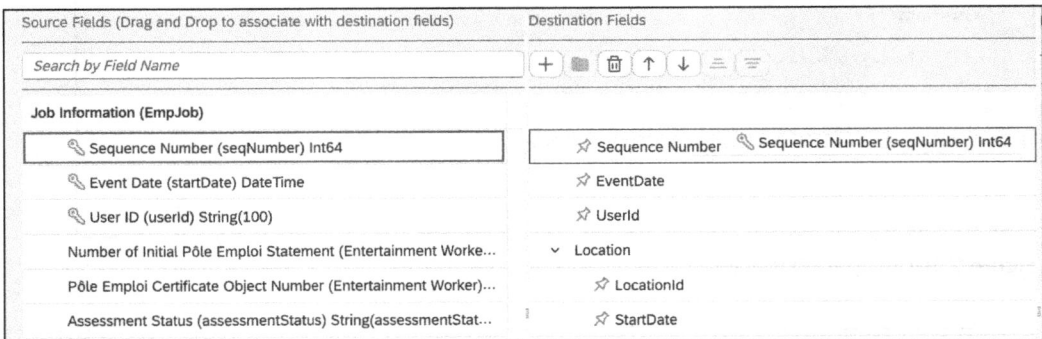

**Figure 11.14** Mapping Fields

You can see the final result in Figure 11.15. On the right, you'll also see a **Preview** section with data from the SAP SuccessFactors instance. Make sure all the fields are mapped correctly and then click **Next**.

Optionally, you can define the **Response Fields** (see Figure 11.16), which are the fields that are returned by SAP Integration Suite as a response. You can do this either manually or by uploading an existing JSON file example.

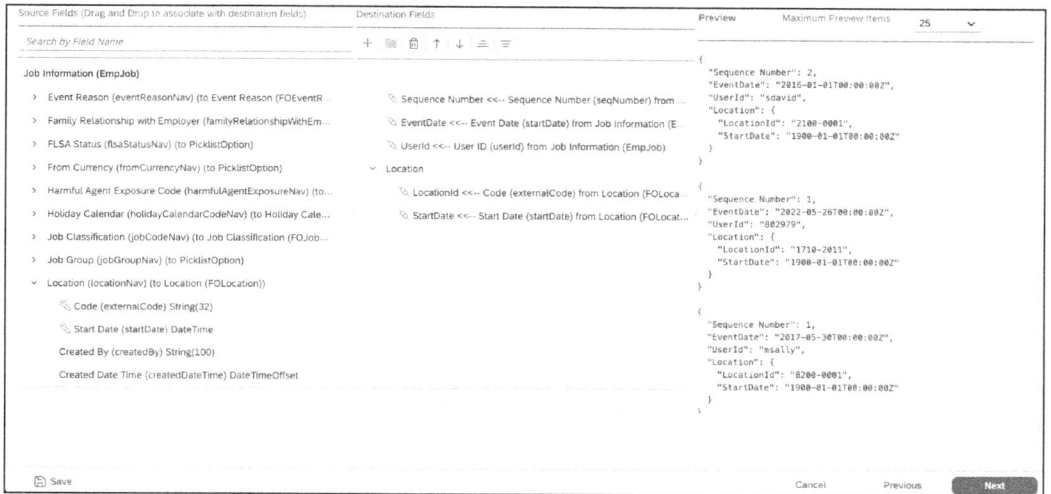

**Figure 11.15** Final Field Mapping with Preview

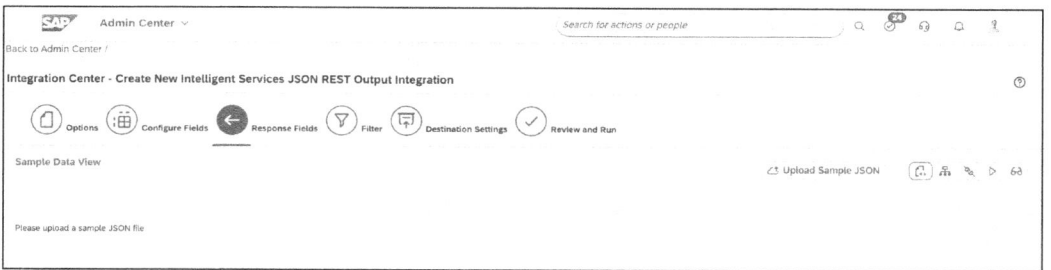

**Figure 11.16** Defining Response Fields

Click on **Upload Sample JSON** to open the dialog box (see Figure 11.17) where you can select the file. Finally, click on **Upload** to upload the JSON file and close the dialog.

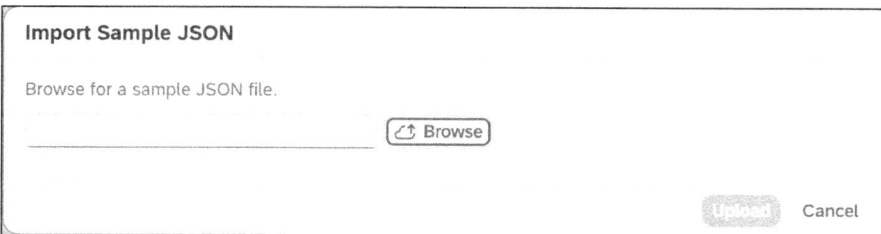

**Figure 11.17** Importing Sample Response JSON

The next step is to define a **Filter**, as shown in Figure 11.18. This lets you process only employees who live in the UK, for example. Along with the advanced filter options, you can create your own **Calculated Filter**, which (as the name suggests) is one that you can't derive directly from a field but that you calculate at runtime.

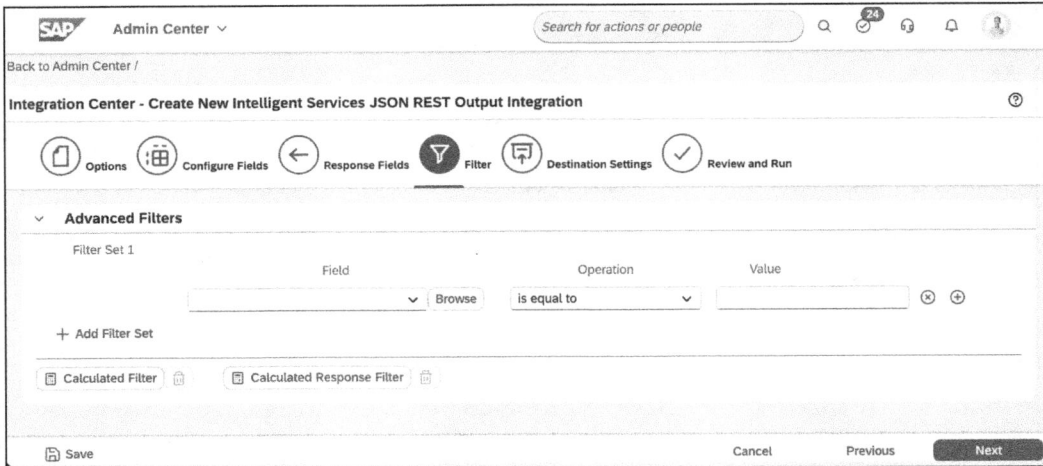

**Figure 11.18** Configuring Filters

In the next step, **Destination Settings**, you must define the connection to the target system. As we mentioned earlier, we recommend that you create and deploy the iFlow in advance (before you get to this step) in Cloud Integration. That will give you an endpoint that can be reached via HTTP and that you can use at this point. To use it, select the **REST Server Settings** option and enter the endpoint of the iFlow in the **REST API URL** field (see Figure 11.19).

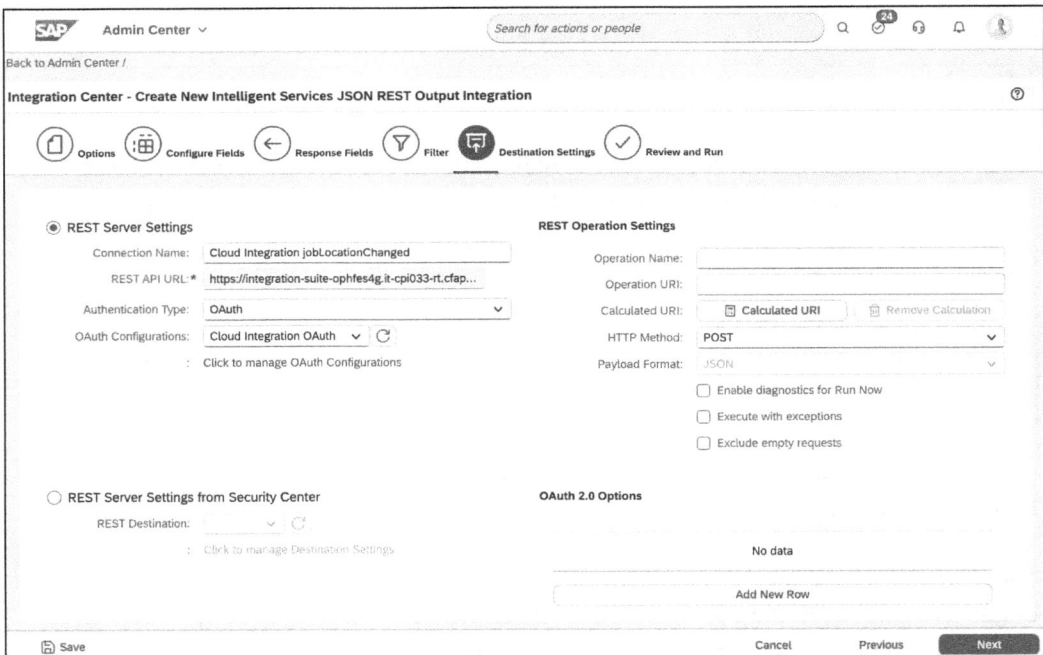

**Figure 11.19** Configuring Destination Settings

Alternatively, you can create a REST destination in the **Security Center**. The settings are the same as those you make directly in the REST **Server Settings** (see Figure 11.20). We recommend that you always create a destination in the **Security Center**, as you can reuse it and it's easier to make changes centrally.

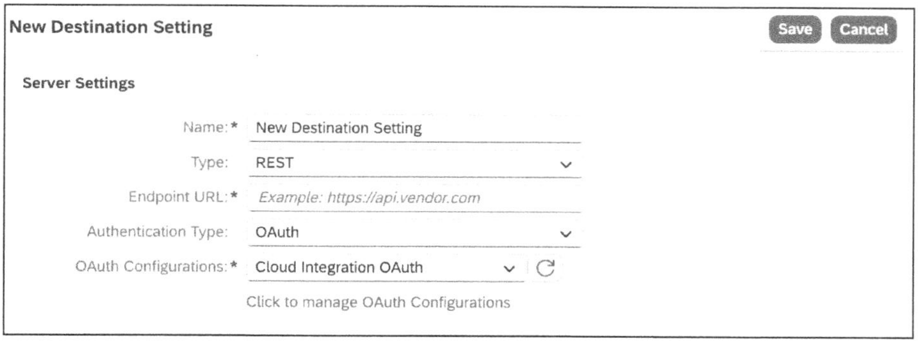

**Figure 11.20** Security Center Destination

After you've completed the destination settings and clicked **Next**, you'll be taken to the **Review and Run** section, where you'll see a summary of all the settings you've made. Make sure that the settings are correct and then click **Save**. Then, by clicking **Go to Intelligent Services Center**, you can return to the **Intelligent Services Center** (see Figure 11.21).

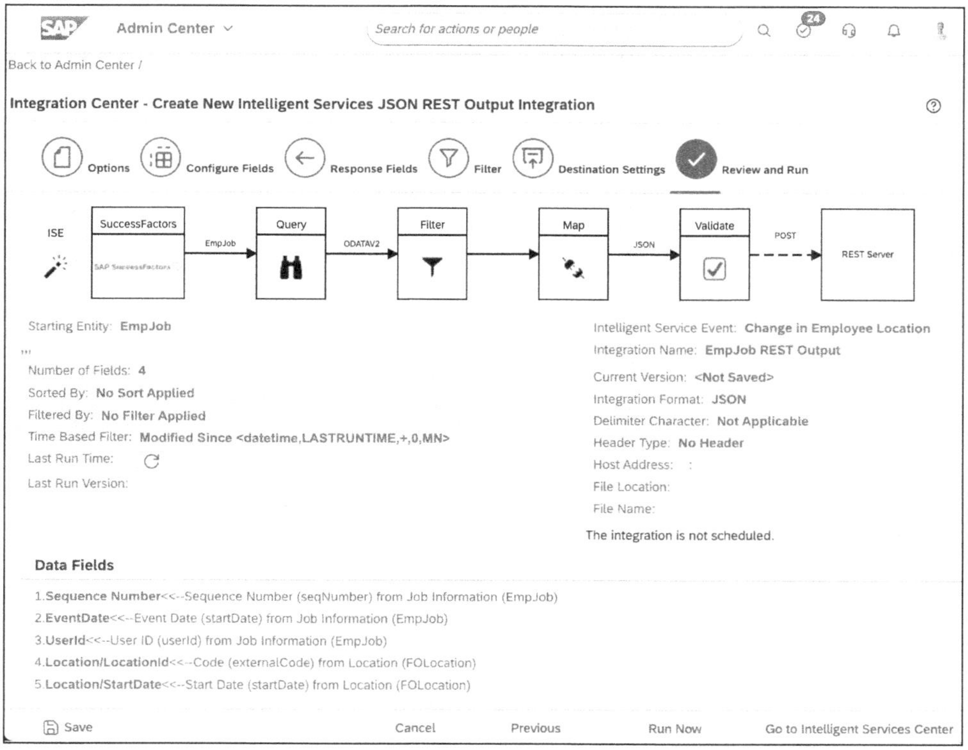

**Figure 11.21** Reviewing Configuration

489

There, you must reestablish the link to the integration you created earlier. To do this, click on the **Integration** entry on the right-hand side in the **Activities** area (see Figure 11.22).

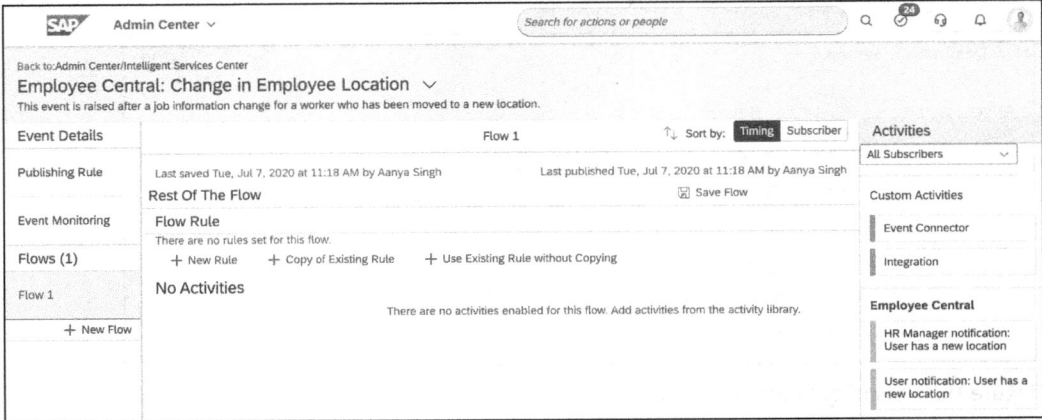

**Figure 11.22** Connecting Flow with Integration

That will open the **Select Integration** dialog box, and in the **My Integrations** section, you'll see the integration you created earlier. Select it and then click **Add Integration** (see Figure 11.23).

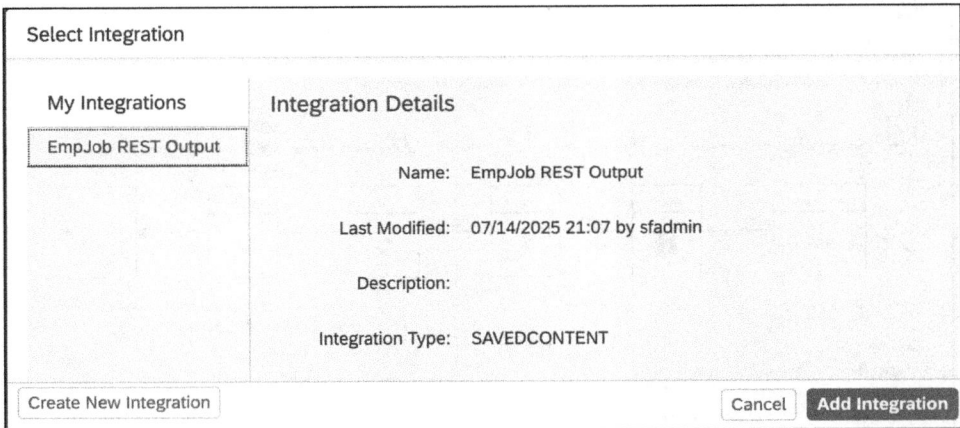

**Figure 11.23** Select Integration

Then, you'll return to the flow in the **Intelligent Service Center**, where you'll see that the flow is now linked to the integration (see Figure 11.24). Click **Save Flow** to save and activate the flow.

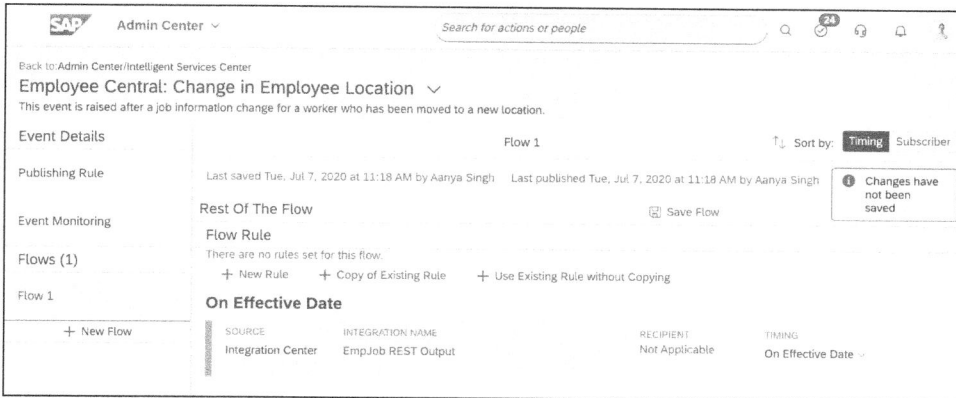

**Figure 11.24** Flow Configuration

A dialog box will then open to inform you that the flow has been successfully saved and published (see Figure 11.25). Close this dialog box by clicking **OK**.

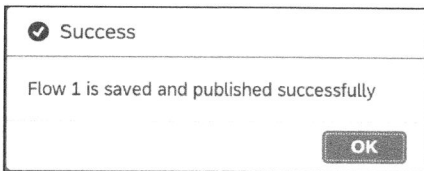

**Figure 11.25** Saving Information

To test the flow, you must make the necessary changes in SAP SuccessFactors. We'll show you how to do this for the flow you previously defined or for the corresponding event. To integrate with carriers, you need to change the employee's location by clicking on the pencil icon in the **Position Information** area, as shown in Figure 11.26.

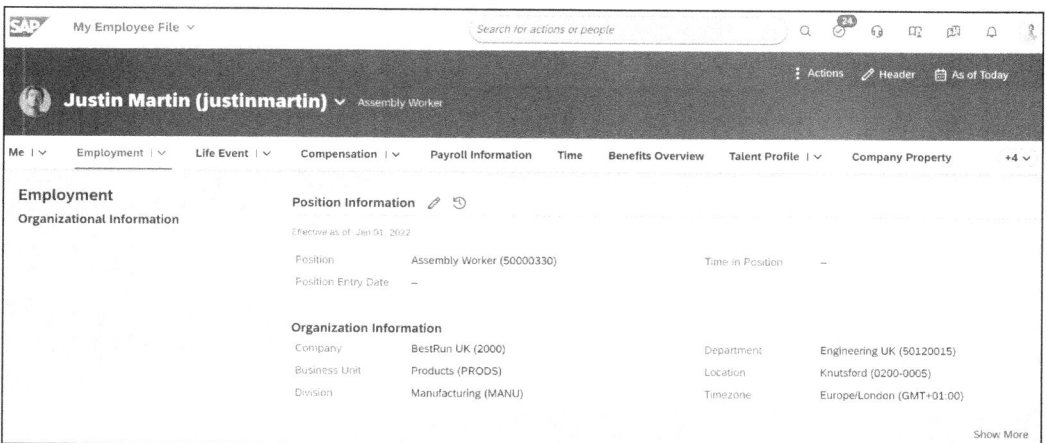

**Figure 11.26** Testing Integration by Changing Location

Now, change the employee's **Location** as shown in Figure 11.27 and click **Save**.

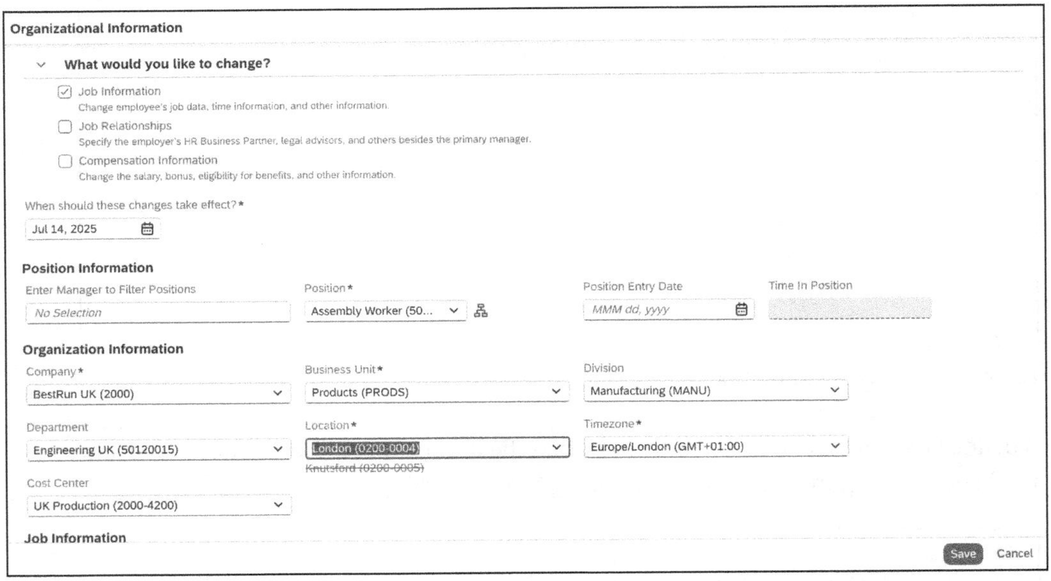

**Figure 11.27**  Changing Employee Location

The change will trigger an approval workflow based on SAP SuccessFactors customizing. A popup window called **Confirm Request** will appear, as shown in Figure 11.28, and there, you can enter a message for the approver. Finally, click **Submit** to send the request to the first approver. Then, by clicking on **Show Workflow Participants**, you can see which approval has been pulled or determined for the selected user.

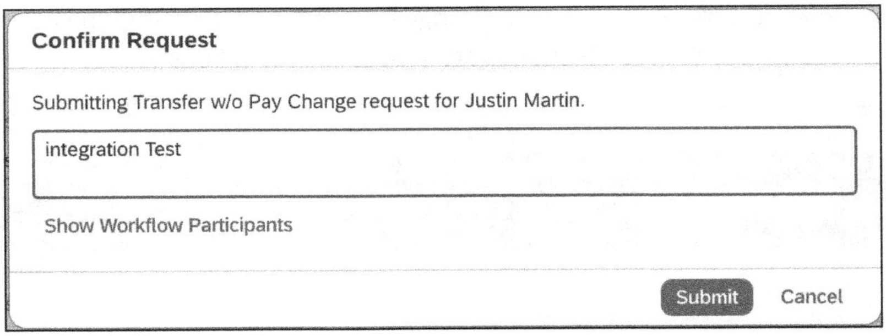

**Figure 11.28**  Confirming Changes

Then, switch to the respective approver and jump to the workflow details (see Figure 11.29). Optionally, you can enter a comment in the **Comment** field and click **Post**.

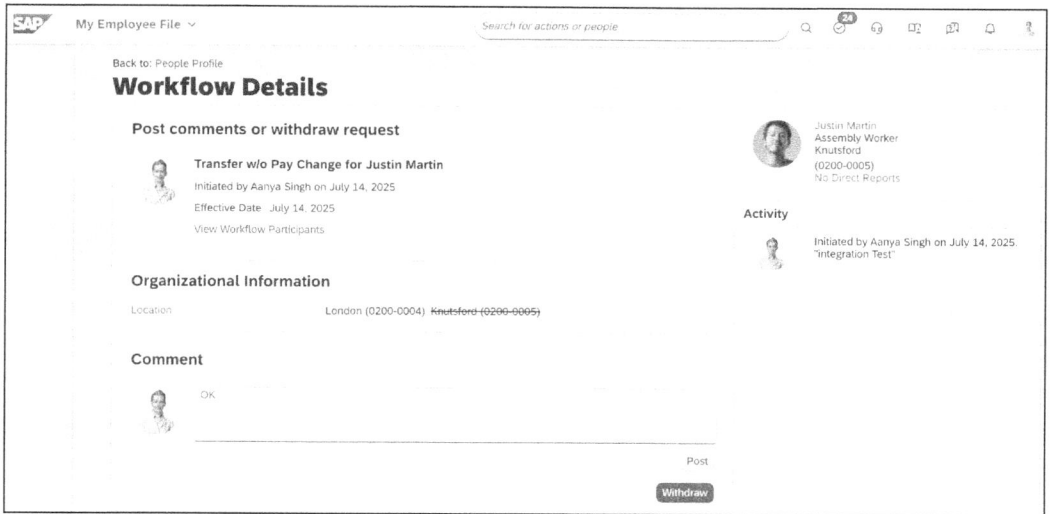

**Figure 11.29**  Workflow Approval

Next, you must now follow the workflow to the last approver, as shown in Figure 11.30. After the last approver has successfully approved the workflow, the integration will be triggered.

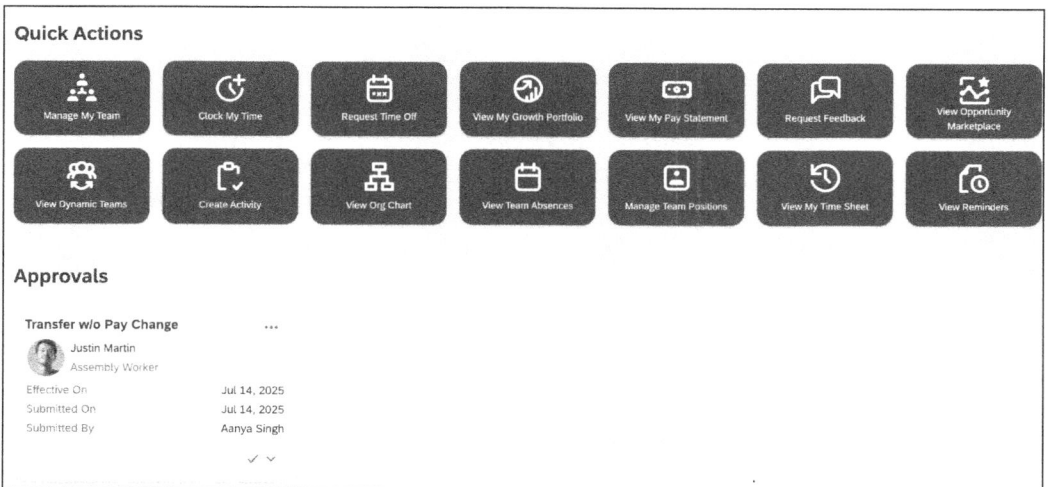

**Figure 11.30**  Final Approval

In the **Intelligent Service Center**, you can check whether the event was actually triggered by opening **Event Monitoring** in the **Event Details** section. Figure 11.31 shows that in our example, the event was triggered. If you don't see an event here, the event may not have been triggered due to the filter settings or because the integration for the flow is not yet active.

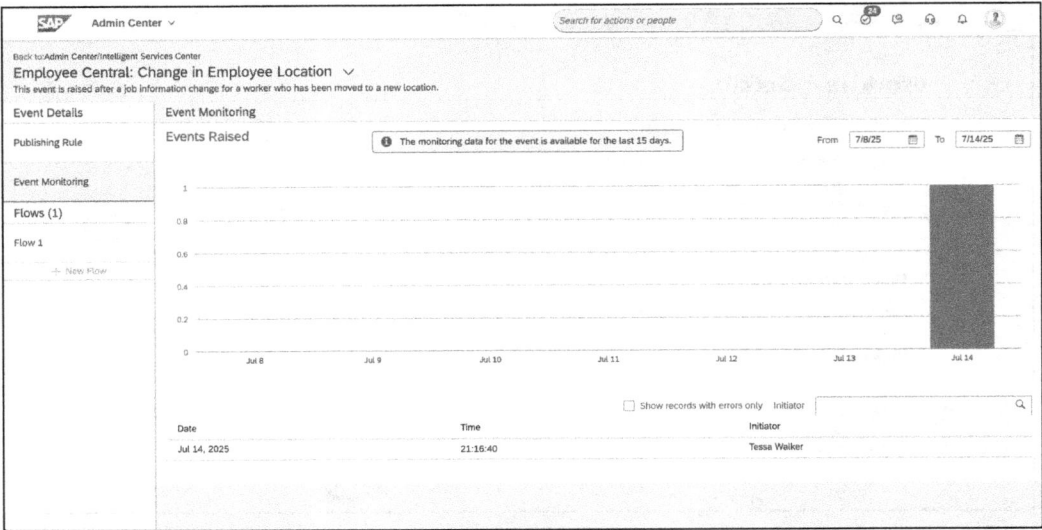

**Figure 11.31** Event Monitoring

You should now also check whether the event has called the iFlow in Cloud Integration. To do this, navigate to **Monitor • Integrations and APIs • Processed Messages** (as shown in Figure 11.32) and search for the messages for the iFlow you want. In our example, you can see that the iFlow has run successfully and has a status of **Completed**.

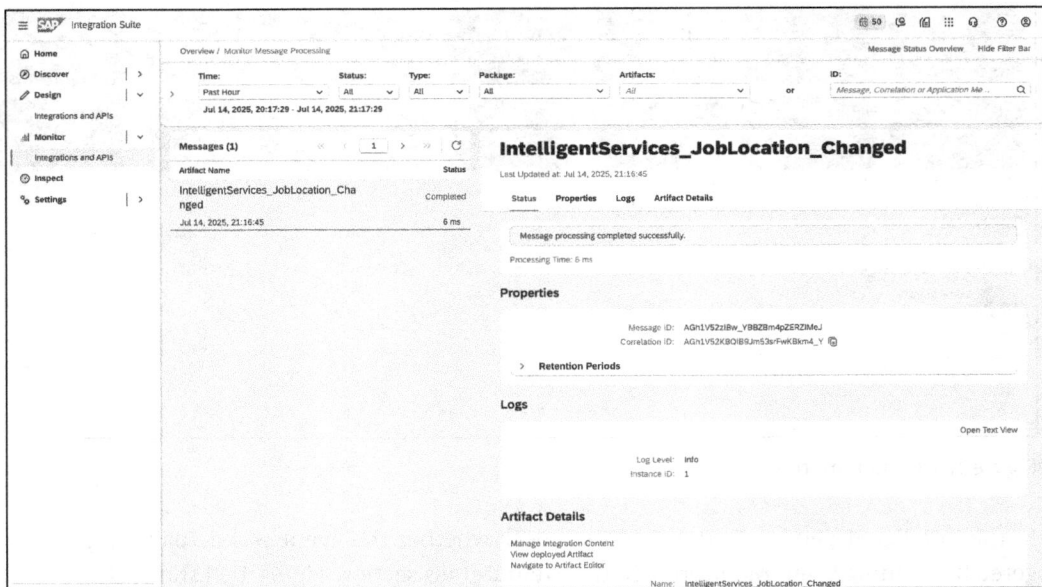

**Figure 11.32** Monitoring Message in Cloud Integration

The approach we've demonstrated in this section assumes that the iFlow has been deployed, is active, and is therefore accessible. However, if maintenance work is currently being carried out on Cloud Integration or if the iFlow is currently being deployed, then it may be temporarily unavailable. This would cause an error in SAP SuccessFactors, and the event would have to be sent again. However, there is a solution for this, which we'll discuss in the next section. The tool of choice for handling it is Event Mesh.

## 11.2    Event Handling in Cloud Integration

As we mentioned earlier, you can also use Event Mesh to send messages from SAP SuccessFactors to Cloud Integration. *Event Mesh* is a capability of SAP Integration Suite, and it must be activated. The Event Mesh capability is named **Manage Business Events**, and you can find the capabilities by navigating to the **Home** section in the side menu (see Figure 11.33).

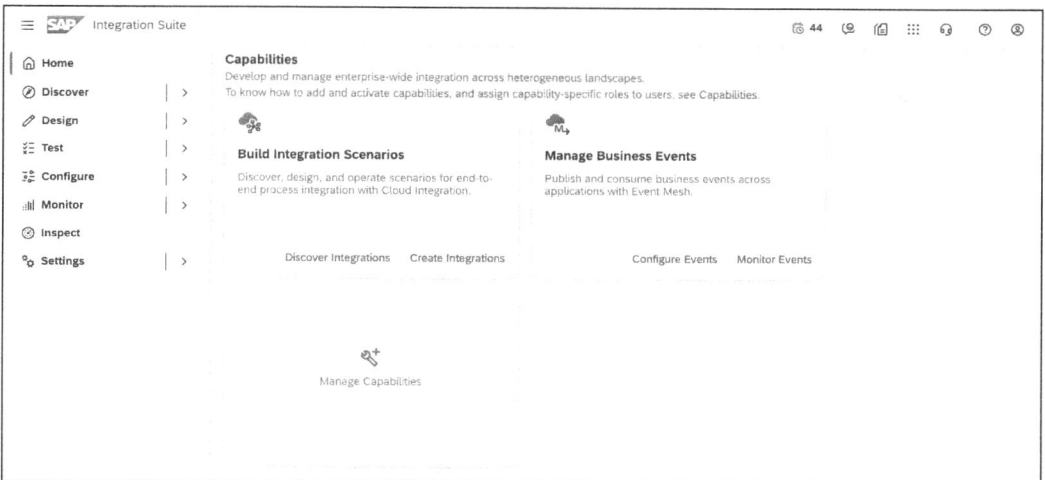

**Figure 11.33** Event Mesh Capability

Here, you need to activate the Event Mesh capability and then activate the event broker once—but you can only do it if you've been assigned the EventMeshAdmin role collection. If you have been, then you can navigate to the **Configure · Event Mesh** area in the side menu and click **Activate** (see Figure 11.34).

After you successfully activate Event Mesh, you should see the success message in Figure 11.35. If an error occurs during activation, you must report an incident to SAP.

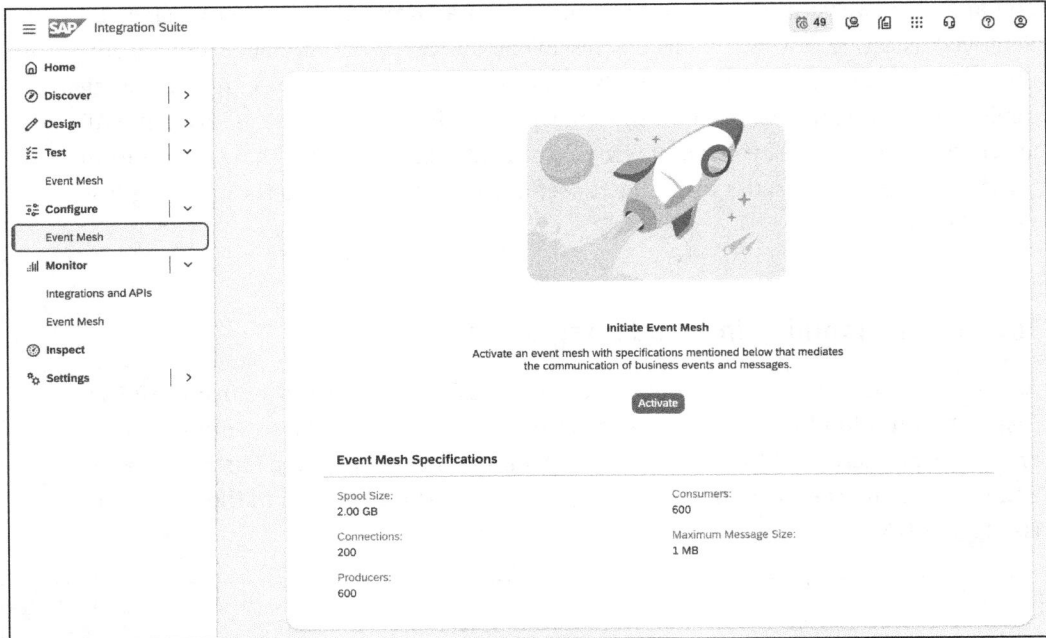

**Figure 11.34** Activating Event Mesh Message Broker

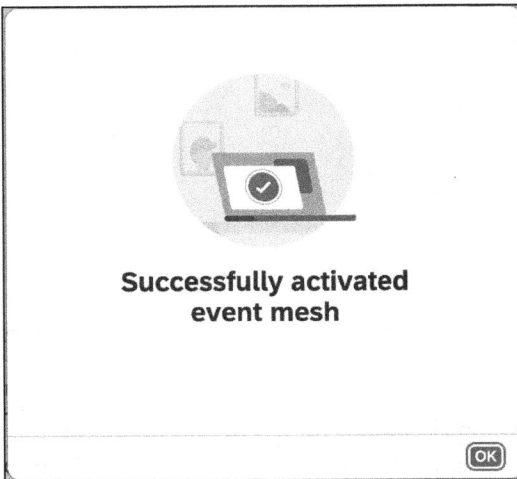

**Figure 11.35** Activation Success Message

After you activate Event Mesh, the details will be displayed in the **Overview** tab, as shown in Figure 11.36. In the **Message Client** area, you'll see a note that will state you must create a message client to publish or consume messages. You'll learn how to activate the message client later in this section.

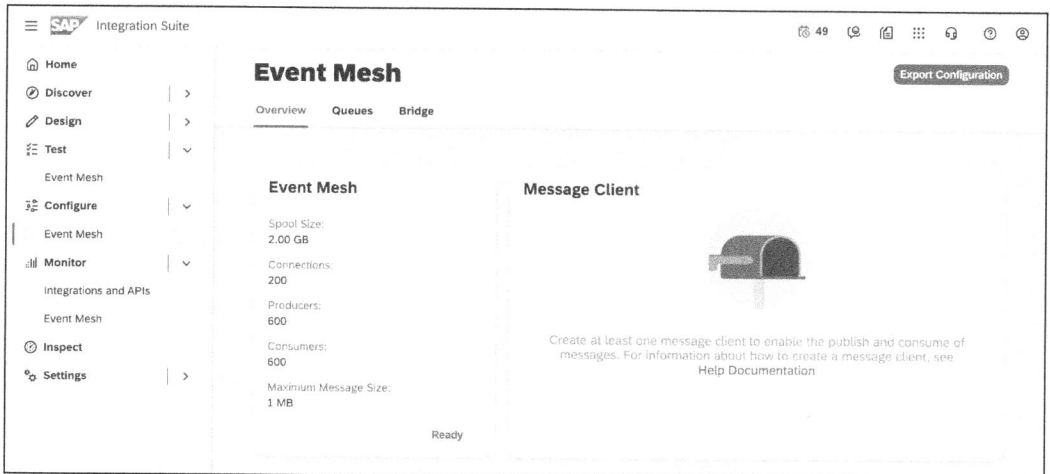

**Figure 11.36**  Event Mesh Overview

On the **Queues** tab, you'll see an overview of the existing queues. Click **Create** to create a new queue (see Figure 11.37).

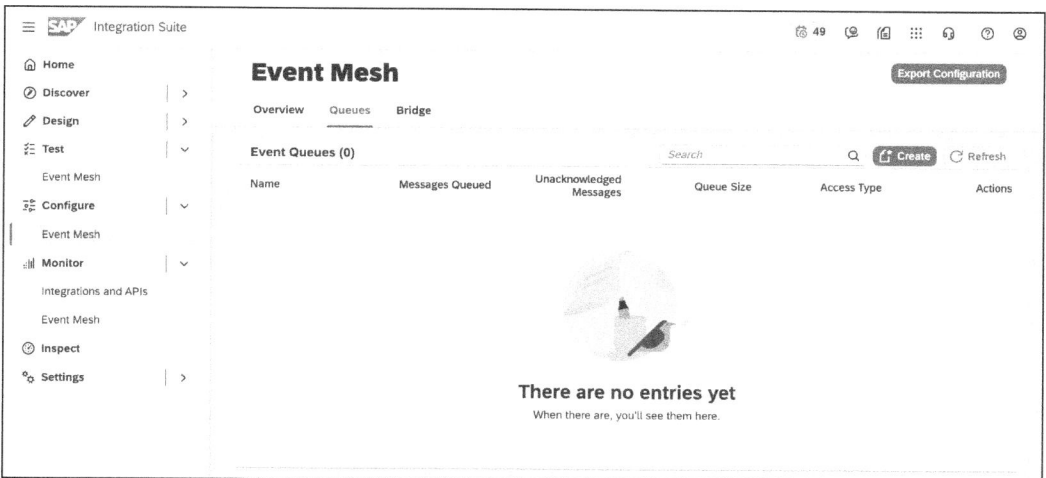

**Figure 11.37**  Event Mesh Queues

In the **Create Queue** dialog, you must configure the queue as shown in Figure 11.38. As you can see, the only mandatory parameter is the **Name**. If you selected a **Namespace** when creating the message client, you can select it here optionally. For the **Access Type** field, you can choose between **EXCLUSIVE** and **NON-EXCLUSIVE**. **EXCLUSIVE** means that only one consumer can access the queue at any given time, and **NON-EXCLUSIVE** means that multiple consumers can access the queue at any given time. Useful information about the other parameters is as follows:

- **Queue Size (in bytes)**
  This is the maximum size of the queue, and the default and he maximum is 1.5 GB.

- **Message Size (in bytes)**
  This is the maximum size of a single message, and the default and the maximum is 1 MB.

- **Max Unacknowledged Messages Per Consumer**
  This is the upper limit for unacknowledged messages per consumer. If this limit is exceeded, no further messages will be delivered.

- **Max Redelivery Count**
  This is the number of retry attempts in case of errors. The default is 0 (which means unlimited attempts until time-to-live expires), and the maximum is 255. When the limit is reached, the message is deleted or moved to a dead message queue (if one has been defined).

- **Dead Message Queue**
  This queue contains messages that have either exceeded the maximum number of delivery attempts or the time-to-live, and to use one, you must create it separately beforehand.

- **Max time-to-live (in seconds)**
  This is the maximum time a message can remain in the queue. The default and the maximum is 604,800 seconds, which equals 7 days. After this time limit expires, the message will be deleted or moved to the **Dead Message Queue**.

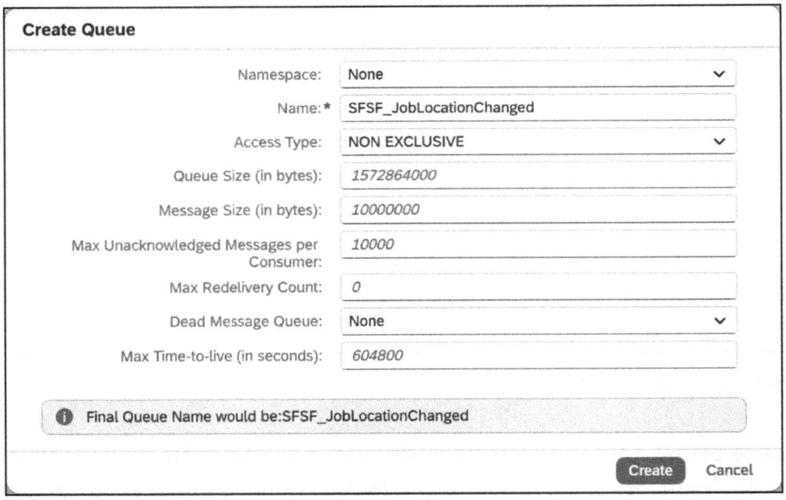

| Create Queue | |
| --- | --- |
| Namespace: | None |
| Name:* | SFSF_JobLocationChanged |
| Access Type: | NON EXCLUSIVE |
| Queue Size (in bytes): | 1572864000 |
| Message Size (in bytes): | 10000000 |
| Max Unacknowledged Messages per Consumer: | 10000 |
| Max Redelivery Count: | 0 |
| Dead Message Queue: | None |
| Max Time-to-live (in seconds): | 604800 |
| ⓘ Final Queue Name would be:SFSF_JobLocationChanged | |
| | Create    Cancel |

**Figure 11.38**  Creating Event Mesh Queue

After you create the queue, you should see the properties of the queue you just created, as shown in Figure 11.39. By clicking on the **Edit** button, you can adjust the properties—which is particularly important if, for example, you want to adjust the queue site or set up a dead letter queue at a later stage.

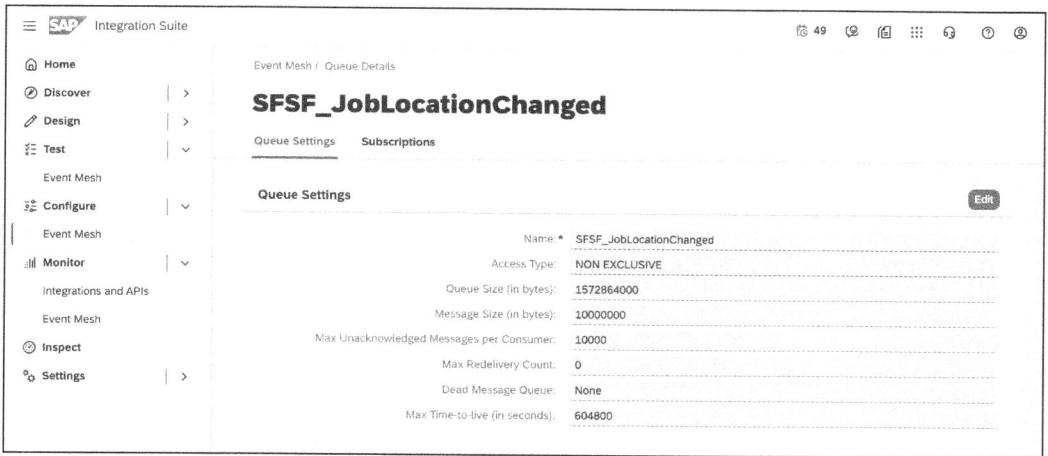

**Figure 11.39** Event Mesh Queue Settings

To ensure that the queue can actually be used, you must create a message client. To do this, you must assign it to the subaccount as a service plan by selecting the **SAP Integration Suite, Event Mesh** service and then the **message-client** plan (see Figure 11.40). Then, click **Add 1 Service Plan** and save your entries.

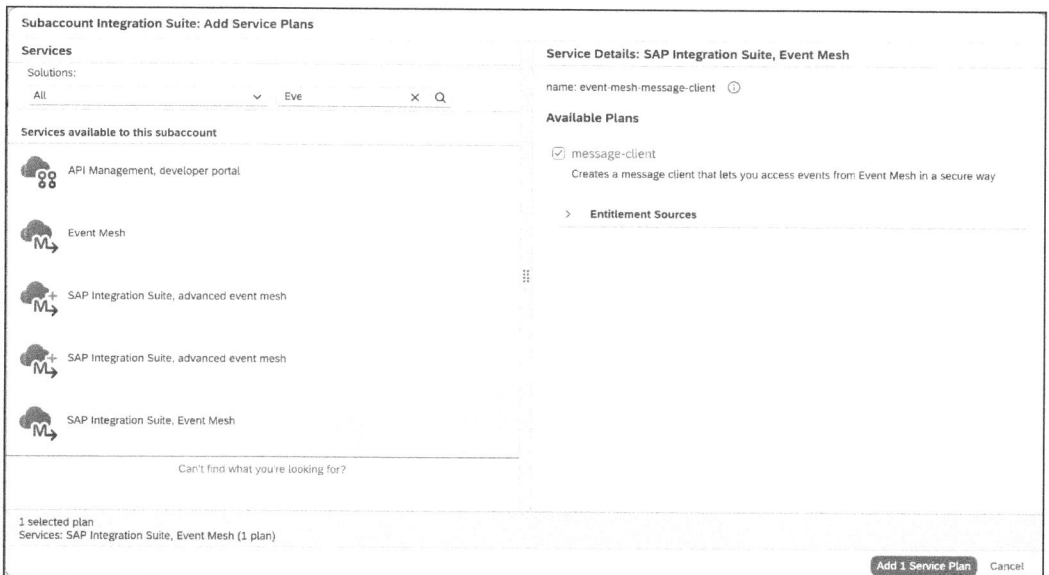

**Figure 11.40** Event Mesh Message Client Service Plan

The next step is to instantiate the service. To do this, navigate to the **Services · Instances and Subscriptions** area in the side menu of the SAP BTP cockpit for the subaccount you want. Then, click the **Create** button (see Figure 11.41).

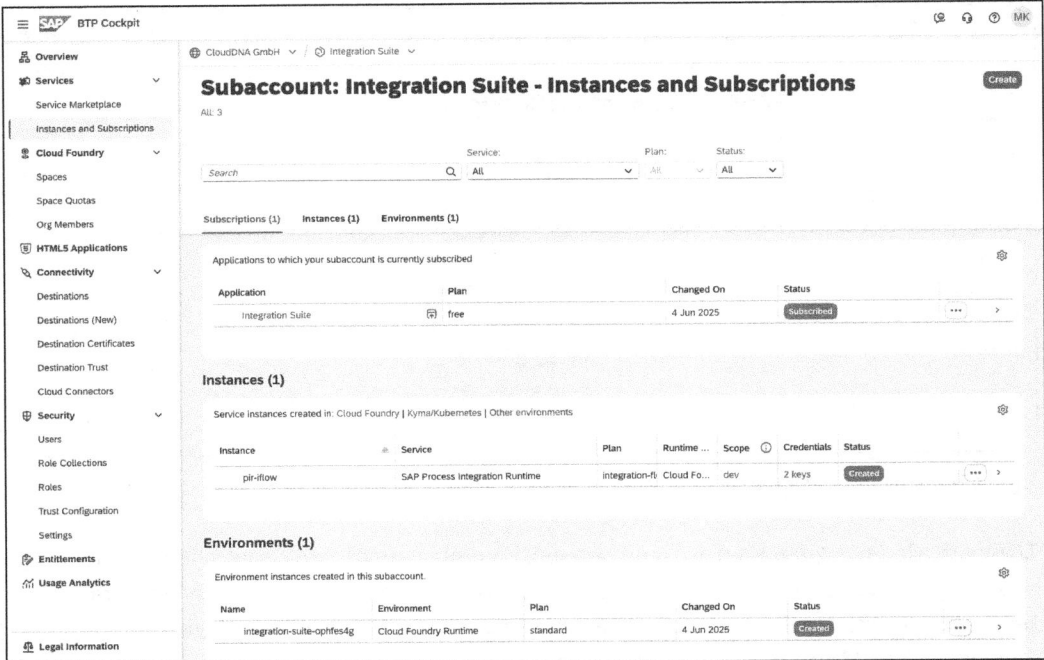

**Figure 11.41**  Creating Service Instance

Then, in the **New Instance or Subscription** popup, select **SAP Integration Suite, Event Mesh** as the **Service** and the **message-client** as the **Plan** (see Figure 11.42). Select **Cloud Foundry** as the **Runtime Environment**, assign the **Space** you want, assign an **Instance Name**, and click **Next**.

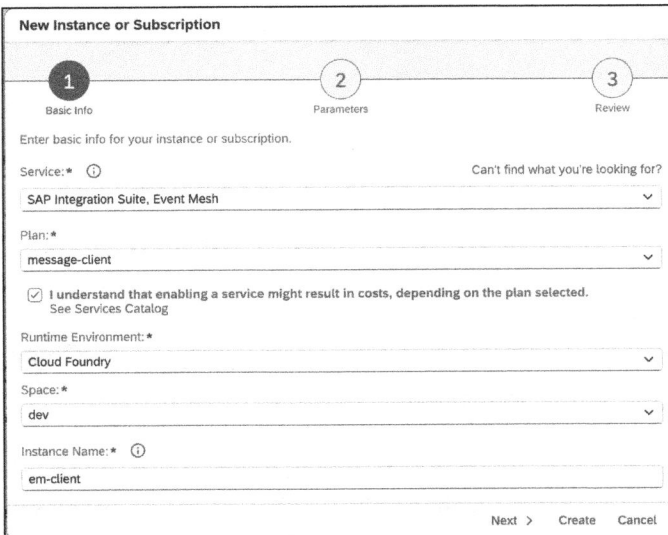

**Figure 11.42**  Specifying Service Instance Details

You can then make adjustments in the **Parameters** step, as shown in Figure 11.43—although usually, you'll only adjust the namespace here. Then, click **Next**.

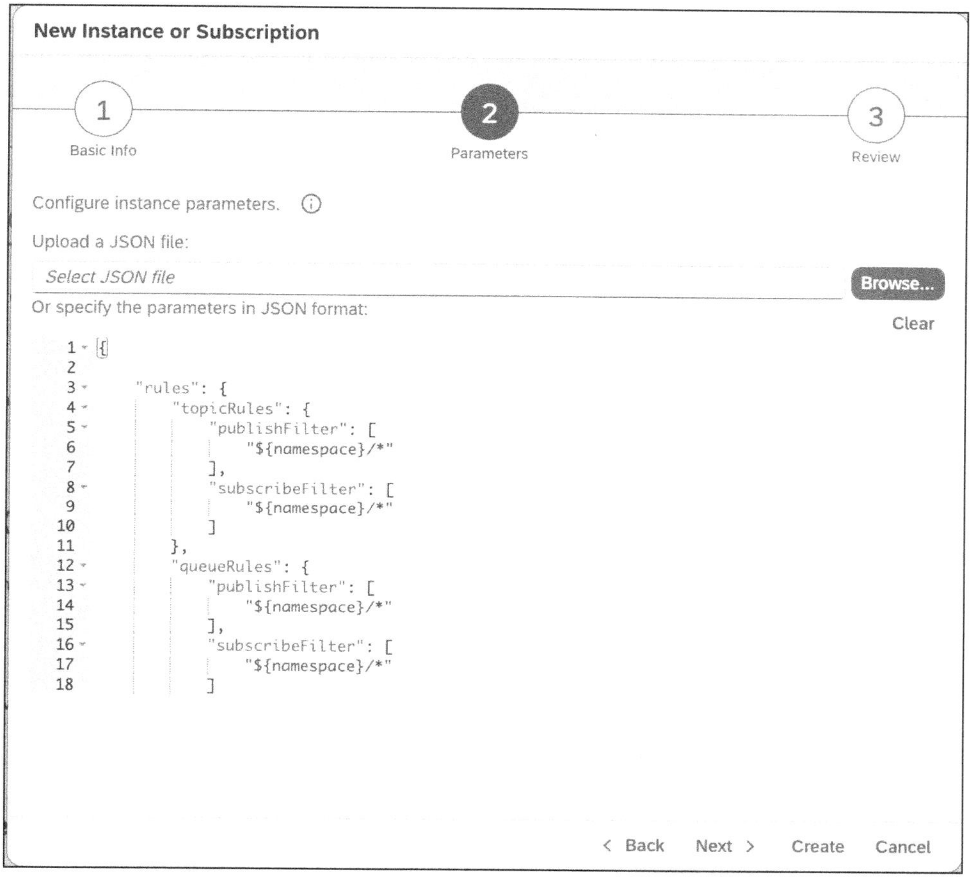

**Figure 11.43** Configure Service Parameters

Finally, in the **Review** step, you'll see a summary of the settings you've made. Check these and then click **Create** to create the service instance (see Figure 11.44).

You'll then see the previously created instance of **message-client** in the **Instances** area, as shown in Figure 11.45. Click on the three dots and select **Create Service Key** to get a service key for this instance.

As shown in Figure 11.46, the service key contains the credentials and endpoints that are required to address the event mesh. The AMQP protocol and REST are supported.

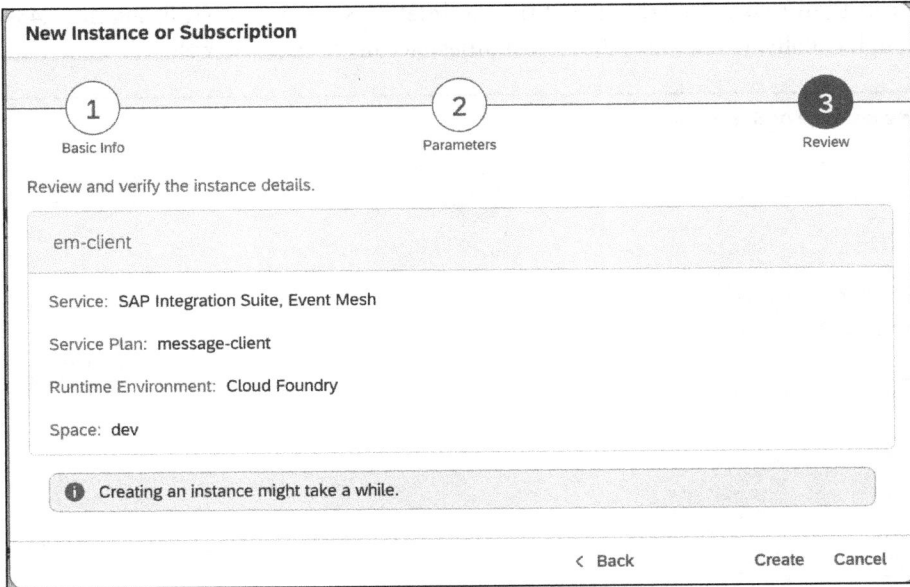

**Figure 11.44** Service Configuration Review

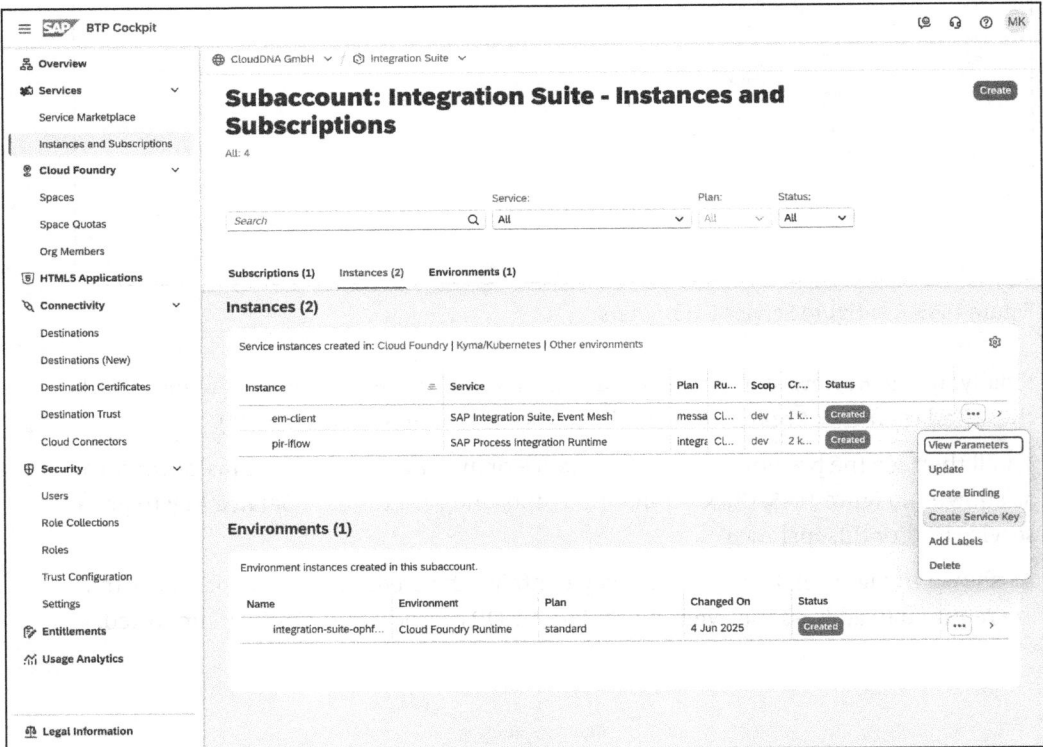

**Figure 11.45** Creating Service Key for Event Mesh Client

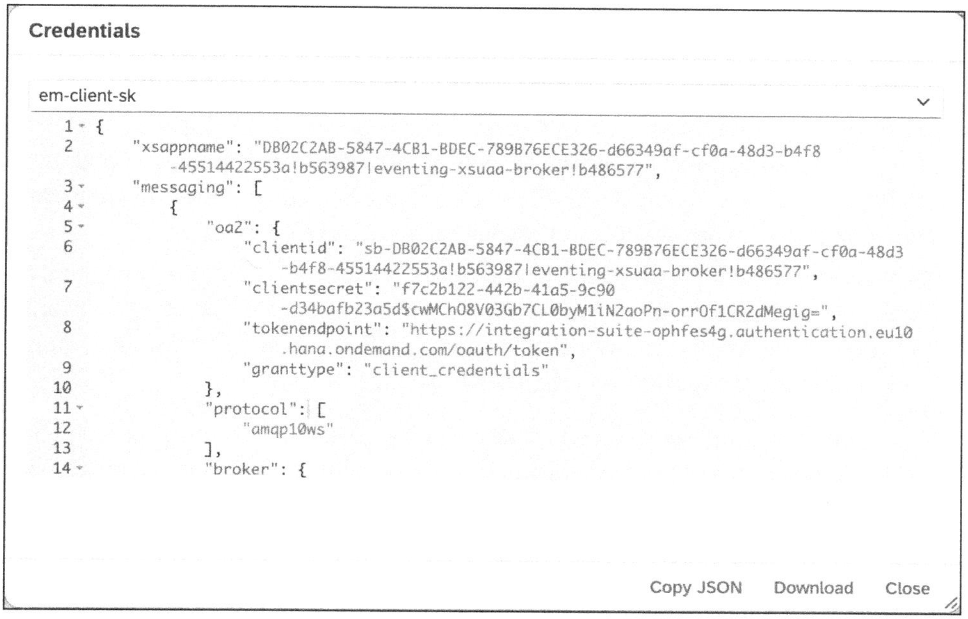

**Credentials**

em-client-sk

```
 1 ⌄ {
 2        "xsappname": "DB02C2AB-5847-4CB1-BDEC-789B76ECE326-d66349af-cf0a-48d3-b4f8
                -45514422553a!b563987|eventing-xsuaa-broker!b486577",
 3 ⌄      "messaging": [
 4 ⌄          {
 5 ⌄              "oa2": {
 6                      "clientid": "sb-DB02C2AB-5847-4CB1-BDEC-789B76ECE326-d66349af-cf0a-48d3
                            -b4f8-45514422553a!b563987|eventing-xsuaa-broker!b486577",
 7                      "clientsecret": "f7c2b122-442b-41a5-9c90
                            -d34bafb23a5d$cwMChO8V03Gb7CL0byM1iN2aoPn-orrOf1CR2dMegig=",
 8                      "tokenendpoint": "https://integration-suite-ophfes4g.authentication.eu10
                            .hana.ondemand.com/oauth/token",
 9                      "granttype": "client_credentials"
10              },
11 ⌄          "protocol": [
12                  "amqp10ws"
13              ],
14 ⌄          "broker": {
```

Copy JSON     Download     Close

**Figure 11.46** Service Key Credentials

REST is used to connect SAP SuccessFactors to Event Mesh, so you can continue with the configuration of the REST endpoint as described in Section 11.1.

## 11.3   Summary

In Section 11.1, we explained how to create and configure event triggers in SAP Success-Factors that call an HTTP endpoint of an iFlow. This approach connects SAP Success-Factors events directly to Cloud Integration by exposing an iFlow endpoint and registering it as the consumer of the event. This method is straightforward and often sufficient for smaller or less complex scenarios, but it does have some limitations. The direct HTTP call tightly couples the sender system with the iFlow, it requires the endpoint to be always available, and it places the responsibility for retries and error handling largely on the iFlow itself.

The most elegant and scalable option for calling Cloud Integration is using Event Mesh, as demonstrated in Section 11.2. Event Mesh reliably publishes events from SAP Success-Factors (or any other SAP or non-SAP system) to a central event broker, and cloud Integration can then subscribe to these events asynchronously. This decoupled approach has clear advantages.

- It provides resilience and buffering, which are key. Event Mesh reliably handles message persistence and retries to ensure that no event is lost, even in the event of temporary unavailability of the iFlow or target system.

- It provides loose coupling, which is essential. Systems no longer need to know each other directly. SuccessFactors publishes events once, and multiple consumers—including Cloud Integration, custom applications, or analytics pipelines—can subscribe independently.

- Event Mesh is the clear choice for scalable solutions because it handles high event volumes with ease. Its capabilities include queueing, load distribution, and parallel processing, which ensure efficient management of large volumes.

- It provides extensibility, which is also essential. Event Mesh seamlessly triggers various downstream scenarios—including onboarding in SuccessFactors, provisioning in Microsoft Entra ID, and notifications in a ticketing system—without requiring any additional configuration in the source system.

You should start simple with direct HTTP event triggers, and as your landscape grows more complex or business processes become more critical, you should move to an event-driven architecture with Event Mesh. This is the recommended best practice.

# Chapter 12

# Working with Event-Based Architectures

*In today's fast-paced digital landscape, businesses are expected to respond to change in real time—such as by reacting to customer interactions, supply chain events, and financial transactions. Traditional integration models, while reliable, often fall short in providing the agility and scalability that are required by modern enterprises. As organizations accelerate their digital transformation journey, the need for more responsive, flexible, and loosely coupled integration paradigms becomes clear. This is where event-driven architecture comes into play.*

This chapter explores the concept of *event-driven architecture (EDA)* within the context of SAP BTP—which is SAP's innovation platform for building and integrating intelligent enterprise solutions. With SAP BTP's support for asynchronous communication patterns, cloud-native services, and real-time data flows, EDA has become an essential architectural approach for integrating SAP systems with internal and external applications.

Section 12.1 highlights the advantages of event-driven integration and outlines why organizations are moving away from synchronous, tightly coupled systems toward more scalable and loosely coupled models. It delves into how EDA supports agility, extensibility, and real-time responsiveness in SAP landscapes, particularly when you're working with services like Event Mesh and SAP Cloud Application Programming Model.

Section 12.2 examines the differences between point-to-point (P2P) and event-driven architectures, comparing their design principles, scalability implications, and use case applicability. You must understand these differences to understand when and why EDA is a better fit for modern, distributed integration scenarios.

Finally, Section 12.3 presents best practices for asynchronous integration and offers practical guidance on designing event-driven systems within SAP BTP. Topics include idempotency, durable subscriptions, event enrichment, security, and the strategic use of Cloud Integration. These practices are intended to help developers and architects build resilient and maintainable solutions that can evolve with business needs.

Whether you're building side-by-side extensions for SAP S/4HANA, integrating with third-party services, or orchestrating cloud-native applications, this chapter provides foundational knowledge and actionable guidance to help you embrace event-driven integration in SAP BTP with confidence.

## 12.1   Advantages of Event-Driven Integration

In today's hyperconnected digital enterprise, the ability to respond to business changes instantly has become a competitive necessity. Enterprises operate in dynamic environments where customer interactions, supply chain updates, and operational events occur continuously across a complex landscape of systems. Traditional integration approaches—which often rely on synchronous, tightly coupled architectures—are increasingly challenged by the demands for flexibility, resilience, and real-time responsiveness. This is especially true in SAP-centric environments where systems such as SAP S/4HANA, SAP SuccessFactors, and SAP Ariba must interact fluidly with custom-built extensions, partner ecosystems, and third-party applications.

EDA offers a modern integration paradigm that directly addresses these challenges. By shifting the focus from tightly orchestrated request-response interactions to loosely coupled, event-based communication, EDA enables systems to publish and react to business events as they happen. This decoupling between producers and consumers not only improves system agility but also allows new applications and services to be introduced or updated independently, without disrupting existing integrations. Within SAP BTP, this model aligns perfectly with the microservices architecture and cloud-native principles that SAP promotes for building scalable, resilient, and modular applications.

At the core of EDA is the notion of treating business events as first-class citizens. These events—such as a sales order being created, a supplier being updated, and an invoice being posted—represent significant moments in a business process. By emitting these events as they occur, systems can notify other components without requiring knowledge of who is consuming the data or how it will be processed.

In SAP BTP, this pattern is made possible through services like Event Mesh, which provides a managed messaging backbone for secure, asynchronous communication between event producers and consumers. This allows for seamless integration between SAP S/4HANA and side-by-side extensions or even non-SAP systems, in a way that is scalable, loosely coupled, and reactive.

Beyond its technical advantages, EDA provides significant business value. It supports faster innovation by allowing IT teams to build and deploy new services without impacting core systems. It also reduces the operational overhead associated with managing complex point-to-point integrations, and it enhances visibility into business operations by enabling event tracing and analytics. For example, when EDA is combined with SAP's enterprise event enablement framework, business events can be exposed in a standardized format and consumed by applications that are built in SAP Cloud Application Programming Model or other technologies. This creates a foundation for building intelligent, event-driven enterprises where business decisions can be automated, triggered, and scaled in real time based on actual business events—not delayed data synchronization. As digital transformation efforts continue to accelerate, embracing event-driven integration is becoming not just a technical evolution but a strategic imperative for SAP customers.

The key benefits of EDA, which we'll discuss further throughout this section, are as follows:

- **Decoupling of systems**
  Events decouple the sender (producer) from the receiver (consumer) to enable the independent development, deployment, and scaling of services. This is particularly beneficial in SAP BTP's multiservice and microservice environments.

- **Real-time responsiveness**
  Business events—such as a sales order creation or a supplier change—can trigger downstream processes without delay and thus significantly improve business process efficiency.

- **Scalability and flexibility**
  New consumers can subscribe to existing event streams without modifying the producing applications. This supports extensibility scenarios in SAP S/4HANA and SAP BTP, such as building side-by-side extensions.

- **Improved maintainability**
  Changes in one system have minimal impact on others, and that leads to reduced regression testing and improved maintainability of integrations.

- **Better alignment with cloud-native principles**
  Event-driven design aligns well with the microservices and serverless paradigms that are offered in SAP BTP. This fosters resilient and scalable cloud-native solutions.

## 12.1.1 Decoupling of Systems

One of the most powerful architectural advantages of adopting EDA is the *decoupling of systems*—a design principle that eliminates the tight dependencies between producers and consumers of data. In traditional integration models, the sender of a message often needs to know where the receiver is, how it works, and whether it's available at the time of communication. This tight coupling creates brittle systems that are difficult to scale, upgrade, and change independently. On the other hand, with EDA, the producer simply emits an *event*—a fact that something has happened—without any concern for who will receive it or what will be done with it. The event is then routed through an intermediary such as Event Mesh, which manages delivery to one or more interested consumers. This separation of concerns not only simplifies system interactions but also supports a far more modular and maintainable architecture.

The immediate advantage of decoupling is evident in the development and deployment lifecycle. You can deploy and build services independently, without needing synchronized releases or preagreed interface designs. For example, in an SAP BTP environment, a development team that's working on a customer feedback application can subscribe to `SalesOrder.Created` events without requiring any changes or coordination with the team that's responsible for SAP S/4HANA Sales. This independence accelerates delivery cycles, allows parallel development tracks, and minimizes the risk of regression issues that are caused by changes in downstream or upstream systems. Furthermore, if the team needs to replace or scale up one consumer service due to increased load, it can do so without impacting the producer or any other subscriber. This agility is particularly beneficial in SAP BTP's multiservice and microservice environments, where loosely coupled components must be orchestrated to deliver integrated yet individually manageable business capabilities.

From a scalability and long-term evolution standpoint, decoupling supports innovation without disrupting the core. You can add new features or integrations by simply subscribing to existing events, which leaves producers untouched. This is invaluable in SAP landscapes, where the core ERP system must remain stable while innovation happens at the edges—in mobile apps, analytics tools, AI models, and partner solutions. For instance, once a `BusinessPartner.Changed` event is emitted by SAP S/4HANA, you can introduce a new compliance-checking service without modifying or even notifying the original system. The ability to evolve each service independently, at its own pace, drastically reduces integration complexity and opens the door to more adaptive, event-aware business processes. In a constantly changing digital landscape, this level of architectural flexibility isn't just beneficial—it's a competitive necessity.

### 12.1.2   Real-Time Responsiveness

In today's dynamic enterprise landscape, speed of response is often the difference between operational efficiency and costly delays. One of the most immediate and impactful advantages of EDA is its ability to enable *real-time responsiveness* across distributed systems. Rather than relying on scheduled jobs, batch processing, or manual triggers, EDA allows systems to react instantly to business events—such as a new sales order, a supplier update, or a stock level change. These events, when emitted from systems like SAP S/4HANA, can be captured and processed by other services without delay, which drastically reduces the latency between data generation and business reaction. For example, when a `SalesOrder.Created` event is published, a downstream fulfillment system can immediately initiate packing and shipping processes. The result isn't just faster operations but a more synchronized, intelligent enterprise that's capable of responding in real time to both risks and opportunities.

The immediate advantage of real-time responsiveness is its impact on business process efficiency and customer satisfaction. In traditional models, a newly created order might sit idle until a batch job runs or an integration tool polls for updates—and that can introduce delays that can accumulate across departments. But with EDA, downstream processes are notified the moment an event occurs so that the next steps can be automatically executed. For instance, a `Supplier.Changed` event can instantly trigger a vendor compliance check, update pricing models, and refresh purchasing agreements—all without human intervention or lag time. In the context of SAP BTP, this means that side-by-side extensions, microservices, and analytics dashboards can operate on near real-time data, without taxing the core ERP system or waiting for data replication. The resulting improvements in speed, accuracy, and decision-making create immediate business value and elevate user experiences—both internally and externally.

Moreover, real-time responsiveness fosters a shift from reactive to proactive operations. By building workflows that listen to relevant business events and act instantly, enterprises can preempt issues rather than merely responding to them. For example, an event like `InventoryThresholdBreached` can automatically initiate replenishment workflows or alert procurement teams before stockouts occur. Likewise, a `CustomerComplaintReceived` event can trigger a sentiment analysis service and notify account managers before the issue escalates. SAP BTP's event-driven capabilities—especially when powered by Event Mesh—allow these events to propagate across services reliably and securely. This approach not only enhances operational agility but also creates a foundation for intelligent automation, predictive insights, and continuous process optimization. In essence, real-time responsiveness transforms business events from passive data points into active triggers for innovation and efficiency.

### 12.1.3    Scalability and Flexibility

A central advantage of EDA is the *scalability and flexibility* it brings to enterprise integration landscapes. Traditional integration models often require significant coordination between producers and consumers—every new connection introduces more complexity, risk, and dependency. In contrast, EDA enables systems to scale by allowing new consumers to subscribe to existing event streams without any need to modify the producing application. This means that once an event, such as `PurchaseOrder.Approved` or `BusinessPartner.Updated`, is published by a system like SAP S/4HANA, it can be consumed by any number of applications or services independently. The producer emits the event once, and multiple consumers—running on SAP BTP or elsewhere—can process that event based on their own requirements and logic. This one-to-many distribution model unlocks a scalable integration pattern that accommodates growing business needs without introducing tighter coupling.

The immediate advantage of this model is its support for agile extensibility, which is particularly valuable in side-by-side extension scenarios. Organizations often need to enhance core enterprise resource planning (ERP) functionality without disrupting existing operations or customizing SAP S/4HANA directly. EDA enables this by allowing you to use SAP BTP to build new applications that simply listen to relevant events that are emitted by the core system. For instance, you can build a new compliance monitoring tool that can subscribe to Supplier.Changed events without requiring any adjustments to the supplier master module in SAP S/4HANA. Similarly, you can build a customer loyalty app that can consume SalesOrder.Completed events and trigger reward calculations in SAP BTP. You can develop, deploy, and maintain these extensions independently to accelerate innovation while preserving the integrity and upgradeability of the digital core. This loose coupling ensures that adding functionality doesn't create a fragile chain of interdependent services but instead enhances the overall architecture's adaptability.

Beyond extensibility, this model supports horizontal scaling and resilience. Because event consumers operate independently, you can deploy multiple instances of the same consumer in parallel to handle increased event volumes. For example, during peak times such as end-of-quarter processing or major sales campaigns, services can scale out to consume and process events concurrently—without affecting other consumers or the event producer. This flexible scaling behavior is intrinsic to event brokers like Event Mesh, which manage message queues and delivery semantics automatically. Moreover, the fact that consumers can be added, removed, or replaced without touching the producer makes EDA ideal for evolving technology landscapes, where you may need to introduce new tools and platforms frequently. In essence, EDA provides a foundation for building robust, adaptable, and future-proof integration ecosystems in SAP S/4HANA and SAP BTP environments.

### 12.1.4   Improved Maintainability

One of the persistent challenges in traditional integration landscapes is the fragile nature of tightly coupled systems. When one application changes, you often need to review, retest, and even refactor all the systems that depend on it. This leads to heavy regression testing cycles, increased coordination between teams, and prolonged deployment timelines. EDA offers a robust solution to this problem by significantly improving *maintainability*. By decoupling the producer and consumer through asynchronous communication, it makes changes in one system—whether they be structural, behavioral, or even changes to its availability—have minimal or no direct impact on the others. A producer can evolve independently, as long as it continues to publish events in the agreed format. Consumers, too, can be updated, scaled, or replaced without disrupting the event source. This separation of responsibilities allows for more modular and sustainable systems that are easier to understand, test, and maintain over time.

The immediate advantage of this decoupled model is a dramatic reduction in the scope and risk of change. For example, when a team working on SAP S/4HANA needs to adjust the logic for how a `SalesOrder.Created` event is generated, it can make and deploy this change without coordinating with every consuming system. As long as the event schema remains consistent, all downstream services that consume the event—such as shipping, invoicing, and analytics—will continue to function without interruption. This contrasts sharply with point-to-point integrations, in which such a change might require schema revalidation, interface adjustments, and retesting across all consuming systems. In SAP BTP, where innovation often happens on the edges in side-by-side extensions, this ability to isolate and control changes streamlines delivery and improves confidence in deploying updates quickly.

Beyond enabling isolated deployments, EDA simplifies the integration landscape itself. With traditional approaches, each integration often brings its own custom logic, error handling, and interdependencies, which accumulate into a maintenance-heavy "spaghetti architecture." By contrast, event-driven systems centralize and standardize communication through an event broker like Event Mesh to create a *uniform pattern* for publishing and consuming events. This not only makes the architecture cleaner but also enhances observability, governance, and troubleshooting. Developers can focus on the specific business logic within each consumer, rather than managing complex interface agreements. As a result, onboarding new developers becomes easier, documentation becomes clearer, and operational support becomes less reactive. Over time, this leads to more predictable integrations, shorter resolution times, and lower total cost of ownership—which are core outcomes of improved maintainability that SAP architects and developers strive for in every solution.

### 12.1.5 Better Alignment with Cloud-Native Principles

As organizations modernize their enterprise landscapes and migrate to the cloud, aligning integration strategies with *cloud-native principles* becomes essential for achieving scalability, resilience, and agility. EDA is a natural fit for these principles. By design, EDA embraces loose coupling, stateless interactions, asynchronous communication, and horizontal scalability—all of which are hallmarks of cloud-native systems. In SAP BTP, where microservices and serverless applications are first-class citizens, EDA provides a communication backbone that allows distributed components to react to business changes in real time, without tight dependencies. Rather than forcing services to interact synchronously or poll APIs for updates, EDA allows services to subscribe to relevant events and respond only when necessary. This results in systems that are more responsive, efficient, and elastic—which are core qualities that define successful cloud-native solutions.

The immediate advantage of this architectural alignment is the ability to build and deploy independently scalable services that respond to business events without bottlenecks or blocking calls. For example, consider a SAP Cloud Application Programming Model–based microservice that is deployed on SAP BTP and listens to `Product-Stock.Updated` events. In a traditional model, this service might need to call multiple upstream systems or be triggered via scheduled jobs—but with EDA, the service simply reacts to the event stream as it arrives and scales out automatically if more messages come in during peak hours. Similarly, a serverless function (such as one that's deployed with SAP's function-as-a-service model) can be triggered by specific events like Invoice.Paid and execute only when needed, and that reduces idle compute time and optimizes resource usage. This model promotes efficiency and responsiveness—which are two vital characteristics of cost-effective, cloud-native application design.

Moreover, this event-driven model supports the composability and fault tolerance that cloud-native strategies rely on. Because services are built as autonomous units that subscribe to events independently, they can fail, restart, and scale without impacting the rest of the system. For instance, if a consumer service that processes CustomerRegistered events fails temporarily, Event Mesh can buffer those events until the service is back online to ensure that no data is lost. This resiliency contrasts sharply with that of tightly coupled, synchronous integrations, where the failure of one component can bring down an entire workflow. This flexibility to compose services around event flows—rather than orchestrated procedures—means that enterprises can evolve their architecture organically and integrate new capabilities over time without large-scale refactoring. As SAP continues to promote extensibility through side-by-side innovation on SAP BTP, event-driven design becomes not just a best practice but a strategic imperative. It ensures that the building blocks of tomorrow's enterprise are modular, reactive, and fully cloud-native by design.

In summary, the advantages of EDA extend far beyond technical improvements—they represent a strategic shift toward more agile, resilient, and scalable enterprise systems.

## 12.2   Differences Between Point-to-Point and Event-Driven Architectures

To successfully adopt modern integration practices within today's enterprise IT landscapes, you need to understand the architectural evolution from traditional P2P systems to EDA. P2P models have long been the foundation of system communication, particularly in SAP and other enterprise environments, where integrations were typically built as direct, synchronous connections between two systems. These setups are relatively straightforward to implement and control, but they often come with limitations in flexibility, scalability, and long-term maintainability. As businesses grow and the demand for real-time responsiveness increases, these limitations become bottlenecks, particularly when changes to one system cascade into disruptions across the entire integration chain.

In contrast, EDA represents a significant paradigm shift by decoupling producers and consumers through asynchronous, message-based communication. Systems emit events to a centralized broker without needing to know who consumes them or how they are processed. This enables multiple consumers to act independently on the same event, so it promotes modularity, fault tolerance, and extensibility. By adopting EDA, organizations can support highly dynamic business requirements while minimizing interdependencies between systems.

Choosing one of these architectures isn't just a matter of technical preference—it's a strategic decision that impacts system agility, resilience, and the ability to innovate. This section offers a detailed comparison of the key features of these two architectures to help clarify when each one is appropriate and why EDA is increasingly becoming the preferred approach in modern SAP BTP landscapes.

The P2P integration model is characterized by a direct and tightly coupled communication channel between two systems, which we call System A and System B in Figure 12.1. In this setup, System A acts as the sender or initiator, while System B is the recipient of the message. Communication in this model is often synchronous, meaning that System A sends a request and waits for an immediate response from System B before proceeding. This synchronous nature creates a time-based dependency between the two systems: both must be online, reachable, and responsive at the moment of interaction. The message flow is entirely controlled by System A, and no further data exchange or reaction occurs unless it initiates another request. As a result, P2P systems are relatively simple to implement but quickly become limiting as the number of systems and required interactions grows.

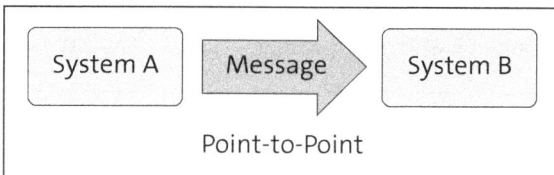

**Figure 12.1** P2P System Architecture with Message Flow

One of the primary traits of this architecture is that error handling is immediate but also rigid. If System B is unavailable or returns an error, System A receives a response right away and must determine how to handle it. While this might seem beneficial from a control perspective, it introduces challenges in resilience and recovery. For example, if System B crashes mid-processing, the message might be lost unless retry logic is explicitly built into System A. Moreover, because the systems are directly linked, any schema change, service disruption, or authentication issue in System B can directly cause System A to fail. There's also little buffer or tolerance for downtime. This model provides no inherent queuing or retry mechanism, meaning there's no persistence of the message once the communication attempt has failed. Thus, without additional error-handling mechanisms such as logging, retry queues, and compensating transactions, there's a high risk of information loss.

Additionally, Figure 12.1 highlights a fundamental limitation of P2P: the message is only transmitted when System A initiates it. In other words, System B is entirely passive in this flow and can't process or react to anything until System A takes action. This unidirectional, demand-based flow limits the architecture's ability to support real-time responsiveness or multiparty distribution. In dynamic enterprise environments— where multiple systems may need to respond to the same event or action—the P2P model falls short. For instance, if a sales order is created in System A, only System B will be aware of it in this setup. Any additional systems that need this information must either be directly integrated with System A or be updated manually. This creates a tightly woven and brittle integration fabric that becomes increasingly hard to maintain, scale, and evolve. Over time, P2P systems can devolve into a *spaghetti architecture*— complex, opaque, and prone to failure with even minor changes.

In an EDA, the flow of information is fundamentally different from the tightly coupled nature of P2P systems. As illustrated in Figure 12.2, EDA introduces an intermediary—a *broker*—between producers and consumers. Producers emit *events*, which represent a business occurrence (e.g., a sales order being created, a payment being completed), without targeting any specific receiver. These events are sent to the broker, such as Event Mesh, which acts as a distribution hub. From there, any number of consumers can subscribe to relevant topics and respond as needed. This design removes direct dependencies between sender and receiver, making the communication asynchronous and loosely coupled. Unlike the synchronous communication in P2P models, producers in EDA don't wait for acknowledgment or success from consumers. They continue their processes independently, thus increasing overall system efficiency and scalability.

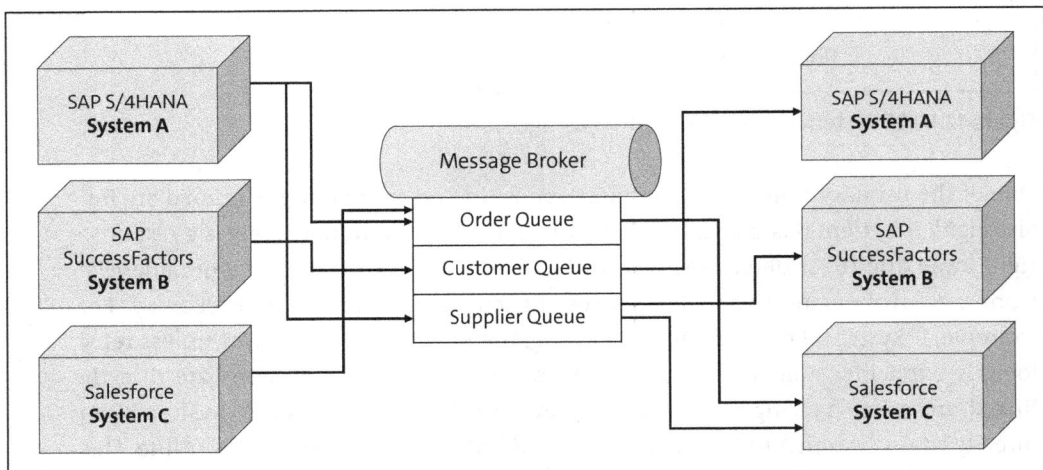

**Figure 12.2** Event-Driven System Architecture with Message Flow

One of the biggest advantages of this approach is its resilience and flexibility in handling errors and failures. Because events are transmitted through a broker, they can be persisted, retried, or redirected with built-in features such as durable queues and DLQs.

This significantly reduces the likelihood of information loss during outages and consumer downtime. For instance, if a consumer service is temporarily unavailable, the event remains in the broker's queue and is delivered once the service is back online. This decoupled model removes the need for immediate response or retry logic at the producer level, as seen in P2P systems. It also allows multiple consumers to process the same event independently, thus enabling parallel workflows such as analytics, auditing, and automation to trigger off a single business action—all without modifying the producer or disrupting other services.

Another key strength of EDA lies in its event-centric, push-based flow. Unlike P2P models, where the consumer remains idle until the sender actively initiates communication, event-driven consumers are reactive—they listen for events and act only when relevant messages are published. This design fosters real-time responsiveness and supports more dynamic, distributed system topologies. For example, a `Customer.Created` event published by SAP S/4HANA could be simultaneously consumed by a CRM system, a loyalty app, and a marketing automation tool—each of which could have its own logic and data requirements. None of these systems need to request the event or know about each other's existence, and that not only simplifies integration but also supports more agile development and deployment practices. You can introduce new services by simply subscribing them to existing events, and that enables innovation without impacting core systems or existing workflows. As a result, event-driven systems better align with modern cloud-native principles and the demands of a real-time digital enterprise.

Figure 12.1 and Figure 12.2 provided a visual snapshot of how P2P and EDA differ in their fundamental information flow. While the simplicity of direct communication in P2P systems is easy to grasp, it also reveals P2P's limitations in scalability and resilience. In contrast, EDA emphasizes decoupling and asynchronous processing to enable multiple systems to independently respond to business events in real time. However, understanding these differences visually is only the beginning. To make informed architectural decisions—especially in complex enterprise scenarios—you need to explore how each approach handles core integration aspects such as coupling, scalability, message flow, maintenance, error handling, and use case suitability. Table 12.1 breaks down these features in detail, highlighting how each architecture addresses key integration challenges and where one may be better suited than another.

| Features | Point-to-Point Integration | Event-Driven Architecture |
|---|---|---|
| Coupling | Tight (via direct communication). | Loose (via event brokers). |
| Scalability | Limited—each new integration requires new logic. | High—multiple consumers can subscribe to the same event. |
| Message Flow | Synchronous or directly asynchronous. | Asynchronous by default. |

**Table 12.1** Comparison of P2P and Event-Driven Architectural Features

| Features | Point-to-Point Integration | Event-Driven Architecture |
|---|---|---|
| Maintenance | Harder, with increasing endpoints. | Easier, due to decoupling. |
| Error Handling | It's usually handled synchronously. | It requires retries, DLQs, and monitoring. |
| Use Cases | Data replication and immediate system-to-system calls. | Real-time updates, microservices, IoT events, and side-by-side extensions. |

**Table 12.1** Comparison of P2P and Event-Driven Architectural Features (Cont.)

Let's dive into each of these features in more detail:

- **Coupling**

  In integration architecture, *coupling* is the degree of dependency between systems. Tightly coupled systems rely heavily on each other to function, and they often require knowledge of endpoints, schemas, and service availability. Loose coupling, by contrast, encourages modularity and separation of concerns, so it enables systems to operate independently and evolve without affecting each other directly. The goal of modern enterprise architecture is to reduce coupling wherever possible to increase agility, resilience, and scalability.

  Let's look at how coupling differs in P2P and EDA:

  - **P2P**

    In P2P, coupling is inherently tight. One system sends a request directly to another system ad expects a response or result. This direct relationship means that any changes to the target system—such as a schema update, endpoint migration, or performance variation—can impact the source system. As the number of integrations grows, these interdependencies multiply and create a fragile and complex network that is difficult to manage and scale. Maintenance becomes time-consuming and error-prone, especially during upgrades or migrations.

  - **EDA**

    EDA promotes loose coupling through asynchronous communication. In EDA, producers emit events without knowing which systems will consume them. Events are published to a broker (like Event Mesh), which decouples the producer and consumer. Consumers can subscribe to events they care about and react to them independently. This separation allows for greater flexibility and easier integration of new systems or services without modifying the existing architecture. EDA thus significantly enhances maintainability, adaptability, and innovation across the enterprise.

- **Scalability**

  *Scalability* is the degree to which a system or architecture can handle increasing loads, users, or events without compromising performance. In integration scenarios, scalability is essential for ensuring that the enterprise can grow and evolve while maintaining the responsiveness and reliability of business processes. An architecture's scalability is often tested during periods of high business activity, such as product launches, seasonal sales, and large data migrations.

Let's look at how scalability differs in P2P and EDA:

- P2P

   In P2P architectures, scalability becomes a challenge because each new integration requires a custom connection between systems. This linear growth in connections increases complexity and introduces potential bottlenecks. Scaling the architecture often involves duplicating integration logic, managing multiple interface endpoints, and ensuring data consistency across various systems. The effort needed to scale P2P integrations is high, so P2P is less suitable for modern, high-growth environments.

- EDA

   EDAs are inherently more scalable than P2P architectures. By decoupling event producers from consumers, EDA allows multiple systems to consume the same event without impacting the producer. You can add new consumers by simply subscribing them to the event stream, with no need to make changes to the producer or existing consumers. This publisher-subscriber model supports horizontal scaling and allows enterprises to expand services and capabilities rapidly. Event Mesh enables this kind of architecture by offering reliable, high-throughput messaging infrastructure that's suitable for global, cloud-scale operations.

- **Message flow**

   Message flow defines how data or events are transmitted between systems, including via timing, delivery guarantees, and processing logic. The flow of information affects the latency, reliability, and structure of integration patterns, and it's a key consideration when you're selecting an appropriate architecture. Let's look at how message flow differs in P2P and EDA:

   - P2P

      In P2P systems, message flow is typically synchronous or semi-synchronous. One system sends a request and waits for a response, often with retry or timeout logic in case of failure. This approach is suitable for immediate, transactional operations where both systems must be available simultaneously. However, it creates dependencies that can lead to bottlenecks and reduced system availability—and failures or slow responses in the receiving system can cascade, thus impacting upstream systems and user experience.

   - EDA

      In EDA, message flow is asynchronous. Events are published to an intermediary (such as Event Mesh) and stored until they are consumed, while consumers process events at their own pace, allowing for time decoupling between systems. This model enables systems to continue operating even if a consumer is temporarily

**12**

unavailable, so it enhances overall resilience. Asynchronous flow also facilitates complex event processing, buffering, and message queuing, which are essential for modern, distributed enterprise applications.

- **Maintenance**
  *Maintenance* involves updating, debugging, scaling, and refactoring integration logic over time. The ease or difficulty of maintenance is heavily influenced by the architecture in place. Architectures that are flexible and loosely coupled are generally easier to maintain and adapt as business requirements evolve. Let's look at how maintenance differs in P2P and EDA:

  - **P2P**
    With P2P integrations, maintenance becomes increasingly burdensome as the number of interfaces grows. Each connection is tightly bound and must be individually monitored, tested, and updated. Any change in a system may necessitate changes in all connected systems, resulting in high regression testing costs and potential downtime. Documentation and governance become crucial, yet they're often hard to maintain as the integration landscape scales.

  - **EDA**
    In contrast, EDAs offer significant maintenance benefits. Since systems are decoupled, changes to one component rarely require changes in others. You can add new features or consumers without impacting producers, and event brokers provide centralized monitoring, logging, and analytics capabilities that simplify issue tracking and resolution. In SAP BTP, tools such as Event Mesh and Cloud Integration offer visibility into event flows, subscriber status, and error queues, making it easier to maintain and evolve the integration architecture without disrupting operations.

In SAP BTP, a typical P2P integration might involve Cloud Integration flows that directly push data from one system to another. In contrast, EDA uses event queuing systems such as Event Mesh and JMS to publish business events (e.g., from SAP S/4HANA, from Producer Flow) to which multiple consumers—like SAP Cloud Application Programming Model–based applications, non-SAP services, and consumer flows—can subscribe and simultaneously process information.

## 12.3    Best Practices for Asynchronous Integration

Adopting EDA within SAP BTP involves more than implementing new tools—it requires a fundamental shift in mindset and architectural thinking. EDA emphasizes asynchronous, loosely coupled interactions in which systems react to business events rather than relying on direct, synchronous calls. This approach challenges traditional integration habits and demands deliberate, forward-thinking design. To help guide you in this transition, the following section outlines essential best practices for building resilient and scalable event-driven solutions in SAP BTP. These include designing with loose coupling in mind, using standardized event formats like CloudEvents, ensuring idempotency in consumers, and implementing durable subscriptions with retry logic and DLQs. Additionally, it explores how to enrich events via APIs, when to use Cloud Integration judiciously, and how to secure event channels by using SAP BTP's OAuth and role-based access controls. Together, these practices form a practical blueprint for working effectively with EDA in enterprise-grade environments.

### 12.3.1   Design with Loose Coupling in Mind

In an event-driven architecture, one of the foundational principles is designing for loose coupling between producers and consumers. Loose coupling is an architectural approach in which system components are independent and unaware of each other's internal workings. This separation allows for greater flexibility, autonomy, and adaptability across the IT landscape. Unlike tightly coupled systems, in which the failure or change of one component can ripple across dependent services, loosely coupled systems promote resilience and allow components to evolve independently. This concept is especially important in SAP BTP scenarios, where different microservices, SAP and non-SAP systems, and custom extensions must interact in real time without introducing brittle dependencies. Designing with loose coupling in mind ensures that event producers only focus on publishing business-relevant facts without having to manage, track, or adapt to the behavior of any downstream consumers.

A key enabler of loose coupling is the use of messaging middleware—such as Event Mesh—that acts as an intermediary between producers and consumers. Rather than messages being sent by producers directly to known receivers, events are published to a broker that categorizes them by topic and handles delivery to all interested subscribers. This decouples the sender from the receiver, not only at the interface level but also in terms of time and implementation. For example, when an SAP S/4HANA system emits an event like `SalesOrder.Created`, it has no awareness of whether that event will be consumed by a workflow extension, a reporting engine, or a legacy billing system. Event Mesh ensures that any subscriber that's listening to the sales order/created topic receives the event. This topic-based routing creates a highly dynamic integration landscape, where new consumers can be added or removed at any time without requiring changes to the producer. This is particularly powerful in agile development environments, where new features are deployed iteratively and with varying timelines.

Event Mesh also provides scalability and extensibility by enabling multiple consumers to subscribe to the same event without interfering with one another. A single `Business-Partner.Changed` event, for instance, might be relevant to a customer notification service, a marketing automation tool, and a data lake ingestion pipeline—and each of these services can consume the event and process it based on its own context and requirements. Furthermore, each consumer can use durable queues, retry logic, and filtering mechanisms to tailor how it handles the event stream. The producer, meanwhile, remains unaffected by how many consumers exist or what they do with the event, which is essential in large-scale enterprise landscapes. This pattern allows developers to innovate faster because they can build new business capabilities by simply plugging into the existing Event Mesh without rewriting or redeploying source systems.

In addition to technical enablement, designing for loose coupling requires thoughtful governance and clear communication patterns. You must design event topics and schemas in a way that expresses intent without exposing internal implementation details. You should assign event names that represent business occurrences (e.g., `OrderShipped`, `InvoiceCancelled`, `MaterialStockAdjusted`) and avoid referencing specific consumer workflows or system behaviors. Define events in a reusable, domain-oriented manner so they can support multiple use cases beyond their initial implementation. SAP BTP offers tooling such as SAP Business Application Studio and Cloud Integration to document and expose event definitions by using standards like AsyncAPI. By fostering a culture of publishing well-governed, self-explanatory, and versioned event contracts, enterprises can ensure long-term maintainability and composability of their solutions. Ultimately, designing with loose coupling in mind not only reduces integration friction but also empowers organizations to adapt swiftly to changing business needs without being hindered by architectural rigidity.

### 12.3.2   Use Standardized Event Formats

In any event-driven integration strategy—particularly across large enterprise landscapes such as those found in SAP environments—you must maintain consistency in how you structure events. Without standardization, each team or system may define event formats differently, which can lead to increased integration effort, misinterpretation of event payloads, and a proliferation of incompatible consumers. Standardized event formats not only reduce complexity but also increase reusability, automation potential, and system interoperability. They serve as a common contract between producers and consumers that allows different teams and applications to build and maintain integrations independently. SAP recognizes this necessity and promotes the adoption of industry-standard formats, such as the CloudEvents specification, across its event-driven offerings in SAP BTP, including Event Mesh and Cloud Integration.

*CloudEvents* is an open standard from the Cloud Native Computing Foundation (CNCF) that defines a common metadata structure for event messages, regardless of the source or payload format. This metadata includes fields—such as ID, source, type, subject, and time—that provide essential context about the event without requiring the consumer to parse the full payload. By using CloudEvents, SAP enables a uniform structure that simplifies event parsing and routing, supports observability, and enhances the developer experience. Whether the event is emitted from S/4HANA, SAP SuccessFactors, or a custom application that's built on SAP BTP, the CloudEvents wrapper ensures that consumers can rely on consistent metadata. For example, a `BusinessPartner.Created` event would carry a type that indicates the nature of the event and a source that denotes the originating system or domain. Consumers can subscribe, filter, and react based on this metadata before they even examine the payload, which streamlines logic and reduces coupling.

SAP goes a step further by offering predefined business events in standardized formats, through catalogs such as SAP Business Accelerator Hub and the event catalogs that are embedded in Event Mesh and Cloud Integration. These catalogs expose domain-specific events like `SalesOrder.Changed`, `PurchaseRequisition.Released`, and `BusinessPartner.Deleted`, which follow consistent naming conventions and data schemas. By leveraging these predefined events, developers can skip the design phase of the event structure entirely and focus on building consumers. This saves time, reduces design errors, and aligns with SAP's intelligent enterprise strategy. Furthermore, these events are typically aligned with SAP's internal data models (e.g., CDS views, BAPIs), and that ensures semantic integrity and traceability back to core business processes. For instance, when consuming a `BusinessPartner.Changed` event from SAP S/4HANA, developers can rely on the fact that the payload fields reflect the structure of the underlying business partner master data. That enhances data accuracy and trust across integrated systems.

Using standardized formats also supports the broader goals of enterprise integration governance. It allows for automated validation of event payloads, compatibility checking during schema evolution, and even tooling support for mock testing or simulation. Integration teams can define generic processing frameworks based on known standards, and that enables reuse across projects. For example, a logging service can ingest all events that conform to CloudEvents and store them in a data lake for analytics, regardless of the originating domain. Similarly, an audit mechanism can inspect the type and time fields of every event to ensure traceability and compliance. By adopting standards, SAP BTP customers also ensure compatibility with non-SAP systems, cloud-native applications, and external partners, since CloudEvents and other open standards are supported across many ecosystems (including AWS, Azure, and Google Cloud). In essence, standardized event formats form the backbone of scalable, reliable, and maintainable event-driven architectures in the SAP world.

### 12.3.3   Implement Idempotency

One of the fundamental design principles in any event-driven system—especially within architectures that are built on SAP BTP—is ensuring that consumer applications are idempotent. *Idempotency* is a property that an operation has if executing the operation multiple times has the same effect as executing it once. This is critically important in systems like Event Mesh that use at-least-once delivery models and in which events may be delivered more than once due to retries, network errors, and broker failovers. Without idempotency, repeated processing of the same event can result in data corruption, duplicate transactions, inflated inventory, and incorrect billing. A lack of idempotency is often a root cause of inconsistencies in distributed systems, and it can be particularly damaging in business-critical SAP scenarios such as financial postings, order management, and HR updates.

The challenge lies in the fact that in event-driven systems, consumers are inherently decoupled from producers and can't rely on delivery guarantees alone. While synchronous request-response systems often assume a one-time, confirmed execution, asynchronous environments make no such promise. Events may arrive twice, out of order, or after long delays. For example, consider a scenario where a `PaymentReceived` event is published by an external payment gateway and consumed by an SAP S/4HANA–based financial extension that's built on SAP BTP. If this consumer records a payment and updates the customer's account balance each time it sees the event, duplicated events could result in the balance being updated multiple times incorrectly. Therefore, designing this consumer to detect and ignore duplicate events ensures data integrity regardless of the number of deliveries. This requires consumers to persist event identifiers (such as a message ID or transaction reference) and verify whether a given event has already been processed before they act on it.

Event Mesh and similar brokers support metadata, such as message IDs and timestamps, that can assist with idempotency, but the implementation responsibility rests with the consumer. Idempotent design strategies typically involve the use of unique transaction identifiers that are embedded in the event payload or metadata. Upon receiving an event, the consumer queries its local database to check whether the event's ID has already been processed. If it has, then the consumer logs the duplicate and safely discards the message, and if it hasn't, then the consumer proceeds with the business logic and marks the event as processed. This pattern is especially effective when paired with transactional data stores like SAP HANA, where event IDs can be stored in a dedicated table with constraints that prevent duplicates. In some scenarios, consumers may also leverage upsert operations or conditional logic to ensure that repeated updates result in no additional side effects. For example, a consumer that reacts to `OrderStatusChanged` events might only update the order status if the incoming status is different from the current one.

It's also important to note that idempotency doesn't mean simply ignoring duplicate messages. Rather, it means producing the same outcome no matter how many times the operation is performed. For some processes, this may require compensating actions, state reconciliation, or business logic that accommodates partial failures. In distributed SAP landscapes, particularly those involving integrations with cloud services, legacy systems, or mobile applications, developers must expect and design for nondeterministic conditions. This is especially true for extensions that are built on SAP BTP with technologies such as SAP Cloud Application Programming Model, in which consumers subscribe to events from Event Mesh or Cloud Integration. Here, incorporating idempotency into business services helps ensure that event processing remains reliable even under heavy load or unstable connectivity. Ultimately, by treating idempotency as a design cornerstone rather than a post-facto patch, SAP architects and developers can build event-driven solutions that are robust, consistent, and ready for real-world complexity.

### 12.3.4   Implement Durable Subscriptions and Retry Logic

In any event-driven architecture—especially one that operates across complex enterprise landscapes, like SAP BTP—resilience is a non-negotiable design requirement. Systems must be built not only to process events under ideal conditions but also to handle transient failures, outages, and downtime gracefully. A core strategy for achieving this resilience is the implementation of *durable subscriptions*, which ensure that a consumer doesn't lose any events, even if it goes offline temporarily. Without such a mechanism, critical events that are published during a downtime window could be lost forever, which can lead to inconsistent business states and missed transactions. For example, if an extension service that subscribes to `PurchaseOrder.Changed` events is temporarily unavailable during a deployment, all events published in that timeframe could vanish unless the broker has persisted them. Event Mesh addresses this issue through *durable queues*, which retain messages until they are acknowledged by the consumer, regardless of whether the consumer is currently online.

Durable subscriptions are especially vital in scenarios involving mission-critical business processes, such as financial approvals, logistics operations, and employee onboarding workflows. Imagine an SAP SuccessFactors integration that listens to `Employee.Created` events to automatically provision IT assets. If this integration is briefly offline and lacks a durable subscription, new employee records could be missed, and that could trigger cascading operational problems. On the other hand, with durable subscriptions in place, Event Mesh holds the events in a queue until the consumer comes back online and confirms successful receipt. This decouples system availability from message delivery and thus allows each component to operate with autonomy and tolerance for temporary outages. Additionally, Event Mesh supports message persistence settings and queue configurations that allow developers to tailor retention periods, priority levels, and acknowledgment models to specific use cases.

However, durability alone doesn't solve all reliability concerns. Event-driven systems must also incorporate robust retry logic to handle transient errors during message processing. Not all failures are fatal—many result from temporary database locks, expired tokens, throttled APIs, or brief network issues. In such cases, retrying the operation after a short interval can lead to successful recovery without any manual intervention. Consumers should implement exponential backoff strategies, capped retries, and circuit breakers to prevent the overloading of downstream systems or getting stuck in endless retry loops. SAP BTP services, such as Cloud Integration and SAP Cloud Application Programming Model, provide support for retry policies either natively or through middleware orchestration. For instance, an SAP BTP microservice that fails to write an invoice event to a third-party tax service can retry a few times with increasing delays before flagging the issue for manual review.

When all retries fail, the system must provide a final safeguard: a DLQ, which is a special-purpose queue that stores messages that could not be successfully processed, even after all retries were exhausted. You can review, reprocess, and rerout these failed messages manually or through automated reconciliation workflows, and incorporating DLQs into your architecture ensures you won't lose any data without knowing about it and that your teams can investigate errors transparently. For example, if a GoodsMovement. Confirmed event repeatedly fails to update a warehouse management system due to malformed data, the event will be sent to a DLQ, where a support team can inspect it. Event Mesh allows developers to configure DLQs for each subscription queue, and they can set thresholds and filters to direct unprocessable events appropriately. This pattern adds an essential layer of resilience and observability that ensures that even in the face of failure, no event will fall through the cracks unnoticed.

Together, durable subscriptions, intelligent retry mechanisms, and DLQs form a three-pillar strategy for reliable event delivery in SAP-based event-driven systems. These features work in concert to prevent data loss, provide fault isolation, and enable recovery from unexpected scenarios. By combining Event Mesh capabilities with best practices from enterprise integration design, organizations can build event-driven solutions on SAP BTP that are not only performant but also dependable and fault-tolerant.

The key takeaway is to proactively plan for failure—not as a rare exception but as a normal operational condition. This mindset will help you develop system designs that handle the messiness of the real world with grace to ensure business continuity and trust in event-driven integrations.

### 12.3.5   Enrich Events When You Need To

In modern event-driven systems, particularly within SAP BTP, the question of how much data to include in an event payload isn't just a technical consideration—it's a strategic architectural decision. The guiding principle is to keep event messages light-weight by emitting only the essential information you need to identify a business occurrence, rather than embedding the entire business object or process context. This approach, which is often called *event notification*, contrasts with *event-carried state transfer*, in which the full dataset is transmitted within the event. SAP advocates the former pattern in many scenarios because it fosters scalability, decouples services more cleanly, and aligns well with SAP's API-first strategy. Rather than sending a complete SalesOrder object when an order is created, for example, SAP recommends emitting an event like SalesOrder.Created with a minimal payload—perhaps just the sales order ID, timestamp, and a source system reference.

The rationale for this design choice lies in its flexibility and performance. Sending minimal data in event messages makes the system more efficient and less error-prone, especially as business objects evolve. A small, well-structured event can be routed quickly by Event Mesh or other brokers, resulting in lower latency and reduced payload sizes. More importantly, this design choice shields consumers from breaking changes if the business object schema changes, since the event structure remains stable. Consumers who require detailed business information can retrieve it on demand through dedicated APIs, which are typically exposed via SAP's OData or REST interfaces. This pattern—which is often called *event sourcing* or an *event reference pattern*—lets consumers decide what context they need and when. For example, a notification service that receives a `BusinessPartner.Changed` event with only a partner ID can then use that ID to call an SAP API like `/sap/opu/odata/sap/API_BUSINESS_PARTNER` and retrieve the full object, including the updated fields.

This approach also supports more dynamic and targeted use cases, especially in extension and integration scenarios on SAP BTP. Imagine a partner-developed application that's built on SAP Build or SAP Cloud Application Programming Model and that listens to `Material.StockUpdated` events. Rather than relying on bulky messages that contain full material master data and stock information (which might be overkill for most cases), the app receives a small event with a material ID and movement type. From there, it can query SAP S/4HANA with standard APIs to retrieve exactly the details it needs—perhaps material availability, storage location, or inventory batch information. This separation of concerns makes the producer system more stable and allows the consumer to shape its logic independently. If the app requires new fields in the future, the event format will not have to change at all—only the API call will have to change to preserve backward compatibility. SAP's suite of APIs, which are documented via SAP Business Accelerator Hub, ensures that this retrieval process is robust, secure, and aligned with enterprise data governance practices.

Furthermore, enriching events through downstream data retrieval promotes better observability and compliance. It enables event messages to act as triggers rather than data carriers, so it gives system administrators a clear separation between signal and state. Logs, audits, and tracing tools can differentiate between the occurrence of an event and the actual data processing that follows. This design also simplifies security, as sensitive fields (such as personal information and financial data) are not transmitted in the event stream but only exposed via authenticated API calls under appropriate access control. From a development perspective, this pattern encourages the reuse of APIs across multiple consumers and avoids the need for producers to manage complex data filtering or transformation logic. It aligns tightly with SAP's strategic direction toward modularity, reusability, and microservice-based architecture within SAP BTP.

### 12.3.6   Use Cloud Integration Judiciously

As enterprises embrace event-driven integration patterns on SAP BTP, you need to make thoughtful decisions about when and how to use integration tools such as Cloud Integration. While Cloud Integration offers powerful capabilities for connecting SAP and non-SAP systems, orchestrating multistep workflows, and transforming messages between formats and protocols, not every event-driven use case benefits from such orchestration. Overusing Cloud Integration where it isn't required can introduce unnecessary latency, increase operational complexity, and add an avoidable maintenance burden. The principle here is simple: use Cloud Integration when orchestration, transformation, or connectivity is essential—and otherwise, opt for direct, lightweight event handling through Event Mesh or consuming services.

Cloud Integration is ideal for scenarios where incoming events must be enriched, transformed, routed, or passed across protocols and environments. For example, imagine a `SalesOrder.Confirmed` event from SAP S/4HANA that must be transformed into a JSON payload, enriched with customer metadata from a third-party CRM, and then sent to an external logistics partner via an HTTP API. In this case, Cloud Integration is the perfect tool—it can sequence each of these steps, apply data mappings, and ensure that the correct headers, formats, and delivery rules are applied. This is also true in B2B integrations, where formats like EDI, AS2, and IDoc need to be bridged to cloud APIs, or when data must be passed between legacy on-premise systems and cloud-native applications. The ability to design, visualize, and monitor such integrations in one place provides visibility and governance that are indispensable for many enterprise use cases.

However, many modern event-driven use cases don't involve these complexities. In fact, inserting Cloud Integration into the middle of a simple publish-subscribe event flow can act as an antipattern. Consider a scenario in which a `ProductCreated` event from SAP S/4HANA needs to trigger a price simulation in an SAP Cloud Application Programming Model–based service that's deployed on SAP BTP. The event could be directly routed through Event Mesh to the SAP Cloud Application Programming Model application, which subscribes to a specific topic like product/created. No mapping, orchestration, or protocol conversion is needed—only event reception and processing logic are needed within the application. Introducing Cloud Integration in this case would add another layer of configuration, deployment, and monitoring, without delivering real business value. In such cases, the simplicity of direct event consumption enhances agility, reduces development and operational overhead, and improves system responsiveness.

To make effective decisions, architects and developers must evaluate each integration use case against its core requirements. Is data transformation or enrichment required before the event reaches its destination? Are there multiple target systems that must be reached in a coordinated manner? Is protocol bridging—such as converting IDocs to REST or vice versa—involved? If the answer to any of these questions is yes, Cloud Integration is likely appropriate—and if not, direct event routing through Event Mesh should be the preferred option. This decision framework supports a more modular, scalable, and maintainable architecture, where each component performs a well-defined role. Additionally, SAP's strategic direction promotes this modularity—SAP BTP is designed to host decoupled services that can be independently scaled, updated, and reused. The Event Mesh service enables this by decoupling producers and consumers to ensure real-time responsiveness and resilience without imposing orchestration unless absolutely necessary.

### 12.3.7  Secure Your Event Channels

Security is a foundation of any integration strategy, and it becomes even more critical in event-driven architectures where data is transmitted asynchronously across potentially distributed systems. As organizations adopt SAP BTP and begin building integrations with Event Mesh and Cloud Integration, ensuring secure communication among event producers, brokers, and consumers isn't optional—it's mandatory. Having unsecured or improperly secured event channels can lead to unauthorized access, data leakage, system misuse, and compliance violations. Since events often represent sensitive business changes such as invoice payments, sales orders, and employee updates, you must handle them with the same level of confidentiality, integrity, and authenticity that you'd use to handle synchronous API communications. In SAP BTP, you achieve this with a combination of OAuth 2.0 for authentication, role-based access control for authorization, and encrypted communication protocols to secure messaging infrastructure end-to-end.

SAP BTP provides a comprehensive identity and access management framework that integrates seamlessly with event-driven services like Event Mesh. All producers and consumers that interact with Event Mesh must authenticate via secure tokens—typically using the OAuth 2.0 client credentials flow. This ensures that only trusted applications with valid credentials can publish or consume messages. Credentials are managed through SAP BTP's service bindings and instance authorizations, and access tokens are generated and validated by the platform's identity provider (e.g., SAP ID service or an integrated corporate identity provider). For example, an SAP Cloud Application Programming Model–based application that consumes `BusinessPartner.Changed` events from Event Mesh must first obtain a valid token via OAuth, which it must then include in its subscription or polling requests. If the token is invalid or expired, access will be denied. This protects the event broker from abuse, enforces application boundaries, and supports compliance with enterprise-level identity governance.

You handle authorization through fine-grained role-based access control, in which you define what specific actions an authenticated identity is allowed to perform. You assign roles to application instances and users, and this can govern operations such as publishing to specific topics, subscribing to message queues, or managing event schemas. Event Mesh allows you to assign different roles to producers, consumers, and administrators, to ensure that a microservice that produce customer events can't inadvertently

12

consume or delete messages that are intended for another team. Role-based access control is especially useful in multitenant or multiteam SAP BTP landscapes, where services are built and operated by different business units. For example, you may adhere to the principle of least privilege by giving a logistics service permission to publish events like GoodsMovement.Confirmed but not to subscribe to financial events like PaymentPosted. You can configure these permissions through the SAP BTP cockpit or programmatically via APIs, so they're suitable for both development and production environments.

Beyond authentication and authorization, encryption plays a vital role in securing event channels at the transport and data levels. All communication between producers, Event Mesh, and consumers is encrypted via HTTPS and TLS to prevent eavesdropping and tampering during transmission. Furthermore, the Event Mesh service ensures that messages stored in queues or temporarily buffered in transit are encrypted at rest with SAP BTP's managed key infrastructure. This protects sensitive data, even if the underlying infrastructure is compromised. In addition to platform-level protections, we encourage developers to encrypt particularly sensitive payloads—such as personally identifiable information (PII) and financial data—within the application layer by using symmetric or asymmetric encryption techniques. Together, these practices ensure that the event-driven architecture upholds confidentiality, integrity, and availability (CIA) and aligns with SAP's and customers' strict compliance requirements, including GDPR, the Health Insurance Portability and Accountability Act (HIPAA), and System and Organization Controls 2 (SOC 2).

Ultimately, securing your event channels isn't just about protecting data—it's about establishing trust in your architecture. Secure channels enable developers and architects to confidently scale integrations, expose services across business domains, and support hybrid or multicloud deployments without compromising control. SAP BTP's security model is designed to make this achievable by integrating identity, access, and encryption into the core platform services. However, it's the responsibility of architects and developers to implement these controls consistently, audit them regularly, and update them as the system landscape evolves. For event-driven solutions to be truly enterprise grade, security must be baked in from the very beginning—not bolted on after deployment. By leveraging OAuth 2.0, role-based access control, and end-to-end encryption, SAP customers can build event-driven systems that are not only innovative and scalable but also secure and compliant by design.

## 12.4   Summary

EDA represents a transformative shift in the way enterprise systems communicate, scale, and adapt to changing business conditions. In contrast to traditional P2P models, in which integrations are tightly coupled and demand synchronous, direct communication, EDA enables loosely coupled, asynchronous interactions via events. This architectural shift isn't only technical but strategic—it realigns system design with the demands of agility, innovation, and cloud-native principles. Within SAP BTP, EDA empowers organizations to move from rigid, linear integrations toward responsive, scalable, and modular ecosystems. Systems emit events without needing to know who consumes them, while services and extensions react in real time, independently, and without friction. This decoupling provides the bedrock for modern integration strategies across hybrid landscapes, and it supports SAP's vision for the intelligent enterprise.

Event-Driven Architecture offers a powerful, future-proof model for integrating and extending SAP systems, particularly within the flexible and service-oriented environment of SAP BTP. It supports agility by decoupling dependencies, delivers speed through real-time responsiveness, and allows enterprises to innovate without disrupting existing operations. Combined with best practices around governance, security, and resilience, EDA enables the construction of intelligent, adaptable ecosystems that can evolve alongside the business. As companies face increasing demands for real-time insights, cloud scalability, and operational agility, adopting EDA isn't just a modernization effort—it's a strategic step toward becoming a truly intelligent enterprise. The insights and practices outlined in this chapter provide you with a blueprint for adopting event-driven thinking across your SAP landscape and unlocking the full potential of SAP BTP's event-native capabilities.

12

# Chapter 13

# Connecting Cloud Integration with Event Mesh

*Modern IT landscapes increasingly require flexible and decoupled communication models to meet requirements for scalability, real-time processing, and system integration. While classic P2P integrations are sufficient in many cases, they quickly reach their limits in highly dynamic, distributed system architectures. EDAs offer an elegant approach here: systems no longer react synchronously to requests but subscribe to relevant events and process them when they occur. They're loosely coupled, robust, and scalable.*

Event Mesh is the central messaging and eventing platform within SAP BTP and Cloud Integration that supports precisely this architectural paradigm. It provides a cloud-based message broker that enables reliable, asynchronous communication across system boundaries. Event Mesh is open-protocol, standards based, and seamlessly integrated with other SAP services—in particular, Cloud Integration.

This chapter is dedicated to the structured connection of Cloud Integration to Event Mesh. Section 13.1 looks at the role of the central message broker, and it explains how messages are exchanged between publishers and subscribers. Section 13.2 explains routing strategies, message handling, and quality criteria for the processing of events. Section 13.3 presents best practices for event handling, and Section 13.4 pays particular attention to the practical implementation of integration scenarios with SAP S/4HANA—from both the perspective of the event source and the perspective of the consuming systems.

The aim of this chapter is to convey the technical basics, design principles, and proven patterns for using the Event Mesh in the context of Cloud Integration. You should gain a complete overview of how the components interact and learn how to build, operate, and scale event-based integrations with SAP technologies—from configuring queues and subscriptions to processing and error handling in iFlow.

## 13.1  Central Message Broker

At the heart of every event-driven architecture is a mechanism for the reliable and scalable transmission of messages: the *message broker*. Within SAP BTP, this role is performed by Event Mesh, which acts as a central mediation system between the components of an IT landscape that react to events and generate them. Producers (publishers) and consumers (subscribers) are decoupled so that they can work independently of each other without having to be aware of each other.

The message broker thus ensures that messages (i.e., the representations of events) are reliably transported, buffered, and delivered to the right recipients. In Event Mesh, this is done with a clear, logical construct: queues, topics, and subscriptions form the backbone of the communication architecture.

Event Mesh follows the *publisher-subscriber model* (pub/sub). In this model, a system publishes an event without knowing which other system will consume it. At the same time, any number of subscribers can "subscribe" to a specific event—regardless of when or how this event was generated. Event Mesh temporarily stores these messages in a queue until they have been picked up by the subscribers.

A central concept here is topic orientation. It is expressed via *topics*, which are logical categories or addresses for certain types of events (e.g., sales/order/created). Publishers send messages to topics, while subscribers register for one or more of the topics. This creates a loose coupling with a high degree of flexibility.

Event Mesh is based on open messaging standards such as AMQP 1.0 (Advanced Message Queuing Protocol), which allows a large number of different systems—both SAP and non-SAP—to be integrated. Communication is encrypted via TLS, whereby you can control both authentication and authorization via SAP Identity Management or your own OAuth2 instances.

A central feature of the message broker is its ability to buffer messages. This is particularly important in asynchronous scenarios in which publishers and subscribers are active at different times. For example, if an SAP S/4HANA system publishes an order confirmation as an event but the processing target system is currently not available, the message is temporarily stored in the event mesh queue. Depending on the configuration and quality of service settings, Event Mesh guarantees either at-least-once or exactly-once delivery.

Above all, a central message broker shows its strengths in complex integration landscapes with several systems, like the following:

- Distributed SAP and non-SAP systems in hybrid scenarios
- Processing business transactions with high message volumes (e.g., logistics, orders, IoT)
- Integration of SAP S/4HANA events with cloud-based microservices
- Scaling of message processing through parallel subscriptions

In conjunction with Cloud Integration, Event Mesh serves as a link between the event source and processing logic. The iFlows can be registered as subscribers to specific topics, and they can thus react to new events (e.g., for transformation, validation, or forwarding).

Using Event Mesh as a central message broker has a number of decisive advantages:

- **Loose coupling**
  Systems don't need to communicate synchronously or know each other.

- **Scalability**
  Messages can be processed in parallel by several recipients.

- **Reliability**
  Messages are not lost, even if systems are temporarily unavailable.

- **Asynchrony**
  Events are processed as soon as the systems are ready, so there's no blocking of processes.

- **Flexibility**
  You can connect new consumers at any time without changing existing systems.

With its central message broker, the Event Mesh forms the heart of event-based architectures within the SAP world. It provides a highly available, standardized, and scalable infrastructure for the secure exchange of messages among systems. For Cloud Integration, this means a strategic expansion: it can evolve from a pure orchestration tool into an intelligent, reactive processing core—which is a decisive step on the way to modern, decoupled IT architectures.

## 13.2   Routing and Message Processing Strategies

The introduction of a central message broker such as Event Mesh creates the basis for a scalable, decoupled message architecture. However, the real intelligence only comes from the targeted processing and routing of messages. This phase determines the level of efficiency, robustness, and control with which events are routed through the integration landscape. Routing and processing strategies are therefore essential parts of an event-driven architecture with Cloud Integration.

The aim of routing is structured event processing, and when a message is received via Event Mesh, Cloud Integration must make the following decisions:

- Which systems should receive the message?
- How should the message be processed, transformed, or enriched?
- What rules apply for forwarding and possible filtering?

Cloud Integration performs these tasks with its iFlows, which react to events as subscribers. Routing strategies determine the flow direction and content of data processing.

You can establish an initial routing logic when you're designing the topics in the Event Mesh. You should also give topics names that are semantically unambiguous and that map hierarchies. Here are some examples of good topic names:

- `customer/created`
- `customer/updated`
- `sales/order/created`
- `sales/order/confirmed`

You can make iFlows subscribe to individual topics or entire topic group. You can also use wildcards (e.g., `sales/order/*`) to allow flexible subscriptions with *low redundancy*, which means that a single iFlow can process all order events while another only responds to confirmed messages. This first form of routing therefore already takes place before entering the iFlow through the topic structure and subscription logic.

After it's delivered by Event Mesh, the message is transferred to an iFlow in Cloud Integration. The content processing begins within this iFlow, where the following types of routing and processing can take place:

- **Message content–based routing**
  In a content-based router, a decision is made on the basis of message elements as to how further processing is to take place. For example, a `Region = EMEA` attribute will route to a different endpoint than `Region = APAC`.

- **Dynamic routing**
  The route is not defined statically but is calculated at runtime using expressions (e.g., with a Groovy script or XSLT). This method offers maximum flexibility but requires more development effort than other methods.

- **Multicast/parallel processing**
  A message can be distributed to several destinations at the same time—for example, for simultaneous archiving and further processing. This improves performance and ensures redundancy.

- **Exception-based routing**
  Depending on the type of error that occurs (e.g. timeouts, mapping errors), an alternative route or error handling logic can be run through.

You must control the processing of events beyond pure routing. We suggest that you use one of these strategies, which have proven themselves in practice:

- **Transformation**
  Message formats are converted into the target system format in iFlows. For example, this strategy can employ XML-to-JSON conversions, structural mapping functions, or Groovy scripts.

- **Enrichment**
  Information that's missing in the event is supplemented via system calls or database access (e.g., customer master data via OData service).

- **Persistence**
  Important messages can be temporarily stored in the data store or filed in the secure store for traceability.

- **Deduplication and idempotency**
  Repeated events (e.g., through retry mechanisms) must not trigger multiple processing, so this strategy involves implementing message ID–based filters via Cloud Integration.

- **Prioritization**
  Messages can be sorted according to urgency—for example, by separating critical events into separate queues or flows.

Also, asynchronous communication naturally brings challenges like these:

- Systems will be temporarily unavailable.
- Messages may be incorrect or incomplete.
- Network problems will delay delivery.

Therefore, you should implement a well-thought-out error-handling system for your iFlows. It should include the following:

- Repetition mechanisms (retry policies)
- Storage of erroneous messages for analysis
- Alerting via email, ServiceNow, or monitoring tools
- Automated notifications to development teams

Cloud Integration offers standardized modules such as exception subprocesses, error end events, and exception-based routing.

You should also adhere to the following principles when designing routing strategies:

- **Loose coupling**
  Your iFlows should be as generic as possible and independent of the source system.

- **Reusability**
  You should use a modular structure to enable use in multiple scenarios.

- **Scalability**
  You should implement separation according to subject areas or message types.

- **Transparency**
  You should make each processing path traceable, and document it.

- **Maintainability**
  You should test and version your dynamic routes well.

Let's look at an example: an incoming customer/created event contains the customer-Type attribute, and a decision is made in the iFlow via context-based routing:

- customerType = B2C → Processing by marketing system
- customerType = B2B → Transfer to CRM system
- Other types → Storage in data store + alert to monitoring

A central logging module can also be connected via multicast.

## 13.3   Best Practices for Event Handling

Event-driven architectures not only offer technical flexibility but also open up completely new possibilities for system integration. To ensure that you can use these advantages sustainably in productive operation, you must implement clearly structured, stable, and comprehensible event handling from the outset. In the interaction between Event Mesh and Cloud Integration, this means above all clean design, consistent routing, robust error handling, and systematic monitoring.

A key objective when processing events is fault tolerance. Because events are transmitted asynchronously and don't have to be processed immediately, they're naturally more susceptible to problems—for example during delivery or when target systems can't be reached. You must therefore design iFlows in such a way that they don't abort but instead recognize the situation and react appropriately. You should therefore take the following approaches that have proven themselves in practice:

- Use retry mechanisms in iFlows for automatic repetition in the event of temporary errors.
- Use exception subprocesses for structured error handling within Cloud Integration.
- Implement persistent storage of important message content in the data store for later processing or tracking.
- Configure DLQs or alternative routes for targeted handling of undeliverable messages.

In addition to error handling, the clarity of the structure plays a decisive role. You must implement well-thought-out naming and hierarchization of topics to clearly assign messages, subscribe to them in a targeted manner, and process them efficiently. Well-structured topics—such as sales order/created and customer/blocked—facilitate routing and enable the use of wildcards for group processing. Standardized naming conventions not only promote readability but also form the basis for subsequent governance.

You should also design the message content itself with care. An event should contain all the essential information that's required for target processing—but no more than necessary. In practice, the use of so-called *enriched events* has proven successful. In addition to the business identifier (such as an order number), enriched events contain meta information such as a timestamp, a source system, an event version, or correlating object IDs. Detailed data that's not available at the moment the event is generated can be enriched in iFlow via service calls if required. This keeps the event size small and transport optimized without sacrificing relevance.

Another proven strategy is the differentiation of event types. It makes distinctions among business events (e.g., `Invoice.Approved`), technical events (e.g., `System.Offline`), status events (e.g., `Order.Confirmed`) and change events (e.g., `Product.Updated`). This clear classification facilitates prioritization in processing and correct classification in monitoring.

Events follow the "fire-and-forget" paradigm: there is no direct feedback as to whether and how they have been processed by the recipient. Publishers should therefore focus exclusively on sending correctly, while the responsibility for further processing lies with the recipient. This loose coupling is a great advantage, but it requires a change in thinking: consistency is no longer achieved immediately, but it will be eventually.

Another key success factor is the visibility of processes and message flows. In reality, it can be difficult to keep track of processed, waiting, or faulty messages—especially with complex event topologies. You should therefore take the following targeted measures to ensure transparency and traceability:

- Use the integrated monitoring tools of Cloud Integration (e.g., for message status and progress monitoring).
- Create an event activity log or dashboard for a cross-system view of incoming and processed events.
- Integrate with SAP Cloud ALM or the SAP Alert Notification service for systematic monitoring and escalation.
- Use external application performance management (APM) or logging tools (e.g., Splunk, Dynatrace) for central evaluation and reporting. (This is optional.)

Security also plays a central role in event handling. All communication among the publisher, message broker, and subscriber should be TLS-encrypted. You should also use established procedures such as OAuth 2.0 and SAML for authentication and authorization, validate payloads in the iFlow to catch structural or security-critical errors at an early stage, and comply with client separation and role-based access control.

Also, don't underestimate the organizational aspect of event architecture—especially when you have larger project teams or heterogeneous system landscapes. To have a successful event architecture, you need governance structures like the central release and documentation of topics, rules for the versioning of events, and control over which roles are allowed to publish or process new events. Without such a set of rules, you run the risk of having your architecture become confusing and difficult to maintain.

Finally, event handling should also be an integral part of your development and test processes. You can simulate events with mocking tools, validate iFlows with automated tests, and integrate them into CI/CD pipelines. You should also use a consistent event data model—for example with JSON schema or XSD—to detect structural errors at an early stage and guarantee processing reliability.

**13**

## 13.4   Event Mesh Integration for SAP S/4HANA

As the central ERP system in many companies, SAP S/4HANA not only provides business-critical functions but also a constant source of business-relevant events. Whenever customer master data is updated, orders are triggered, or deliveries are booked, a potentially processable event takes place. To make these events usable from SAP S/4HANA, SAP offers a structured eventing model in combination with Event Mesh, which bridges the gap between the ERP world and modern cloud integration.

In the following sections, we'll provide an introduction to Event Mesh integration with SAP S/4HANA and then dive into the practical instructions for setting it up.

### 13.4.1   Overview

When SAP S/4HANA was released, it introduced *business events*—which are standardized events that are triggered as soon as certain business processes occur. Business events are not to be confused with technical system notifications—they refer directly to business statuses such as *Business Partner Created*, *Sales Order Confirmed*, and *Invoice Rejected*.

Business events can be activated via defined trigger points (e.g., BAdIs, CDS views, change documents) and will be available for publication to a central messaging system thereafter. You can use the SAP Enterprise Messaging Framework (or, in newer versions, Event Mesh directly) for the connection to Event Mesh.

Before you can send events from SAP S/4HANA, you must prepare the backend system for eventing by doing the following:

- Activate the event enablement add-ons if it's an on-premise system.
- Configure an outbound event channel that will instantiate the connection to Event Mesh.
- Activate specific event topics within the system's event catalog.

In the SAP Fiori launchpad of the SAP S/4HANA system, you can use the Business Event Enablement app to view which events are available and activated. Many of these events will be predefined, but you can supplement them with customer-specific extensions.

The actual technical integration takes place via a service instance in SAP BTP, which grants access to Event Mesh. This creates a *messaging client* whose parameters (client ID, secret, broker URL) are registered in the backend system. You establish the connection via the AMQP protocol, and you can secure it with TLS and set it up with OAuth authentication.

Once you've connected it, SAP S/4HANA can send directly to Event Mesh—or more precisely, to certain topics that you previously created manually or that were created by the system itself. These topics follow a configurable naming scheme like */sap/s4/beh/businesspartner/v1/BusinessPartner/Changed/v1* or */sap/s4/beh/businesspartner/v1/BusinessPartner/Created/v1*.

Once an event has been published from SAP S/4HANA to Event Mesh, it's ready for further processing in Cloud Integration. Here, an iFlow can be registered as a subscriber to the corresponding topic. The iFlow processes the incoming message, extracts the relevant data—typically a reference ID such as a SalesOrderId—and then performs the necessary steps like data enrichment, making a system call, triggering a follow-up process, or even simple logging.

You must also know the structure of the event payload and correctly model it in the iFlow. SAP provides some structured JSON models that you can use for this purpose, or you can analyze the event format by using test messages and monitoring tools.

The integration of SAP S/4HANA events into the event mesh opens up numerous fields of application:

- **Cross-system process chains**
  A delivery release in S/4HANA automatically triggers the creation of a shipping order in a third-party logistics system.

- **Real-time reporting**
  New orders are forwarded as an event and integrated into a reporting tool or dashboard.

- **Notifications**
  Events such as price changes and customer blocking trigger automatic emails or alerts.

- **Application decoupling**
  Specialist departments and microservices only subscribe to the events that are relevant to them, without directly addressing S/4HANA.

These examples show that the event-based architecture achieves not only technical efficiency but also business autonomy and scalability.

When you're integrating SAP S/4HANA with Event Mesh, you should follow a few key recommendations that have proven their worth:

- **Implement clear topic design**
  Use consistent and descriptive topic names to facilitate routing and subsequent monitoring.

- **Keep event size minimal**
  The event payload should contain key information but not excessive detailed data.

- **Implement idempotent processing**
  Design iFlows in such a way that they don't process events received more than once twice.

- **Implement error logging**
  Save faulty events and record them in the monitoring system in a traceable manner.

- **Actively use monitoring**
  Use monitoring in the S/4HANA system (in event logs), Event Mesh, and Cloud Integration.

### 13.4.2   Communication with Cloud Integration and Event Mesh with SAP S/4HANA

In this section, you'll configure Cloud Integration to consume events that are sent from an on-premise SAP S/4HANA system with Event Mesh. This setup will allow you to enable event-driven integration scenarios, where business events that are triggered in SAP S/4HANA (such as the creation of a sales order or the change of a business partner) can be sent to the cloud and processed asynchronously.

Before you can begin designing and deploying event-driven flows in Cloud Integration, you need to make sure that two essential capabilities are activated: **Event Mesh** and **Cloud Integration** with message queues. As shown in Figure 13.1, navigate to your SAP Integration Suite instance and ensure that both **Event Mesh** and **Cloud Integration** are listed as **Active** under your service capabilities. If not, click the **Add Capabilities button** in the upper right-hand corner and activate them accordingly. You need to perform this step to support event routing and message handling between SAP S/4HANA and your iFlows in the cloud.

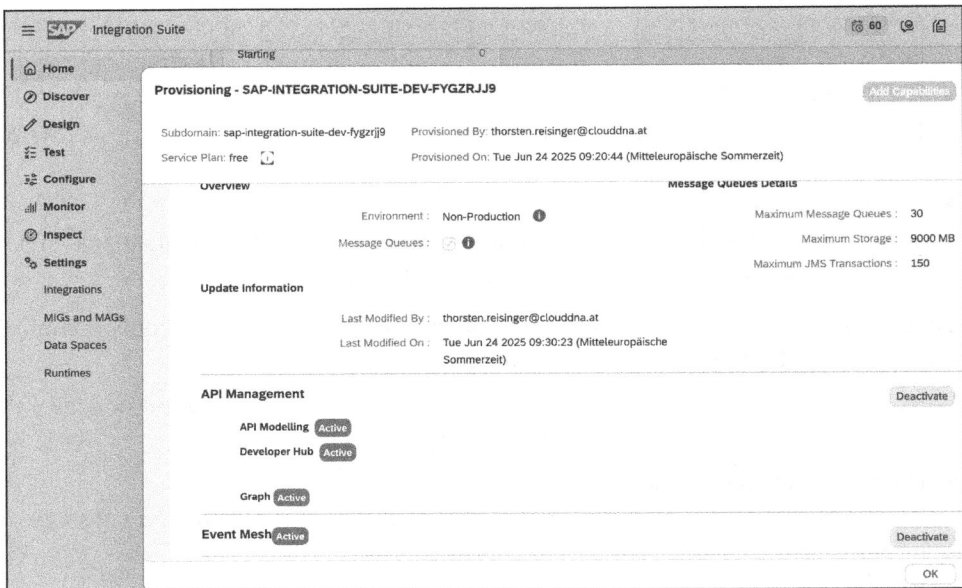

**Figure 13.1** Activating Event Mesh and Cloud Integration

The next step in the setup for consuming events in Cloud Integration is to create a new Event Mesh instance (see Figure 13.2). Navigate to the **Service Marketplace**, search for "SAP Integration Suite, Event Mesh" in your SAP BTP subaccount, and begin the creation process by clicking **Create**. In the service creation dialog, choose the **message-client** plan, check the box to accept the cost-related terms, and select **Cloud Foundry** as the **Runtime Environment**. Then, select your Cloud Foundry **Space** and enter an **Instance Name**. This instance will enable your Cloud Integration to interact with Event Mesh and thus receive event messages from your SAP S/4HANA system.

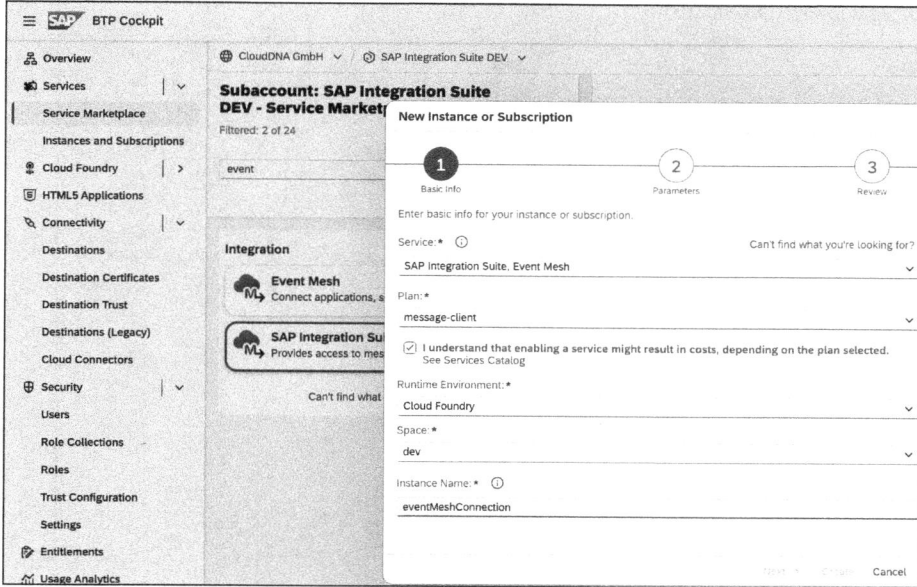

**Figure 13.2** Creating Event Mesh Instance

Next, you need to define the configuration parameters for the Event Mesh instance by using a JSON structure. You do this in the **Parameters** tab during the instance creation process. You can either upload a *.json* file or paste the configuration directly into the editor, as shown in Figure 13.3 and Figure 13.4.

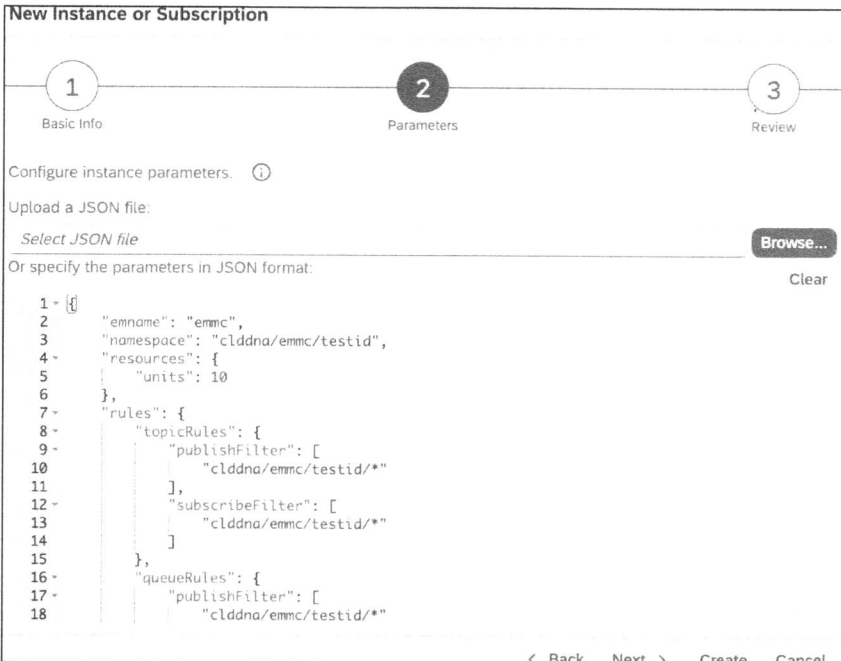

**Figure 13.3** First Part of JSON That Defines Event Mesh

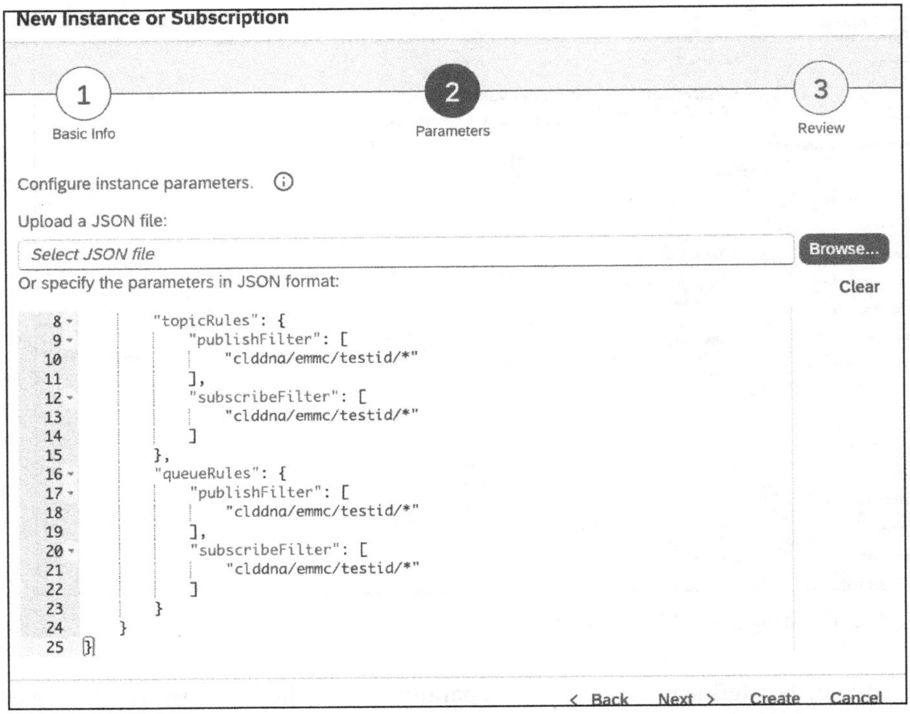

**Figure 13.4** Second Parto of JSON That Defines Event Mesh

Here's what each field in the JSON means:

- **emname**
  This is the name of the message client. It must be unique within the subaccount, and it's used to identify this client across the Event Mesh setup.

- **namespace**
  This defines a unique scope for your message client, and it's not generated automatically. You must manually provide a valid namespace with exactly three segments, using the following format: orgName/clientName/uniqueId. The total length must not exceed 24 characters.

- **resources**
  This section specifies the resource allocation. You must define units to determines how many Event Mesh resources the client can consume (in this case, enter "10").

- **rules**
  This section defines access privileges for topics and queues.
  - Under topicRules, the publishFilter and subscribeFilter define which topics the message client can publish to or subscribe from.
  - Similarly, queueRules define queue access, and the publishFilter here determines which queue messages the client can send to.

---

**Important**

The namespace you use here should match the namespace of the owner message client, to ensure that permissions are granted correctly.

---

Once you've entered the correct parameters, click **Create** to create the instance.

Once you've completed these steps correctly, your screen should look like the one shown in Figure 13.5. You should see the Event Mesh instance you just created in the **Instances** section with a status of **Created**. On the right side, under **Service Keys**, you'll find the option to manage service keys for this instance.

Then, proceed by clicking **Create** in the **Service Keys** section. This will allow you to generate a new service key that's essential for accessing the Event Mesh instance from external applications such as Cloud Integration and SAP S/4HANA.

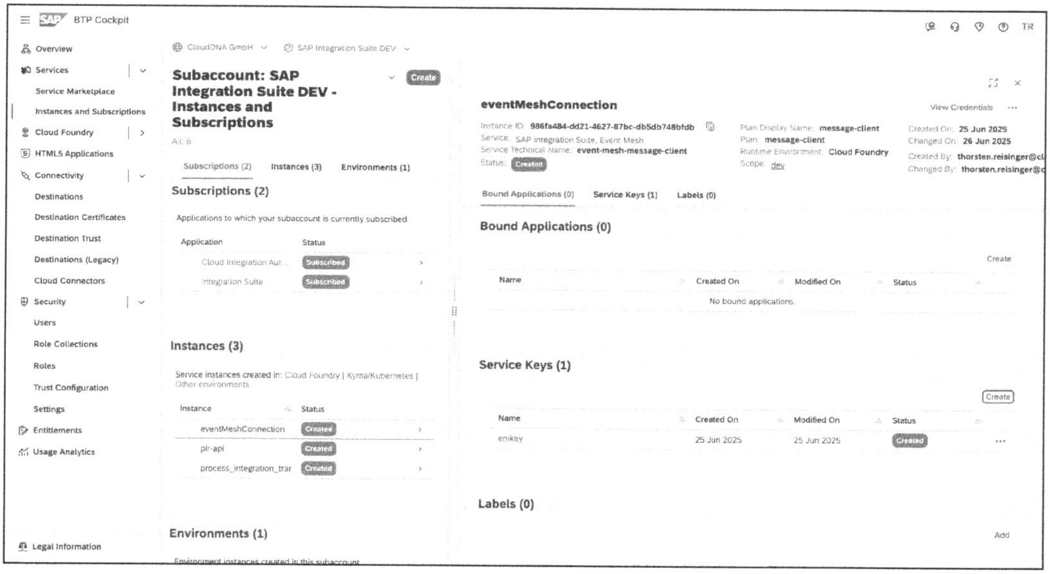

**Figure 13.5**  Created Event Mesh Instance

In the next step, you'll be prompted to create a new service key. Enter a meaningful **Service Key Name** like "emkey" to help you identify its purpose later. You don't need to configure any binding parameters at this stage, so you can leave the JSON editor empty (`{}`), as shown in Figure 13.6. Then, simply click on **Create** at the bottom right to generate the credentials that are required to authenticate and securely connect to your Event Mesh instance from other systems like SAP S/4HANA and Cloud Integration.

After you successfully create the service key, a credentials JSON will be displayed as shown in Figure 13.7. You should save this service key because it contains important details like the `clientid`, `clientsecret`, `tokenendpoint`, and messaging protocol information.

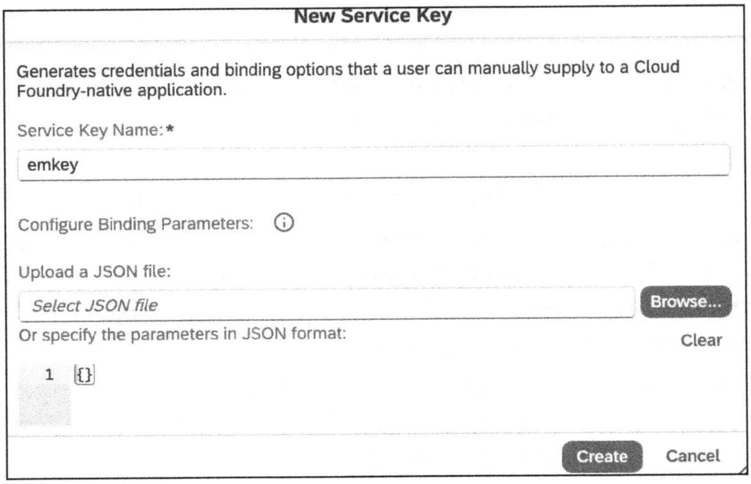

**Figure 13.6**  Creating Event Mesh Service Key

You'll need these credentials later when you're configuring the connection between SAP S/4HANA and Cloud Integration to securely send and receive events through Event Mesh. You can either download the JSON or copy it by using the options at the bottom of the screen shown in Figure 13.7.

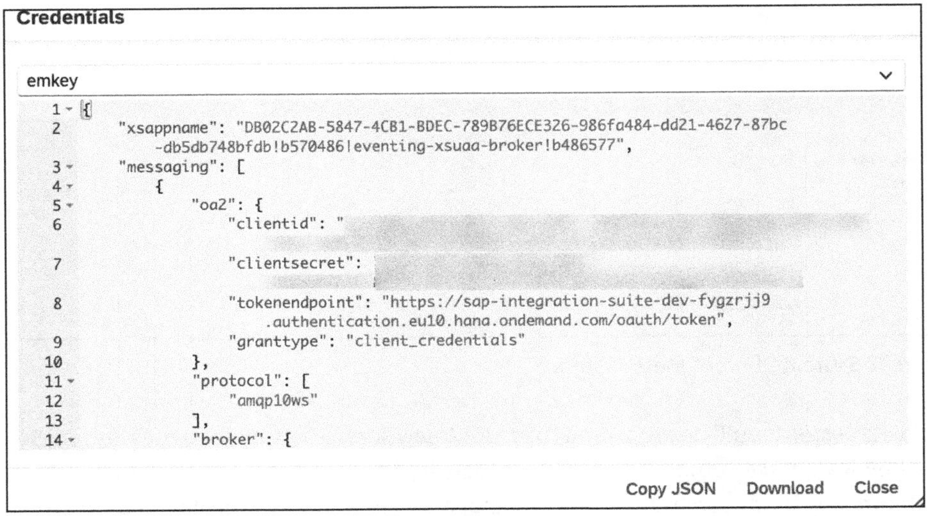

**Figure 13.7**  Created Event Mesh Service Key

Next, switch over to your SAP S/4HANA system and open Transaction /IWXBE/CONFIG (see Figure 13.8), which is used to configure outbound event communication channels. On the **Channel Configuration** screen, you'll create a new channel with the previously saved service key by choosing the **via Service Key** option. Select **SAP Event Mesh** as the channel type to proceed with the setup.

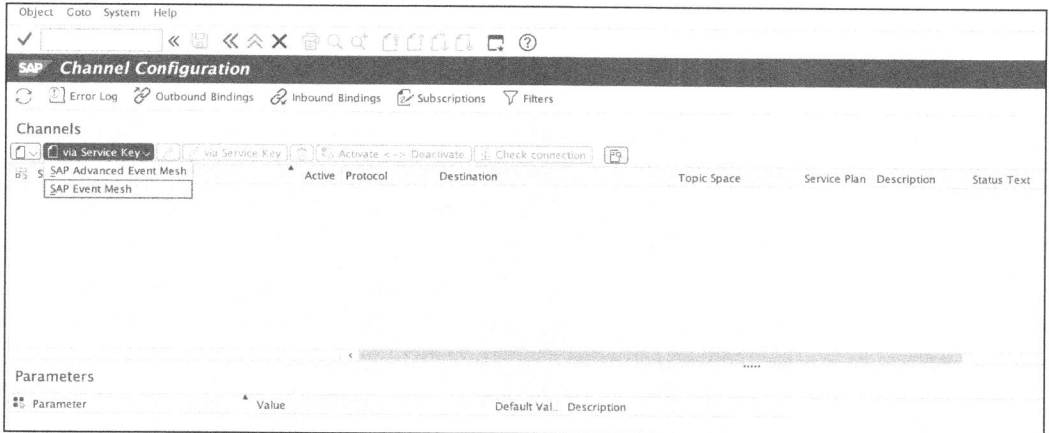

**Figure 13.8** Event Mesh Transaction

Then, you need to complete the channel configuration in your SAP S/4HANA system (see Figure 13.9). Start by entering a name for the **Channel** and providing a meaningful **Description**.

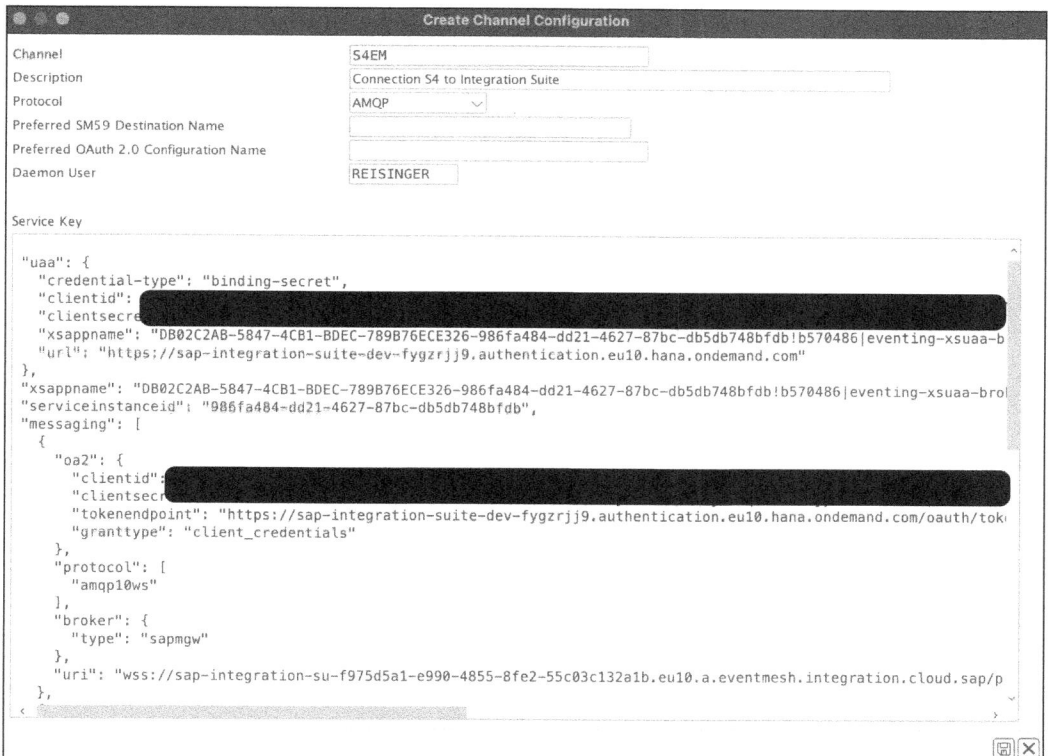

**Figure 13.9** Creating SAP S/4HANA to Event Mesh Configuration

Select **AMQP** for the for the **Protocol**, and you can leave the fields for **Preferred SM59 Destination Name** and **Preferred OAuth 2.0 Configuration Name** empty since they'll be generated automatically at the end of the process.

Next, go to the **Service Key** section and paste the *entire* service key JSON that you previously generated in the BTP cockpit. Once you've filled everything in, click **Save** to create the channel. This will automatically fill in the fields you left empty and establish the connection between your SAP S/4HANA system and Cloud Integration using Event Mesh.

Then, you'll be redirected back to the overview screen, where you'll see the channel you've just created. To enable it, simply select the channel entry and click the **Activate** button in the toolbar (see Figure 13.10). You have to perform this step to make the connection between your SAP S/4HANA system and Cloud Integration via Event Mesh fully operational.

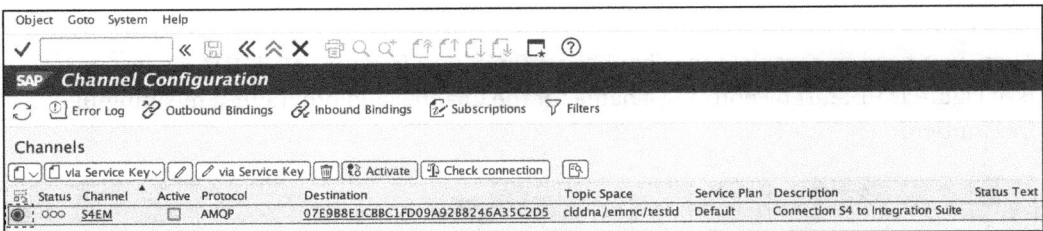

**Figure 13.10**  Activating Channel

The next step is to validate the connection to the channel. Click the **Check connection** button in the toolbar, and if you've configured everything correctly, you'll see a success message at the bottom of the screen confirming that the connection test for your channel was successful (see Figure 13.11). Then, switch to the **Outbound Bindings** section by clicking the corresponding button in the upper left-hand corner of the screen.

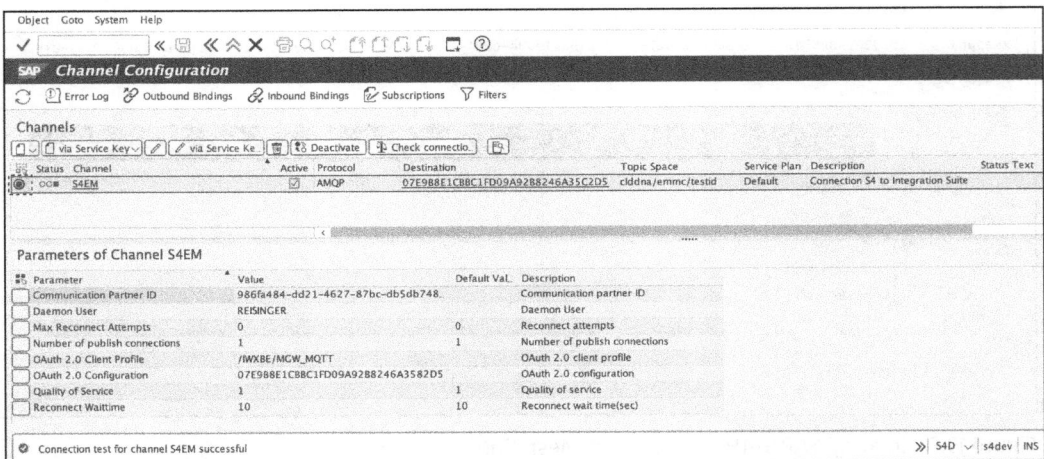

**Figure 13.11**  Checking Connection to S4EM Chanel

That will take you to a view where you can configure which events should be sent from SAP S/4HANA via the previously created Event Mesh channel (see Figure 13.12). Under **Outbound Bindings of Channel <YourChannelName>,** click the document icon to create a new outbound binding. This binding represents the actual event you want to trigger (e.g., a change in a business partner).

Then, you'll see a popup titled **Create Outbound Binding**, where you can enter or search for the event topic you want. You can also use the value help button on the right-hand side to browse and select from available event topics. Once you've selected the appropriate topic, confirm your choice to complete the outbound binding creation.

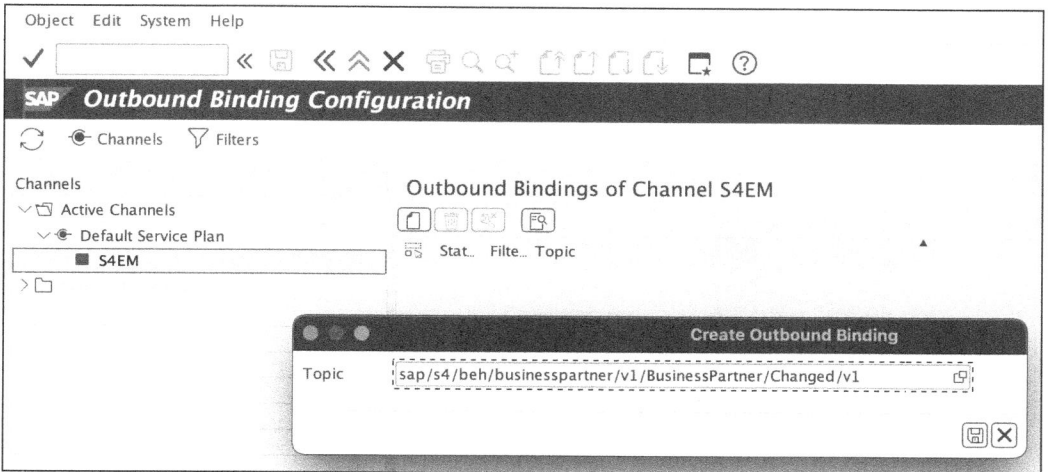

**Figure 13.12** Configuring Outbound Binding for Channel S4EM

After you've created the outbound binding, it should appear under **Outbound Bindings of Channel <YourChannelName>**, as shown in Figure 13.13. You should see the topics you've configured—in this case, **BusinessPartner/Changed** and **BusinessPartner/Created**—along with their status indicators.

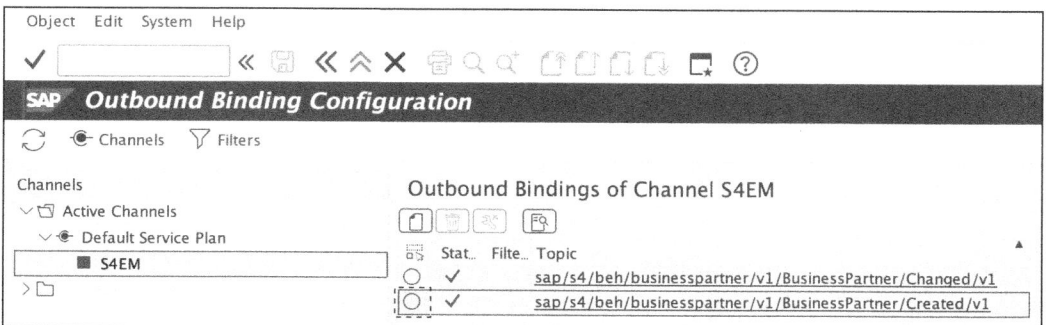

**Figure 13.13** S4 Created Outbound Bindings

Next, switch back to SAP Integration Suite and navigate to **Configure** and then **Event Mesh** in the left-hand menu. There, you'll see an overview of your Event Mesh environment, including the **Message Client that** you previously created on the screen in Figure 13.5, with its corresponding **Namespace** and **Message Client ID** (see Figure 13.14). Click on the message client entry to access its detailed configuration and manage the queues and subscriptions associated with it.

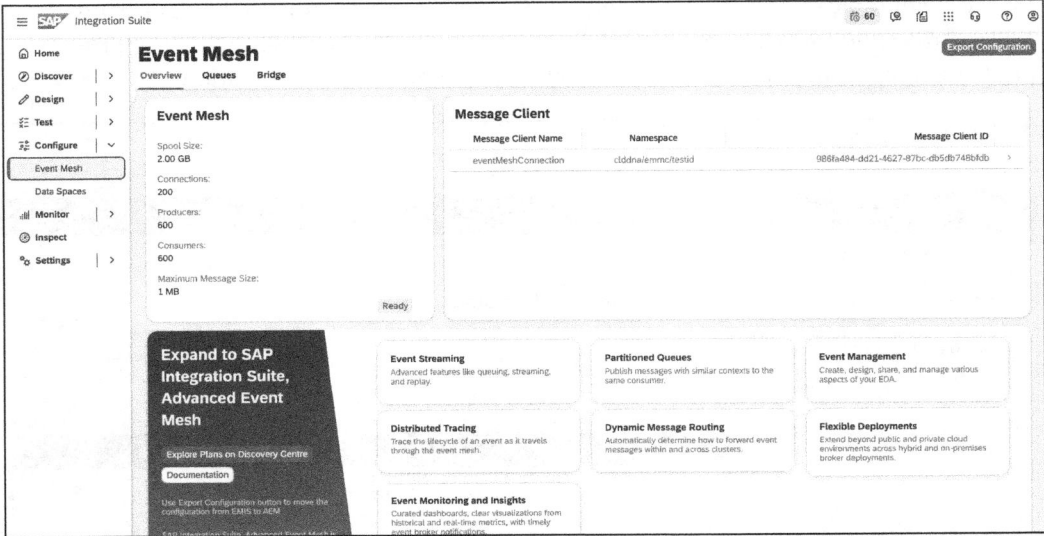

**Figure 13.14** Event Mesh Environment Overview

Once you're in the message client view, switch to the **Queues** tab—and since there are currently no queues, go ahead and click on the **Create** button in the top right-hand corner to define a new event queue (see Figure 13.15). You'll use this queue to receive and store events that are published by your SAP S/4HANA system.

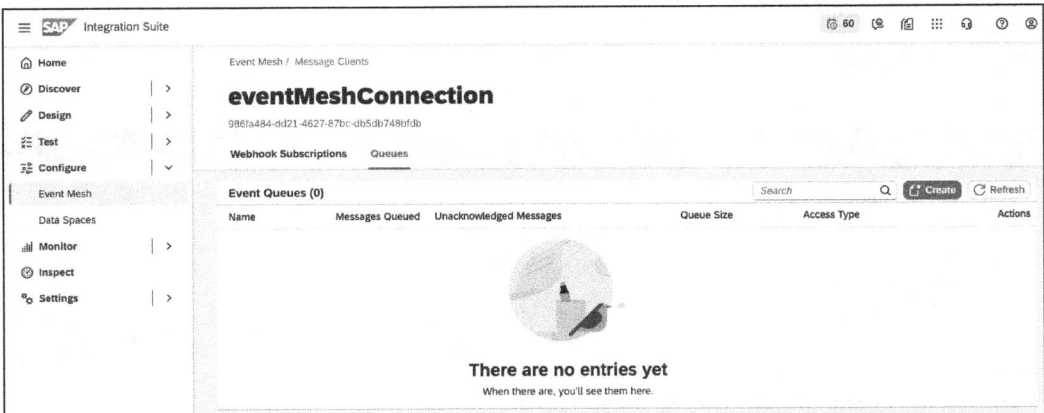

**Figure 13.15** Creating Message Queue

In the next step, you create the actual event queue. You must enter a **Name** for it, but the **Namespace** will be prefilled based on your message client (see Figure 13.16).

---

**Important**

If you leave the remaining fields blank, the default values (displayed in gray) will be applied automatically by the system, and you won't need to modify them unless you require specific behavior (e.g., redelivery, time-to-live, size limits).

---

When you're ready, click **Create** to finalize your queue.

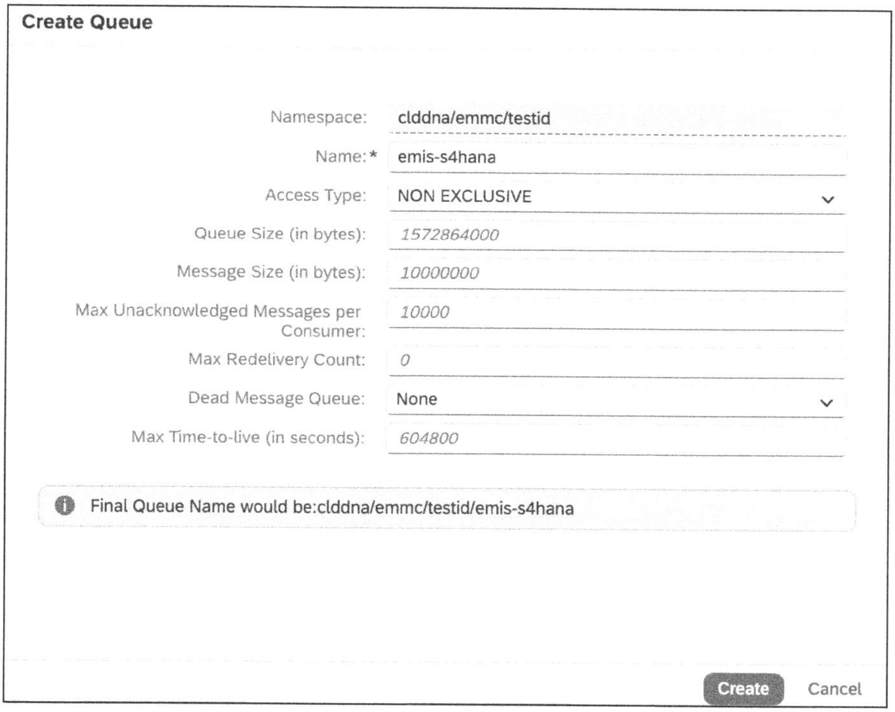

**Create Queue**

| | |
|---|---|
| Namespace: | clddna/emmc/testid |
| Name:* | emis-s4hana |
| Access Type: | NON EXCLUSIVE |
| Queue Size (in bytes): | 1572864000 |
| Message Size (in bytes): | 10000000 |
| Max Unacknowledged Messages per Consumer: | 10000 |
| Max Redelivery Count: | 0 |
| Dead Message Queue: | None |
| Max Time-to-live (in seconds): | 604800 |

Final Queue Name would be:clddna/emmc/testid/emis-s4hana

Create    Cancel

**Figure 13.16** Configuring Your Message Queue

This is your created event queue—in this example, it's *clddna/emmc/testid/emis-s4hana*. Next, click on the queue entry to open its configuration (see Figure 13.17). You need to add a subscription to link the queue to the events coming from your SAP S/4HANA system.

Then, switch to the **Subscriptions** tab (if it's not already selected) and click the **Create** button on the right side to add a new subscription for this queue (see Figure 13.18). This will allow the queue to receive messages from events that are triggered in the SAP S/4HANA system.

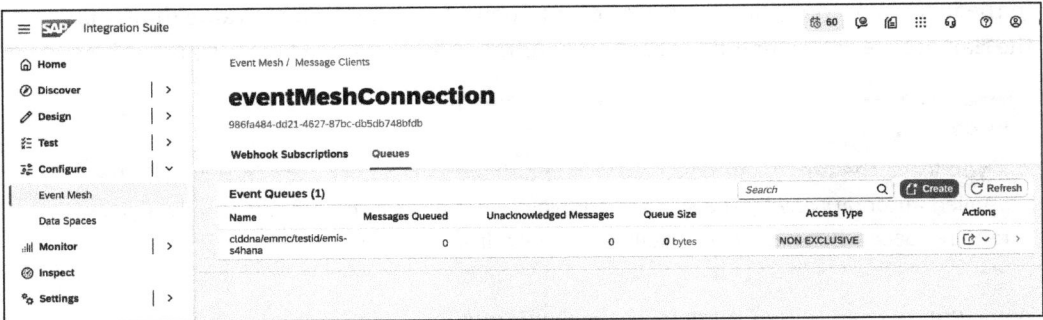

**Figure 13.17** Created Message Queue

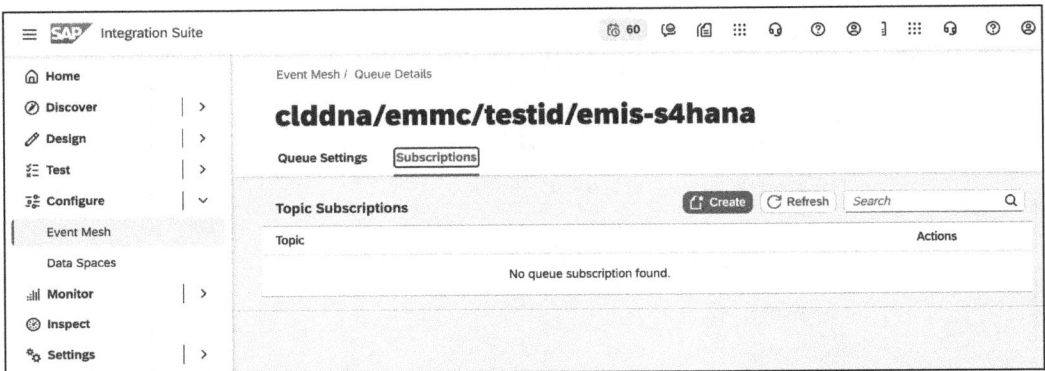

**Figure 13.18** Creating Subscription

Next, configure the topic subscription for the message queue as shown in Figure 13.19. You can enter a wildcard pattern like "<namespace>/*" to subscribe to all events that are sent to the queue via the configured outbound bindings. This is useful if you want to catch everything that comes from the SAP S/4HANA system without filtering.

However, it may often be better for you to narrow down the subscription to a specific event. For instance, if you only want to receive events that are related to changes in business partners, you can define the topic as "<namespace>/ec/sap/s4/beh/business-partner/v1/BusinessPartner/Changed/v1" instead. Everything you enter after "/ec/" in this topic will reflect the exact outbound binding topic that you configured in the SAP S/4HANA system. This approach allows you to subscribe only to the events you are interested in and thus ensure a more efficient and focused message flow.

Once you've successfully created the topic, your configuration screen should look like Figure 13.20. You'll see the full topic string listed under **Topic Subscriptions**, and that will confirm that the message queue is now subscribed and ready to receive events that are published to the matching topic(s).

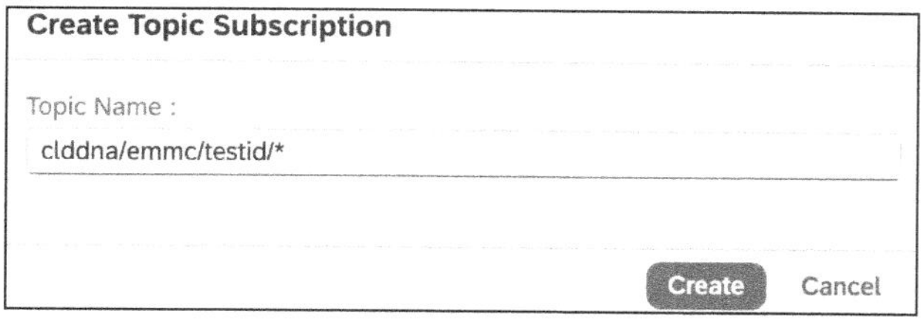

**Create Topic Subscription**

Topic Name :

clddna/emmc/testid/*

Create    Cancel

**Figure 13.19**  Configuring Topic Subscription

With the subscription in place, you can switch back to your SAP S/4HANA system, launch Transaction /IWXBE/CONFIG again to go back to the **Channel Configuration** interface, and click on the **Subscriptions** tab at the top of the screen. Then, click on the **Create** icon to open the **Create Channel Subscription** dialog window.

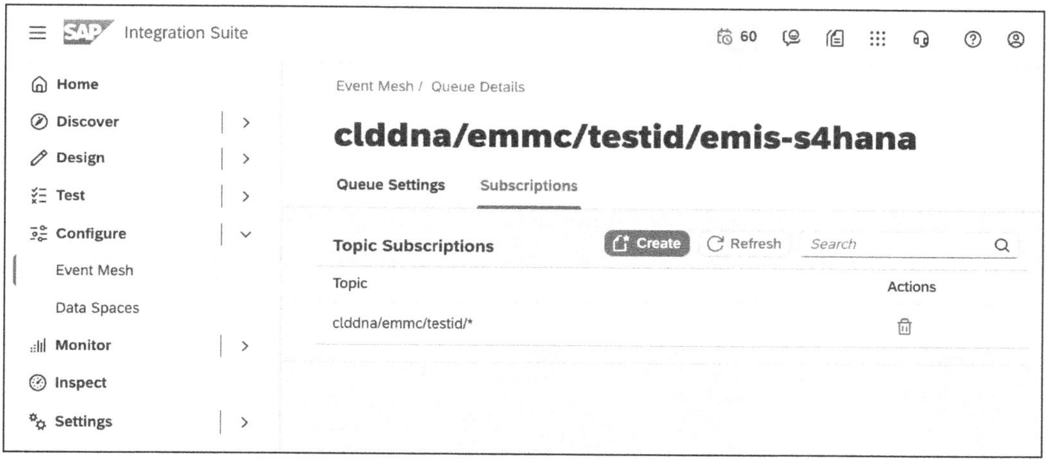

≡   SAP  Integration Suite                                              📅 60   ☺   📁   ⠿   🎧   ⑦   ⊚

🏠 **Home**

⊘ **Discover**            |   >              Event Mesh /  Queue Details

✏ **Design**              |   >

⅀ **Test**                |   >              # clddna/emmc/testid/emis-s4hana

⧆ **Configure**           |   ∨              **Queue Settings**    Subscriptions

   Event Mesh

   Data Spaces                                **Topic Subscriptions**        ⌐ Create   ↻ Refresh    Search                    🔍

.ıll **Monitor**          |   >              Topic                                                        Actions

⊘ **Inspect**                                clddna/emmc/testid/*                                          🗑

⚙ **Settings**            |   >

**Figure 13.20**  Created Topic Subscription

Then, in the **Address** field, enter the full path to the event queue you previously created in Cloud Integration (see Figure 13.21). This address must exactly match the queue name in the correct namespace, as we showed earlier during the Event Mesh configuration. Once you've entered this address, confirm it by clicking the checkmark icon to finalize the subscription. Performing this step ensures that events published to the subscribed topics will be forwarded to the specified queue, so it enables your integration scenario to function end-to-end.

Next, switch back to SAP Integration Suite to configure secure access to Event Mesh. In the navigation pane, go to **Monitor • Integrations and APIs**, click on **Security Material**, click on **Create,** and choose **OAuth2 Client Credentials**. A popup window will appear, and you'll need to enter the required values there, based on the service key you previously created. Focus specifically on the section in the service key where the protocol is *amqp10ws*.

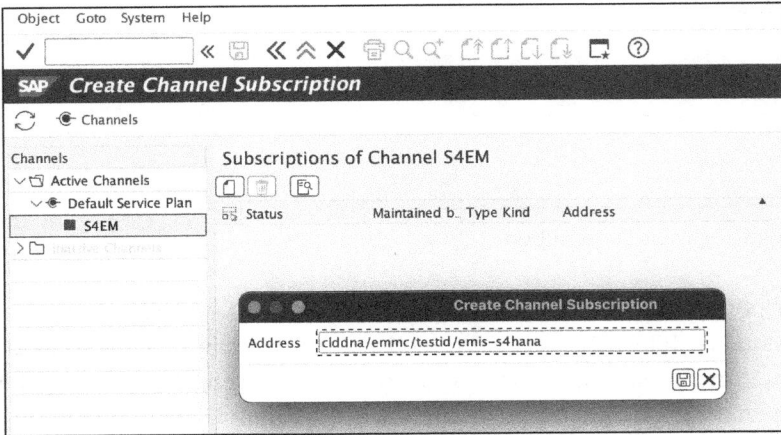

**Figure 13.21** Configuring S4 Subscription Channels

Enter a **Name** for the credentials, and for the **Token Service URL**, use the `tokenendpoint` from the service key. For the **Client ID** and **Client Secret**, take the corresponding values from the service key as well. Set the **Client Authentication** method to **Send as Request Header** and the **Content Type** to **application/json**. You can leave all other fields empty unless your setup explicitly requires you to fill them in. Once you've input all information correctly, click on **Deploy** to save the configuration (see Figure 13.22).

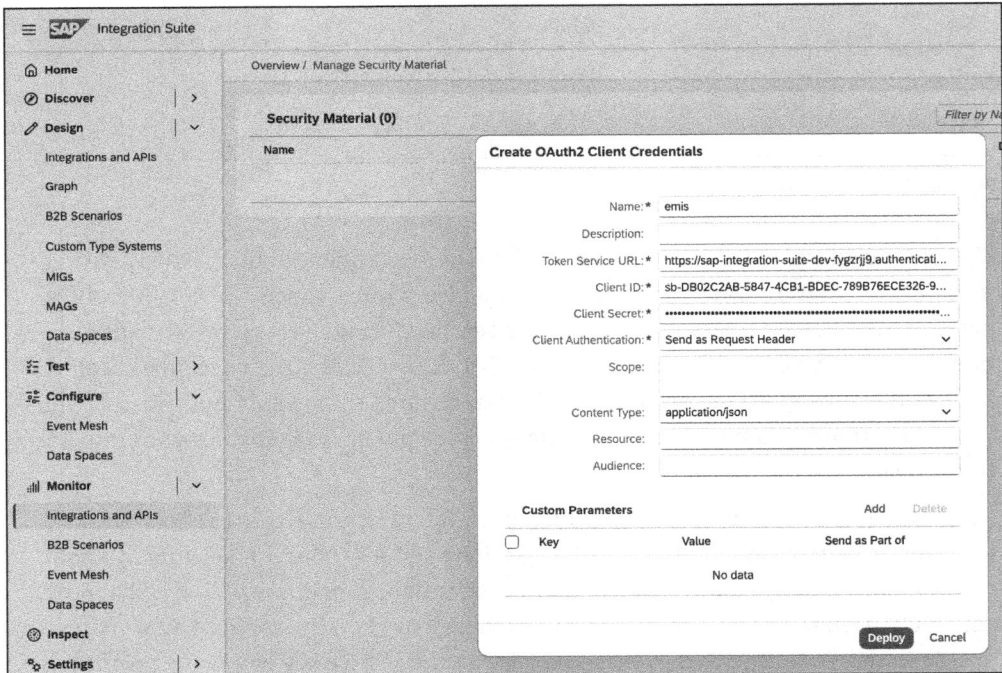

**Figure 13.22** Creating Event Mesh Security Material

Once you've completed the security configuration, the next step is to consume the events from the queue by creating a new iFlow. In Cloud Integration, navigate to the **Design** section and create a new iFlow. Once the iFlow is open, add a **Sender** component and select the **AMQP** adapter as the **Adapter Type** (see Figure 13.23). This adapter will enable the iFlow to connect to Event Mesh and receive messages from the configured queue.

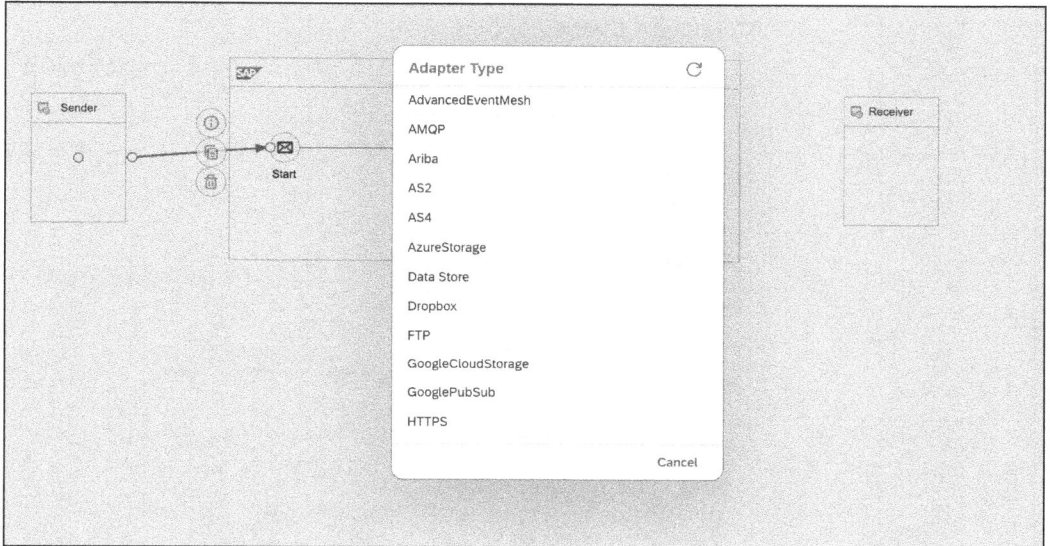

**Figure 13.23** Sender AMQP Protocol

In the AMQP sender adapter configuration within the iFlow, you must fill in several fields to establish a proper connection to Event Mesh. In the **Host** field, you must enter the URI taken from the service key, specifically from the section where the protocol is defined as *amqp10ws*. This ensures that the iFlow will connect to the correct Event Mesh endpoint by using the appropriate protocol (see Figure 13.24).

In the **Port** field, enter "443" because it's the default port that's used for secure Web-Socket and HTTPS communication and it's required for AMQP over TLS. For the **Path**, enter "/protocols/amqp10ws" because it's the mandatory path for AMQP communication over WebSocket in Event Mesh and it should not be modified. Set the **Proxy Type** to **Internet**, which specifies that the iFlow will communicate over the public internet and is necessary for accessing cloud-based Event Mesh instances.

Then, check the box for **Connect with TLS** to ensure that the connection is encrypted with TLS, which is required for secure communication with the Event Mesh service. Leave the **Disable Reply-To** box checked to tell the system not to expect a reply-to address in the incoming AMQP messages, which is appropriate for event-driven scenarios where messages are typically one-way.

In the **Authentication** field, select **OAuth2 Client Credentials** to indicate that the iFlow will authenticate using a previously configured OAuth2 client credential. Finally, in the **Credential Name** field, enter the name you created before on the screen shown in Figure 13.22.

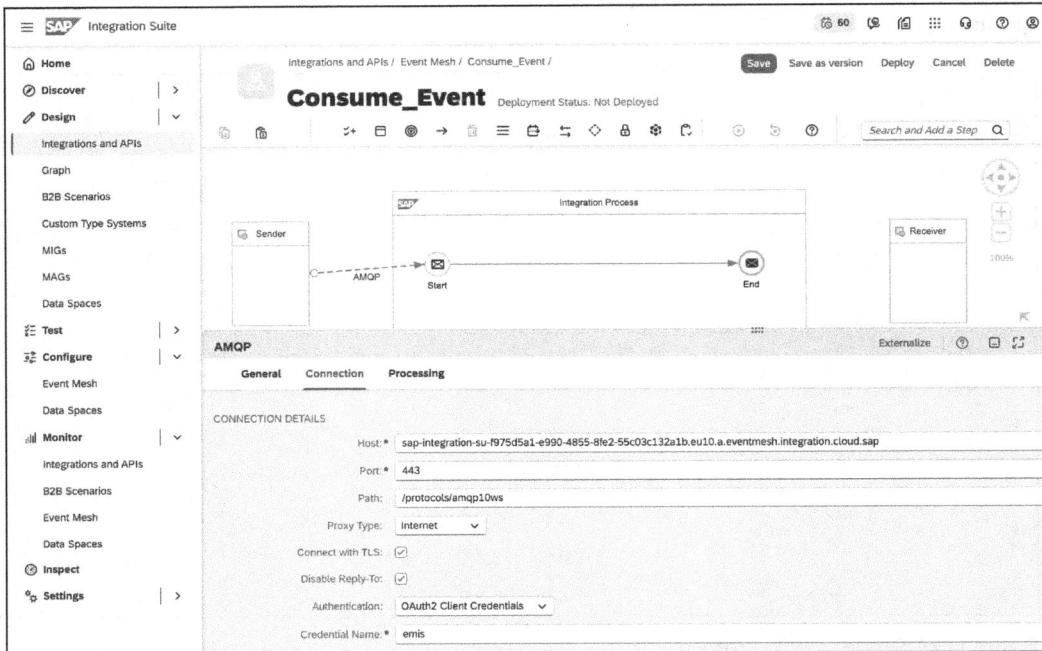

**Figure 13.24** Configuring Sender Event Mesh Connection

Then, in the **Processing** tab of the AMQP sender adapter, you must enter the name of the event queue that you created earlier in Event Mesh. This ensures the iFlow will listen to the correct queue for incoming events. Enter a name in the **Queue Name** that exactly matches the full name of your queue, including the namespace. This will bind the sender adapter directly to the Event Mesh queue so that all published messages will be consumed by this iFlow (see Figure 13.25).

After the **Queue Name**, you can configure additional processing options. You can enter a value in the **Number of Concurrent Processes** field to define how many parallel processes will consume messages from the queue. For example, you can enter a value of 1 to make messages be processed sequentially. You can also enter a value in the **Max. Number of Prefetched Messages field to** determine how many messages the adapter should fetch in advance from the queue. The default value is 5 and is typically sufficient for testing or low-volume scenarios.

If you check the **Consume Expired Messages** box, the adapter will process messages even after they've exceeded their time-to-live. In most cases, this box should remain unchecked unless your situation requires it.

Under **Retry Details**, you can set the **Max. Number of Retries to** controls how many times the system will attempt to redeliver a message in case of a failure. You can use the **Delivery Status After Max. Retries** field to specify what will happen to a message after all retry attempts fail, and you can choose from options like **REJECTED** (to discard the message) and **SUSPENDED** (to hold the message for manual review).

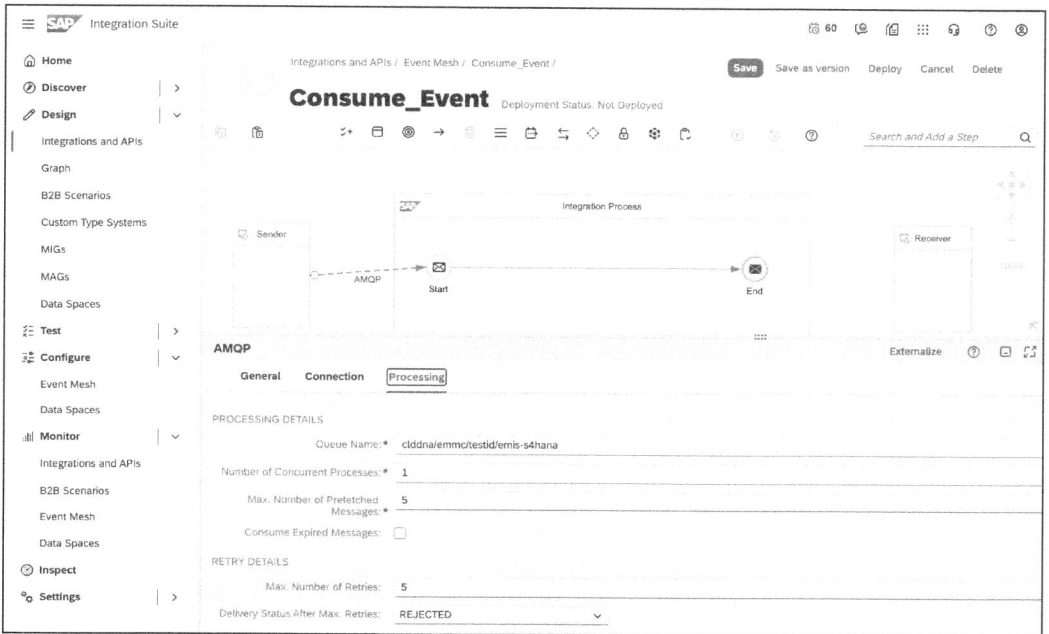

**Figure 13.25** Configuring Event Mesh Sender Processing

Once you've successfully deployed the iFlow and completed all configurations correctly, you'll see the **Polling Information** section appear in the iFlow's monitoring view. This section will confirm that the adapter is actively polling the configured Event Mesh queue, and it will display the full connection string, including the host, port, protocol path, and exact queue name. Additionally, the **Consumption Status** will show as **Successful** to indicate that the flow is connected to Event Mesh and ready to consume messages in real time (see Figure 13.26).

Once you have everything configured and the iFlow is actively consuming messages, you can perform a functional test (see Figure 13.27). To do this, you simply need to change a business partner within your SAP S/4HANA system—in this example, we modified the salutation of an existing business partner and clicked **Save**. This action will trigger the outbound event defined in the system configuration and should result in a message being sent to the configured Event Mesh queue. If you've set up everything correctly, this message will be picked up by your iFlow and processed accordingly.

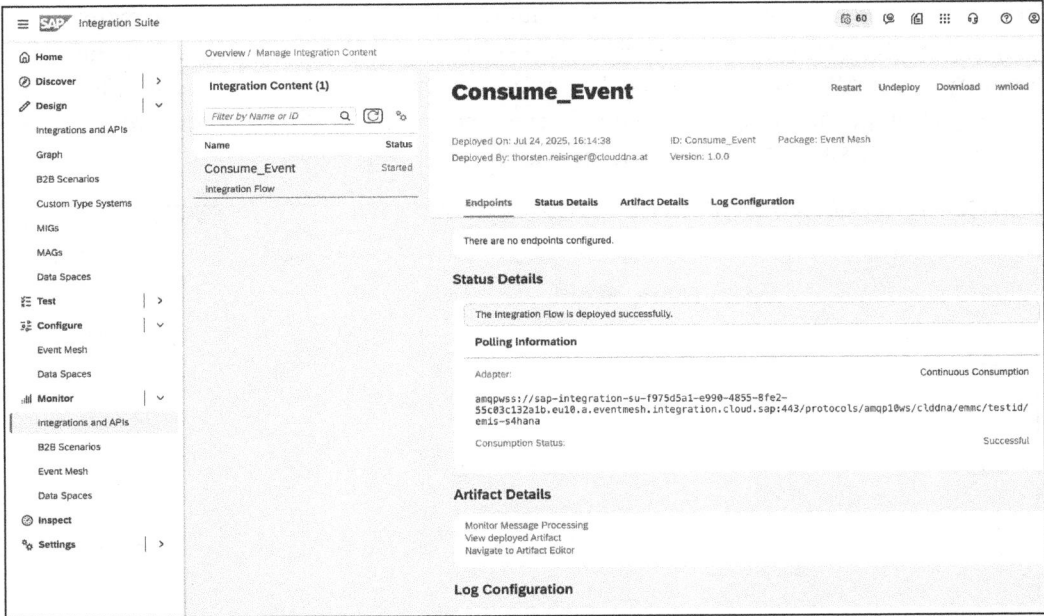

**Figure 13.26** Deployed Event Mesh iFlow

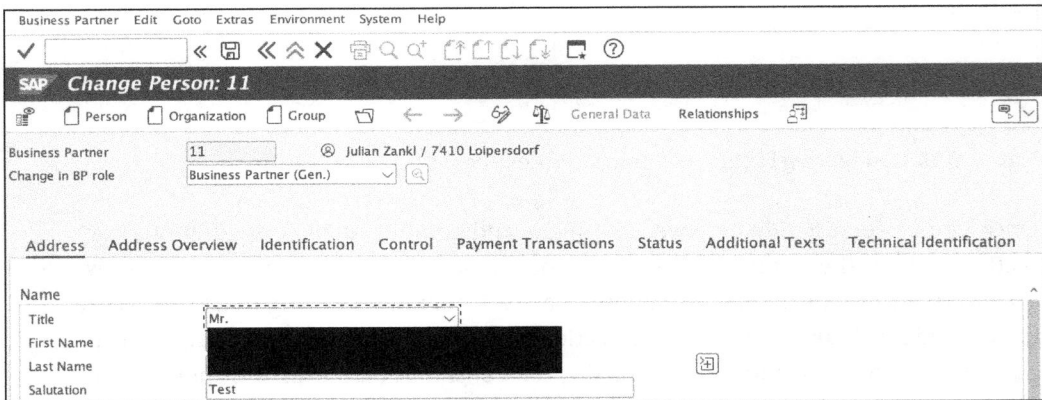

**Figure 13.27** Changing Business Partner

If you've completed the entire configuration successfully, you'll see the event appear automatically in the **Monitor Message Processing** section of your Cloud Integration. In our example, the business partner change triggered an event, which was then consumed and processed by the deployed iFlow. The monitoring screen confirms that the event was received and handled, with the status showing as **Completed** and the processing time being listed. This confirms the end-to-end setup from SAP S/4HANA event emission to Event Mesh queue consumption and iFlow execution (see Figure 13.28).

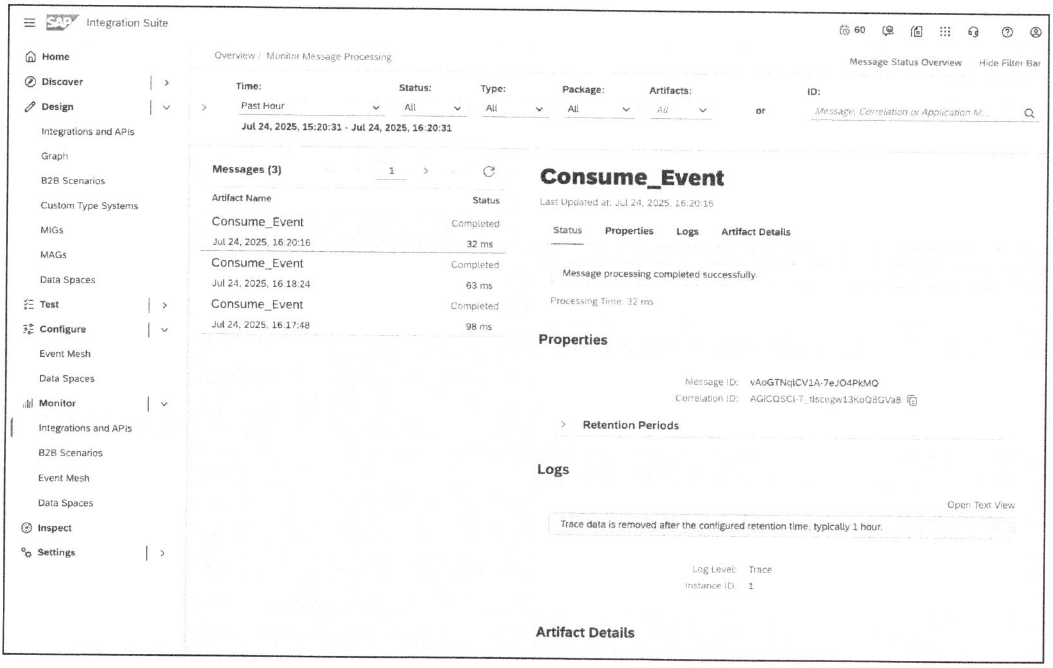

**Figure 13.28**  Consumed Event

Figure 13.29 shows the actual payload that's received from the event that was triggered in the SAP S/4HANA system. In this example, the payload indicates that a change has occurred to a business partner with an **ID** of **11**. The event includes metadata such as the type, source, timestamp, and content type, along with a data section that identifies the specific affected business partner. This payload structure conforms to the CloudEvents specification and can be processed further within the iFlow to trigger downstream actions.

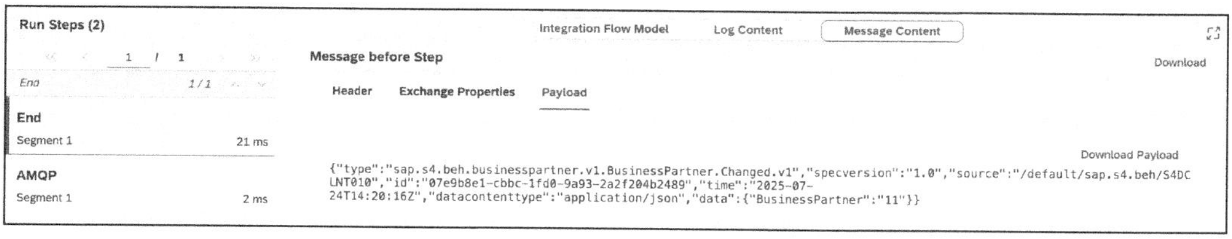

**Figure 13.29**  Consumed Event Payload

## 13.5   Summary

In this chapter, we laid out the foundation for implementing event-driven architectures in SAP landscapes through the structured connection between Cloud Integration and Event Mesh. We focused not solely on feature exploration or technical experimentation but also on establishing a robust, loosely coupled, and scalable messaging framework to enable real-time, asynchronous system interactions based on standardized enterprise events.

We started walking you through the configuration by having you set up a central message broker via Event Mesh. We emphasized the conceptual model behind event-driven systems, particularly the publisher-subscriber (pub/sub) paradigm. Once you had this messaging foundation in place, we addressed routing and message processing strategies within Cloud Integration. By subscribing iFlows to specific topics and applying content-based routing, transformation logic, and error handling within those flows, you can process messages in a structured, traceable, and intelligent way. In parallel, we introduced best practices to support stable event handling across system landscapes. These included retry mechanisms, exception processing, structured topic hierarchies, and consistent event naming conventions.

In the second half of the chapter, we focused on the technical integration of SAP S/4HANA with Event Mesh. We had you configure the SAP S/4HANA system, acting as a producer of business events, to publish standardized messages such as business partner changes and order confirmations. Once you'd successfully published events into Event Mesh, we demonstrated a complete consumption scenario using Cloud Integration. To conclude the setup, we had you execute a fully integrated test scenario in which a business partner change in SAP S/4HANA triggered an outbound event that was routed through Event Mesh, subscribed to by an Cloud Integration queue, and processed by a deployed iFlow.

Therefore, this chapter established the operational, technical, and architectural foundation for event-driven SAP integration. Its configurations and patterns empower development teams to implement scalable, resilient, and modular solutions that respond to business events in near real time across hybrid system landscapes. By completing the steps outlined in this chapter, you've established an event-driven integration layer that's not only technically functional but also scalable, auditable, and business-ready. This foundation will enable you to connect transactional systems like SAP S/4HANA with modern, cloud-native applications in a way that supports real-time responsiveness and future-oriented integration strategies.

# The Authors

**Martin Koch** is the managing director of CloudDNA GmbH, an SAP partner in Austria. He and his team conduct training for SAP and have developed four of their own training courses on the topics of SAPUI5, SAP Fiori, cloud integration, and cloud security, which are listed in the SAP training catalog.

**Thorsten Reisinger** is an integration developer and solution architect at CloudDNA GmbH. In this role, he supports companies with implementing complex integration scenarios in SAP Integration Suite and with developing secure, scalable architectures. With many years of experience in SAP Business Technology Platform (SAP BTP) projects, he is responsible for the design and implementation of user, authorization, and security concepts. In addition, he has in-depth expertise in identity management and IT security. His work is characterized by a practical approach that combines technical know-how with a deep understanding of business processes.

**Marc Urschick** is a development lead at CloudDNA GmbH.

# Index

**Interested in reading more?**

Please visit our website for all new book
and e-book releases from SAP PRESS.

**www.sap-press.com**